Diseases of Field Crops: Diagnostics and Management

Vaibhav Kumar Singh • Jameel Akhtar •
Krishna Pratap Singh
Editors

Diseases of Field Crops: Diagnostics and Management

Springer

Editors
Vaibhav Kumar Singh
Division of Plant Pathology
ICAR-Indian Agricultural Research Institute, Pusa
New Delhi, India

Jameel Akhtar
Division of Plant Quarantine
ICAR-National Bureau of Plant Genetic Resources
New Delhi, India

Krishna Pratap Singh
Department of Plant Pathology
G. B. Pant University of Agriculture and Technology
Pantnagar, Uttarakhand, India

ISBN 978-981-97-6159-3 ISBN 978-981-97-6160-9 (eBook)
https://doi.org/10.1007/978-981-97-6160-9

This Springer imprint is published by the registered company Springer Nature Singapore Pte Ltd.
The registered company address is: 152 Beach Road, #21-01/04 Gateway East, Singapore 189721, Singapore

Foreword

As Editor-in-Chief of *Food Security—The Science, Sociology and Economics of Food Production and Access to Food*, I am fortunate enough to have a viewpoint of what the literature related to food security worldwide looks like. Over the past 4 years, I have been surveying nearly 4000 submissions to the journal—every day, two or three submissions reach me. These manuscripts may cover any aspect of food security, from the production of food, to its use and its nutritional value.

One point is very clear: food production today is downplayed, by the scientific community and by the policy environment alike. This happens everywhere in the world, even where food is scarce. Food production is taken for granted. Ironically, it is downplayed by the same policymakers who speak of climate change, of the world population, of the loss of agricultural land, of the necessity of sustainable practices, and of the collapse of global biodiversity.

Better understanding plant diseases, improving insights in plant disease management, developing disease management methods that do not only address the problem at hand but also consider other associated constraints, and making so that these management methods do not have negative impact on ecosystem health and resilience: these are critical questions that scientists must address now. This is urgent. This calls upon open-minded scientists that have both strong disciplinary roots, and also are open to new views and interdisciplinary thinking.

Plant diseases do not happen by accident; plant disease epidemics even less so. These happen in contexts which make them possible. There have been plant pathogens (and parasites of living organisms) since the beginning of life on Earth. So, why is it, then, that plant diseases would be such a problem today, especially in agriculture? The reasons must be found, I think, in the overall context where agriculture takes place: not only just the ecology of plant production, and the technologies used to generate harvests, but also the economic, social, political, and policy environment of agricultural production. Having a minimum understanding of these contextual elements are, in my view, paramount in better understanding plant diseases and in better managing them.

Plant pathology is at the forefront of technology development. Plant pathology was long seen as a very dusty area of science, full of complicated taxonomy words, where scientists were first of all fascinated by the diversity of plant pathogens, their forms, their lifestyles, their sophisticated ways to survive and specialise to host plants. Not anymore. Plant pathology today involves a whole array of technologies, from molecular, to remote sensing, to machine-learning. To many scientists and students, this is a sign of being "modern", an indication of "progress".

Charles Darwin stated: "one cannot be a good theorician if one is a good observer". To me, technology could be a Trojan horse, and for two reasons. One is that our human eyes, these wonderful organs strongly connected to our brains, perfected over million years, remain the best observation instruments: one cannot dispense of observation with one's own eyes; there cannot be a good plant pathologist who does not spend some time in the field, looking at plants. Could there be a good medical doctor who refuses to see one's patient? The other is that technology drives us, gives us the impression that everything is possible through technique, that technology will fix anything, always. This is a dangerous fallacy. Solutions will come not from technology, but from the science, and from the scientist's ability to decide which technique is best to achieve a given objective.

I am very honoured by the Editors to write this foreword to their book, *Diseases of Field Crops: Diagnostics and Management*. This book offers the reader with a very diverse and fascinating sample of what plant pathology has best to offer today. It will provide inspiration to scientists and students on the specific questions addressed by each chapter, and also, to the broader question of how plant pathology could become even better, and better serve society and citizens. These are urgent questions for all of us.

Plant Pathology, GB Pant University of Agriculture and Technology
Pantnagar, India

Serge Savary

Preface

In the vast expanse of agricultural science, the battle against plant diseases is a perpetual journey marked by relentless inquiry and strategic innovation. We the editors, Dr. Vaibhav Kumar Singh, Dr. Jameel Akhtar, and Dr. Krishna Pratap Singh, take immense pride in presenting this comprehensive treatise, *Diseases of Field Crops: Diagnostics and Management*. This collaborative effort represents a pinnacle of scientific exploration, offering a reservoir of knowledge essential for researchers, practitioners, academicians, and students immersed in the intricate world of plant pathology.

Diagnosis and Detection of Plant Pathogens of Field Crops: In the first section, 'Diagnosis and Detection of Plant Pathogens of Field Crops', our distinguished contributors delve into a diverse array of topics, showcasing the multidimensional nature of modern plant pathology. From the precise diagnostics of major fungal pathogens of rice to cutting-edge applications of machine learning and artificial intelligence in plant disease detection, each chapter underscores the evolving landscape of diagnostic methodologies. Additionally, chapters exploring the multi-locus gene characterization of phytoplasmas, the diagnosis of plant parasitic nematodes, and the identification of seed-borne pathogens offer a nuanced understanding of the challenges posed to field crops. This reveals the depth and diversity of our contributors' expertise.

Management of Field Crop Pathogens/Diseases: The second section, 'Management of Field Crop Pathogens/Diseases', unfolds with a compendium of 12 chapters that offer insightful perspectives on the dynamic strategies employed in the field to address critical issues in contemporary agriculture. From Bakanae disease management in rice to the revolutionary CRISPR technology's application in plant disease management, the authors elucidate pioneering approaches. Integrated management strategies for soybean diseases, proactive measures against major fungal diseases of pulse crops, and the exploration of nanotechnology for enhanced disease protection reflect the authors' dedication to mitigating losses and advancing sustainable practices in agriculture. Emerging threats like *Rhizoctonia* and rice bakanae disease prompt a thorough examination of biointensive-based management approaches.

Editors' Vision: Navigating the diverse and evolving landscape of plant pathology, we, the editors, Dr. Vaibhav Kumar Singh, Dr. Jameel Akhtar, and Dr. Krishna Pratap Singh, are privileged to guide this scholarly journey and view this treatise not merely as a compilation of chapters but as a pivotal contribution to the scientific discourse. It is our sincere belief that the intersection of knowledge presented in this book will not only inform but inspire future research. Our esteemed contributors, experts in their respective fields, have seamlessly woven a tapestry of insights that will undoubtedly influence the trajectory of future research in plant diseases and management.

Concluding Remarks: In a world where agricultural sustainability is paramount, understanding and managing plant diseases are at the forefront of ensuring global food security. We extend our deepest gratitude to the esteemed contributors whose collective wisdom and dedication have shaped this volume. As editors, we aspire for this treatise to transcend its pages, becoming a catalyst for innovation, collaboration, and transformative advancements in plant pathology.

We express our deepest gratitude to all the contributors whose dedication and expertise have enriched this volume. It is our sincere hope that this treatise becomes a cornerstone for future research, paving the way for sustainable and resilient agricultural practices globally.

New Delhi, India — Vaibhav Kumar Singh
New Delhi, India — Jameel Akhtar
Pantnagar, India — Krishna Pratap Singh

Contents

Editors and Contributors

About the Editors

Vaibhav Kumar Singh is Agricultural Research Service Scientist currently working as Senior Scientist at the Division of Plant Pathology, ICAR-Indian Agricultural Research Institute, Pusa, New Delhi, India. He has received his B.Sc. (Agri.) from DDU University, Gorakhpur; M.Sc. (Agri.) in Plant Pathology from C.S. Azad University of Agriculture and Technology, Kanpur and Ph.D. in Plant Pathology from G.B. Pant University of Agriculture and Technology, Pantnagar, UK, India. His areas of research interest are genetic resistance, pathogenic variability, host–pathogen interaction, molecular diagnostics, biological control, epidemiology, and disease management. Dr. Singh has made outstanding contribution in the field of wheat pathology and genetics under wheat improvement programme as well as wheat rust pathology programme of IARI by characterizing a number of rusts-resistant wheat genotypes including slow rusting adult plant resistance. He is effectively involved in the development of about 38 high yielding, durable rusts resistance, multiple diseases resistance, and quality wheat varieties, which have been widely accepted by the farmers in different parts of the country. His contribution to research has won him national awards like 'Nanaji Deshmukh ICAR Outstanding Team Award-2019', besides other prestigious awards including NESA Eminent Scientist of the Year Award-2024, Dr. R.S. Paroda Award–2014, Dr. S. Nagarajan Memorial Award–2021, Outstanding Scientist Award–2019, Young Educator Award–2019, Young Scientist Award–2015, Vikash Mitra Samman–2022, and Krishak Mitra Samman–2016. He has been conferred with the Fellowship of Indian Society of Genetics and Plant Breeding, Andaman Sciences Association, and ISEE. He has done much collaborative research with international organization like CIMMYT and BISA. He also trained at international level at KARI-Kenya, BARI-Bangladesh, CIMMYT-Mexico, INIAF-Bolivia and USDA-ARS, FDWSR Unit-USA. Have more than 58 publications in peer-reviewed high impact factor journals of repute. He has written four books and handled five research projects financed by ICAR-NFAS, ICAR-AMAAS, NFSM-DAC, MoA&FW, and DBT.

Jameel Akhtar graduated from C.S. Azad University of Agriculture and Technology, Kanpur, U.P. and obtained B.Sc. (Ag & AH) degree in 1998 and Master and Doctoral degree in Plant Pathology in 2000 and 2003, respectively, from G.B. Pant University of Agriculture and Technology, Pantnagar, US Nagar, Uttarakhand. He joined regular service in Birsa Agricultural University, Ranchi, Jharkhand as Jr. Scientist-cum-Assistant Professor in July 2004 and worked at Ranchi Agricultural College, Ranchi. After serving BAU, Ranchi for 6 1/2 years, he joined ICAR-National Bureau of Plant Genetic Resources, New Delhi as Senior Scientist in the Division of Plant Quarantine on December 31, 2010 and currently working as Principal Scientist. He has more than 75 peer-reviewed publications in various national and international journals and delivered many poster and oral presentations in numerous national/international conferences and symposia, supervised three M.Sc. (Ag) students for their thesis research in Plant Pathology. Also, involved in teaching various PG/Ph.D. courses of post-graduate school of ICAR-IARI, New Delhi. He has been conferred 18 awards and honours including Fellow of Indian Phytopathological Society, New Delhi in 2020 and Society of Plant Protection Sciences, New Delhi in 2009, Young Scientist Award by UP Council of Science and Technology, Govt. of Uttar Pradesh, Lucknow in 2009, Smt. Kavuri Sarada Memorial Award by the Plant Protection Association of India in 2014, etc. and member of editorial board of various national/international journals and life member of nine academic societies of India.

Krishna Pratap Singh graduated from B.R.D. PG College of Gorakhpur University, Deoria, U.P. and obtained B.Sc. (Ag) degree in 1980 and Master and Doctoral degree in Mycology and Plant Pathology in 1984 and 1990, respectively, from the Institute of Agriculture Sciences, Banaras Hindu University, Varanasi. He joined service as Assistant Professor at G.B. Pant University of Agriculture and Technology in Nov 1992 and worked at College of Forestry, Ranichauri in various capacities viz. Assistant Professor, Associate Professor, Professor, Co-coordinator, UUHF Entrance, Director, Communication and Publication, UUHF and Officer In-charge, Discipline of Plant Pathology, published more than 70 research papers in various national and international journals, including three edited books, one authored book, five laboratory manuals, technical bulletins, book chapters, number of popular articles and presented many research papers and posters in various national and international seminars/symposia/workshops/conferences, etc. He has supervised five Ph.D. and three PG students for their thesis research in Plant Pathology. He has been serving as a member, editorial board of Applied Microbiology (UK), Plant Disease Research, Ludhiana, Indian Journal of Plant Pathology, Lucknow and Indian Phytopathology, India. He is a life member of eight scientific professional societies and Fellow of four scientific societies including Indian Phytopathological Society, Indian Society of Hill Agriculture (ISHA), Indian Society of Mycology and Plant Pathology, and International College of Nutrition (FICN). He has been awarded with prestigious Prof. KPV Menon and Sharda Lele, Plant Pathology Leadership awards. Presently, he is working as Professor, Plant Pathology at GBPUAT, Pantnagar.

Contributors

Rashmi Aggarwal Fungal Molecular Biology Laboratory, Division of Plant Pathology, ICAR-Indian Agricultural Research Institute, New Delhi, India

Amit Ahuja Division of Nematology, ICAR-Indian Agricultural Research Institute, Pusa, New Delhi, India

Jameel Akhtar ICAR-National Bureau of Plant Genetic Resources, Pusa Campus, New Delhi, India

T. Aravind Department of Plant Pathology, M. S. Swaminathan School of Agriculture, Centurion University of Technology and Management, Paralakhemundi, Odisha, India

Alka Arora ICAR-Indian Agricultural Statistics Research Institute, New Delhi, India

Bishnu Maya Bashyal Fungal Molecular Biology Laboratory, Division of Plant Pathology, ICAR-Indian Agricultural Research Institute, New Delhi, India

Pooja Bhatt Department of Plant Pathology, College of Agriculture, Govind Ballabh Pant University of Agriculture and Technology, Pantnagar, India

Nuclear Agriculture and Bio-Technology Division, Bhabha Atomic Research Centre, Trombay, Mumbai, India

W. Tampakleima Chanu ICAR Research Complex for NEH Region, Manipur Centre, Imphal, India

Chandan Kumar Deb ICAR-Indian Agriculture Statistical Research Institute, New Delhi, India

H. Chandrajini Devi ICAR Research Complex for NEH Region, Manipur Centre, Imphal, India

K. Sarda Devi ICAR Research Complex for NEH Region, Manipur Centre, Imphal, India

O. Priyoda Devi ICAR Research Complex for NEH Region, Manipur Centre, Imphal, India

Manisha Dev Department of Plant Pathology, College of Agriculture, Govind Ballabh Pant University of Agriculture and Technology, Pantnagar, India

S. C. Dubey ICAR-National Bureau of Plant Genetic Resources, Pusa Campus, New Delhi, India

Pooja Garg ICAR-National Institute for Plant Biotechnology, Pusa, New Delhi, India

Kavita Gupta ICAR-National Bureau of Plant Genetic Resources, Pusa Campus, New Delhi, India

Nitika Gupta Advanced Centre for Plant Virology, Division of Plant Pathology, ICAR-Indian Agricultural Research Institute, New Delhi, India

Pham Thi Thu Ha Genomic Research Institute and Seed, Ton Duc Thang University, Ho Chi Minh City, Vietnam

Md Ashraful Haque ICAR-Indian Agriculture Statistical Research Institute, New Delhi, India

S. N. Islam ICAR-Indian Agriculture Statistical Research Institute, New Delhi, India

Rajni Jain ICAR-National Institute of Agricultural Economics and Policy Research, New Delhi, India

Samyak Jain Shiv Nadar University, New Delhi, India

Himani Jeena Department of Plant Pathology, College of Agriculture, Govind Ballabh Pant University of Agriculture and Technology, Pantnagar, India

B. Jeevan Division of Crop Protection, ICAR-Vivekanand Parvatiya Krishi Anusandhan Sansthan, Almora, Uttarakhand, India

Abhijeet S. Kashyap ICAR-National Bureau of Agriculturally Important Microorganisms, Maunath Bhanjan, Uttar Pradesh, India

Anamika Kashyap ICAR-National Institute for Plant Biotechnology, Pusa, New Delhi, India

Malemnganbi Keisham ICAR Research Complex for NEH Region, Manipur Centre, Imphal, India

P. Krishnan Division of Agricultural Physics, ICAR-Indian Agricultural Research Institute, New Delhi, India

Ashish Kumar Gupta ICAR-National Institute for Plant Biotechnology, Pusa, New Delhi, India

Sujata Kumari ICAR-National Institute for Plant Biotechnology, Pusa, New Delhi, India

Ravindra Kumar ICAR-Indian Institute of Wheat and Barley Research, Karnal, Haryana, India

Mansi Machal Wheat Pathology Laboratory, Division of Plant Pathology, ICAR-Indian Agricultural Research Institute, New Delhi, Delhi, India

Albert Maibam ICAR Research Complex for NEH Region, Manipur Centre, Imphal, India

Raghavendra Reddy Manda Wageningen University and Research, Wageningen, The Netherlands

Nazia Manzar ICAR-National Bureau of Agriculturally Important Microorganisms, Maunath Bhanjan, Uttar Pradesh, India

Sudeep Marwaha ICAR-Indian Agricultural Statistics Research Institute, New Delhi, India

K. K. Mishra Division of Crop Protection, ICAR-Vivekanand Parvatiya Krishi Anusandhan Sansthan, Almora, Uttarakhand, India

Raj K. Mishra Division of Crop Protection, ICAR-Indian Institute of Pulses Research, Kanpur, Uttar Pradesh, India

Sapam Monteshori ICAR Research Complex for NEH Region, Manipur Centre, Imphal, India

Sajad Un Nabi ICAR-Central Institute of Temperate Horticulture, Srinagar, Jammu, India

Yalavarthi Nagaraju ICAR-National Bureau of Agriculturally Important Microorganisms, Maunath Bhanjan, Uttar Pradesh, India

Sapna Nigam ICAR-Indian Agricultural Statistics Research Institute, New Delhi, India

Chhattar Pal Regional Sericultural Research Station, Central Silk Board, Jammu, India

Sonika Pandey Division of Crop Protection, ICAR-Indian Institute of Pulses Research, Kanpur, Uttar Pradesh, India

Pankaj Division of Nematology, ICAR-Indian Agricultural Research Institute, Pusa, New Delhi, India

Snehalatha Pasupuleti Division of Nematology, ICAR-Indian Agricultural Research Institute, Pusa, New Delhi, India

Aakash V. Patel Wheat Pathology Laboratory, Division of Plant Pathology, ICAR-Indian Agricultural Research Institute, New Delhi, India

Neeta Pathaw ICAR Research Complex for NEH Region, Manipur Centre, Imphal, India

Praveen S. Patted ICAR-Indian Institute of Seed Science, Maunath Bhanjan, Uttar Pradesh, India

H. Prashanth Babu Division of Genetics, ICAR-Indian Agricultural Research Institute, Pusa, New Delhi, Delhi, India

S. T. Prashantha Fungal Molecular Biology Laboratory, Division of Plant Pathology, ICAR-Indian Agricultural Research Institute, New Delhi, India

H. Rajashekara Division of Crop Protection, ICAR-Vivekanand Parvatiya Krishi Anusandhan Sansthan, Almora, Uttarakhand, India

Gaurav Rakhonde Department of Plant Pathology, University of Agricultural Sciences, GKVK, Bengaluru, Karnataka, India

Govind Pratap Rao Division of Plant Pathology, ICAR-Indian Agriculture Research Institute, Pusa Campus, New Delhi, India

Mahesh Rao ICAR-National Institute for Plant Biotechnology, Pusa, New Delhi, India

Madem Gurivi Reddy Department of Plant Pathology, S.V. Agricultural College, Tirupati, Acharya N. G. Ranga Agricultural University, Tirupati, Andhra Pradesh, India

S. S. Roy ICAR-Central Citrus Research Institute, Nagpur, Maharashtra, India

Sonam Sah ICAR-National Institute of Abiotic Stress Management, Baramati, Maharashtra, India

Jyotsana Sanasam ICAR Research Complex for NEH Region, Manipur Centre, Imphal, India

Devender Sharma Crop Improvement Division, ICAR-Vivekananda Parvatiya Krishi Anusandhan Sansthan, Almora, Uttarakhand, India

Jyoti Sharma ICAR-National Institute for Plant Biotechnology, Pusa, New Delhi, India

Parul Sharma Punjab Agricultural University, Ludhiana, Punjab, India

Susheel Kumar Sharma Advanced Centre for Plant Virology, Division of Plant Pathology, ICAR-Indian Agricultural Research Institute, New Delhi, India

Krishna Pratap Singh Department of Plant Pathology, College of Agriculture, Govind Ballabh Pant University of Agriculture and Technology, Pantnagar, India

RN. Singh ICAR-National Institute of Abiotic Stress Management, Baramati, Maharashtra, India

Sanjay Singh Uttar Pradesh Council of Agricultural Research (UPCAR), Lucknow, Uttar Pradesh, India

Shweta Singh ICAR-Indian Institute of Sugarcane Research, Lucknow, Uttar Pradesh, India

Th. Surjit Singh ICAR Research Complex for NEH Region, Manipur Centre, Imphal, India

Vaibhav Kumar Singh Wheat Pathology Laboratory, Division of Plant Pathology, ICAR-Indian Agricultural Research Institute, Pusa, New Delhi, India

Vishal Singh Somvanshi Division of Nematology, ICAR-Indian Agricultural Research Institute, Pusa, New Delhi, India

Seweta Srivastava School of Agriculture, Lovely Professional University, Phagwara, Punjab, India

A. R. N. S. Subbanna Division of Crop Protection, ICAR-Vivekanand Parvatiya Krishi Anusandhan Sansthan, Almora, Uttarakhand, India

Sunil Kumar Sunani ICAR-IIPR Regional Station, Bhubaneswar, Khordha, Khordha, Odisha, India

Kumari Surbhi Department of Plant Pathology, College of Agriculture, Govind Ballabh Pant University of Agriculture and Technology, Pantnagar, India

Gohar Taj Department of Molecular Biology and Genetic Engineering, College of Basic Sciences and Humanities, GBPUAT, Pantnagar, Uttarakhand, India

Kunal Tanwar ICAR-National Institute for Plant Biotechnology, Pusa, New Delhi, India

Shikha Tripathi ICAR-National Institute for Plant Biotechnology, Pusa, New Delhi, India

Sandhya Upadhyay Department of Molecular Biology and Genetic Engineering, College of Basic Sciences and Humanities, GBPUAT, Pantnagar, Uttarakhand, India

Baby Wangkhem ICAR Research Complex for NEH Region, Manipur Centre, Imphal, India

Gaurav Yadav Fungal Molecular Biology Laboratory, Division of Plant Pathology, ICAR-Indian Agricultural Research Institute, New Delhi, India

Tabalique Yumkhaibam ICAR Research Complex for NEH Region, Manipur Centre, Imphal, India

Part I

Diagnosis and Detection of Plant Pathogens of Field Crops

1 Diagnostics of Major Fungal Pathogens of Rice

Sunil Kumar Sunani, Bishnu Maya Bashyal, S. T. Prashantha, Gaurav Kumar Yadav, and Rashmi Aggarwal

Abstract

Rice (*Oryza sativa* L.) stands as one of the foremost cereals cultivated across diverse agro-climatic regions globally. However, the rice crop faces numerous challenges from both biotic and abiotic factors. Fungal diseases, in particular, pose a significant impediment to rice production, affecting the crop at various growth stages. Therefore, prompt detection and accurate identification of pathogens are essential for effectively managing the disease. Conventional methods for detecting fungal pathogens of rice typically involve isolating the presumed pathogen on a suitable nutrient medium, followed by microscopic examination of spores and spore-bearing structures, and identification based on taxonomic characteristics. These processes are time-consuming and labor-intensive and can yield inconsistent results, often requiring significant expertise in fungal taxonomy. In contrast, the identification of certain rice fungal pathogens through their biochemical and physiological characteristics offers greater certainty compared to isolation-based methods. Indeed, nucleic acid–based techniques, such as polymerase chain reaction (PCR), along with its variations like real-time PCR and loop-mediated isothermal amplification (LAMP), offer rapidity and consistency in results. Notably, they exhibit enhanced specificity, sensitivity, and reliability. These advanced methods facilitate the detection, differentiation, and quantification of rice fungal pathogens, whether present in symptomatic or asymptomatic rice plants, as well as in seeds.

S. K. Sunani
ICAR-IIPR Regional Station, Bhubaneswar, Khordha, Khordha, Odisha, India

B. M. Bashyal (✉) · S. T. Prashantha · G. K. Yadav · R. Aggarwal
Fungal Molecular Biology Laboratory, Division of Plant Pathology, ICAR—Indian Agricultural Research Institute, New Delhi, India

V. K. Singh et al. (eds.), *Diseases of Field Crops: Diagnostics and Management*,
https://doi.org/10.1007/978-981-97-6160-9_1

Keywords

Rice · Blast · Sheath blight · Bakanae · Brown spot · False smut · Sheath rot · Diagnostics · LAMP · PCR · Nucleic acid

1.1 Introduction

Rice (*Oryza sativa* L.) holds significant importance as a cereal crop globally, serving as a staple food for over 60% of the world's population. Rice covers the world's largest sown area of 167.13 mha approximately with an annual production of 782.00 mmt (FAO Stat. 2018) thus constituting 37% of the total area under food grains. Indeed, rice productivity faces challenges from various plant pathogens and abiotic factors. Fungal diseases pose a significant threat to rice production, affecting the crop at nearly every growth stage (Savary et al. 2012). These fungal pathogens can manifest as different species, strains, varieties, races, or biotypes, each varying in virulence levels. Fungal pathogens can enter rice crops through multiple avenues, including infected seeds or propagules present in soil, water, air, natural vectors, and alternative host plant species. The pathogen also develops newer more virulent strains due to the development of new cultivars and the use of fungicides, other changing agri-practices, and changing global climatic conditions. Various studies suggest that the pathogen coevolves with the newly developed cultivars hence overcoming resistance and causing more severe diseases (Lo Iacono et al. 2012). Additionally, the evolution of fungal strains with the use of newly developed fungicides becomes a matter of major concern for the farmers and the administrators (Hahn 2014). Hence, it is imperative to maintain constant and regular monitoring of emerging new strains that exhibit higher infection potential or increased resistance to agrochemicals. This proactive approach is essential for preventing potential damage in the future. Pathogen detection and disease diagnosis may be required for numerous purposes: such as (a) to determine the presence and quantity of the pathogens; (b) to evaluate the efficacy of the application of physical, cultural, biological, or chemical methods of containing the pathogens; (c) seed certification for plant quarantine and certification programs; (d) to detect and recognize new pathogens strains or fungicide resistant strain in order to manage them; (e) to study taxonomic and evolutionary relationships of rice pathogens; (f) to evaluate the level of disease incidence and resulting yield loss; (g) to evaluate pathogen infection in plant materials used in breeding programs; and (h) to study the pathogenesis and gene functions. Hence, an appropriate knowledge of disease assessment protocols and the associated tools is a prerequisite in order to implement effective disease management strategies. The chapter presents a comprehensive review of rice diseases and existing disease detection methodology focusing on visual and nucleic acid–based techniques.

1.2 Rice Blast

Rice blast is an important disease in the rice crop around the world. It causes a 10–30% loss in rice production (Talbot 2003). *Magnaporthe oryzae* is an ascomycete which belongs to the family Magnaporthaceae. The pathogen infects and causes blast disease in more than 50 cultivated plant species including rice, wheat, barley, and finger millet (Ou 1980). The Commonwealth Mycological Institute (CMI) has documented its presence in 85 countries all over the world. Blast disease occurs in varying environmental conditions and different types of rice ecosystems.

1.2.1 Conventional Approach to Diagnose *Magnaporthe oryzae*

The rice blast is categorized as leaf blast, collar blast, node blast, and neck blast. In leaf blasts, the lesions on the leaf blade are spindle-shaped or elliptical with greenish-gray centers and brown margins. Later on, the lesions enlarge, coalesce, and eventually kill the leaves under favorable conditions. The size, color, and shape of the lesions may vary depending on the prevailing climatic conditions (Padmanabhan 1974; Rao and Manibhushan 1994). Collar blast occurs when the pathogen infects the collar region, potentially leading to the complete death of the leaf blade. Additionally, the pathogen can infect the stem node, causing it to turn blackish and become fragile, a condition known as node blast. The panicle neck can also become infected, exhibiting a grayish-brown lesion that causes the panicle to bend or fall over in severe cases. Furthermore, the pathogen is responsible for brown lesions on the panicle branches and spikelets. After observing the symptoms, the isolation of the pathogen from the diseased plant part using an appropriate growth medium can be done to further study the morphological and cultural characteristics of the pathogen for identification.

On potato dextrose agar, the blast fungus develops a grayish colony with a circular smooth margin and a concentric ring pattern. The mycelium of *M. oryzae* is septate, and both the nuclei within the mycelium and the spores are haploid. In laboratory conditions, the sexual or teleomorphic stage of the rice blast pathogen can be induced by pairing isolates of opposite mating types, although it has not been observed in field conditions. It forms fusiform-shaped ascospores with three septa, contained within specialized structures called perithecia, where the asci are unitunicate. The fungus is classified as heterothallic, featuring a bipolar mating system where mating is governed by two different alleles at a single locus, alongside additional genes that regulate the sexual cycle. The asexual stage of *M. oryzae* is known as *Pyricularia oryzae*, which prolifically produces spores called conidia on lesions and in culture on conidiophores. Conidiophores of *M. oryzae* typically form clusters around each stoma, although solitary occurrences are infrequent, and they possess 2–4 septa. The basal part of the conidiophores is swollen and tapers toward a lighter apex. The conidia are three-celled and are produced at the apex of a conidiophore. They are typically pyriform or obclavate, measure 20–22 × 10–12 μm, and are 2–3 septate, translucent, and slightly darkened (Gashaw et al. 2014).

1.2.2 Nucleic Acid–Based Diagnosis of Rice Blast

Kuribayashi (1928) initially documented the seed transmission of *M. grisea* in Japan, later observed in various other regions worldwide. Infected rice seeds serve as a means to introduce the fungus into new areas. Therefore, employing disease-free seeds is crucial in integrated rice blast disease management (Sattar and Savitri 1999). Early detection of *M. grisea* in rice seeds is imperative. Hence, a PCR-based diagnostic assay was devised for identifying *M. grisea* in infested rice seeds (Chadha and Gopalakrishna 2006). Primers were designed using the nucleotide sequence of an infection-specific gene, mif23. This PCR-based seed assay has the capability to detect *M. grisea* in rice seed lots with as low as 0.2% infestation and exhibits a sensitivity of 20 μg of genomic DNA.

A swift and highly accurate detection technique was devised, utilizing real-time quantitative analysis, employing a Taqman probe tailored to MHP1, a unique gene responsible for encoding HYDROPHOBIN, which plays a crucial role in virulence. The newly developed method demonstrated remarkable sensitivity, capable of detecting fungal DNA at levels as low as 1 pg (Su'udi et al. 2013). Additionally, primers targeting the 18s–28s region of ribosomal DNA were tailored for the detection of *M. oryzae*, exhibiting a sensitivity of 6.9×10^{-5} ng genomic DNA (Sun et al. 2015). The efficacy of ribosomal DNA in detecting *M. grisea* has previously been confirmed by Qi and Yang (2002). The advancement led to the creation of the LAMP technique, allowing for the early detection of pathogens with a sensitivity of 10 pg of pure genomic DNA (Li et al. 2019). In designing the LAMP primers, the Alb1 superfamily hypothetical protein MGG_04322, involved in ribosome and melanin biogenesis, was selected as the target. To quantify rice blast fungus conidia and enhance sensitivity, they developed a quantitative LAMP (q-LAMP) detection system. This assay facilitated rapid detection (within 35 min) of fungal spores, achieving a sensitivity of 3.2 spores per milliliter.

1.3 Sheath Blight of Rice

Sheath blight of rice is another important disease of rice caused by a soil-borne fungus *Rhizoctonia solani* AG-1 IA, belonging to the phylum Basidiomycota, family Ceratobasidiaceae. The teleomorph is *Thanatephorus cucumeris* (A.B. Frank) Donk. It causes significant damage to crops, and up to 50%, yield losses have been observed under most conducive environments (Sridevi et al. 2008).

1.3.1 Conventional Approach to Diagnose *Rhizoctonia solani*

The infected rice plant was observed for its typical sheath blight symptoms. The symptoms on the leaf sheath manifest as circular, oval, or ellipsoid water-soaked spots with a greenish-gray color. With the progression of the disease, these spots enlarge and merge, forming larger lesions characterized by grayish-white centers surrounded by irregular margins or outlines ranging from tan to dark brown. When the infection spreads to leaf blades, it results in irregular lesions with margins displaying hues of dark green, brown, or yellow-orange. These lesions can proliferate extensively, merging on either partial or entire leaf blades, resembling a pattern similar to rattlesnake skin. The infected tissues disrupt the transportation of water and essential mineral nutrients to the leaves, flag leaves, and panicles. As the pathogen advances from initial sheath infections, it can infect the culms, weakening them significantly. This leads to lodging and the collapse of tillers. In favorable conditions, the fungus develops white, web-like hyphae on the sheaths and leaves, aiding in the transmission of the disease from one leaf to another and leading to infections in neighboring plants. During the boot to heading stages, sclerotia of the fungus typically form within the infected sheaths and leaves. Initially appearing white, these sclerotia gradually darken to brown or dark brown. Loosely attached, they easily detach from the plants at maturity.

In field conditions, sexual reproduction is not frequently observed in *R. solani*. The perfect stage is *Thanatephorus cucumeris*, which forms a thin, mildew-like growth on soil, leaves, and infected sheaths located just above the ground line. Under conditions of high humidity, it forms basidia with four sterigmata, each bearing a single basidiospore.

R. solani lacks the production of asexual spores or conidia, relying instead on plant infection through vegetative mycelia or hyphal threads formed from germinating sclerotia. When young, the mycelium appears white, but as it matures, it transitions to a yellowish or light brown hue. It produces long multinucleate cells that grow approximately perpendicular to the main hypha, exhibiting a slight constriction at the junction between the main hypha and its branches. Hyphal cells are separated by septa containing pores, facilitating the movement of mitochondria, nuclei, and cytoplasm between cells.

1.3.2 Nucleic Acid–Based Diagnosis

The infection of rice due to different species, such as *R. solani*, *R. oryzae*, and *R. oryzae-sativae*, of sheath disease is not easily distinguished at the early stages of lesion development. Therefore, a PCR-based method relying on variation in ITS region was used to generate specific primers to differentiate *Rhizoctonia* species present in rice (Johanson et al. 1998). Matsumoto et al. (2002) used 28S ribosomal DNA of *R. solani* and developed a primed PCR assay for the detection and identification of *R. solani* isolates in AG 1-IA, AG 1-IB, and AG 1-IC.

Further, the detection and quantification of *Rhizoctonia solani* AG-1 IA were accomplished by real-time PCR with 1 pg sensitivity (Sayler and Yang 2007). Real-time PCR is specific and sensitive for detecting and quantifying sheath blight pathogen in infected rice tissue before mycelia and sclerotia are visible, even before symptoms appear. It can evaluate the resistance level of rice cultivars and can be performed in the field using portable machines. A more efficient LAMP assay is designed to detect the pathogen in plant tissues and soil samples (Patel et al. 2015).

The ITS sequence was used to design LAMP primers, for detecting most anastomosis groups of *R. solani* with improved sensitivity and could distinguish *R. solani* and *R. zeae*, from *R. oryzae*. Additionally, for field-based applications to identify the pathogen in infected plant samples, seeds, vegetative cuttings, and soil samples, the LAMP protocol was integrated into a lateral flow detection (LFD) system using generic anti-biotin and anti-fluorescein antibodies. The LAMP reactions utilized biotin-labeled primers, which were then hybridized with a fluorescein amidite (FAM)-labeled hybridization probe and subsequently detected using the LFD. This LAMP-LFD method is simple and rapid, offering comparable performance to quantitative real-time PCR (Patel et al. 2015).

1.4 Brown Spot of Rice

The brown spot of rice, caused by *Bipolaris oryzae* (teleomorph, *Cochliobolus miyabeanus*), has been documented in various rice-growing countries, including Japan, China, Burma, Sri Lanka, Bangladesh, Russia, North America, Philippines, Saudi Arabia, Iran, Africa, South America, Australia, Malaya, and Thailand (Ou 1985a, b; Khalili et al. 2012). In India, also, it affects the rice productivity in all rice-growing states (Frinking 1985; Ou 1985a, b). The severity of the disease is increased under low-input management and in dry/direct seeded rice cultivation, particularly in the states of Bihar, Orissa, Assam, Chhattisgarh, Madhya Pradesh, Jharkhand, and West Bengal.

It causes a yield loss of 12% under moderate infection and is negligible at lower infection ratings; however, 30–45% yield loss occurs during heavy infection as it significantly reduces the number of tillers and grains (Ou 1985a, b). Grain yield losses vary according to rice cultivars and the stage of infection (Kulkarni et al. 1980).

1.4.1 Conventional Approach to Diagnose *B. oryzae*

The pathogen damages the crop from the seedling to the milk stage, manifesting symptoms as minute spots on the coleoptile, leaf blade, leaf sheath, and glume. The symptoms are distinguished by small circular brown spots with gray or whitish centers, typically cylindrical or oval in shape, resembling sesame seeds and often surrounded by a yellow halo. As the infection advances, the merging of spots becomes apparent, ultimately resulting in the drying up of the leaf.

Under favorable conditions, discolored and shriveled grains, marked by black or dark brown spots, develop on the glumes. Additionally, dark brown conidiophores and conidia emerge on these spots, imparting a velvety appearance. In severe instances, grayish mycelial growth becomes apparent between the sheath and stalk. Dry rot symptoms may also be observed as the infection progresses upward. The brown spot pathogen forms brown or grayish-brown spots on the neck region, contrasting with the blackening observed in cases of neck blast (Sunder et al. 2005).

The pure culture appears gray to olive or black in color. Under microscopic examination, sporophores are observed as thick, erect structures, geniculate, and dark olivaceous at the base, gradually lightening toward the tip. They emerge through stomata in clusters of 3–5. Conidia, typically 5–10 septate, exhibit an arrangement where the oldest conidium is situated toward the base. They are slightly curved and widest in the middle. Completely matured conidia display a fuliginous or brownish hue and germinate with two polar germ tubes, one originating from each of the thin-walled regions. In less mature stages, subhyaline spores may produce germ tubes from intermediate segments. In India, the size of conidiophores and conidia from *Bipolaris oryzae* isolates has been noted to range from 70–175 × 5.6–7 μm to 45–106 × 14–17 μm, respectively (Ou 1985a, b).

1.4.2 Nucleic Acid–Based Diagnosis of Brown Spot of Rice

Identification of the correct species of *Bipolaris* solely based on morphological characteristics can be challenging and is not commonly practiced. Nowadays, molecular characterization using DNA fingerprinting has emerged as a faster and more reliable method for identifying fungi. A prevalent method for identifying plant pathogenic fungi involves analyzing the internal transcribed spacer (ITS) sequences of ribosomal DNA (rDNA). Archana et al. (2011) developed a PCR-based diagnostic technique to detect *Bipolaris oryzae* in rice. They employed species-specific primers *BoVf* and *BoVr*, which exclusively targeted DNA from all *B. oryzae* isolates without amplifying DNA from other species. The sensitivity of PCR detection ranged from 100 ng μL^{-1} to 1 fg μL^{-1} of *B. oryzae* DNA. This method offers a diagnosis within one day, contrasting with the cultural method, which requires 10 days.

1.5 Bakanae Disease of Rice

The emergence of the bakanae disease, attributed to *Gibberella fujikuroi* (*Fusarium fujikuroi* Nirenberg), is posing significant challenges to rice cultivation in India, Japan, Thailand, and Taiwan, as documented by various studies (Bashyal et al. 2016a, 2017; Kini et al. 2002; Saremi 2005; Webster and Gunnell 1992). The northwestern region of India is affected by *Fusarium fujikuroi*, as noted by Fiyaz et al. (2016). Bakanae disease significantly diminishes both the quality and yield of basmati rice, as evidenced by studies conducted by Bashyal et al. (2016a, b) and Gupta et al. (2015). Although the bakanae disease of rice was initially described in Japan in 1828 (Ito and Kimura 1931), the specific *Fusarium* species associated with it could not be identified at that time. Originally, *G. fujikuroi* was viewed as a single species by early authorities like Snyder and Hansen (1945). However, subsequent research revealed that this classification actually comprises multiple distinct species, now collectively known as the *G. fujikuroi* species complex (GFSC).

In India, the bakanae disease is associated with three *Fusarium* species: *Fusarium fujikuroi*, *F. proliferatum*, and *F. verticillioides*, as documented by Bashyal and

Aggarwal (2013) and Bashyal et al. (2016b). Among these, *F. fujikuroi* is identified as the primary cause of this disease in the field (Bashyal et al. 2016a). The pathogen is primarily transmitted through seeds but can also persist in soil and infected plant debris (Ou 1985a, b). Consequently, plants can become infected either through contaminated seeds or after transplantation in the main field. The main reservoir of inoculum for this disease is rice seeds, resulting in yield loss, quality decline, and germination issues. Therefore, the most effective strategy for managing bakanae disease is to use healthy seeds.

1.5.1 Conventional Approach to Diagnose *F. fujikuroi*

The infected rice plant was visually examined for symptoms, as bakanae disease presents various manifestations such as seedling rot before emergence, seedling blight, crown rot, root decay, plant stunting, abnormal elongation, sterility, and the formation of empty panicles. These symptoms occur under diverse climatic conditions worldwide, as documented by Sun and Snyder (1981a, b), Ou 1985a, b), Webster and Gunnell (1992), and Desjardins et al. (2000). Bakanae disease affects all parts of the rice plant, including the roots, crowns, stems, leaf sheaths, and panicles. While some infected plants may perish during the crop's growth period, others manage to survive until maturity. These surviving plants often exhibit tall, slender, and lanky tillers, accompanied by white, empty, chaffy panicles. Panicles typically collapse under their own weight, causing plants to succumb before reaching maturity (Webster and Gunnell 1992). On the damaged grains, at the junction of the lemma and palea, one can observe the pink sporodochia of the pathogen. In India, the emergence of adventitious roots from lower nodes of the stem was documented, leading to the naming of the disease as foot rot (Thomas 1931).

In terms of morphology, *F. fujikuroi* initially produces white mycelia, which darkens to gray-violet or magenta over time. While sporodochia are generally absent, if they do appear, they are pale orange in color. In instances where sporodochia are present, they contain macroconidia. However, numerous isolates exhibit minimal sporodochia formation or completely lose this capability after undergoing repeated subculturing. Macroconidia are moderately slender, with a tapered apical end and a poorly developed basal cell containing three to five septa. Microconidia, meanwhile, are either club-shaped or oval, featuring a flattened base and zero to one septum. Polyphialides give rise to false heads and short to medium chains, which may proliferate and commonly develop into monophialides. Notably, chlamydospores are absent in this process (Leslie and Summerell 2006). Mating populations (MPs) are utilized to address taxonomic complexities and differentiate species. Within each MP, the mating pattern is heterothallic, controlled by two alleles at a single mating–type locus. Members of the same MPs are sexually fertile with one another but not with members of different MPs. *F. fujikuroi* stands out as a significant pathogen characterized by considerable variability in pathogenicity and a broad host range. Additionally, many *Fusarium* species exhibit various formae speciales (f. sp.) and physiological races (Zemankova and Lebeda 2001). Sun and Snyder

(1981a, b) successfully generated perithecia by laboratory crosses of *F. fujikuroi* strains. This led to the identification of four reproductively isolated groups within *G. fujikuroi*, labeled as A, B, C, and D. Mating group C comprised genetically interfertile strains originating from rice (Hseieh et al. 1977; Kuhlman 1982). Mating population C (MP-C), the anamorph of *Fusarium fujikuroi*, was initially identified among rice strains originating from Taiwan (Hseieh et al. 1977). Additionally, it has been implicated as the causative agent of bakanae disease in Italy (Amatulli et al. 2010; Dal Pra et al. 2010). Mating population A (MP-A), the anamorph of *F. verticillioides* (synonym *F. moniliforme*), and mating population D (MP-D), the anamorph of *Fusarium proliferatum*, have both been isolated from rice originating in Asia. Furthermore, MP-D has been isolated from rice crops in Africa, Australia, and the USA (Amoah et al. 1995; Desjardins et al. 1997). Group C is primarily restricted to rice, while the other groups are associated with hosts such as sugarcane, pine, rye, wheat, asparagus, corn, or sorghum (Puhalla and Spieth 1985). Kuhlman (1982) identified a fourth mating population, designated as D, and classified four varieties of G. fujikuroi—moniliformis, subglutinans, fujikuroi, and intermedia—based on mating population, ascospores, perithecia size, phialide type, and microconidial formations.

1.5.2 Nucleic Acid–Based Diagnosis

The fungal pathogen of bakanae was identified initially by real-time polymerase chain reaction (PCR) (TaqMan) assays using species-specific TEF primers (Amatulli et al. 2012). The developed method could detect pathogens at early stages from infected rice tissues and seeds. Carneiro et al. (2017) introduced a more efficient and sensitive real-time PCR targeting the elongation factor-1 alpha (EF-1α) region to identify *F. fujikuroi*, achieving a detection threshold as low as 27.5 fg of DNA. This assay facilitated the detection of bakanae disease in asymptomatic tissues during the initial stages of rice growth. Additionally, Ortega et al. (2018) devised an enhanced and swift loop-mediated isothermal amplification (LAMP) assay for *F. fujikuroi*. This LAMP assay targeted the elongation factor-1 alpha sequence of *F. fujikuroi*, achieving a detection threshold ranging from 100 to 999 fg of DNA. Thereafter, Rong et al. (2018) utilized the intergenic spacer (IGS) region of the nuclear ribosomal operon in *F. fujikuroi* to design a LAMP assay. The assays produced results within 70 min at 62 °C, requiring only 67 pg/μL DNA for detection. A visible color change from purple to sky blue was observed solely in the presence of the DNA from the targeted pathogens. This change occurred by adding hydroxynaphthol blue (HNB) to the reaction system before amplification, as illustrated in Fig. 1.1.

A specific LAMP assay using a gibberellin biosynthesis pathway gene NRPS31 was established by Zhang et al. (2019). Given the seed-borne nature of the pathogen and the challenges associated with morphological identification, there is a pressing need to develop a rapid detection assay for the effective management of bakanae (Sunani et al. 2019). PCR- and LAMP-based assays were devised for detecting *F. fujikuroi*. Utilizing the unique sequences (NRPS31 gene) of *F.*

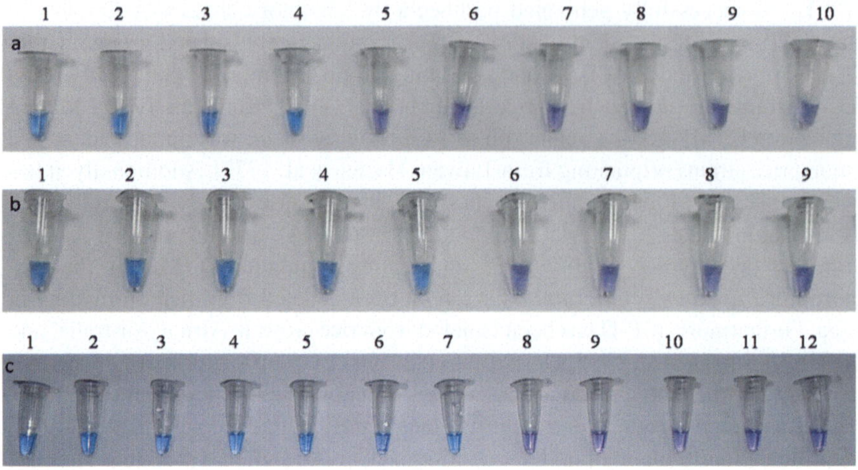

Fig. 1.1 Visual detection of LAMP reaction using HNB dye: The color changes from violet (negative reaction) to sky blue (positive reaction). (**a**) Specificity of the LAMP assay. Tubes from left to right (1) *Fusarium fujikuroi* (F250); (2) *F. fujikuroi* (F322); (3) *F. fujikuroi* (F308); (4) *F. fujikuroi* (F309); (5) *F. proliferatum*; (6) *F. verticillioides*; (7) *F. thapsinum*; (8) *Rhizoctonia solani*; (9) *Bipolaris oryzae*; (10) NTC (non-template control). (**b**) Sensitivity of the LAMP assay for detection of *Fusarium fujikuroi.* (1) 100 ng; (2) 10 ng; (3) 1 ng; (4) 100 pg; (5) 10 pg; (6) 1 pg; (7) 100 fg; (8) 10 fg; (9) negative control. (**c**) Validation of LAMP assay in *F. fujikuroi* inoculated rice seedlings. (1, 2) inoculated seedlings of Pusa Basmati 1121; (3, 4) suspected seeds of Pusa Basmati 1121; (5) diseased plant of Pusa Basmati 1509; (6, 7) suspected plants collected from rice field; (8–12) disease-free rice plants

fujikuroi retrieved from GenBank accession KY953210, a specific marker, BMFf1900, was established for amplifying *F. fujikuroi* with a sensitivity of 10 pg. For accurate quantification of the inoculum, real-time primers (NRPS31RT-F and NRPS31RT-R) were formulated, increasing the assay's sensitivity to 10 fg. These PCR and LAMP assays have been verified in field conditions, seed testing, and the assessment of rice genotype resistance. The developed diagnostics are efficient, rapid, and highly sensitive for detecting *F. fujikuroi*. They prove valuable for assessing seed health, confirming varietal resistance, and monitoring pathogen inoculum levels (Sunani et al. 2019).

1.6 False Smut of Rice

The rice disease known as false smut, attributed to *Ustilaginoidea virens* (Cooke) (Takahashi), is a common disease affecting rice crops worldwide. Its first occurrence in India was reported in the Tirunelveli District of Tamil Nadu State (Cooke 1878). Initially perceived as a sign of a bountiful harvest and regarded as a minor ailment due to its sporadic occurrence, this disease is now on the rise. The increased incidence can be attributed to the widespread adoption of high fertilizer-responsive cultivars and hybrids, extensive use of nitrogenous fertilizers, and notable shifts in climatic conditions. Since 2001, the disease has been observed in severe forms in the primary rice-growing states of India such as Uttar Pradesh, Uttaranchal, Haryana, Punjab, Jammu and Kashmir, Karnataka, Andhra Pradesh, Tamil Nadu, Puducherry Bihar, Jharkhand, Gujarat, and Maharashtra (Dodan and Singh 1996; Mandhare et al. 2008).

The disease leads to both qualitative and quantitative losses, with reductions in grain yield due to factors such as chaffiness, decreased test weight, and sterility of spikelets adjacent to smut balls. Yield losses have been estimated to vary from 0.2 to 50% across different regions of the country (Dodan and Singh 1996). Additionally, a recent proposal suggests *Villosiclava virens* as the new name for the teleomorph of the false smut fungus (Tanaka and Tanaka 2008).

1.6.1 Conventional Approach to Diagnose False Smut

The symptoms caused by *U. virens* are observed only after flowering. The usual symptom induced by *U. virens* in rice involves the development of a white fungal mass emerging from the inner space of a spikelet during the initial stages of infection. This mass then transforms into a light-yellow smut ball before ultimately turning greenish-black in color. The mature smut ball's outer layer is rich in chlamydospores and often coated with sclerotia in autumn (Ikegami 1961, 1963). Its lifecycle includes both sexual (ascospores) and asexual (chlamydospores) stages (Biswas 2001). In its pure culture, the fungus grows slowly, displaying two distinct colony types: one characterized by green, hardened sclerotia-like features after 2 weeks of incubation and the other by white, spreading mycelia evident after 3 weeks of incubation.

The utilization of a light microscope enables the observation of morphological features such as spined chlamydospores, germinating chlamydospores, hyphae, and secondary conidia, which aids in the identification of isolates as *U. virens*. The chlamydospores are round to elliptical and measure from 3 to 5 μm in diameter (Kim and Park 2007). Irregularly curved spines with a length of 200–500 nm can be observed on the surface of chlamydospores (Kim and Park 2007). The mature false smut ball occasionally generates sclerotia, characterized by being black, horseshoe-shaped, irregularly oblong, or flat, measuring from 2 to 20 mm in length (Fu et al. 2012). Scanning electron microscopy reveals a rough outer sclerotial wall and a compactly intertwined interior of hyphae within the sclerotia (Fu et al. 2012).

1.6.2 Nucleic Acid–Based Diagnosis

U. virens-specific primers and corresponding nested primers have been designed based on the ribosomal DNA internal transcribed spacers (ITS1 and ITS2) and the 5.8S rRNA gene (Zhou et al. 2003) and effectively used for the detection of *U. virens* in various tissues of rice and associated grasses, expediting the timely prediction of false smut disease in fields (Zhou et al. 2006). Nested PCR has been devised for the detection of *U. virens* in rice panicles, both pre- and post-heading. Findings reveal the pathogen's presence before heading (Ashizawa and Kataoka 2005). Real-time PCR-based assays, known for their enhanced sensitivity over conventional nested PCR, have been developed and effectively utilized to detect the presence of *U. virens*. This detection has been achieved in soil (Ashizawa et al. 2010), rice samples (Li et al. 2013), and air samples collected around susceptible rice plants in fields (Zheng et al. 2012).

Tang et al. (2017) devised a nested polymerase chain reaction (PCR) assay employing a comparative genomics approach to detect *U. virens* within the plant host. This method utilized six sets of nested PCR primers with high specificity and sensitivity, capable of detecting 1–10 fg/mL of DNA from field or lab samples. These primers enable the detection of *U. virens* in various rice tissues and even in field conditions, facilitating a deeper understanding of the mechanisms underlying the *U. virens* infection cycle.

Several techniques, including polymerase chain reaction (PCR), nested PCR, and real-time PCR, have been devised for *U. virens* detection. However, the majority of these methods require advanced equipment such as a thermal cycler. An alternative, the LAMP assay, utilizing a specific region of the UvG-β1 gene, has been developed with a sensitivity of approximately 1 pg (Yang et al. 2018).

1.7 Sheath Rot of Rice

Sheath rot of rice, induced by *Sarocladium oryzae*, is prevalent in rice-growing regions worldwide. For a significant period, rice sheath rot was regarded as a minor ailment. However, it has recently garnered attention and emerged as a notable threat to rice production. This shift is attributed to substantial alterations in rice farming practices, including increased plant density, elevated nitrogen fertilizer application rates, the adoption of semi-dwarf and photoperiod-insensitive cultivars, utilization of high-yielding varieties, implementation of efficient water management systems, and advancements in seeding techniques, among others. Consequently, losses attributed to *S. oryzae* can vary widely, ranging from 20 to 85% (Sakthivel 2001).

1.7.1 Identification of *S. oryzae* Based on Morphological and Cultural Characters

The primary symptoms of sheath rot disease include rotting of the uppermost leaf sheaths surrounding the young panicles. These manifest as oblong or irregular lesions, ranging from 0.5 to 1.5 cm, with gray centers and brown margins, or they may appear grayish brown throughout. As the disease progresses, these lesions enlarge and often merge, covering a significant portion of the leaf sheath. Consequently, the young panicles may remain enclosed within the affected sheaths or emerge partially. Additionally, a dense whitish powdery growth may be observed inside the affected sheaths, ultimately leading to the rotting of young panicles (Ou 1985a, b). As a consequence of *S. oryzae* infection, the seeds exhibit characteristics such as being chaffy and discolored. Moreover, the infection affects both the viability and nutritional value of the seeds (Sakthivel 2001; Gopalakrishnan and Valluvaparidasan 2010).

The mycelial colony of *S. oryzae* displayed a compact appearance with restricted growth. Sparse, cottony white aerial mycelium was observed. Additionally, the underside of the plate, corresponding to the culture on the upper side, exhibited orange-colored discoloration. When the colony was consistently exposed to light, it displayed hints of orange coloration. The hyphae exhibited sparse branching and were septate. Conidiophores emerging from the mycelium were slightly thicker than the vegetative hyphae and branched once or twice. Conidia were cylindrical to slightly fusiform, aseptate, and hyaline, measuring 4–7 × 1–2 μm in size, and arranged in slimy heads.

A detailed summary of diagnostics developed for the detection and characterization of major fungal pathogens infecting rice crops is given in Table 1.1.

Table 1.1 Diagnostics developed against major pathogens of rice crop

Disease	Primers	Sequences (5′–3′)	Purpose	Reference
Rice blast	F R	GGCATCGTTAGCGGTCTTC CTACGAGGCTGCATAACGAC	qPCR	Qi and Yang (2002)
	F R	CGTCACACGTTCTTCAACC CGTTTCACGCTTCTCCG	PCR	Harmon et al. (2003)
	F3 B3 FIP(F1c-F2) BIP (B1c-B2)	GCCAGGCCAAGACGGT GCGCACTTTCAGAAATCGC GGCCTCGAGCTGCTTCTTGGACGGAAGGCTTGGGACG ACA AAGGAAGACGATGATGCGGCCCGAATTTTGACACGAC CCCG	LAMP	Li et al. (2019)
Sheath blight of rice	ITS1F ITS2 R ITS3 F ITS4R	TCCGTAGGTGAACCTGCGGGCT GCGTTCTTCATCGATGC GCATCGATGAAGAACGCA TCCTCCGCTTATTGATATGC	PCR	White et al. (1990)
	GMROS-2 R GMROS-6 F	CGGCTCATCCTTGAAAAATCC GAAAGAGAGAGAGGTCGCCTC	PCR	Johanson et al. (1998)
	R. solani AG-common primer (F) *R. solani* subgroup-specific primer (R) AG 1-IA	CTCAAACAGGCATGCTC CAGCAATAGTTGGTGGA	PCR	Matsumoto et al. (2002)
	Rs1F Rs2R	GCCTTTTCTACCTTAATTTGGCAG GTGTGTAAATTAAGTAGACAGCAAATG	Real-time PCR	Sayler and Yang (2007)
	RUBQFRUBQR	GTGGCCAGTAAGTCCTCAGC GAAACGGGACACGACCAAGG	qPCR	Sayler and Yang (2007)

Brown leaf spot	BoVf BoVr	CGCCCATCTTATGCAGTTTCC CGAGTCTCCCAGAAAGAGG	PCR	Matsumoto et al. (2002)
	CmSCD1_44F CmSCD1_302R	CATGTGTGCAGTAAAGTGACTC GTCTTGAGGAGGGGGTT	Taqman real-time PCR	Su'udi et al. (2012)
Bakanae	ef1	ATGGGTAAGGA(A/G)GACAAGAC	PCR	O'Donnell et al. (1998)
	ef2	GGA(G/A)GTACCAGT(G/C)ATCATGTT		
F. fujikuroi	Fuji1F	ACGTGTCAAACTAAACATTCGA	qPCR	Amatulli et al. (2012)
	TEF1R	GCGACAACATACCAATGACG		
F. proliferatum	Proli1F TEF	GTCACGTGTCAAGCAGCGA	qPCR	Amatulli et al. (2012)
	TEF1R	GCGACAACATACCAATGACG		
	TqF2	GGCGCGTTTTGCCCTTTCCT		Carneiro et al. (2017)
	TqR	AGCGGCTTCCTATTGTCGAA		
	FfujiPq	[FAM]-TCACGTGTCAAACTAAA-[NF-MGB]		
	TqF1	CGAGTGATGGGCGCGTTTTG		
	TqR	AGCGGCTTCCTATTGTCGAA		
	TaqFuji	[FAM]-CGCATTGTCACGTGTCAAAC- [NF-MGB]		
	FfujiFq	CACGTGTCAAACTAAACATTCG		
	FfujiRq	GATGGTGATACCACGCTCAC		
	BakanPq	[FAM]-GCCGCTGAGCTCGGTAAGGG-[NF-MGB]		

(Continued)

Table 1.1 (continued)

Disease	Primers	Sequences (5′–3′)	Purpose	Reference
F. fujikuroi	IGS-Ff F3 B3 FIP (F1c + F2) BIP (B1c + B2) LF	AGACACCGTTTTGTTTTGC CTTTTCTCCAACTTCACAGC TCAGCCTACCCTAGACCCTC GCCCTTACCTTTGCTTGA AGCTAGACTACTCTAAGGGTAGGT ACAACTTCCACCAATTGCA ATCCGGCCAAGACCGTC	LAMP	Rong et al. (2018)
F. proliferatum	RED1-Fp-F3 B3 FIP (F1c + F2) BIP (B1c + B2) LF LB	ATCGGCGTTTTCAACTCT TTGGATGCTGACGACATGAA CTATGCCAGAGCCGACGTAGAC CCAAGCATTCCTCCCCCT GACCCCAATGGCTGAGGTACCT CGCGAAGTAGTCAACCATCT CTTTGCAGCTGGGGCTG GTTTAGTTACGCCGCCGCCAAA	LAMP	
	F3 B3 FIP BIP Biotin-FIP FAM-probe	CGAATGTAATGGATGTAACACA CATGATACTCAAACAGGCATG TTCGCTGCGTTCTTCATCGATTCTCATACTAAGTTTCAACAACG TGCGATAAGTAATGTGAATTGCAGATCCAAGGAATACCAAGGAG Biotin-TTCGCTGCGTTCTTCATCGATTCTCATACTAAGTTTCAACAACG FAM-TCGAATCTTTGAACGCACCTT		
	Nrps31 F Nrps31 R NRPS31 RT F NRPS31 RT R F3 B3 FIP (F1c-F2) BIP (B1c-B2)	CCTCATACCGTTTCGCTTGT CGCTACCTACGCATTTCGGG GAGTATCTGAGGAACGAGCTTG CAGTTCATATTACCCCAGTCCG AAGTTCCTCGATGCTGTAGC CTCAGACTGGTGCTTCACAA CGCGGACAACCCAATTTCCGTA-CTCTCTACTGCCTGGGCTC CTGTATGGGCAGTCTTGGACGC-GAAGGGCTGGAAGAGGTTC	PCR qPCR LAMP	Sunani et al. (2019)
False smut of rice	US1-5 US2-5 US3-3 US4-3	CCGGAGGATACAACCAAAAAAACTCT CAATGCATGTCTGAGTGGATTTTTG GCTCCAAGTGCGAGGATAACTGAAT CCAACACCAAGCGCAAGACAGA	PCR	Zhou et al. (2003)

	uvr1497F uvr1586R probe	CCGCTGCCTAAGATAAAGTCC AGGCTCCCGGTTGTTTTTAC GGCCCCAG	qPCR	Ashizawa et al. (2010)
	ITS1-DP2 F ITS4 R	GGATACAACCAAAAAAACTC GCATATCAATAAGCGGAGGA	PCR	Ashizawa et al. (2010)
	ITS1-DP3 F ITS2-DP1 R	TCTTGTGTTTTCCAATGCA TTCAGTTATCCTCGCACTTG	Nested PCR	Ashizawa et al. (2010)
	RFS-F RFS-R Probe	GCGAGAGGATGCTACAATGGA AATCCGCAGCCAAGCCCCTACG GCATACGGGGAAGGTTCAGA	qPCR	Li et al. (2013)
	Uv544F-O Uv544R-O Uv544F-I Uv544R-I Uv858F-O Uv858R-O Uv858F-I Uv858R-I Uv924F-O Uv924R-O Uv924F-I Uv924R-I Uv1384F-O Uv1384R-O Uv1384F-I Uv1384R-I Uv1581F-O Uv1581R-O Uv1581F-I Uv1581R-I Uv5140F-O Uv5140R-O Uv5140F-I Uv5140R-I	TCTTGGCTCCTCGGAAGCTC CATGTCTTGCCCAAAGATGCG AGGTTCTACATGCGTGTTGTGA GCCCTGTGAAACTTGGTGC ACGAGCAGGCAGCAAACACC CCGCCACTTTGCCTGTCGATG TCTTCTTGCGCCTTCTGACA GCTCCTCCCATGCGTTTT TCCTCGTCCCTCCTGTCTT CGCCGAGAAGTATCCATCC CGACCTTGTCCACCTCGACCA TCACCCTGGCTCTCTTGCGAA AGCCCTTCTTTCTCGTCA TATTTCGGACCCGTGTTT CGTTTTCAAGCGGGACCCTA TCCGAGTTATTACCAGACATGC CGCCCTGAAATCGCCCTTTA ACCTGACCGTCTTGGCTCCC TCCTTCTTGTCTCGGCGTCT CCCTTGCCAATGGTCCATGCTT CCAGGCTCAGGTCAACTTCA CGCAGACCTTTCCCTTGC TCGGCACGCACCAGACACGTT TGGAGCCAGCGGAAACATGCAAA	Nested PCR	Tang et al. (2017)

(Continued)

Table 1.1 (continued)

Disease	Primers	Sequences (5′–3′)	Purpose	Reference
	F3	GCTCCTGGGATTCTTTGGTG	LAMP	Yang et al. (2018)
	B3	GCTCGATCGGGACAACCA		
	FIP (F1c-F2)	TTGTCGCGTTGCAAGGACACGGGCATTCGACTCGGAACAG		
	BIP (B1c-B2)	TTCACTGACATGTGCGCTGACCCGGTTTCGTGGAACGGATT		
	LoopF	CGATTACCGTGCGTGTTGGA		
	LoopB	TCCGCAGTTGAAAATATGG		

1.8 Conclusion

The rice crop suffers from several fungal diseases which are considered a major constraint in rice production. The rapid and accurate diagnosis of rice diseases is a prerequisite in order to manage the disease timely and effectively. There are numerous methods for the diagnosis of rice diseases. Conventional diagnostic methods entail isolating the presumed fungal pathogen from infected plants and identifying them based on taxonomical criteria. While these methods do not necessitate sophisticated equipment, they are time-consuming and labor-intensive and demand substantial knowledge of fungal taxonomy. Nucleic acid–based techniques, including polymerase chain reaction (PCR), real-time PCR, and LAMP, offer greater specificity, sensitivity, rapidity, and reliability for detecting, differentiating, and quantifying rice fungal pathogens. These methods are effective for identifying pathogens in symptomatic as well as asymptomatic rice plants and seeds.

References

Amatulli MT, Spadaro D, Gullino ML, Garibaldi A (2010) Molecular identification of *Fusarium* spp. associated with bakanae disease of rice in Italy and assessment of their pathogenicity. Plant Pathol 59:839–844

Amatulli MT, Spadaro D, Gullino ML, Garibaldi A (2012) Conventional and real-time PCR for the identification of *Fusarium fujikuroi* and *Fusarium proliferatum* from diseased rice tissues and seeds. Eur J Plant Pathol 134:401–408

Amoah BK, Rezanoor HN, Nicholson P, Mac-Donald MV (1995) Variation in the *Fusarium* section Liseola: pathogenicity and genetic studies of *Fusarium moniliforme* Sheldon from different hosts in Ghana. Plant Pathol 44:563–572

Archana B, Udaya Shankar AC, Chandra Nayaka S, Niranjana SR, Mortensen CN, Lund OS, Prakash HS (2011) Development of a PCR-based diagnostic assay for detecting *Bipolaris oryzae*, the brown spot fungus in rice. In: World congress on biotechnology, pp 352

Ashizawa T, Kataoka Y (2005) Detection of Ustilaginoidea virens in rice panicles before and after heading in the field using nested-PCR technique with species-specific primers. Jpn J Phytopathol 71(1):16–19

Ashizawa T, Takahashi M, Moriwaki J, Hirayae K (2010) Quantification of the rice false smut pathogen *Ustilaginoidea virens* from soil in Japan using real-time PCR. Eur J Plant Pathol 128:221–232

Bashyal BM, Aggarwal R (2013) Molecular identification of *Fusarium* spp. associated with bakanae disease of rice in India. Indian J Agric Sci 83(1):72–77

Bashyal BM, Aggarwal R, Sharma S, Gupta S, Rawat K, Singh D, Krishnan SG (2016a) Occurrence, identification and pathogenicity of *Fusarium* spp. associated with bakanae disease of basmati rice in India. Eur J Plant Pathol 144(2):457–466

Bashyal BM, Aggarwal R, Sharma S, Gupta S, Singh UB (2016b) Single and combined effects of three *Fusarium* species associated with rice seeds on the severity of bakanae disease of rice. J Plant Pathol 98(2):405–412

Bashyal BM, Rawat K, Sharma S, Kulshreshtha D, Gopala Krishnan S, Singh AK, Aggarwal R (2017) Whole genome sequencing of *Fusarium fujikuroi* provides insight into the role of secretory proteins and cell wall degrading enzymes in causing bakanae disease of rice. Front Plant Sci 8:2013–2025

Biswas A (2001) False smut disease of rice: a review. Environ Biol 19:67–83

Carneiro GA, Matic S, Ortu G, Garibaldi A, Spadaro D, Gullino ML (2017) Development and validation of a TaqMan real-time PCR assay for the specific detection and quantification of *Fusarium fujikuroi* in rice plants and seeds. Phytopathology 107(7):885–892

Chadha S, Gopalakrishna T (2006) Detection of *Magnaporthe grisea* in infested rice seeds using polymerase chain reaction. J Appl Microbiol 100(5):1147–1153

Cooke MC (1878) Some extra European fungi. Grevillea 7:13–15

Dal Pra MS, Tonti DP, Nipoti P, Alberti I (2010) First report of *Fusarium* audiyazi associated with rice bakanae in Italy. Plant Dis 94:1070–1070

Desjardins AE, Plattner RD, Nelson PE (1997) Production of fumonisin B1 and *moniliformin* by *Gibberella fujikuroi* from rice from various geographic areas. Appl Environ Microbiol 63:1838–1842

Desjardins AE, Manandhar HK, Plattner RD, Manandhar GG, Poling SM, Maragos CM (2000) *Fusarium* species from Nepalese rice and production of mycotoxins and gibberellic acid by selected species. Appl Environ Microbiol 66(3):1020–1025

Dodan DS, Singh R (1996) Recent advances in research on cowpea diseases. Present Status Agric Rev 17(4):227–240

Fiyaz RA, Yadav AK, Krishnan SG, Ellur RK, Bashyal BM, Grover N, Prabhu KV (2016) Mapping quantitative trait loci responsible for resistance to bakanae disease in rice. Rice 9(1):45–54

Frinking HD (1985) Current concepts on fungal diseases of rice. Neth J Plant Pathol 91:14

Gashaw G, Alemu T, Tesfaye K (2014) Morphological, physiological and biochemical studies on *Pyricularia grisea* isolates causing blast disease on finger millet in Ethiopia. J Appl Biosci 74:6059–6071

Gopalakrishnan KA, Valluvaparidasan V (2010) Effect of seed-borne *Sarocladium oryzae*, the incitant of rice sheath rot on rice seed quality. J Plant Protect Res 50:98–102

Gupta AK, Solanki IS, Bashyal BM, Singh Y, Srivastava K (2015) Bakanae of rice—an emerging disease in Asia. J Animal Plant Sci 25:1499–1514

Hahn M (2014) The rising threat of fungicide resistance in plant pathogenic fungi: *Botrytis* as a case study. J Chem Biol 7(4):133–141

Harmon PF, Dunkle LD, Latin R (2003) A rapid PCR-based method for the detection of *Magnaporthe oryzae* from infected perennial ryegrass. Plant Dis 87(9):1072–1076

Hseieh WH, Smith SN, Snyder WC (1977) Mating groups in *Fusarium moniliforme*. Phytopathology 67:1041–1043

Ikegami H (1961) Studies on the false smut of rice VII. Analysis on the occurrence of the smut balls at the different ages. Research bulletin, Faculty of Agriculture, Gifu University 14:27–37

Ikegami H (1963) Occurrence and development of sclerotia of the rice false smut fungus. Research bulletin, Faculty of Agriculture, Gifu University 18:47–53

Ito S, Kimura J (1931) Studies on the bakanae disease of the rice plant. Report of Hokkaido Prefectural Agricultural Experiment Stations 27:1–95

Johanson A, Turner HC, McKay GJ, Brown AE (1998) A PCR-based method to distinguish fungi of the rice sheath-blight complex, *Rhizoctonia solani*, *R. oryzae* and *R. oryzae-sativae*. FEMS Microbiol Lett 162(2):289–294

Khalili E, Sadravi M, Naeimi S, Khosravi V (2012) Biological control of rice brown spot with native isolates of three *Trichoderma* species. Braz J Microbiol 43:297–305

Kim KW, Park EW (2007) Ultrastructure of spined conidia and hyphae of the rice false smut fungus Ustilaginoidea virens. Micron 38(6):626–631

Kini KR, Leth V, Mathur SB (2002) Genetic variation in *Fusarium moniliforme* isolated from seeds of different host species from Burkina Faso based on random amplified polymorphic DNA analysis. J Phytopathol 150(4–5):209–212

Fu R, Ding L, Zhu J, Li P, Zheng AP (2012) Morphological structure of propagules and electrophoretic karyotype analysis of false smut Villosiclava virens in rice. J Microbiol 50:263–269

Kuhlman EG (1982) Varieties of *Gibberella fujikuroi* with anamorphs in *Fusarium* section Liseola. Mycologia 74:759–768

Kulkarni S, Ramakrishnan K, Hegde RK (1980) Incidence of brown leaf spot of rice caused by *Dreschslera oryzae* (Breda de Haan) Subram. and Jain under different agroclimatic conditions of Karnataka. Mysore J Agric Sci 14(3):321–322

Kuribayashi K (1928) Studies on overwintering, primary infection and control of rice blast fungus, Pyricularia oryzae. Jpn J Phytopathol 2(2):99–117

Leslie JF, Summerell BA (2006) *Fusarium* laboratory workshops—A recent history. Mycotoxin Res 22(2):73–74

Li H, Ni DH, Duan YB, Chen Y, Li J, Song FS, Yang JB (2013) Quantitative detection of the rice false smut pathogen *Ustilaginoidea virens* by real-time PCR. Genet Mol Res 12(4):6433–6441

Li L, Zhang SY, Zhang CQ (2019) Establishment of a rapid detection method for rice blast fungus based on one-step loop-mediated isothermal amplification (LAMP). Plant Dis 103(8):1967–1973

Lo Iacono G, Van den Bosch FN, Paveley N (2012) The evolution of plant pathogens in response to host resistance: factors affecting the gain from deployment of qualitative and quantitative resistance. J Theor Biol 304:152–163

Mandhare VK, Gawade SB, Game BC, Padule DN (2008) Prevalence and incidence of bunt and false smut in paddy (*Oryza sativa* L.) seeds in Maharastra. Agric Sci Dig 28:292–294

Matsumoto M (2002) Trials of direct detection and identification of Rhizoctonia solani AG 1 and AG 2 subgroups using specifically primed PCR analysis. Mycoscience 43(2):185–189

O'Donnell K, Kistler HC, Cigelnik E, Ploetz RC (1998) Multiple evolutionary origins of the fungus causing Panama disease of banana: concordant evidence from nuclear and mitochondrial gene genealogies. Appl Biol Sci 95:2044–2049

Ortega SF, Tomlinson J, Hodgetts J, Spadaro D, Gullino ML, Boonham N (2018) Development of loop-mediated isothermal amplification assays for the detection of seedborne fungal pathogens *Fusarium fujikuroi* and *Magnaporthe oryzae* in rice seed. Plant Dis 102:1549–1558

Ou SH (1980) Pathogen variability and host resistance in rice blast disease. Annu Rev Phytopathol 18:167–187

Ou SH (1985a) Bakanae disease and foot rot. In: Rice diseases survey. Kew, Commonwealth Mycological Institute, pp 262–272

Ou SH (1985b) False Smut (Green Smut). In: Ou SH (ed) Rice disease, 2nd edn. Commonwealth Mycology Institute, Famham Royal, pp 307–311

Padmanabhan SY (1974) Control of rice diseases in India. Indian Phytopathol 22:396–398

Patel JS, Brennan MS, Khan A, Ali G (2015) Implementation of loop-mediated isothermal amplification methods in lateral flow devices for the detection of *Rhizoctonia solani*. Can J Plant Pathol 37(1):118–129

Puhalla JE, Spieth PT (1985) A comparison of heterokaryos is and vegetative compatibility among varieties of *Gibberella fujikurai* (*Fusarium monoliforme*). Exp Mycol 9:39–47

Qi M, Yang Y (2002) Quantification of *Magnaporthe grisea* during infection of rice plants using real-time polymerase chain reaction and northern blot/phosphoimaging analyses. Phytopathology 92(8):870–876

Rao KM, Manibhushan R (1994) Rice blast disease. Daya Publishing House, New Delhi

Rong Z, Yuan Y, Ye W, Wang X, Zheng X (2018) Rapid diagnosis of rice bakanae caused by *Fusarium fujikuroi* and *F. proliferatum* using loop-mediated isothermal amplification assays. J Phytopathol 166(4):283–290

Sakthivel N (2001) Sheath rot disease of rice: current status and control strategies. In: Sreenivasaprasad S, Johnson R (eds) Major fungal diseases of rice: recent advances. Springer, Dordrecht, pp 271–283

Saremi H (2005) *Fusarium*, biology, ecology and taxonomy. Jihad Daneshgahi, Ferdossy Mashhad University, Iran, p 152

Sattar MA, Savitri H (1999) Disease free seed production of rice. In: Choudhury RK, Guar A. Technical bulletin no. 10th edn. Indian Agricultural Research Institute, New Delhi

Savary S, Ficke A, Aubertot JN, Hollier C (2012) Crop losses due to diseases and their implications for global food production losses and food security. Food Secur 4:519–537

Sayler RJ, Yang Y (2007) Detection and quantification of *Rhizoctonia solani* AG-1 IA, the rice sheath blight pathogen, in rice using real-time PCR. Plant Dis 91(12):1663–1668

Snyder WC, Hansen HM (1945) The species concept in *Fusarium* with reference to *discolor* and other sections. Am J Bot 32:657–666

Sridevi G, Parameswari C, Sabapathi N, Raghupathy V, Veluthambi K (2008) Combined expression of chitinase and β-1, 3-glucanase genes in Indica rice (*Oryza sativa* L.) enhances resistance against *Rhizoctonia solani*. Plant Sci 175:283–290

Su'udi M, Park J, Kang W, Park S, Hwang D, Ahn I (2012) Quantification of rice brown leaf spot through Taqman real-time PCR specific to the unigene encoding *Cochliobolus miyabeanus* SCYTALONE DEHYDRATASE1 involved in fungal melanin biosynthesis. J Microbiol 50:947–954

Su'udi M, Kim J, Park JM et al (2013) Quantification of rice blast disease progressions through Taqman real-time PCR. Mol Biotechnol 55(1):43–48

Sun SK, Snyder WC (1981a) The bakanae disease of the rice plant. In: Nelson PE, Toussoun TA, Cook RJ (eds) *Fusarium*: diseases, biology and taxonomy. The Pennsylvania State University Press, University Park, pp 104–113

Sun SK, Snyder WC (1981b) The bakanae disease of the rice plant. In: Nelson PE, Toussoun TA, Cook RJ (eds) *Fusarium*: disease, biology and taxonomy. The Pennsylvania State University Press, University Park, pp 104–113

Sun G, Liu J, Li G, Zhang X, Chen T, Chen J, Pan H (2015) Quick and accurate detection and quantification of *Magnaporthe oryzae* in rice using real-time quantitative polymerase chain reaction. Plant Dis 99(2):219–224

Sunani SK, Bashyal BM, Rawat K, Manjunatha C, Sharma S, Prakash G, Aggarwal R (2019) Development of PCR and loop mediated isothermal amplification assay for the detection of bakanae pathogen *Fusarium fujikuroi*. Eur J Plant Pathol 154(3):715–725

Sunder S, Singh R, Dodan DS, Mehla DS (2005) Effect of different nitrogen levels on brown spot (*Drechslera oryzae*) of rice and its management through host resistance and fungicides. Plant Dis Res 20(2):111–114

Talbot NJ (2003) On the trail of a cereal killer: exploring the biology of *Magnaporthe grisea*. Annu Rev Microbiol 57(1):177–202

Tang J, Zheng L, Jia Q, Liu H, Hsiang T, Huang J (2017) PCR markers derived from comparative genomics for detection and identification of the rice pathogen *Ustilaginoidea virens* in plant tissues. Plant Dis 101:1515–1521

Tanaka E, Tanaka C (2008) Phylogenetic study of clavicipitaceous fungi using acetaldehyde dehydrogenase gene sequences. Mycoscience 49(2):115–125

Thomas KM (1931) A new paddy disease in Madras. Madras Agric J 19:34–36

Webster RK, Gunnell PS (1992) Compendium of rice diseases. The American Phytopathological Society Press, St. Paul, p 62

White TJ, Bruns T, Lee SJWT, Taylor JL (1990) Amplification and direct sequencing of fungal ribosomal RNA genes for phylogenetics. In: Innis MA, Gelfand DH, Sninsky JJ, White TJ (eds) PCR protocols: a guide to methods and applications. Academic, London, pp 315–322

Yang X, Al-Attala MN, Zhang Y, Zhang AF, Zang HY, Gu CY, Li YF (2018) Rapid detection of *Ustilaginoidea virens* from rice using loop-mediated isothermal amplification assay. Plant Dis 102(9):1741–1747

Zemankova M, Lebeda A (2001) *Fusarium* species, their taxonomy, variability and significance in plant pathology. Plant Prot Sci 37(1):25–42

Zhang SY, Dai DJ, Wang HD, Zhang CQ (2019) One-step loop-mediated isothermal amplification (LAMP) for the rapid and sensitive detection of *Fusarium fujikuroi* in bakanae disease through *NRPS31*, an important gene in the gibberellic acid bio-synthesis. Sci Rep 9:3726

Zheng J, Zhang Z, Jiang H, Wang Y, Chai R, Qiu H, Mao X, Wang J, Du X, Lan Z, Sun G (2012) Establishment and preliminary application of a quantitative real-time PCR method for detection of conidia of *Ustilaginoidea virens*. Chin J Rice Sci 26:500–505

Zhou YL, Izumitsu K, Sonoda R, Nakazaki T, Tanaka E, Tsuda M, Tanaka C (2003) PCR-based specific detection of *Ustilaginoidea virens* and *Ephelis japonica*. J Phytopathol 151(9):513–518

Zhou YL, Xie XW, Wang S, Pan YJ, Liu XZ, Yang H, Xu JL, Li ZK (2006) Detection of *Ustilaginoidea virens* from rice field and plants by nested PCR. J Agric Biotechnol 14:542–545

Recent Developments in the Detection and Diagnosis of Fungal-Bacterial Plant Pathogens: From Lab Assay to Field Detection

2

Yalavarthi Nagaraju, Praveen S. Patted, Nazia Manzar, and Abhijeet S. Kashyap

Abstract

Diseases are the great cause of the food crisis, demanding continuous monitoring and screening to minimize the damage. The untiring importunity of diseases severely affects food security and quality worldwide. Rapid and accurate measurement of diseases assumes importance, and the efforts in bringing it to reality by the scientific community resulted in numerous technological developments. The technologies are based on immunology and the detection of DNA, RNA, or specific proteins of interest. The versatility of PCR methods finds its way into diagnosis, detection, and interpretations of disease incidence, identifying the pathogen and potential threats, respectively. The demand for ready detection at the field level has sparked the genesis of LAMP assay and microneedle assays; however, they lack sensitivity and specificity sometimes. The highly sensitive methods include real-time PCR and laser techniques which have more practical applicability. The data of DNA methods is supported by the electron microscopy techniques which is indispensable in confirming the diseases. The future of detection and diagnosis relies on the sensitivity of the methods, and high-sensitivity methods have more reliability. In this chapter, we tried to cover the

Y. Nagaraju (✉)
ICAR—National Bureau of Agriculturally Important Microorganisms, Maunath Bhanjan, Uttar Pradesh, India

Microbiology, CSB-Central Sericultural Research and Training Institute, Berhamapore, India
e-mail: yalavarthi.nagaraju@icar.gov.in

P. S. Patted
Seed Science and Technology, ICAR-Central Potato research Institute, Shimla, India

N. Manzar · A. S. Kashyap
ICAR—National Bureau of Agriculturally Important Microorganisms, Maunath Bhanjan, Uttar Pradesh, India

V. K. Singh et al. (eds.), *Diseases of Field Crops: Diagnostics and Management*, https://doi.org/10.1007/978-981-97-6160-9_2

different methods used in the diagnosis and detection of pathogens and their reliability.

Keywords

PCR · LAMP assay · Microscopic techniques · Immunoassays

2.1 Introduction

Constant global pathogen surveillance must be needed to protect human health, food security, bioterrorism, and biodiversity (Fisher et al. 2012). Even while just a few specialist centers have historically been responsible for monitoring the world's disease burden, more effective real-time detection could be achieved by making precise diagnostics available at the point of care (Bissonnette et al. 2010). These tests, however, are limited to a small number of pathogens and may need expensive chemicals, cold chain distribution, specialized tools, and professional staff (Singh and Sidhu 2014). Long-term culture techniques continue to be the primary strategy for monitoring infections in places with low resources because there are not any creative replacements for pricey antibody- and nucleic acid–based assays. Similar to how developments in molecular biology gave rise to antibody diagnostics, the developing area of synthetic biology has the potential to offer novel diagnostic platforms to address global health concerns (Church et al. 2014). Although metabolic engineering has been predominantly used to exploit synthetic biology for the commercial fermentation of industrial and biomedical commodities up to this point (Ro et al. 2006), there are still substantial unexplored possibilities in public health, animal health, and agriculture.

Any living thing can be genetically altered and replicated through biotechnology to create improved products that can be used in various ways (Agrios 2005; Fagwalawa et al. 2013). Combining biotechnology with conventional agricultural methods will enhance our understanding of plants' capacity to identify and defend themselves against pathogens like fungi, and this will help us develop sustainable agriculture (Haggag 2008). It can be used for molecular markers, genetic engineering, tissue culture, and diagnostics (James and Sugar 2003). Chemical control of plant diseases is often non-specific in its effects, killing beneficial organisms and pathogens, and it may have undesirable health, safety, and environmental risks (Manzar et al. 2021a, b). Because of these factors, plant disease control is of great interest to biotechnologists. Biotechnology is an essential discipline for the accurate diagnosis of plant disease. Enzyme-linked immunosorbent assays (ELISA) and nucleic acid techniques identify fungal, viral, and bacterial infections (Kashyap et al. 2021, 2023). There have been several reports on discovering plant diseases (Manzar et al. 2022; Kashyap et al. 2022a, b). For them, a new approach in management practices like advanced plant genetic engineering is required, and many scholars have discussed and presented the application of genetic engineering for plant disease resistance. The different tools of biotechnology and genomic approaches have enhanced the process of identifying plant pathogens with greater accuracy. Plant genes and their products interacting with plant pathogens have been identified and inserted into plants or used as specific molecular markers to improve breeding for disease resistance. Therefore, acquiring noble information from applied biotechnology to plant-pathogen interactions provides new knowledge and approaches to improve plant health, yield, and quality of plants contributing to food (Fagwalawa et al. 2013).

For long-term food security and to stop the spread of invasive pathogens, accurate diagnosis and identification of plant diseases are crucial (Balodi et al. 2017). In this regard, particular attention should be paid to the early detection of pathogens in plant propagation materials (seeds, mother plants, and vegetative plant material) to prevent introducing and further disseminating new infections in a growing area where they were not previously present. In order to make the final decision regarding the most effective management approaches, quick detection and identification of plant diseases utilizing swift, sensitive, and precise methodologies are strongly required. PCR-based isothermal amplification techniques like LAMP, DNA, RNA fingerprinting hybridization technology, and pyro-sequencing are some of the most crucial molecular methods for detecting plant pathogenic diseases (Kang et al. 2010).

The successful management of plant diseases depends on the accurate and efficient detection of pathogens, knowledge of the genetic and pathogenic variability level present in a pathogen population, development of disease-resistant cultivars, and quantification of disease-resistant genes in various epidemiological regions. These factors allow for more informed decisions regarding cultivar selection and how and when chemicals can control the disease. By screening seeds and other planting materials, pathogen detection and identification are crucial for halting the spread of disease. It is also required to put quarantine restrictions into practice. Additionally, it plays a more general role in many fields of plant pathology research, such as disease epidemiology, investigations into the links between diseases that cause yield loss, and the development of new disease prevention tactics. Almost every area of plant pathology depends on accurately detecting pathogens or conditions, and visual interpretation is typically the initial step in traditional methods of illness diagnosis. The diagnosis may then be confirmed using laboratory identification techniques like selective media and microscopy, and these techniques are often the least expensive, most accessible, and most suitable. For instance, look at Stowell and Gelernter's (2001) most current assessment on diagnosing turf grass diseases.

However, conventional procedures have several shortcomings, so researchers seek new diagnostic approaches. Traditional techniques typically need specialized microbiological knowledge that might take years to develop. Utilizing general methods that relatively untrained personnel may learn fast and readily is necessary. When a quick, high-throughput diagnosis is needed, it is unacceptable for cultivation procedures to take days or weeks to complete. The conclusions are not always clear-cut, for instance, when many diseases or physiological states might generate the same symptoms. It could be challenging to distinguish between closely related creatures using only morphological traits. Discriminating between populations of the same pathogen with specific characteristics, such as fungicide resistance, toxin

Fig. 2.1 Techniques used for the identification of plant pathogens

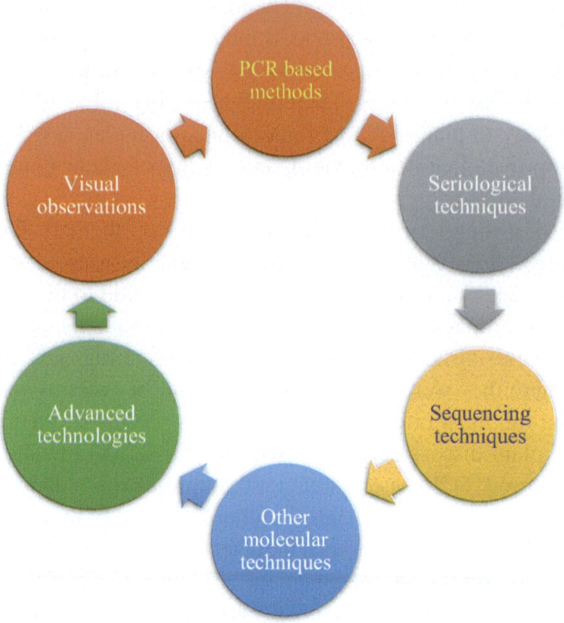

production, or variations in virulence, may also be necessary. Traditional approaches might not be sensitive enough (for instance, when the diagnosis of presymptomatic infection is required); hence, during the past 10 years, much work has been put into creating new techniques for locating and classifying plant pathogens. Recent developments in molecular biology methods, notably the advent of many DNA-based markers, have profoundly altered plant protection tactics and more conventional techniques for disease diagnosis and breeding resistant strains. This article's primary objective is to highlight the broad uses of biotechnological approaches for plant disease detection, diagnosis, and control and to give an overview of proteomics research in bacterial diseases (Fig. 2.1).

Most work has been put into creating technologies using nucleic acids and antibodies for diagnostics (Duncan and Torrance 1992; Ward and Bej 2006; Martin et al. 2000; McCartney et al. 2003). This chapter addresses the methodology and tactics employed in these brand-new diagnostic tests and how they are used in plant pathology. This covers research on viruses, phytoplasmas, and bacteria, albeit fungi have received more attention than the other categories. We also contrast the different approaches and talk about potential future advances.

2.2 Advanced Techniques

2.2.1 Microneedle

With this technique, DNA may be extracted from a sample in less than a minute instead of many hours, and the isolated DNA is immediately available for amplification without further sample purification processes. Despite the current microneedle design's lack of specificity in the isolation of nucleic acids and other intracellular analytes, the combination of efficient microneedle patch techniques and miniature DNA amplification assays (Koo et al. 2013) opens the door to simple, quick, and deployable plant pathogen amplification and diagnosis. The polyvinyl alcohol (PVA) used to make microneedle patches is powerful enough to rupture the cell walls of plant leaf tissues and release pathogenic or plant DNA (Paul et al. 2019).

2.2.2 Real-Time Microchip PCR System

In the last 10 years, microchip PCR systems have become popular to miniaturize traditional PCR systems and save costs and operating time. Several portable real-time microchip PCR systems have also been reported; however, they all require different accessory devices. Here, we describe a standalone real-time microchip PCR system comprising a battery, a microcontroller to manage the entire thermocycling process, a small fluorescence detector to detect amplified DNA, and a PCR reaction chamber microchip with an integrated thin-film heater. The total system measures $25 \times 16 \times 8$ cm^3 and weighs 843 g. The single PCR run on the disposable microchip uses 110 mAh of power and only needs an 8 L sample volume. A DNA extraction process was created for field operations, particularly without liquid nitrogen, chemicals, or other substantial lab equipment. Six distinct fungal and bacterial plant diseases were successfully detected with a 100% success rate using the designed real-time microchip PCR equipment and the DNA extraction methodology, with a detection limit of 5 ng/8 μL of the sample (Koo et al. 2013).

2.2.3 Plant Wearables

A cutting-edge advancement in agricultural diagnostics, plant wearables attach flexible sensor devices directly to plant tissues like leaves for continuous monitoring using micro-electro-mechanical systems (MEMS) and ultrathin, ultra-lightweight nanosensors. For instance, Oren et al. (2018) created a roll-to-roll manufacturing technique for a wearable sensor based on graphene that can track water evaporation from plant leaves. Graphene's electrical resistance changes in various humidity settings serve as the foundation for the sensing process. Im et al. proposed a wearable plant drought stress sensor based on a different approach. The fluctuation of printed 100-nm-think gold electrode capacitances on a flexible polyimide (PI) was recorded in real time to measure local humidity.

2.2.4 Nanopore Sequencing

The third-generation sequencing (TGS) method, known as the nanopore sequencing platform, employs a motor protein to move a single-stranded DNA or RNA through a protein or solid-state nanopore to produce nucleotide-dependent electrical current signals that make pathogenic polynucleotides easy to analyze. Nanopore sequencing technology outperforms earlier sequencing techniques with real-time and extensive readings of sequence data, rapid running times, scalable genome mapping, and low sample loadings. Recent studies using nanopore sequencing devices have centered on diagnosing plant diseases. Recent studies using nanopore sequencing devices have centered on diagnosing plant diseases. In order to diagnose various plant bacteria, viruses, fungi, and phytoplasma, such as *P. digitatum* in lemon and *S. lycopersicum* in tomato, Chalupowicz et al. (2019) successfully established a standard protocol using a handheld sequencing system created by Oxford Nanopore Technologies (named "MinION").

The MinION is a portable nanopore sequencing device that may be used by just plugging it into a regular USB connection which began the initial beta testing MinION in 2014. The MinION technology detects DNA strand which passes through nanopores at a certain speed based on size, detecting the translocation of the DNA molecule with single-nucleotide accuracy. The system utilizes consumable flow cells with immobilized nanopores, and each flow cell can now produce between 5 and 10 Gb of sequencing data. The MinION is appropriate for use even in distant areas due to its compact size and simple connectivity (Quick et al. 2016), even used by NASA in 2016.

Most of the device's reported diagnosis applications which have produced encouraging results involve medical, animal viral illness, and even food safety. For instance, it has been proven that the MinION can fully detect pathogens from urine samples in just 4 h without needing culture (Schmidt et al. 2016). Additionally, Oxford Nanopore Technologies' barcoding kits, which enable the detection of up to 96 distinct barcodes in a single sample, can be used for multiplex detections.

2.2.5 Nanosensors

Several nanoparticle-based chemo- or biosensors have been suggested for agricultural diagnostics and made available for purchase. The planned sensory interactions may have transduction mechanisms that allow the analytes to be detected by the nanosensors' optical or electrical outputs. Selective chemical interactions or biospecific recognition components like DNA oligos, antibodies, aptamers, and enzymes might improve the sensor's detecting specificity. Surface-enhanced optical features, such as surface plasmon resonance (SPR), or electron-conductive nanoscale substrates, such as carbon-based nanomaterials (such as graphene or carbon nanotube) as transducers, might be used to increase detection sensitivity (Mukherjee et al. 2016). Nanoparticles made of metal or metalloids, quantum dots, and array-based nanosensors are the three primary groups of nanosensors or nanosensors.

Nanoparticles are made of metal or a metalloid. A famous family of nanosensors uses inorganic nanomaterials like Ag, Au, Si, and metal oxide nanoparticles to detect and measure molecular targets of interest using recognition ligands like antibodies or DNA oligos (Darr et al. 2017; Wu and Lin 2013; Narware et al. 2023). Semiconductor nanocrystals called quantum dots (QDs) have enormous promise as optical nanosensors (Gao et al. 2018). Because of their unique and advantageous photophysical characteristics, QDs have successfully been biosensors for plant imaging and disease diagnostics (Wang et al. 2016).

The QDs' tiny size (1–10 nm) makes them perfect fluorescent imaging contrast agents for in-planta transit and quick uptake, where the fluorescent signals may be easily collected to trace their distributions in the live system. The array-based sensor assemblies, modeled after the human olfactory system and comprising many chromophores or artificial nanomaterials, are one type of cutting-edge nanosensor capable of multiplexing and discriminating between distinct analytes. The term "electronic nose" (e-nose) refers to a typical type of array-based sensor that uses electronic transducers rather than chemical sensors in the array (Geng et al. 2019; Röck et al. 2008; Wang et al. 2016). Due to its cross-reactivity and capacity to analyze chemical fingerprints, the array-based technique is instrumental in differentiating pretty similar analyte combinations. Chemometrics may be used to assess the multidimensional output data of sensor arrays in a supervised or unsupervised fashion, producing extraordinarily accurate and prognostic findings (Cuypers and Lieberzeit 2018). Instead of e-nose signals, array-based chemical sensors (Askim et al. 2013) can produce optical readout (such as colorimetric or fluorescent), which is simpler to comprehend and analyze.

Additionally, the creation of biosensors for disease detection has included quantum dots (QDs) (Frasco and Chaniotakis 2009). They have been employed for disease detection via the fluorescence resonance energy transfer (FRET) mechanism, which explains energy transfer between two light-reactive molecules, because of their distinctive and beneficial optical features (Algar and Krull 2008). To detect plant diseases like *Candidatus*, *Phytoplasma aurantifolia*, and witches' broom disease of lime (WBDL), many QD-FRET-based sensors have been created (Ca. *P. aurantifolia*). The immunosensor developed had a detection limit of 5 ca. *P. aurantifolia*/L, high sensitivity, and specificity of 100% (Rad and Mohsenifar 2012). It was also noted that QD-FRET might identify the disease vectors. For instance, the beet necrotic yellow vein virus causes the most damaging disease affecting sugar beets, rhizomania (BNYVV). The sole known vector of BNYVV for plant transmission, *Polymyxa betae* (Keskin), has been successfully identified using a QD-FRET-based sensor (Safarpour et al. 2012).

2.2.6 Thermography

Thermography makes it possible to visualize the variations in surface temperatures between plant leaves and canopies. Thermographic cameras capture the emitted infrared radiation, which can then examine the color difference. Previous studies have shown that phytopathogens would impact the water loss in plants controlled by stomata (Chaerle et al. 2007; Lindenthal et al. 2005; Oerke et al. 2006). The volume of water that transpired may be measured without being influenced by external temperature changes, and the associated disease can be tracked using thermographic imaging (Oerke et al. 2006). Several research groups have documented the temperature variations brought on by plant pathogen infection. For direct disease detection in plants, the thermographic detection of plant diseases may also be scaled up (Lindenthal 2005). A potential technique to track the diversity of soilborne pathogen infections is thermography. However, thermography's practical utility for disease monitoring is restricted due to its excellent sensitivity to changes in ambient circumstances during measurements. Furthermore, thermographic detection is not disease-specific, so it cannot be utilized to determine the kind of infection or differentiate between diseases with similar thermographic patterns.

2.2.7 Fluorescence Imaging

According to this method, which measures the chlorophyll fluorescence on the leaves as a function of the incident light, changes in the fluorescence parameters can be used to analyze pathogen infections based on modifications to the photosynthetic apparatus and reactions that transport electrons during photosynthetic processes (Bürling et al. 2012, Bürling et al. 2011; Kruckenberg et al. 2011). Temporal and spatial fluctuations in chlorophyll fluorescence were examined to accurately detect leaf rust and powdery mildew infections in wheat leaves at 470 nm (Kuckenberg et al. 2009). Even though fluorescence measurement offers sensitive photosynthetic anomaly detection, this approach has only a limited number of field applications (Chaerle et al. 2009; Cséfalvay et al. 2009).

2.2.8 Hyperspectral Techniques

In a wide range of spectrum between 350 and 2500 nm, hyperspectral imaging may be utilized to gather insightful data on plant health. In large-scale agriculture, hyperspectral imaging is being utilized increasingly for crop disease detection and plant phenotyping. The method is reliable and offers a quick interpretation of the imaging data. Furthermore, hyperspectral imaging cameras make capturing data in three dimensions easier, with *X*- and *Y*-axes for geographical data and *Z* for spectral data, which results in more precise and detailed information regarding plant health over a vast region (Mahlein et al. 2012). By observing the variations in reflectance brought on by the biophysical and biochemical characteristic changes during infection, hyperspectral methods have been extensively employed for plant disease detection. *Phytophthora infestans*, *Venturia inaequalis*, and *Magnaporthe grisea* infections of rice, tomatoes, and apple trees have all been found and reported using hyperspectral imaging techniques (Delalieux et al. 2007; Kobayashi et al. 2001).

2.2.9 Gas Chromatography

Profiling the volatile chemical signature of the diseased plants is an indirect way of identifying plant diseases. Certain volatile organic compounds (VOCs) highly suggestive of the stress encountered by plants may be released due to pathogen infections in plants (Fang et al. 2014). When strawberries are infected with *Phytophthora cactorum*, a fungus that causes crown rot, the affected area of the strawberry plant or fruit releases p-ethylguaiacol and p-ethylphenol as distinctive VOCs (Ellis and Grove 1982; Fang et al. 2014). Additionally, pathogen- and mechanical-induced damage to green leaf plants produces VOC. Examples of green leaf volatiles (GLVs) that have been reported to have been created following mechanical injury to plant leaves or during herbivore invasion include cis-3-hexenol, cis-hexenyl acetate, and hexyl acetate. The kind and nature of the infection might be determined by the profiling of such VOC, which could then be utilized for illness diagnosis and confirmation (Fang et al. 2014). The presence of a particular VOC that is symptomatic of a particular illness might be determined by gas chromatography (GC) analysis of the volatile signature of plants (Jansen et al. 2009; Kashyap et al. 2022a, b). Gas chromatography and mass spectrometry are frequently coupled (GC-MS) to improve the performance of chemical separation and analysis and to discover unidentified compounds in volatile samples (Isidorov et al. 1985; Kesselmeier and Staudt 1999). Due to its high specificity, GC/GC-MS can offer more precise information about plant disease than the optical imaging-based detection techniques covered above. Based on the quantitative data gathered from the VOC sample, it also enables the early diagnosis of diseases. However, GC/GC-MS needs sampling of pre-collected VOC for a more extended period before data processing, which significantly restricts its on-field deployment, unlike the imaging system, which can instantly get the data on-field.

2.2.10 Biospeckle Laser Technique

The "biospeckle" laser technique is a recently developed technology that may detect the presence of harmful fungus on seeds. This method is based on the optical interference phenomena produced when laser light interacts with the seed coat. Laser examination of seeds enables the identification of regions with various activities (Braga et al. 2005; Rabelo et al. 2011). This approach can identify the presence of fungus on seeds since they exhibit biological activity when present.

2.2.11 Videometer Lab Instrument

A multispectral vision system is a recently created tool known as videometer lab equipment that can detect diseased seeds from healthy seeds and is also used to evaluate the color, texture, and chemical content of seed surfaces (Boelt et al. 2018). When separating healthy spinach seeds from seeds infected with *Stemphylium botryosum*, *Cladosporium* spp., *Fusarium* spp., *Verticillium* spp., or *A. alternata*, the characteristics from pictures collected by visible light wavelengths and near-infrared wavelengths were combined successfully (Olesen et al. 2017). Using the Videometer lab device, multispectral imaging technology was used to forecast the quality of castor (*Ricinus communis* L.) seeds based on the color of the seed coat. With 92% accuracy, this method could discriminate between live and dead seeds, indicating its potential use in detecting pathogen-induced seed degeneration.

2.2.12 Lateral Flow

According to De Boer and Lopez (2012), one of the earliest grower-friendly pathogen monitoring techniques is LF, which is based on immune detection and immune-printing kits. *Phytophthora species*, *Ralstonia solanacearum*, and potato virus Y are just a few plant pathogens that can be detected using LFA devices currently on the market. Lateral flow assays (LFAs) are one of the first user-friendly pathogen surveillance approaches, according to De Boer and Lopez (2012). Plant infections that may now be found with LFA devices include *Phytophthora* species, *Ralstonia solanacearum*, and potato virus Y.

2.3 PCR-Based Methods

2.3.1 Isothermal Mediated Loop Amplification

Notomi et al. (2000) published the first report using loop-mediated isothermal amplification (LAMP) for hepatitis B virus detection. The *Bacillus subtilis* polymerase, also known as Bst polymerase, is used in the PCR-based LAMP procedure, which naturally does not require a denaturation phase. The enzyme works by strand displacement activity, enabling template amplification under isothermal circumstances. The reaction may be carried out in a straightforward water bath that can maintain a temperature of 60–65 °C; this is a significant benefit for LAMP-based diagnostics and reduces the complexity and expense of the necessary equipment. In contrast to traditional PCR, LAMP requires a minimum of four primers complementary to six distinct sections of the DNA sequence. A total of 10^9 copies of the target sequence may be amplified using a LAMP reaction's end products, stem-loop DNAs containing multiple inverted repeats of the target, and cauliflower-like structures (Notomi et al. 2000). LAMP is a popular approach for pathogen identification because of its quick findings and reasonable specificity (Yan et al. 2017).

Like traditional PCR analysis, agarose gel electrophoresis may be used to see LAMP-amplified results. The outcomes may be seen with the naked eye in certain instances because of the reaction's large quantity of amplified DNA. The response is recorded using a color change from violet to sky-blue, as described in a 2009 paper by Goto et al. using hydroxy naphthol blue (HNB) as a colorimetric indication for visualizing LAMP findings. Since then, HNB has been incorporated into various investigations that employ LAMP-based plant pathogen detection techniques (Kong et al. 2016). LAMP findings following amplification have also been observed using Picogreen, a DNA intercalating dye. Turbidimetry has also been utilized in LAMP-based research to see and measure template DNA (Mori et al. 2004). Pyrophosphate is produced as a by-product that interacts with the magnesium and produces a white magnesium precipitate pyrophosphate. Mori et al. (2004) created a straightforward gadget that can measure and heat the turbidity of LAMP as the response. For the cutoff point for exceeding a specific according to reports, the turbidity value depends on the starting DNA quantity, enabling them to determine the least amount of DNA.

Plant infections may now be diagnosed on-site utilizing LAMP, real-time LAMP, and RT-LAMP, significantly reducing analysis time. Commercial kits based on LAMP have been validated for identifying food pathogens, including *Salmonella* spp., *Listeria* spp., and *Escherichia coli* (Lim et al. 2017). 3M introduced one such kit. However, these technology-based portable tests for plant pathogen detection are currently being developed. Kogovsek et al. (2015) developed a LAMP detection method for identifying grapevines. The findings confirmed many straightforward DNA extraction techniques, and accurate detection was achieved even when grape crude extracts were used for the detection. Applying the reagent stabilization strategies outlined above might simplify methods like this.

2.3.2 Recombinase Polymerase Amplification

Recombinase polymerase amplification (RPA), similar to LAMP technology, runs at a static temperature (often 37–42 °C), making it most suited for in-field diagnosis. Three key enzymes—a strand displacement polymerase, a recombinase enzyme, and single-stranded DNA-binding proteins—are used with oligonucleotide primers to produce isothermal amplification (Ahmed et al. 2014). A complex is formed between the recombinase enzyme and primers that "scan" the target sequence and shift the template strand in the process. The single-stranded DNA-binding proteins then stabilize the template strand, preventing the shifting of primers. The strand displacement polymerase initiates the target sequence's multiplication (Piepenburg et al. 2006). Since its initial publication, the RPA has been frequently used to diagnose human and animal diseases. Although it has been used to identify fungi, bacteria, and viruses that cause plant diseases, there are few reports of this (Londono et al. 2016). Sensitivity (detection can be done within 10–15 min) is superior to the traditional PCR-based diagnostics.

Terminally labeled RPA amplicons have been used to quickly and efficiently diagnose the plum pox virus and the small cherry virus 2 using high-affinity lateral flow strips (Mekuria et al. 2014). The multiplex detection of plant diseases using recombinase polymerase amplification is achieved by combining the process with surface-enhanced Raman scattering (SERS) tagged nano-tags (Lau et al. 2016).

2.4 Immunodetection Methods

The most prevalent methods for performing immunodetection include ELISA, dot blot immunoassays, and microsphere immunoassays. These methods are labor-intensive and lengthy and need skilled personnel and specialized laboratory equipment (Safenkova et al. 2017). The current trend is to create multiplex immunoassays to facilitate and make immunodetection an inexpensive method for pathogen diagnosis.

2.4.1 Microsphere Immunoassay

For the detection of microorganisms, microsphere immunoassays (MIAs) have become a potential substitute for detecting microorganisms using microspheres that have been studded with various fluorophores. The pathogen antigen is captured by the microspheres conjugated with specific antibodies, and the bound and unbound microspheres are separated using a magnet. This was followed by adding specific antibodies coupled with a reporter fluorophore to the microsphere sample mixture. Multiple analytes may be specifically detected simultaneously by using a Luminex xMAP analyzer to excite the internal microsphere dyes with a red laser and the reporter fluorochrome, collected by the antigen, with a green laser (Charlermroj et al. 2016).

There were no instances of multiplex MIAs being used to identify plant pathogens before 2008. The first reports concerned the creation of a multiplex MIA for detecting *Acidovorax avenae* subsp. *citrulli*, the chilli vein-banding mottle virus (potyvirus), the watermelon silver mottle virus (tospovirus serogroup IV), and the melon yellow spot virus (tospovirus) (Charlermroj et al. 2016). Compared to ELISA, the multiplex MIA in the latter investigation allowed for more sensitive identification of the infections. Another benefit of MIAs is that the tests only take around an hour instead of a typical ELISA's 4 h (Charlermroj et al. 2016). The method is user-friendly for pathogen detection, wherein 100 parameters may be examined at a stretch, even though portable instruments for MIAs have not yet been created (De Boer and Lopez 2012). Agricultural firms may eventually have their diagnosis station, eliminating the need to ship samples to specialized labs with the proper personnel training and adequately calibrated equipment for fluorescence reading. For small-scale farmers, technology expenses are still expensive.

2.4.2 Aptamer-Based Diagnosis

Aptamers, a single-stranded ssDNA or RNA with 100-mer ligands, were chosen to bind specific target molecules. SELEX (systematic evolution of ligands by exponential enrichment) is a methodology that is typically used to identify these ligands. Target molecules are screened against a random pool of oligonucleotides (typically about 1000 sequences) to identify aptamers. Once the final aptamers have been chosen, these oligonucleotides are sequenced and employed in the subsequent SELEX rounds (normally 12–16 rounds) (Sefah et al. 2010). Only the functional aptamers that attach to the target molecule remain after numerous washes and remove all nonbinding oligonucleotides.

Most diagnostic uses for aptamers depend on conformational changes brought on by ligands. Different techniques, such as dye binding or fluorescence quenching, can be used to identify these changes (Bunka and Stockley 2006). Due to their tiny size, chemical stability, and low cost, aptamers have been shown to have an affinity to their targets that is equivalent to or even higher than that of frequently used monoclonal antibodies (Hamula et al. 2016). Even though aptamers are feasible and promising antibodies to replace, they are utilized for food security or the detection of human diseases. The identification of plant pathogens has not been the subject of much research in the field of plant pathology. There have been reports of fumonisin (a toxin generated by *Fusarium* species) detection aptamers for fungus, although this research took a food safety approach, as aforementioned (McKeague et al. 2010).

2.5 Other Methods

2.5.1 Dielectrophoresis

When a polarizable particle moves as a result of an alternating current (AC) electric field, the process is known as dielectrophoresis (DEP) (Adekanmbi and Srivastava 2016). Particles travel toward high- or low-field density areas depending on their force. At different AC frequencies, the target particle and the suspension medium undergo changes in their permittivity and conductivity, which are measured by the Clausius-Mossotti (CM) factor. The particle is pushed toward the highest or weakest electric field according to different CM parameters. Pohl and Hawk (1966) initially described dielectrophoresis (DEP) in live cells. Since then, DEP has been widely utilized to manipulate and separate artificial or biological particles (Gagnon et al. 2011). It has been chiefly employed in the diagnostic field to concentrate and separate germs before using conventional or molecular methods to detect them. For instance, DEP has been used to trap cells with varying electrical characteristics, including living and dead cells, at various locations along a microchannel to selectively separate three distinct bacteria (Pysher and Hayes 2007).

2.6 Conclusion

It is common knowledge that the appropriate treatment methods must be applied after a plant disease occurrence. Early detection can also distinguish between a little outbreak and a sizable production loss in the field. Therefore, it is essential to check for infections frequently. As the inoculum concentrations might not be high enough for conventional methods to detect a disease, relying on high-throughput, focused, and sensitive technology is crucial to see a condition in its early stages, frequently when plants are still symptomless. According to research, these parameters are met by most immunologic and nucleic acid–based methods. However, most of these are restricted to specialist labs that need advanced tools and knowledgeable personnel, making the entire process difficult and time-consuming from sample to the outcome. The methods discussed in this chapter will help to advance the development of precise, sensitive, and focused pathogen detection for various applications. The majority of them have been successfully used in research on humans, animals, and food safety, making the subject of plant pathology an untapped one where these technologies have the potential to alter how diseases are currently diagnosed shortly significantly.

References

Adekanmbi EO, Srivastava SK (2016) Dielectrophoretic applications for disease diagnostics using lab-on-a-chip platforms. Lab Chip 16:2148–2167

Agrios GN (2005) Plant pathology. Elsevier

Ahmed A, van der Linden H, Hartskeerl RA (2014) Development of a recombinase polymerase amplification assay for the detection of pathogenic *Leptospira*. Int J Environ Res Public Health 11:4953–4964

Algar WR, Krull UJ (2008) Quantum dots as donors in fluorescence resonance energy transfer for the bioanalysis of nucleic acids, proteins, and other biological molecules. Anal Bioanal Chem 391(5):1609–1618. https://doi.org/10.1007/S00216-007-1703-3

Askim JR, Mahmoudi M, Suslick KS (2013) Optical sensor arrays for chemical sensing: the optoelectronic nose. Chem Soc Rev 42(22):8649–8682. https://doi.org/10.1039/C3CS60179J

Balodi R, Bisht SUNAINA, Ghatak A, Rao KH (2017) Plant disease diagnosis: technological advancements and challenges. Indian Phytopathol 70(3):275–281

Bissonnette L, St-Arnaud M, Labrecque M (2010) Phytoextraction of heavy metals by two Salicaceae clones in symbiosis with arbuscular mycorrhizal fungi during the second year of a field trial. Plant Soil 332:55–67

Boelt B, Shrestha S, Salimi Z, Jorgensen JR, Nicolaisen M, Carstensen JM (2018) Multispectral imaging—a new tool in seed quality assessment? Seed Sci Res 28(3):222–228. https://doi.org/10.1017/S0960258518000235

Braga RA, Rabelo GF, Granato LR, Santos EF, Machado JC, Arizaga R, Rabal HJ, Trivi M (2005) Detection of fungi in beans by the laser biospeckle technique. Biosyst Eng 91(4):465–469. https://doi.org/10.1016/J.BIOSYSTEMSENG.2005.05.006

Bunka DHJ, Stockley PG (2006) Aptamers come of age—at last. Nat Rev Microbiol 4:588–596

Bürling K, Hunsche M, Noga G (2011) Use of blue-green and chlorophyll fluorescence measurements for differentiation between nitrogen deficiency and pathogen infection in winter wheat. J Plant Physiol 168(14):1641–1648. https://doi.org/10.1016/J.JPLPH.2011.03.016

Bürling K, Hunsche M, Noga G (2012) Presymptomatic detection of powdery mildew infection in winter wheat cultivars by laser-induced fluorescence. Appl Spectrosc 66(12):1411–1419

Chaerle L, Leinonen I, Jones HG, van der Straeten D (2007) Monitoring and screening plant populations with combined thermal and chlorophyll fluorescence imaging. J Exp Bot 58(4):773–784. https://doi.org/10.1093/JXB/ERL257

Chaerle L, Lenk S, Leinonen I, Jones HG, van der Straeten D, Buschmann C (2009) Multi-sensor plant imaging: towards the development of a stress-catalogue. Biotechnol J 4(8):1152–1167. https://doi.org/10.1002/BIOT.200800242

Chalupowicz L, Dombrovsky A, Gaba V, Luria N, Reuven M, Beerman A, Lachman O, Dror O, Nissan G, Manulis-Sasson S (2019) Diagnosis of plant diseases using the nanopore sequencing platform. Wiley Online Library 68(2):229–238. https://doi.org/10.1111/ppa.12957

Charlermroj R, Makornwattana M, Grant IR, Elliott CT, Karoonuthaisiri N (2016) Validation of a highthroughput immunobead array technique for multiplex detection of three foodborne pathogens in chicken products. Int J Food Microbiol 224:47–54

Church A, Fish R, Haines-Young R, Mourato S, Tratalos J, Stapleton L Willis C, (2014) UK national ecosystem assessment follow-on: cultural ecosystem services and indicators

Cséfalvay L, di Gaspero G, Matouš K, Bellin D, Ruperti B, Olejníčková J (2009) Presymptomatic detection of *Plasmopara viticola* infection in grapevine leaves using chlorophyll fluorescence imaging. Eur J Plant Pathol 125(2):291–302. https://doi.org/10.1007/S10658-009-9482-7

Cuypers W, Lieberzeit PA (2018) Combining two selection principles: sensor arrays based on both biomimetic recognition and chemometrics. Front Chem 6:268. https://doi.org/10.3389/FCHEM.2018.00268/FULL

Darr JA, Zhang J, Makwana NM, Weng X (2017) Continuous hydrothermal synthesis of inorganic nanoparticles: applications and future directions. Chem Rev 117(17):11125–11238. https://doi.org/10.1021/ACS.CHEMREV.6B00417

De Boer SH, Lopez MM (2012) New grower-friendly methods for plant pathogen monitoring. Annu Rev Phytopathol 50:197–218

Delalieux S, van Aardt J, Keulemans W, Schrevens E, Coppin P (2007) Detection of biotic stress (*Venturia inaequalis*) in apple trees using hyperspectral data: non-parametric statistical approaches and physiological implications. Eur J Agron 27(1):130–143

Duncan JM, Torrance L 1992 Techniques for the rapid detection of plant pathogens (pp xiii+-235)

Ellis MA, Grove G (1982) Fruit rots cause losses in Ohio-strawberries

Fagwalawa LD, Yakasai MT, Kutama AS (2013) Growth and yield parameters of sorghum genotypes as affected by artificial inoculation techniques for screening against head smut in Nigeria. Bayero Journal of Pure and Applied Sciences 6(1):144–151

Fang Y, Umasankar Y, Ramasamy RP (2014) Electrochemical detection of p-ethylguaiacol, a fungi infected fruit volatile using metal oxide nanoparticles. Analyst 139(15):3804–3810. https://doi.org/10.1039/C4AN00384E

Fisher MC, Henk DA, Briggs CJ, Brownstein JS, Madoff LC, McCraw SL, Gurr SJ (2012) Emerging fungal threats to animal, plant and ecosystem health. Nature 484(7393):186–194

Frasco M, Chaniotakis N (2009) Semiconductor quantum dots in chemical sensors and biosensors. Sensors (Basel) 9(9):7266–7286. https://doi.org/10.3390/s90907266

Gagnon N, Barret M, Topp E, Kalmokoff M, Massé D, Masse L, Talbot G (2011) A novel fingerprint method to assess the diversity of methanogens in microbial systems. FEMS Microbiol Lett 325(2):115–122

Gao G, Jiang YW, Sun W, Wu FG (2018) Fluorescent quantum dots for microbial imaging. Chin Chem Lett 29(10):1475–1485. https://doi.org/10.1016/J.CCLET.2018.07.004

Geng Y, Peveler WJ, Rotello VM (2019) Array-based "chemical nose" sensing in diagnostics and drug discovery. Angew Chem Int Ed 58(16):5190–5200. https://doi.org/10.1002/ANIE.201809607

Goto M, Honda E, Ogura A, Nomoto A, Hanaki K (2009) Colorimetric detection of loop-mediated isothermal amplification reaction by using hydroxy naphthol blue. Biotechniques 46:167–172

Haggag WM (2008) Biotechnological aspects of plant resistant for fungal diseases management

Hamula CLA, Peng H, Wang Z, Tyrrell GJ, Li XF, Le XC (2016) An improved SELEX technique for selection of DNA aptamers binding to M-type 11 of Streptococcus pyogenes. Methods 97:51–57

Isidorov VA, Zenkevich IG, Ioffe BV (1985) Volatile organic compounds in the atmosphere of forests. Atmos Environ (1967) 19(1):1–8. https://doi.org/10.1016/0004-6981(85)90131-3

James GM, Sugar CA (2003) Clustering for sparsely sampled functional data. J Am Stat Assoc 98(462):397–408

Jansen RMC, Hofstee JW, Wildt J, Verstappen FWA, Bouwmeester H, van Henten EJ (2009) Induced plant volatiles allow sensitive monitoring of plant health status in greenhouses. Plant Signal Behav 4(9):824–829. https://doi.org/10.4161/psb.4.9.9431

Kang S, Xu Y, You Q, Flügel WA, Pepin N, Yao T (2010) Review of climate and cryospheric change in the Tibetan Plateau. Environ Res Lett 5(1):015101

Kashyap AS, Manzar N, Rajawat MVS, Kesharwani AK, Singh RP, Dubey SC, Pattanayak D, Dhar S, Lal SK, Singh D (2021) Screening and biocontrol potential of Rhizobacteria native to Gangetic Plains and hilly regions to induce systemic resistance and promote plant growth in Chilli against bacterial wilt disease. Plants 10:2125

Kashyap AS, Manzar N, Ahamad F, Tilgam J, Sharma PK, Saxena AK (2022a) First report of root rot disease in green gram (*Vigna radiata*) caused by *Ectophoma multirostrata* in India. Plant Dis. https://doi.org/10.1094/PDIS-11-21-2400-PDN

Kashyap AS, Manzar N, Nebapure SM, Rajawat MVS, Deo MM, Singh JP, Kesharwani AK, Singh RP, Dubey SC, Singh D (2022b) Unraveling microbial volatile elicitors using a transparent methodology for induction of systemic resistance and regulation of antioxidant genes at expression levels in chili against bacterial wilt disease. Antioxidants (Basel) 11:404. https://doi.org/10.3390/antiox11020404

Kashyap AS, Manzar N, Meshram S, Sharma PK (2023) Screening microbial inoculants and their interventions for cross kingdom management of wilt disease of solanaceous crop—A step towards sustainable agriculture. Front Microbiol 14:1174532. https://doi.org/10.3389/fmicb.2023.1174532

Kesselmeier J, Staudt M (1999) Biogenic volatile organic compounds (VOC): an overview on emission, physiology and ecology. J Atmos Chem 33(1):23–88. https://doi.org/10.1023/A:1006127516791

Kobayashi T, Kanda E, Kitada K, Ishiguro K, Torigoe Y (2001) Detection of rice panicle blast with multispectral radiometer and the potential of using airborne multispectral scanners. Phytopathology 91(3):316–323. https://doi.org/10.1094/PHYTO.2001.91.3.316

Kogovsek P, Hodgetts J, Hall J (2015) LAMP assay and rapid sample preparation method for on-site detection of flavescence doree phytoplasma in grapevine. Plant Pathol 64:286–296

Kong X, Qin W, Huang X (2016) Development and application of loop-mediated isothermal amplification (LAMP) for detection of *Plasmopara viticola*. Sci Rep 6:28935

Koo C, MalapiWight M, Kim HS, Cifci OS, Vaughn-Diaz VL, Ma B, Kim S, Abdel-Raziq H, Ong K, Jo YK, Gross DC, Shim WB, Han A (2013) Development of a real-time microchip PCR system for portable plant disease diagnosis. PLoS One 8(12):e82704. https://doi.org/10.1371/JOURNAL.PONE.0082704

Kruckenberg H, Müller T, Freuling C, Mühle RU, Globig A, Schirrmeier H, Buss M, Harder T, Kramer M, Teske K, Polderdijk K (2011) Serological and virological survey and resighting of marked wild geese in Germany. Eur J Wildl Res 57:1025–1032

Kuckenberg J, Tartachnyk I, Noga G (2009) Temporal and spatial changes of chlorophyll fluorescence as a basis for early and precise detection of leaf rust and powdery mildew infections in wheat leaves. Precis Agric 10(1):34–44. https://doi.org/10.1007/S11119-008-9082-0

Lau HY, Wang Y, Wee EJH, Botella JR, Trau M (2016) Field demonstration of a multiplexed point-of-care diagnostic platform for plant pathogens. Anal Chem 88:8074–8081

Lim MC, Park JY, Park K, Ok G, Jang HJ, Choi SW (2017) An automated system for separation and concentration of food-borne pathogens using immunomagnetic separation. Food Control 73:1541–1547

Lindenthal M (2005) Visualisierung der Krankheitsentwicklung von Falschem Mehltau an Gurken durch *Pseudoperonospora cubensis* mittels Thermografie. https://bonndoc.ulb.uni-bonn.de/xmlui/handle/20.500.11811/2198

Lindenthal M, Steiner U, Phytopathology HD (2005) Effect of downy mildew development on transpiration of cucumber leaves visualized by digital infrared thermography. Am Phytopath Soc 95(3):233. https://doi.org/10.1094/PHYTO-95-0233

Londono MA, Harmon CL, Polston JE (2016) Evaluation of recombinase polymerase amplification for detection of begomoviruses by plant diagnostic clinics. Virol J 13:48

Mahlein AK, Oerke EC, Steiner U, Dehne HW (2012) Recent advances in sensing plant diseases for precision crop protection. Eur J Plant Pathol 133(1):197–209. https://doi.org/10.1007/S10658-011-9878-Z

Manzar N, Kashyap AS, Sharma PK, Saxena AK (2021a) First report of leaf spot of maize caused by *Curvularia geniculata* in India. Plant Dis 105:4155. https://doi.org/10.1094/PDIS-03-21-0637-PDN

Manzar N, Singh Y, Kashyap AS, Sahu PK, Rajawat SMV, Bhowmik A, Sharma PK, Saxena AK (2021b) Biocontrol potential of native *Trichoderma* spp. against anthracnose of great millet (*Sorghum bicolour* L.) from Tarai and hill regions of India. Biol Control 152:104474

Manzar N, Kashyap AS, Goutam RS, Rajawat MVS, Sharma PK, Sharma SK, Singh HV (2022) Trichoderma: advent of versatile biocontrol agent, its secrets and insights into mechanism of biocontrol potential. Sustain For 14(19):12786

Martin C, Roberts D, van der Weide M, Rossau R, Jannes G, Smith T, Maher M (2000) Development of a PCR-based line probe assay for identification of fungal pathogens. J Clin Microbiol 38(10):3735–3742

McCartney HA, Foster SJ, Fraaije BA, Ward E (2003) Molecular diagnostics for fungal plant pathogens. Pest Management Science: formerly Pesticide Science 59(2):129–142

McKeague M, Bradley CR, De Girolamo A, Visconti A, Miller JD, Derosa MC (2010) Screening and initial binding assessment of fumonisin B1 aptamers. Int J Mol Sci 11:4864–4881

Mekuria TA, Zhang S, Eastwell KC (2014) Rapid and sensitive detection of little cherry virus 2 using isothermal reverse transcription recombinase polymerase amplification. J Virol Methods 205:24–30

Mori Y, Kitao M, Tomita N, Notomi T (2004) Real-time turbidimetry of LAMP reaction for quantifying template DNA. J Biochem Biophys Methods 59:145–157

Mukherjee A, Majumdar S, Servin AD, Pagano L, Dhankher OP, White JC (2016) Carbon nanomaterials in agriculture: a critical review. Front Plant Sci 7:172. https://doi.org/10.3389/FPLS.2016.00172/XML/NLM

Narware J, Singh SP, Manzar N, Kashyap AS (2023) Biogenic synthesis, characterization, and evaluation of synthesized nanoparticles against the pathogenic fungus Alternaria solani. Front Microbiol 14:1159251. https://doi.org/10.3389/fmicb.2023.1159251

Notomi T, Okayama H, Masubuchi H (2000) Loop-mediated isothermal amplification of DNA. Nucleic Acids Res 28:63

Oerke EC, Steiner U, Dehne HW, Lindenthal M (2006) Thermal imaging of cucumber leaves affected by downy mildew and environmental conditions. J Exp Bot 57(9):2121–2132. https://doi.org/10.1093/JXB/ERJ170

Olesen AS, Lohse L, Boklund A, Halasa T, Gallardo C, Pejsak Z, Belsham GJ, Rasmussen TB, Bøtner A (2017) Transmission of African swine fever virus from infected pigs by direct contact and aerosol routes. Vet Microbiol 211:92–102

Oren E, Klingeman W, Gazis R, Moulton J, Lambdin P, Coggeshall M, Hulcr J, Seybold SJ, Hadziabdic D (2018) A novel molecular toolkit for rapid detection of the pathogen and primary vector of thousand cankers disease. PLoS One 13(1):e0185087

Paul R, Saville AC, Hansel JC, Ye Y, Ball C, Williams A, Chang X, Chen G, Gu Z, Ristaino JB, Wei Q (2019) Extraction of plant DNA by microneedle patch for rapid detection of plant diseases. ACS Nano 13:6540. https://doi.org/10.1021/ACSNANO.9B00193/SUPPL_FILE/NN9B00193_SI_002.AVI

Piepenburg O, Williams CH, Stemple DL, Armes NA (2006) DNA detection using recombination proteins. PLoS Biol 4:204

Pohl HA, Hawk I (1966) Separation of living and dead cells by dielectrophoresis. Science 152:647–649

Pysher MD, Hayes MA (2007) Electrophoretic and dielectrophoretic field gradient technique for separating bioparticles. Anal Chem 79:4552–4557

Quick J, Loman NJ, Duraffour S, Simpson JT, Severi E, Cowley L, Bore JA, Koundouno R, Dudas G, Mikhail A, Ouédraogo N (2016) Real-time, portable genome sequencing for Ebola surveillance. Nature 530(7589):228–232

Rabelo GF, Enes AM, Braga JRA, Dal Fabbro IM (2011) Frequency response of biospeckle laser images of bean seeds contaminated by fungi. Biosyst Eng 110(3):297–301. https://doi.org/10.1016/J.BIOSYSTEMSENG.2011.09.002

Rad F, Mohsenifar A (2012) Detection of Candidatus *Phytoplasma aurantifolia* with a quantum dots fret-based biosensor. J Plant Pathol 94(3):525–534

Ro DK, Paradise EM, Ouellet M, Fisher KJ, Newman KL, Ndungu JM, Ho KA, Eachus RA, Ham TS, Kirby J, Chang MC (2006) Production of the antimalarial drug precursor artemisinic acid in engineered yeast. Nature 440(7086):940–943

Röck F, Barsan N, Weimar U (2008) Electronic nose: current status and future trends. Chem Rev 108(2):705–725. https://doi.org/10.1021/CR068121Q/ASSET/CR068121Q.FP.PNG_V03

Safarpour H, Safarnejad MR, Tabatabaei M, Mohsenifar A, Rad F, Basirat M, Shahryari F, Hasanzadeh F (2012) Development of a quantum dots FRET-based biosensor for efficient detection of *Polymyxa betae*. Can J Plant Pathol 34(4):507–515. https://doi.org/10.1080/07060661.2012.709885

Safenkova IV, Zaitsev IA, Varitsev YA, Byzova NA, Drenova NV, Zherdev AV, Dzantiev BB (2017) Development of a lateral flow immunoassay for rapid diagnosis of potato blackleg caused by Dickeya species. Anal Bioanal Chem 409:1915–1927

Schmidt K, Mwaigwisya S, Crossman LC, Doumith M, Munroe D, Pires C, Khan AM, Woodford N, Saunders NJ, Wain J, O'grady J (2016) Identification of bacterial pathogens and antimicrobial resistance directly from clinical urines by nanopore-based metagenomic sequencing. J Antimicrob Chemother 72(1):104–114

Sefah K, Shangguan D, Xiong X, O'Donoghue MB, Tan W (2010) Development of DNA aptamers using cell-SELEX. Nat Protoc 5:1169–1185

Singh Y, Sidhu HS (2014) Management of cereal crop residues for sustainable rice-wheat production system in the Indo-Gangetic plains of India. Proc Indian Natl Sci Acad 80(1):95–114

Stowell LJ, Gelernter WD (2001) Diagnosis of turfgrass diseases. Annu Rev Phytopathol 39(1):135–155

Wang P, Lombi E, Zhao FJ, Kopittke PM (2016) Nanotechnology: a new opportunity in plant sciences. Trends Plant Sci 21(8):699–712. https://doi.org/10.1016/J.TPLANTS.2016.04.005

Ward LN, Bej AK (2006) Detection of Vibrio parahaemolyticus in shellfish by use of multiplexed real-time PCR with TaqMan fluorescent probes. Appl Environ Microbiol 72(3):2031–2042

Wu SH, Lin HP (2013) Synthesis of mesoporous silica nanoparticles. Chem Soc Rev 42(9):3862–3875. https://doi.org/10.1039/C3CS35405A

Yan M, Xu L, Jiang H, Zhou Z, Zhou S, Zhang L (2017) PMA-LAMP for rapid detection of *Escherichia coli* and Shiga toxins from viable but non-culturable state. Microb Pathog 105:245–250

Recent Advances in the Detection of Plant Viruses and Viroids in Field Crops

3

Neeta Pathaw, Susheel Kumar Sharma, Nitika Gupta, Shweta Singh, Devender Sharma, Parul Sharma, Sajad Un Nabi, Jyotsana Sanasam, K. Sarda Devi, Baby Wangkhem, O. Priyoda Devi, Th. Surjit Singh, Albert Maibam, Sapam Monteshori, H. Chandrajini Devi, Malemnganbi Keisham, Tabalique Yumkhaibam, W. Tampakleima Chanu, and S. S. Roy

Abstract

Crop plants are susceptible to infection by a number of viruses and viroids, which can only be controlled by early detection along with accurate diagnosis. Initial diagnosis of viral diseases was based on biological assays for the identification followed by serological and cytopathological detection methods. Enzyme-linked immunosorbent assay (ELISA) has emerged as a highly favored technique

N. Pathaw · J. Sanasam · K. S. Devi · B. Wangkhem · O. P. Devi · T. S. Singh · A. Maibam · S. Monteshori · H. C. Devi · M. Keisham · T. Yumkhaibam · W. T. Chanu
ICAR Research Complex for NEH Region, Manipur Centre, Lamphelpat, Imphal, India

S. K. Sharma (✉) · N. Gupta (✉)
Advanced Centre for Plant Virology, Division of Plant Pathology, ICAR—Indian Agricultural Research Institute, New Delhi, India

S. Singh
ICAR—Indian Institute of Sugarcane Research, Lucknow, Uttar Pradesh, India

D. Sharma
Crop Improvement Division, ICAR—Vivekananda Parvatiya Krishi Anusandhan Sansthan, Almora, Uttarakhand, India

P. Sharma
Punjab Agricultural University, Ludhiana, Punjab, India

S. U. Nabi
ICAR—Central Institute of Temperate Horticulture, Srinagar, Jammu and Kashmir, India

S. S. Roy
ICAR—Central Citrus Research Institute, Nagpur, Maharashtra, India

V. K. Singh et al. (eds.), *Diseases of Field Crops: Diagnostics and Management*, https://doi.org/10.1007/978-981-97-6160-9_3

for its versatility; however, with the advent of polymerase chain reaction (PCR), nucleic acid–based techniques has evolved as a specific and accurate tool for the detection of viruses and virus-like pathogens. In recent years, variants of PCR like reverse transcription-PCR (RT-PCR), real-time PCR, multiplex PCR, and isothermal nucleic acid amplification techniques were found advantageous in the detection of pathogens owing to their sensitivity and versatility as compared to ELISA. The use of next-generation sequencing (NGS) for the discovery of novel viruses or technologically advanced biosensor devices has revolutionized the field of viral disease diagnosis. This chapter reviews various methods of viral detection in field crops from the early to the most advanced technologies, although some are established protocols others are still under development.

Keywords

Detection · Serological detection · Nucleic acid–based detection · Isothermal detection · Next-generation sequencing

3.1 Introduction

Viruses are tiny, obligate intracellular parasites with very small genomes. Owing to the simple structure, they are however unable to reproduce or replicate without a host. They consist of single or double-stranded nucleic acid (DNA or RNA) as their genetic material packaged by capsid proteins with or without a lipid envelope. Viruses and viroids cause infection across the domains of the living world, giving rise to many important diseases in plants. They lead to the significant yield and quality losses in many agricultural crops due to which they are ranked as the second most economically important pathogen (Vidaver and Lambrecht 2004; Jeong et al. 2014). Economic impact due to viral infection in plants has been estimated to be about several billions (Hull 2002; Jeong et al. 2014; Islam et al. 2017); however, the absolute damage to the crops in terms of revenue is difficult to predict as the rate of infection depends on several factors like prevailing environmental conditions, viral strain, plant variety, and the stage of infection (Strange 2005). Viruses are known to modify host cellular metabolism by redirecting nutrients and altering the host physiology to suit their growth and development resulting in multiple symptoms that mimic other diseases caused due to various biotic or abiotic conditions (Lievens et al. 2005). While most of the plants do show typical symptoms like mosaic, stunting, curling, puckering, and mottling, others, however, are symptomless which could be attributed to the virus being present in the latent form (Nabi et al. 2018). Therefore, detection based on the symptom expression is not a preferred choice of diagnostics of plant viral diseases. Viruses are also endowed with a high rate of mutation; this combined with molecular events like recombination and exchange of genetic materials between closely related species leads to the development of diverse and aggressive strains (García-Arenal et al. 2001). While changing weather patterns have brought about an emergence of newer viral strains or a resurgence of the milder viruses as potent ones, factors like rising international trade in agricultural stocks viz. seeds and plantlets have led to an increased likelihood of introducing newer pathogens into the uncharted territory (Massart et al. 2017). Therefore, early detection and identification remains the most important method to control the diseases caused by these vital pathogens. From the first virus (tobacco mosaic virus) to be identified a century ago to the most recent ones, several techniques have been established and utilized for detecting these potent pathogens. Over the years, several methods including biological, cytological, immunological, and molecular as well as more recent ones that include the use of high-throughput sequencing and computing methods have been utilized to develop more robust detection techniques (Yadav and Khurana 2016; Rani et al. 2019). Although most of the studies on standardization of diagnostic methods are focused particularly on horticultural crops, this chapter reviews the various diagnostic techniques from the classical to the more advanced technologies that have been developed and utilized for the identification of viruses in major field crops. A list of important viral diseases in field crops is highlighted in Table 3.1.

Table 3.1 List of important viral diseases in field crops

Crops	Viruses
Cereal crops	
Rice	Rice tungro bacilliform virus (RTBV), rice tungro spherical virus (RTSV), rice black–streaked dwarf disease (RBSDV), Southern rice black–streaked dwarf virus (SRBSDV), rice stripe virus (RSV), rice dwarf virus (RDV), rice gall dwarf virus (RGDV), rice ragged stunt virus (RRSV), rice grassy stunt virus (RGSV), rice yellow stunt virus (RYSV)
Maize	Maize streak virus (MSV), maize dwarf mosaic virus (MDMV)
Wheat	Wheat dwarf India virus (WDIV), wheat streak mosaic virus (WSMV), Triticum mosaic virus (TriMV), wheat American striate mosaic virus (WASMV), wheat spindle streak mosaic virus (WSSMV), barley yellow dwarf virus (BYDV)
Barley	Barley stripe mosaic virus (BSMV), barley yellow dwarf virus (BYDV)
Millets	Maize streak virus (MSV), maize dwarf mosaic virus (MDMV), sugarcane mosaic virus (SMV)
Pulses	
Cowpea	Cowpea aphid-borne mosaic virus (CABMV), BCMV-blackeye cowpea mosaic strain (BCMV-BlCM), cowpea chlorotic mottle virus (CCMV), CMV, cowpea golden mosaic virus (CGMV), cowpea mild mottle virus (CPMMV), Southern bean mosaic virus (SBMV)
Green gram and black gram	Mungbean yellow mosaic virus (MYMV), mungbean yellow mosaic India virus (MYMIV), bean common mosaic virus (BCMV)
Chickpea	Chickpea chlorotic dwarf virus (CpCDV)
Pigeon pea	Pigeon pea sterility mosaic virus (PPSMV)
French bean	Bean common mosaic virus (BCMV)

(continued)

Table 3.1 (continued)

Crops	Viruses
Cash crops	
Sugarcane	Sugarcane bacilliform viruses (SCBV), sugarcane yellow leaf virus (SCYLV), sugarcane streak mosaic virus (SCSMV)
Sugarbeet	Beet ringspot virus (BRV), beet necrotic yellow vein virus (BNYVV)
Tobacco	Tobacco mosaic virus (TMV), tobacco leaf curl Pusa virus (TbLCPuV), papaya leaf curl virus (PaLCV), radish leaf curl virus (RaLCV)
Oilseeds	
Soybean	Soybean mosaic virus (SbMV), soybean dwarf virus (SbDV), tobacco ring spot virus (TRSV), cowpea severe mosaic virus (CPSMV), peanut bud necrosis virus (PBNV), tobacco streak virus (TSV), mungbean yellow mosaic virus (MYMV), bean golden mosaic virus (BGMV)
Sesame	Broad bean wilt virus-2 (BBWV-2), cowpea aphid-borne mosaic virus (CABMV)
Groundnut/ peanut	Peanut clump virus (PCV), peanut yellow mosaic virus (PeYMV), groundnut bud necrosis virus (GBNV), groundnut rosette assistor virus (GRAV), groundnut rosette virus (GRV), tomato spotted wilt virus (TSWV), peanut bud necrosis virus (PBNV), Indian peanut clump virus (IPCV), peanut mottle virus (PeMoV), peanut stripe virus (PStV), cucumber mosaic virus (CMV)

3.2 Detection Methods

Historically, the discovery of new plant viruses is often the output of etiological investigations of economically important plant diseases, and the follow-up of the identification and characterization of the pathogen is the development of a reliable detection method. The ability to produce a robust and effective detection method is therefore an essential aspect of managing plant viral diseases. In many cases, viral infection in plants can remain systemic throughout its lifespan with few exceptions like blueberry necrotic ring blotch virus (BNRBV) which although causes disease does not move systemically (Robinson et al. 2016). The detection and identification of viruses and viroids rely on a wide range of techniques based on the biological, morphological, and intrinsic properties of viruses. Figure 3.1 summarizes the different techniques used for the diagnosis of diverse plant viruses. The following sections provide a compendium on the various techniques used particularly for the diagnosis of viruses and viroids in agricultural crops.

3.2.1 Detection Based on Biological Properties

Biological assay as a tool was once the preferred choice for the detection and identification of viral diseases in plants. The detection was based primarily on the development of symptom expression on indicator plants that are either a natural host or a susceptible plant that has been mechanically inoculated using sap, grafting, budding, or vector-mediated methods. Biological assays were highly recommended in some species as they allowed bio-amplification of the pathogen when the titer of the

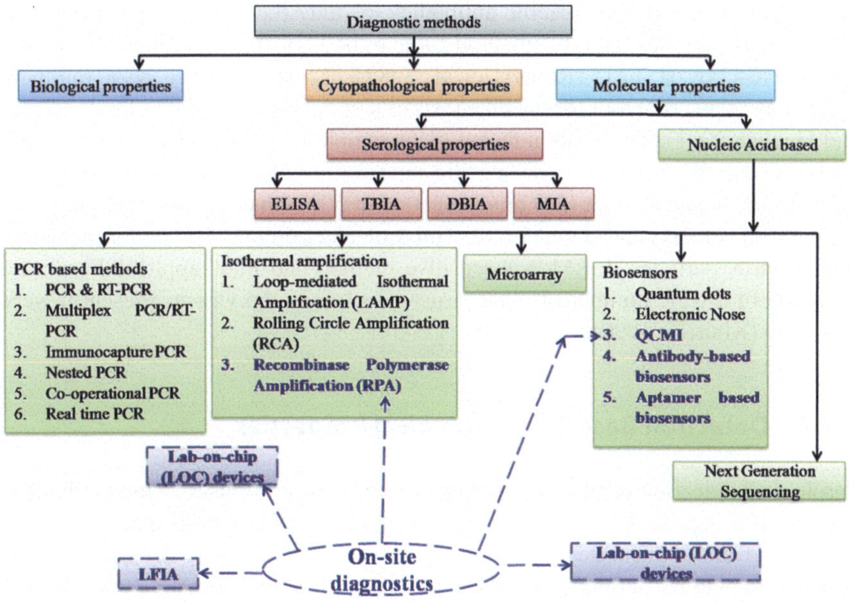

Fig. 3.1 Techniques used for the detection and diagnosis of viruses and viroids in plants

pathogen was too low for detection, particularly useful as they increased the concentration of the viral titer in plant tissues and hence allowed easy detection and characterization by other serological and molecular methods. With the advent of other advanced and less labor-intensive methods, however, this once essential technique is now not a routinely used method for the detection of viruses in plants.

3.2.2 Detection Based on Cytopathological or Physical Properties

Viruses are unique, and every individual virus has specific physical and anatomical features. Therefore, the basic physical features like shape, size, and other anatomical properties of individual virions were employed for the development of a vital viral detection method in plants using the electron microscopy (EM) technique (Bernd and Günther 2009). The detailed structural study of the viruses utilizes a negative leaf-dip staining or epidermal strip technique for the detection of the shape and size of the individual viruses with the commonly used negative strains such as ammonium molybdate, sodium phosphotungstate, or uranyl acetate. EM is a rapid detection technique for the diagnosis of specific viruses (Harris 2007; Wild 2008). Currently, various microscopic studies are also used in combination with immuno-based detection like immuno-electron microscopy (IEM). The use of virus-antibody interaction in electron microscopy was first demonstrated by Anderson and Stanley 1941 to detect tobacco mosaic virus (TMV). Derrick (1973) used a combination of electron microscopy with specific antibodies to study the morphological observation of virus particles, and this method came to be known as immunosorbent electron microscopy (ISEM) (Roberts et al. 1982). ISEM has since been routinely utilized as a detection tool for the diagnosis of a broad range of plant viruses (Naidu and Hughes 2001). The virus infected plant sample (sap) is coated onto a copper grid treated earlier with the virus-specific antiserum followed by washing and staining with the negative stain (2% uranyl acetate in 50% ethanol) and air drying. The copper grid is then viewed under a transmission electron microscope for the detection of virus particles. ISEM is a sensitive method and finds applicability in the detection of a wide number of plant viruses, including the viruses present in very low titers (Khurana 1990).

3.2.3 Detection Based on Serological Properties

Serological detection of plant viruses employs the use of virus-specific antibodies. The past few decades have seen a breakthrough in the field of molecular diagnostics with the advent of accurate and rapid methods based on antibody detection, in particular monoclonal antibodies and enzyme-linked immunosorbent assay (Kohler and Milsten 1975; Clark and Adams 1977). The surface of viruses has various proteins that can be used for detection and diagnosis purposes. Most plant viruses have coat protein which is used to study pathogenicity as well as for detection through immunological response by a generation of antibodies against the protein. Such kinds of techniques have now become routine diagnostic tools, with methods like enzyme-linked immunosorbent assay (ELISA), dot-blot immunobinding assay (DBIA), and tissue blot immunobinding assay (TBIA).

3.2.3.1 Enzyme-Linked Immunosorbent Assay (ELISA)

ELISA method is a simple, versatile, and robust technique, which is performed on a polystyrene plate coated with specific antibodies (double-antibody sandwich-ELISA: DAS-ELISA). The microtiter plate for ELISA is coated with the primary antibody followed by the infected crude sap or purified protein that is further detected with the help of a secondary antibody conjugated with a reporter molecule. The plant virus detection is confirmed through a colorimetric estimation wherein the colored product formed is estimated through an ELISA reader. An alternative to DAS-ELISA is the direct antigen coated-ELISA (DAC-ELISA) where the infected plant sap or the protein is firstly pre-coated into the polystyrene plate and then detected by primary and secondary antibodies, respectively (Clark and Adams 1977). ELISA as a technique is quite sensitive and can be used for the detection of many samples at one time (Vemulapati et al. 2014) with the requirement of a very small amount of antibody and is a semi-automatic process (Naidu and Hughes 2001). Bajet et al. (1985) have reported the development of a double-antibody sandwich (DAS)-ELISA for the detection of rice tungro disease caused by rice tungro bacilliform virus (RTBV) and rice tungro spherical virus (RTSV) in infected greenhouse plants. Also, polyclonal antisera raised against

RTBV and RTSV have been used for the development of a simple, rapid diagnostic using DAS-ELISA for field surveys (Nath et al. 2000). ELISA has also been used for the detection of a wide variety of viruses including tobacco ringspot virus (TRSV), soybean mosaic virus (SMV), cowpea aphid-borne mosaic virus (CABMV), and cucumber mosaic virus (CMV) (Lister 1978; Ojuederie et al. 2009; Sun et al. 2001). Although a sensitive method, it sometimes gives false positives because of cross-reactivity or non-specific reactions (Kfir and Genthe 1993) with strains of closely related viral species which are not differentiated using ELISA (Boonham et al. 2014).

3.2.3.2 Dot-Blot Immunobinding Assay (DBIA)

The dot-blot immunobinding technique is a modification of ELISA wherein the dot assay is done on a nitrocellulose membrane. The sample-laden membrane is further incubated in a test antibody followed by a peroxidase-conjugated secondary antibody to the primary antibody and observed for the colour development of 4-chloro-1-naphthol. DBIA has very high sensitivity and can be used to detect very low titer (nanogram quantities) viruses as a very small volume is required in the process (Powell 1987; Abd El-Aziz et al. 2019). This technique in some viral species was found to be superior to both ELISA and immunosorbent electron microscopy methods. It could even detect cowpea mosaic virus (CPMV) in nanogram quantities (Zheng et al. 1988). Berger et al. (1985) used DBIA to detect pictogram quantities of potyviruses using a dot immunobinding assay technique.

3.2.3.3 Tissue Blot Immunobinding Assay (TBIA)

Tissue blot immunoassay is based on the principle of ELISA except for using nitrocellulose and nylon membranes as the platform for detection (Webster et al. 2004). The technique requires a larger volume of virus concentration and the use of a specific antibody to eliminate any false result. TBIA is found to be superior as a technique for its cost-effectiveness, reliability, and high sensitivity as compared to ELISA and hence used extensively for the detection of many viral diseases like bean yellow mosaic virus (BYMV), bamboo mosaic virus (BoMV), papaya ringspot virus (PRSV), cymbidium mosaic virus (CyMV), tomato spotted wilt virus (TSWV), and sweet potato feathery mottle virus (SPFMV) (Bove et al. 1988; Lin et al. 1990; Webster et al. 2004; Makkouk and Kumari 2006; Eid et al. 2008; Shang et al. 2011; Hančević et al. 2012). Al-Khalaf et al. (2009) have reported the use of tissue blot immunoassay on imprinted nitrocellulose membranes for the detection of bean yellow mosaic virus (BYMV) in faba bean, lentil, and pea plants. International Centre for Agricultural Research in the Dry Areas (ICARDA) has developed commercial detection kits based on polyclonal antibodies that can diagnose a wide variety of legume-infecting viruses like cucumber mosaic virus (CMV), bean yellow mosaic virus (BYMV), alfalfa mosaic virus (AMV), pea seed-borne mosaic virus (PSbMV), and chickpea chlorotic stunt virus (CpCSV) including some based on monoclonal antisera like bean leaf roll virus (BLRV), soybean dwarf virus (SbDV), and beet western yellow virus (BWYV) (Katul 1992; Kumari et al. 2006, 2008; Makkouk and Kumari 1996).

3.2.3.4 Microsphere Immunoassay (MIA)

ELISA remains the most utilized or modified method for the detection of viruses; however, a recent technology using microspheres in the immunoassay has been developed for the detection of microorganisms (Charlermroj et al. 2013). Microsphere immunoassay employs the use of different sets of fluorescence-coded microspheres in which each bead comprises a combination of dyes coupled with target-specific antibodies linked to a fluorophore. Bergervoet et al. (2008) reported the use of this bead-based technology in detecting diverse potato-infecting viruses in the same reaction. MIA utilizing paramagnetic beads instead of conventional beads has resulted in high specificity and lower background reaction. This technique has been used to simultaneously detect potato virus Y (PVY), potato virus X (PVX), and potato leaf roll virus (PLRV) in potato leaf extracts with equal sensitivity and specificity (Selvarajan and Balasubramanian 2016). Microsphere-based immunoassay has been able to detect four plant pathogens together with equal sensitivity viz. a fruit blotch bacterium: *Acidovorax avenae* subsp. citrulli (Aac), a potyvirus: chilli veinal mottle virus (ChiVMV), a orthotospovirus serogroup IV: watermelon silver mottle virus (WSMoV), and a orthotospovirus: melon yellow spot virus (MYSV) (Charlermroj et al. 2013). The application of this technique for the detection of viruses infecting field crops however remains to be explored.

3.2.4 Detection Based on Nucleic Acid

The discovery of a thermostable *Taq* polymerase enzyme revolutionized the use of modern molecular techniques in laboratories where polymerase chain reaction (PCR) could be used in the detection of viral pathogens based on the genetic information of viruses. Molecular detection methods have evolved as a simplified alternative to costly and time-consuming serological methods. It is used routinely in laboratories worldwide nowadays owing to its high accuracy and sensitivity. Molecular detection methods include techniques such as PCR, reverse transcriptase-PCR (RT-PCR), multiplex PCR, and quantitative PCR (qPCR) along with many modifications to these techniques. These molecular techniques have many advantages over biological or serological techniques, as any viral genomic region can be targeted to develop the detection tools.

3.2.4.1 PCR and RT-PCR

PCR as a technique is used for amplification, i.e., the creation of millions of copies of the same DNA sequence in a short period. This technique being highly accurate and sensitive has effectively been used for the detection and identification of many viruses and is presently the base of all modern detection methods (Webster et al. 2004; Lopez et al. 2008). PCR proceeds in three basic steps: (1) denaturation of double-stranded DNA at a temperature of 94 °C or above, (2) annealing of primers at 50–75 °C (depending on the melting temperature of primers), and (3) elongation at 72 °C (Makkouk and Kumari 2006). It has been used to identify and characterize many important viruses like rice tungro bacilliform virus (RTBV), wheat dwarf

India virus (WDIV), sugarcane bacilliform virus (SCBV), and mungbean yellow mosaic virus (MYMV) (Dasgupta et al. 1996; Mandal et al. 1997; Singh et al. 2009; Kumar et al. 2012; Karuppaiah et al. 2013; Rao et al. 2014). RNA viruses require the conversion of RNA into cDNA employing a reverse transcriptase enzyme before the regular PCR in a process known as RT-PCR (Lopez et al. 2008; Webster et al. 2004). RT-PCR is an accurate, inexpensive, and specific technique as compared to the serological methods; therefore, it has been extensively used to detect many important viruses of pulses like cowpea aphid-borne mosaic virus (CABMV), pigeonpea sterility mosaic virus (PPSMV), and soybean mosaic virus (SMV) (Colinet et al. 1998; Kumar et al. 2003; Ha et al. 2008; Elbeaino et al. 2013; Mishra et al. 2015). Both PCR and RT-PCR have also been used to detect several viruses in crops like groundnut and mustard such as turnip yellow virus (TuYV), groundnut bud necrosis virus (GBNV), peanut mottle virus (PeMoV) and peanut clump virus (PCV) (Naidu et al. 2000; Umamaheswaran et al. 2003; Sharma et al. 2013; Soumya et al. 2014).

3.2.4.2 Multiplex PCR/RT-PCR

Multiplex PCR is a technique that can detect two or more viruses/viroids together (Lopez et al. 2008; Webster et al. 2004). This method allows the detection of many viruses altogether in a particular host and is extremely useful for the detection of different viral pathogens in case of mixed infections. Multiplex PCR technique requires the use of several primers targeting specific regions of the genome of multiple viruses in a single reaction. The specific primers used in this method are designed to lack self-complementarity and have the same annealing temperature enabling it to be used in a single amplification along with the amplified products differing in size for effective detection through agarose gel visualization. This technique has an edge over other conventional methods of detection as many viruses can be detected in a single reaction making it cost and time-effective (Bertolini et al. 2003; Roy et al. 2005; Deb and Anderson 2008). Periasamy et al. (2006) have reported on the standardization of a multiplex RT-PCR-based detection protocol for the simultaneous detection of rice tungro bacilliform virus (RTBV) and rice tungro spherical virus (RTSV), the causal agents of the devastating rice tungro disease in India. Multiplex RT-PCR technique has also been used to detect four different viruses in garlic and for the simultaneous detection of four viruses along with a greening bacterium in citrus (Majumder and Baranwal 2014; Meena and Baranwal 2015). This technique apart from multiple detection of different viruses in crops has also been used in combination with high-throughput sequencing techniques in detecting both known and novel viruses/viroids (Villamor et al. 2016; Jones et al. 2017).

3.2.4.3 Immunocapture PCR (IC-PCR)

This technique is a combination of nucleo- and immuno-based detection techniques and involves the use of antibodies to entrap virus particles followed by amplification using virus-specific oligonucleotide. The process requires adsorption of virus particles on the antibody-coated tubes followed by heating in the presence of non-ionic surfactant Triton X-100 prior to PCR. IC-PCR helps in concentrating the virus particles due to antibody-enabled entrapment, thus making it highly sensitive even for low-titer viruses. Subba Reddy et al. (2011) reported the development of a duplex-immunocapture-reverse transcriptase-PCR (D-IC-RT-PCR) for the detection and distinction of naturally infecting sugarcane streak mosaic virus (SCSMV) and sugarcane mosaic virus (SCMV). For the detection, a common reverse primer (5′-d(T)18(AGC)-3′) was used along with a viral genome-specific forward primer to detect 1.4 kb (SCSMV) and 0.9 kb (SCMV) amplicons, respectively. Both viruses could be detected up to the dilution of 10^{-4} in D-IC-RT-PCR (Subba Reddy et al. 2011).

3.2.4.4 Nested PCR

Nested PCR is a variant of polymerase chain reaction which is used when the viral titer is very low, which otherwise using the conventional method results in low amplification and hence difficult to detect (Webster et al. 2004). This method requires a second amplification step using the primary amplicon product. The nested PCR technique has been used to detect several viruses infecting cowpea plantations in West Africa like cowpea mottle virus (CPMoV), Southern cowpea mosaic virus (SCPMV), tombusvirus-like viruses, and mycotymovirus (Palanga et al. 2016). This technique has also been used for detecting viruses like prunus necrotic ringspot virus (PNRSV), prune dwarf virus (PDV), PPV, and citrus tristeza virus (CTV) (Olmos et al. 1999; Helguera et al. 2001, 2002; Adkar-Purushothama et al. 2011).

3.2.4.5 Co-operational PCR (Co-PCR)

Co-operational PCR requires a second amplification step and unlike the nested PCR can be performed in the same reaction tube (Martos et al. 2011). This technique requires the use of an external along with three internal primers for the reaction and requires diluted samples for PCR (Pantaleo et al. 2001; Olmos et al. 2002; Capote et al. 2009). Olmos et al. (2002) reported the detection of cherry leaf roll virus (CLRV) utilizing this technique resulting in a higher detection sensitivity of 100times more than RT-PCR.

3.2.4.6 Real-Time PCR (qPCR)

Real-time PCR (rt-PCR) or quantitative PCR (qPCR) is a real-time closed-tube technique which is an accurate, sensitive and highly specific method of detection. It is used for the detection of even small quantities of target DNA and requires no gel electrophoresis for the confirmation of the amplification product. The qPCR requires the use of labeled primers and results in the accumulation of fluorescent amplicons in real time. At present, there are a variety of fluorogenic dyes that are used to bind to any dsDNA in real-time method and hence are rendered one of the simplest and economical ways of quantification. Boonham et al. (2004) used the TaqMan assay for the detection of potato spindle tuber viroid that is ~1000 times more sensitive as compared to chemiluminescent assays. This technique has also been reportedly employed for the detection and quantitation of wheat yellow mosaic virus (WYMV) in wheat plants (Liu et al. 2013). A single-step reverse transcription-real-time PCR assay was used for monitoring the changes in viral density with subsequent effects on symptom expression of Southern rice black–streaked dwarf virus (SRBSDV) infecting rice (Matsukura et al. 2013). The real-time technique was also used to routinely analyze and differentiate closely related viruses like rice black–streaked dwarf virus (RBSDV) and Southern rice black–streaked dwarf virus (SRBSDV) (Sharma and Dasgupta 2012; Matsukura et al. 2013).

3.2.4.7 Microarray

Microarray is a technique that evolved from southern blotting which utilizes glass as a support medium instead of nitrocellulose or nylon membrane. Though initially developed for the detection of differential expression of mRNA, it was further developed for accurate diagnosis of plant viruses without the need for amplification. Oligochips used in the oligonucleotide tray comprise thousands of specific 25–75 bp nucleotide single-stranded synthesized DNA probes spotted onto the glass support which hybridizes with the viruses extracted from plant samples. The major drawback of this technique is the cost and requirement of advanced automated machinery required for spotting and analysis that has to be hosted in a clean environment. The other requirement is the construction of oligonucleotide probes that are highly specific to the target DNA (Dugat-Bony et al. 2012). Microarray technique has been used efficiently for the detection of tomato infectious chlorosis virus (TICV), tomato mosaic virus (TMV), pepino mosaic virus (PepMV), and cucumber mosaic virus (CMV) (Tiberini et al. 2010), potato virus X (PVX), potato virus A (PVA), potato leafroll virus (PLRV), potato virus Y (PVY), and potato virus S (PVS) (Bystricka et al. 2005; Lee et al. 2003). A microarray chip for the detection of wheat streak mosaic virus was generated using an SNP-based single-base primer extension (Yadav and Khurana 2016).

3.2.4.8 Isothermal Amplification

The isothermal amplification technique depends on a non-thermal separation of dsDNA and just like the conventional PCR uses specific primers and polymerases to amplify the target DNA. The amplification step can be done in an isothermal condition without the need for thermal cycling and hence does not require the use of costly or power-consuming machines making it cheaper and effective. Owing to its cost-effectiveness, simplicity, and robust utility, it is easily used as an on-site detection technique (Boonham et al. 2014). There are different variants of this technique that have been developed for the diagnosis of plant viruses like loop-mediated isothermal amplification (LAMP), rolling circle amplification (RCA), and recombinase polymerase amplification (RPA).

Loop-Mediated Isothermal Amplification (LAMP)

Loop-mediated isothermal amplification method is a novel detection technique that overcomes the cost barrier of a conventional PCR while being a highly specific and sensitive method of viral detection in plants (Notomi et al. 2000). This technique can amplify the target DNA at very low copy number in less time. The reaction comprises a specific *Bst* DNA polymerase (high processivity and strand displacement activity) along with three sets of primers each showing complementarity to their specific genomic regions in the template. The reaction is executed under a constant temperature between 60 and 65 °C for 30–60 min. It exhibits high tolerance to various biological contaminants, thus avoiding false negatives as a result of the inactivation of the polymerase enzyme mostly faced in conventional PCR (Kaneko et al. 2007). LAMP has been used to detect several plant viruses such as wheat yellow mosaic virus, tomato yellow leaf curl virus, plum pox virus, banana bunchy top virus, and zucchini yellow mosaic virus (Kuan et al. 2014; Fukuta et al. 2003; Varga and James 2006; Peng et al. 2012; Ahmadi et al. 2013). Le et al. (2010) reported the use of reverse transcription-LAMP (RT-LAMP) for the detection of nine rice viruses in Asia. This technique has also been used to detect and differentiate rice black streaked dwarf virus (RBSDV) and southern rice black streaked dwarf virus (SRBSDV) from host plants and insect vectors (Zhou et al. 2012). Recently the use of LAMP detection method combined and converted into a lateral flow dipstick (LFD) device has been used as potential on-site or field-based detection method. Tomlinson et al. (2013) have reported the use of LAMP detection products using the lateral flow device for the simultaneous detection of the amplification products of cassava brown streak virus (CBSV), Ugandan cassava brown streak virus (UCBSV) and cytochrome oxidase (a plant internal control gene).

Rolling Circle Amplification (RCA)

The rolling circle amplification technique amplifies circular genomes using the *Bacillus subtilis* bacteriophage *Phi29* DNA polymerase. It can be performed using both sequence-specific or random hexamer primers at an isothermal temperature. As the RCA reaction involves the use of random primers for amplification, hence no prior sequence information is required, and it can amplify novel circular viral genomes. RCA-amplified product is digested using restriction enzymes before sequencing for the confirmation of the viral species. Restriction fragments of the RCA product using specific restriction enzymes can also be utilized as a characteristic profile for the screening of a large number of circular viruses such as begomoviruses and badnaviruses. RCA is a highly sensitive method for the detection of integrated and episomal viral sequences of badnaviruses (James et al. 2011; Baranwal et al. 2014; Sharma et al. 2015).

3.2.5 Biosensor

Recent developments in the field of biomaterials have paved the way for the development of a wide variety of sensor-based techniques for detecting microorganisms such as bacteria, viruses, and fungi (Fang and Ramasamy 2015). Depending on the sensor principle, the analytes can be further detected using electrical, chemical, optical, vibration, or magnetic signal-based sensors. The use of nanomaterial-based matrices as transducers along with the use of biomolecules for recognition like

DNA, enzymes, and antibodies has further enhanced the limit of detection of these techniques. Various nanomaterials are currently used for the construction of biosensors like quantum dots, metal or metal oxide, carbon, and polymeric nanoparticles. Biosensor holds great promise for the in situ detection of biomolecules such as protein-based biomarkers or viruses that can be used to detect specific viruses at a field or to identify emerging viruses as well as fluctuations in the population of emerging ones.

3.2.5.1 Quantum Dots (QDs)

Quantum dot (QD) comprises tiny (a few nm in dm) fluorescent, crystalline particles which have been successfully used for biosensor construction for disease detection in plants. QDs are stable light emitters and due to their robustness can be tuned into different sizes. Due to technological advancement in the field of nanotechnology, QDs as a biomarker have come up as a reliable and precise tool for disease diagnosis. Owing to their photostability and optical sensitivity as biolabels, they can easily be detected using non-sophisticated equipment. QDs with their unique optical properties are used as bioanalyzers using the fluorescence resonance energy transfer (FRET) mechanism (Algar and Krull 2008). Fang and Ramasamy (2015) reported the use of a QD-FRET sensor to detect witches' broom disease (WBDL) caused by *Candidatus Phytoplasma aurantifolia* (Ca. *P. aurantifolia*) in lime. QD-FRET-based antibody sensor has also been developed for the rapid and accurate detection of the vector *Polymyxa betae* (Keskin) specific to beet necrotic yellow vein virus (BNYVV) in sugarbeet (Safarpour et al. 2012). Although this technique has shown much success in the laboratories, the on-site detection using QD based sensors are yet to be explored.

3.2.5.2 Electronic Nose or E-Nose (EN)

E-Nose (EN) is an instrument comprising an array of electronic chemical sensors that are specifically able to recognize profiles of volatile chemical signatures of diseased plants. Plants release a wide variety of volatile organic compounds (VOCs) under stress. However, many plants do not show any symptoms until the late stage of infection. Hence, it is highly valuable to detect the disease at its onset to develop appropriate management strategies. Many studies for the detection and analysis of plant VOCs are undertaken successfully with invasive methods like gas chromatography-mass spectrometry (GC-MS) and proton transfer reaction mass spectrometry (PTR-MS), but these are bulky and expensive instruments requiring skilled manpower and do not work in a real-time fashion. EN is a non-invasive method that can work in a real-time fashion and even detect trace amounts of VOCs. The system is coupled with a variety of sensor arrays that transform the VOC signals into an electronic signal, the specific pattern of which is then modulated by conditioning circuits and saved as data sets to be read and analyzed for pattern recognition analysis (Cui et al. 2018). This technique has been used successfully for the early detection of bacterial diseases of chilli and papaya (Prak Chang et al. 2014). Hararika et al. (2020) have reported the use of EN as a diagnostic device for detecting citrus tristeza virus (CTV) in Khasi mandarin orange.

3.2.6 Next-Generation Sequencing (NGS)

The past few decades have seen unprecedented technological advancement in the field of sequencing techniques that have led to the development of newer methods for the detection and characterization of viruses. The NGS technique unlike the conventional mode of detection provides a powerful alternative for the detection of viruses or viroids without any prior sequence information (Prabha et al. 2013; Villamor et al. 2019). NGS or high-throughput or deep sequencing is a highly efficient and rapid technique that has been used to study pathogenomics, particularly the detection and identification of novel viruses/viroids. It is superior to the traditional Sanger chain termination method. It works in a different approach of sequencing by the synthesis of generating massive data of millions to billions of nucleotides in parallel order. Being a culture and sequence-independent approach, NGS can be used for the simultaneous detection of DNA/RNA viruses/viroids even at very low titer. It is performed using different platforms like Illumina, SOLID, Ion torrent, 454 pyrosequencing, Roche 454, PacBio-Illumina, MinION, and Nanopore and template preparations for the simultaneous detection of known as well as novel viruses in a plant sample. NGS can also be supplemented with other techniques like rolling circle amplification (RCA) to target specifically the circular genomes of viruses like badnavirus, begomovirus, geminivirus, and babuvirus (Wyant et al. 2012). Adams et al. (2009) have reported about the enrichment of nucleic acid populations for the screening of Gayfeather mild mottle virus (a novel cucumovirus). Several reports on the study and virome profiling of plants like apple, citrus, and grapevine have been done using NGS (Loconsole et al. 2012; Sidharthan et al. 2020). Elbeaino et al. (2014) have reported the identification of sterility mosaic disease in pigeon pea caused by pigeon pea sterility mosaic virus (PPSMV) using the Illumina platform. Further, platforms like MinION were also efficiently used for the detection of wheat streak mosaic virus (WSMV), barley yellow dwarf virus (BYDV), and Triticum mosaic virus (TriMV) in wheat plants (Fellers et al. 2019) and a novel bipartite begomovirus infecting cowpea (Naito et al. 2019).

3.2.7 CRISPR-based Detection

CRISPR/Cas (Clustered Regularly Interspaced Short Palindromic Repeats/CRISPR-associated genes) is an emerging technology that can be applied for a wide variety of biological applications including plant protection and breeding. This gene-editing technique was developed for the production of viral-resistant plants by targeting the DNA/RNA viral genomes (Abudayyeh et al. 2016; Wolter and Puchta 2018). Advancements in the field of synthetic and sequencing biology lately have led to the development of next-generation diagnostic techniques combining methods like isothermal technologies with CRISPR enzymes like Cas13 for the detection of many human and animal pathogens (Sullivan et al. 2019). Gootenberg et al. (2017) have reported the development of a new diagnostic tool "SHERLOCK" used for the detection of viral RNA and DNA. In this method, the samples are amplified first for enrichment using recombinase polymerase reactions (RPA); in case of DNA, or when the target is RNA, it undergoes a step of reverse transcription (RT) before RPA amplification. The RPA-derived product is then transcribed into RNA using T7 RNA polymerase, wherein these transcripts are targeted and cleaved by a fluorescent quenched Cas13 enzyme that releases photo signals indicating the presence of the target DNA/RNA virus. This technique recently has been used to develop a sensitive detection assay for the application of a rapid and on-site lateral flow "paper strip" detection kit (Gootenberg et al. 2018). A CRISPR-based SHERLOCK method has been used to develop an accurate and rapid detection assay that can also be used on-site detection of white spot syndrome virus in shrimps (Sullivan et al. 2019). Although SHERLOCK has been widely used in the diagnostics of many human and animal viruses, it remains a widely unexplored domain for application in plant virology.

3.2.8 On-site Detection

On-site detection approaches are mainly instrumental, as their working principles are simple to operate and require no sophisticated instrument. These devices are fast, reliable, accurate, and sensitive and used on-field, thus making them highly favorable (Lau et al. 2016).

3.2.8.1 Lateral Flow Immunoassay (LFIA)

Lateral flow immunoassay (LFIA) is a simple, efficient, and rapid detection device that works on the chromatographic principle combined with an immuno-based recognition technique. LFIA works on the interaction between the target virus and the conjugated antibodies in a test strip, which upon interaction produces an immunochemical reaction detectable via a visible coloration in the test and sample lines (Von Lode 2005; Price and Kricka 2007). Typically, LFIA is a paper-based platform consisting of a sample loading spot through which it flows until the immunochemical reaction develops. Currently, a number of different membrane strips are used for this purpose from nylon, nitrocellulose, polyethene, fused silica, or polyethersulfone. Many commercial firms developing these devices are encasing the membrane strips inside a plastic case with only the sample loading and reading window exposed for easy handling. LFIA test strips have been used for the detection of a vast number of viruses (Kusano et al. 2007; Drygin et al. 2009). This immunochromatographic assay has been utilized for the rapid on-field detection of plum pox virus (PPV) and potato virus X (PVX) (Byzova et al. 2010; Drygin et al. 2012). Yoon et al. (2014) have reported the development of a rapid immune-gold strip (RIGS) kit for the on-site detection of tomato spotted wilt virus (TSWV) using the same principle. This technique has also been used to develop an on-farm detection assay of six different viruses viz. groundnut bud necrosis virus (GBNV), large cardamom chirke virus (LCCV), cucumber mosaic virus (CMV), peanut mottle virus (PeMoV), potato virus Y (PVY) and papaya ringspot virus (Selvarajan and Balasubramanian 2016).

3.2.8.2 Recombinase Polymerase Amplification (RPA)

Recombinase polymerase amplification is an isothermal enzymatic process that utilizes recombinase enzyme for the initiation of amplification. This process amplifies the DNA template, wherein the amplification is performed at 37–42 °C in the presence of a specific primer (32–36 nucleotides long), buffers, and three enzymes (recombinase, strand displacing polymerase, and ssDNA binding proteins). RPA begins with the scanning of the DNA with the recombinase enzyme that binds the primers on the cognate sites to open the helical structure that is stabilized by the SSB proteins, and the synthesis occurs via the strand displacing polymerase to form the new strand. RPA has been used for the detection of viruses like cucumber mosaic virus (CMV), plum pox virus (PPV), and banana streak MY virus (BSMYV) (Zhang et al. 2014: Kapoor et al. 2017).

3.2.8.3 Quartz Crystal Microbalance Immunosensors (QCMI)

Quartz crystal microbalance is a technique based on the detection of frequency or vibration changes under vacuum, gas, or liquid conditions in real time (Kurosawa et al. 2006). Recently, QCM combined with the immunogenic reagents resulting in the QCMI technique has shown better sensitivity and detection (Owen et al. 2007). QCMI due to its high sensitivity, maintenance, low cost, and portability offers a better advantage for on-site detection of pathogens (Chen and Tang 2007; Becker and Cooper 2011). This technique has proven successful in the on-site detection of a number of viruses such as turnip yellow mosaic virus (TYMV), tobacco mosaic virus (TMV), and cymbidium mosaic virus (CyMV) (Eun et al. 2002; Dickert et al. 2004; Zan et al. 2012).

3.2.8.4 Antibody-Based Biosensors

Antibody-based immunosensors allow fast detection, high sensitivity, real-time analysis, and potential for quantification. It works on the principle where the antibody-based immune response is identified by a potent transducer which converts the immunogenic interaction and relays the signal for analysis (Fang and Ramasamy 2015). Most of the antibody-based biosensors use amperometric, impedimetric, potentiometric, or conductometric transducers. During the past few years, this technique has been used for the detection of viruses like tobacco mosaic virus, cowpea mosaic virus, lettuce mosaic virus, some orchid viruses, and citrus tristeza virus (Eun et al. 2002, Dickert et al. 2004, Candresse et al. 2007, Fang and Ramasamy 2015, Shojaei et al. (2016).

3.2.8.5 Aptamer-Based Biosensors (Aptasensors)

Aptamers are small 30- to 32-bp-long oligonucleotide or amino acid polymers that are highly specific to the target biomolecules. An aptamer-based strategy has critical focal points over the antibodies utilized due to their cheapness and less generation time (2–3 days). They display high specificity, chemical stability, and cost-effectiveness (Seok et al. 2016). The aptamers are immobilized onto a solid surface to arrest the target biomolecules where the aptasensors help to convert the signal with a transducer. For enhanced sensitivity and specificity, nanomaterials are currently used (Khedri et al. 2018). Balogh et al. (2010) have reported the first-time detection of apple stem pitting virus (ASPV) using the coat protein of the virus using aptamer-based sensors.

3.2.8.6 Lab-on-Chip (LOC) Devices

LOC is a small microfluidic device integrating different laboratory methods onto a single chip like PCR, DNA hybridization, and sequencing leading to pathogen detection. The device comprises various components like the sampling processor for sampling, an amplifier for signal multiplication, a transducer for signal quantification, and software for data analysis (Luka et al. 2015). The last few decades have seen lab-on-chip devices gaining popularity for the diagnostics of pathogens (Mark et al. 2010). This technique has several advantages like low energy and reagent consumption and requires less time for detection. The small size of the lab-on-chip device enables portability and high processivity of samples on the field. At the same time, the requirement of a label-free assay and unstable polymers hinders the on-site detection (Julich et al. 2011). Although a lot of bacterial and fungal pathogens have been detected using the LOC, it has not been used for plant viral diagnosis (Julich et al. 2011).

3.2.8.7 Cell Phone (CP)-Based Devices

These devices combine a modern communication system with sensors to capture target images which are then compared with the stored images of diseased plants. This technique allows communication of the generated data from one place to another. At present, there are many such applications developed where disease detection is based on some morphological changes like color intensity, edge outline, texture, and number of spots in the infected plant samples. Plantix (Google app) developed by the AgTech start-up contains around 55,000 plant images that can detect about 60 plant diseases based on these parameters. Although accurate, these cell phone–based devices still require optimization and refinement before being considered as an alternative to robust and sensitive detection techniques based on molecular or serological methods (Vashist et al. 2015).

3.3 Application of Diagnostics in Crops Plants

3.3.1 Cereal Crops

Rice Tungro Disease: Tungro disease of rice is caused by two viral species, i.e., a double-stranded DNA virus rice tungro bacilliform virus (RTBV) (genus: *Tungrovirus*, family: *Caulimoviriude*) and a single-stranded RNA virus rice tungro spherical virus (RTSV) (genus: *Waikavirus*, family: *Secoviridae*). Identification and molecular characterization of the RTBV strain were initiated by Dasgupta et al. (1996) using PCR which identified a 1.1 kb amplicon of the RTBV isolates. Joshi and Dasgupta (2001) further reported on the Southern blot analysis and sequencing of the restricted fragments for the detection of RTBV. PCR/RFLP technique was used to identify a 30-bp-long fragment as a genetic marker for variability in the RTBV genome of tungro-infected leaf samples of rice from six geographically diverse isolates (Joshi et al. 2003). Coat protein-based primers were used for the detection of RTSV in rice samples from Andhra Pradesh, Orissa, Puducherry, West Bengal, and Tamil Nadu using RT-PCR (Mangrauthia et al. 2012). Periasamy et al. (2006) have reported a multiplex RT-PCR for simultaneous detection of RTBV and RTSV.

Rice Black–Streaked Dwarf Disease: It is caused by a dsRNA virus, rice black–streaked dwarf virus (RBSDV) (genus: *Fijivirus*, family: *Reoviridae*). Rice black–streaked dwarf was first reported in the 1950s in Japan, and it has slowly spread to China and Korea. The dot-blot hybridization method using cDNA clones was used for the detection of RBSDV field isolates collected from Korea and Japan (Isogai et al. 2001). Further, the RT-PCR technique using S10-specific primers from the genomic dsRNA region was used for the detection of RBSDV from diverse locations in Korea (Lee et al. 2005). Wang et al. (2006) used an indirect enzyme-linked immunosorbent assay (ID-ELISA) protocol for screening and detection of RBSDV in dilute wheat leaf extracts. Le et al. (2010) have reported the use of reverse transcription-LAMP (RT-LAMP) for the detection of nine rice viruses in Asia. A multiplex reverse transcription-polymerase chain reaction (mRT-PCR) was developed for the simultaneous detection of rice stripe virus (RSV), rice black–streaked dwarf virus (RBSDV), and rice dwarf virus (RDV) from infected rice fields in Korea. The viruses were detected based on three sets of primers targeting the capsid protein-coding genes amplifying to 703 bp, 485 bp, and 252 bp amplicons, respectively (Cho et al. 2013).

Southern Rice Black–Streaked Dwarf Virus (SRBSDV): It is a plant hopper transmitted persistent virus of rice belonging to genus *Fijivirus* and family *Reoviridae*. Southern rice black–streaked dwarf virus was first reported in the Guangdong Province of China in 2001 and has slowly spread to other South Asian countries like Vietnam. Its major vector is identified as the white-backed planthopper (WBPH; *Sogatella furcifera* Horváth). Whole genome sequencing was employed to identify the new strain, which showed some similarity to RBSDV but represented characteristics of a new species within the genus *Fijivirus*, family *Reoviridae* (Wang et al. 2010). RT-PCR technique was employed to identify and characterize the virus in infected rice fields in Vietnam and China, where gene-specific primers were used for PCR (Hoang et al. 2011; Wu et al. 2013). Wu et al. (2013) reported the development of an mRT-PCR method for the detection of SRBSDV, RBSDV, and rice stripe virus (RSV) from rice plants in Eastern China. A reverse transcription real-time PCR technique was used to study the dynamics and relationship between the infected rice plants and virus acquisition by *S. furcifera* (Matsukura et al. 2013). An one-step RT-PCR technique was developed and utilized, which could distinguish RBSDV from SRBSDV (Zhang et al. 2013). Zhou et al. (2012) reported the development of a reverse transcription loop-mediated isothermal amplification (RT-LAMP) assay for the detection of SRBSDV from rice tissue which was ten times more sensitive than a classical RT-PCR assay.

Wheat Dwarf India Virus (WDIV): WDIV is a leafhopper-transmitted virus belonging to the genus *Mastrevirus* and family *Geminiviridae*. Wheat infection due to WDIV was first reported in India by Kumar et al. (2012). The PCR-based technique was used to amplify a 2.8 kb DNA fragment using WDIV-specific primers, thus confirming the viral infection. The association of two alphasatellites (leaf curl Multan and Guar leaf curl) and a betasatellite (Ageratum leaf curl) with WDIV showed severe dwarfing symptoms with the subsequent study showing ageratum leaf curl betasatellite as a pathogenic determinant of WDIV infection in wheat (Kumar et al. 2015a, b). Identification of the association was done using PCR technique using the gene-specific primers. PCR and sequencing technique was used to study the WDIV infection in association with ageratum yellow leaf curl betasatellite in wheat from different regions of Pakistan (Kumar et al. 2020).

Wheat Streak Mosaic Virus (WSMV): WSMV is a pathogenic virus belonging to the genus *Tritimovirus* and family *Potyviridae* and is known to infect plants of the Poaceae family. It was first reported in Nebraska in the 1920s and is currently one of the major wheat-infecting pathogens. WSMV is transmitted by an eriophyid mite, *Aceria tosichella* Keifer (Mar et al. 2013). A PCR-based diagnostic was developed for the parallel detection of WSMV and barley yellow dwarf virus (BYDV) in infected wheat samples that could be detected up to a dilution of 1000-fold in plant saps (French and Robertson 1994). Mar et al. (2013) have reported on the identification and characterization of WSMV isolates using an RT-PCR method from the wheat-growing areas of Brazil. Of the six isolates identified, two of the isolates were used for determining the host range of which the isolates were unable to infect any dicots but showed symptoms in *Hordeum vulgare*, *Triticum aestivum*, *Zea mays*, *Secale cereale*, × *Triticosecale*, and *Avena sativa* (Mar et al. 2013). A combination of DAS-ELISA (double-antibody sandwich-enzyme-linked immunosorbent assay), electron microscopy, and reverse transcription-polymerase chain reaction (RT-PCR) techniques were employed to identify WSMV isolates from wheat-growing fields of Lithuania. The isolates were identified by RT-PCR using primers specific to the coat protein gene of WSMV (Urbanavičienė et al. 2015).

Maize Dwarf Mosaic Virus (MDMV): It is a ssRNA virus belonging to the genus *Potyvirus* and family Potyviridae. It was first identified in 1964 in Illinois, USA, and is currently among the most important viral diseases of maize worldwide. The serological technique (ELISA) was used to detect the virus particles in seed samples of infected corns collected during the late 1970s to early 1980s in Illinois. Nault et al. (1979) have reported on the detection of MDMV using ELISA in infected maize crops collected from the coastal valleys of Peru. The PCR-based technique (RT-PCR) was used to confirm the identification of MDMV in symptomatic maize plants from different maize-growing regions of Poland. The method employed the use of forward primers designed from the NIb region and reverse primers from the coat protein region (Trzmiel and Jeżewska 2008). Morales et al. (2014) reported the identification of two major co-infecting viruses: maize dwarf mosaic virus (MDMV) and maize chlorotic dwarf virus (MCDV) using a quantitative reverse transcriptase-PCR (qRT-PCR) technique in the corn-growing regions of Ohio. A reverse transcriptase-PCR (RT-PCR) technique was also reportedly used to detect and differentiate MDMV from closely related species like sugarcane mosaic virus (SCMV) and johnsongrass mosaic virus (JGMV) (Kannan et al. 2018).

3.3.2 Pulses

Mungbean Yellow Mosaic Virus (MYMV): MYMV belonging to the genus *Begomovirus* and family Geminiviridae causes infection in different pulse crops like mungbean, black gram, pigeon pea, soybean, and French bean. Mandal et al. (1997) have reported on the use of PCR technique for the detection of MYMV in symptomatic plants of black gram. Mungbean yellow mosaic India virus (MYMIV) one of the causal agents of yellow mosaic disease along with MYMY infecting several pulse species was detected using Southern blotting and PCR methods (Balaji et al. 2004; Usharani et al. 2005). PCR using overlapping primers was reportedly used in detecting MYMIV and MYMV infection in symptomatic plants of soybean from Madhya Pradesh and the South Indian region, respectively (Girish and Usha 2005). Haq et al. (2011) reported the identification of begomovirus infection in symptomatic black gram plants collected from Tamil Nadu by PCR using abutting primers specific to DNA-A and DNA-B. Sequence comparison of the two components, however, indicated the DNA-A component to be a variant of MYMV, while DNA-B showed higher similarity to MYMIV. A wild relative of black gram *Vigna mungo* var. *silvestris* showing yellow mosaic symptoms confirmed MYMIV infection by PCR using begomovirus coat protein (AV1) gene–specific primers followed by sequencing and sequence analysis (Kamaal et al. 2011). Reddy et al. (2015) reported the detection of MYMIV infection in black gram plant samples collected from regions of Andhra Pradesh using coat protein (CP)-based gene amplification technique. PCR with CP gene–specific primers has also been used to ascertain begomovirus infection in various symptomatic plants of green gram, black gram, cowpea, and horse gram collected from different agronomic regions of Tamil Nadu (Maheshwari et al. 2014). The leaf samples of green gram, black gram, and cowpea detected positive for MYMV infection, while horse gram samples showed positive for horse gram yellow mosaic virus (HYMV) (Maheshwari et al. 2014). RCA-amplified products of begomovirus detected *Vigna hainiana* and *V. Trilobata* confirmed infection with MYMIV (Naimuddin and Pratap 2011).

Cowpea Aphid-Borne Mosaic Virus (CABMV): CABMV is another significant virus belonging to the genus *Potyvirus* and family *Potyviridae* and is known to cause widespread infection in cowpea. It causes leaf deformation, green vein banding, and interveinal chlorosis on the leaf lamina of cowpea. The field sample of cowpeas collected from Rajasthan detected CABMV infection by RT-PCR using gene-specific primers, simultaneously overlapping genome-wide primers that were used to sequence the complete genome of two CABMV isolates (Mishra et al. 2015).

Chickpea Chlorotic Dwarf Virus (CpCDV): It is an ssDNA virus belonging to the genus *Mastrevirus* and family *Geminiviridae* and causes severe stunt disease in chickpea plants. Kanakala et al. (2013a) reported on the identification of chickpea chlorotic dwarf Pakistan virus by RCA amplification and sequencing of linear monomer viral genome showing 99% similarity at the full genome level. Further, chickpea chlorotic dwarf virus (CpCDV) from infected field samples was detected using the dot-blot hybridization technique (Kanakala et al. 2013b).

Pigeonpea Sterility Mosaic Virus (PPSMV): PPSMV belongs to the genus *Emaravirus* (family unassigned) and causes sterility mosaic disease in pigeonpea (*Cajanus cajan*). It is known to be transmitted by eriophyid mites (*Aceria cajani*) (Patil and Kumar 2015). Kumar et al. (2003) reported the use of a 764 bp random cDNA clone from PPSMV to identify and hybridize with RNA isolated from purified virus preparation, based on which RT-PCR was done using gene-specific primers to amplify a 321 bp RNA-3 region of PPSMV. The 276 and 360 bp genomic fragments of PPSMV specific to the RNA-dependent RNA polymerase gene (RdRp) region of the emaravirus genome were detected by PCR using degenerate primers specific to the RdRp conserved region (Elbeaino et al. 2013).

3.3.3 Cash Crops

Sugarcane Bacilliform Viruses (SCBV): SCBV belonging to the genus *Badnavirus* and family *Caulimoviridae* is associated with the freckling disease of sugarcane and infects a number of species like *Saccharum officinarum*, *S. barberi*, *S. sinense*, *S. robustum*, and *Saccharum* interspecific hybrids. The detection of SCBV-infected sugarcane plants for the screening of diseased and healthy plants was done by PCR using primers specific to the RT/RNase H region of the genome or core RT/RNase H region (Singh et al. 2009; Karuppaiah et al. 2013; Rao et al. 2014). Sharma et al. (2015) reported on the use of RCA-concentrated DNA as a template for the sensitive detection of SCBV in field samples.

Sugarcane Yellow Leaf Virus (SCYLV): SCYLV belonging to the genus *Polerovirus* and family *Luteoviridae* is another economically important disease infecting sugarcane. RT-PCR using primers specific to the four ORFs of the SCYLV genome was used for indexing of sugarcane field samples including some tissue cultured samples. Out of the tested positive samples, sequence analysis helped to identify three SCYLV genotypes (CUB, IND, and BRA-PER) associated with the yellow leaf disease of sugarcane (Viswanathan et al. 2008a). Viswanathan et al. (2009) reported the detection of SCYLV-infected sugarcane field plants that were asymptomatic using a robust RT-PCR-based technique. The RT-PCR-based detection using primers specific to 5′-UTR, ORF0, and partial ORF1 region of the SCYLV genome could detect infection in sugarcane cultivars (Chinnaraja et al. 2013).

Sugarcane Streak Mosaic Virus (SCSMV): It belongs to the genus *Poacevirus* and family *Potyviridae*. Viswanathan et al. (2008b) reported the development of an RT-PCR-based detection of SCSMV targeting the N-terminal region of the viral coat protein (CP). This method was successful in indexing the sugarcane samples collected from different regions of India, South Africa, Australia, and the USA, of which 92% tested positive for SCSMV infection (Viswanathan et al. 2008b). Hema et al. (2003) reported the detection and sequencing of two overlapping gene fragments of 3.1 kb comprising the partial CI, complete 6 K, VPg-NIa, and NIb gene, respectively. PCR using overlapping primers was also employed for the characterization of the SCSMV full genome (Parameswari et al. 2013).

Sugarcane Mosaic Virus (SCMV): SCMV belongs to the genus *Potyvirus* and family *Potyviridae*. Immunocapture-RT-PCR technique for the detection of SCMV using coat protein (CP) specific primers was reported (Gaur et al. 2003). Rao et al. (2006) reported the utilization of RT-PCR-based detection of sugarcane mosaic virus (SCMV), sorghum mosaic virus (SrMV), and sugarcane streak mosaic virus (SCSMV). RT-PCR using gene-specific primer pairs was also used to detect the prevalence of SCMV, SCYLV, and SCSMV from different sugarcane-growing regions (Viswanathan et al. 2010). A multiplex detection technique using duplex-immunocapture-PCR (D-IC-PCR) was used for the simultaneous detection of SCMV and SCSMV (Subba Reddy et al. 2011).

Beet Necrotic Yellow Vein Virus (BNYVV): BNYVV belonging to the genus *Benyvirus* and family *Benyviridae* is associated with the rhizomania disease, one of the most destructive diseases in sugar beet (Safarpour et al. 2012). It is transmitted by the vector *Polymyxa betae* (Keskin). Double-antibody sandwich-enzyme-linked immunosorbent assay (DAS-ELISA) and RT-PCR techniques were used to detect BNYVV infection in sugarbeet plants in Turkey, of which RT-PCR was found more suitable for BNYVV detection (Yardımcı and Kılıç 2011). The QD-FRET-based antibody sensor has also been developed for the rapid and accurate detection of *Polymyxa betae* (Keskin), the vector of beet necrotic yellow vein virus (BNYVV) in sugarbeet (Safarpour et al. 2012).

3.3.4 Oilseeds

Soybean Mosaic Virus (SMV): SMV belonging to the genus *Potyvirus* and family *Potyviridae* is the most important viral pathogen infecting soybean worldwide. It was first reported from Connecticut, USA, in 1915 (Gardner and Kendrick 1921). The virus is known to cause severe infection when co-infected with cowpea mosaic virus (CPMV), bean pod mottle virus (BPMV), and tobacco ringspot virus (TRSV) (Cui et al. 2011; Hwang et al. 2011). Wang et al. (2005) reported the development of a rapid and sensitive RT-PCR method for the detection of SMV in dry soybean seeds which employed the use of an improved SDS-phenol-chloroform procedure for RNA extraction. Ahangaran et al. (2009) reportedly used different serological methods (DAS-ELISA, DAC-ELISA, TBIA, and DIBA) to compare and identify SMV from the soybean-growing regions of Iran. Of the different methods tested, TBIA was found the most sensitive and rapid for field testing and surveys. Detection of SMV was done further using primers specific to the coat protein region of the virus using techniques like RT-PCR and immunocapture-RT-PCR (IC-RT-PCR) (Ahangaran et al. 2009). Naghavi et al. (2008) further reported the use of immunocapture-reverse transcription-PCR (IC-RT-PCR) using primer pairs SMV-CPr and SMV-CPf specific to the coat protein region to detect and identify SMV in soybean plants collected from Lorestan Province, South West of Iran. An RT-PCR technique was employed for the detection and diagnosis of SMV symptomatic plants using specific primers CIF/CIrev and POT1/POT2 from soybean plants collected from farmers and experimental fields from different districts of Akola, Maharashtra (Nandakishor et al. 2017).

Soybean Dwarf Disease: This disease is caused by soybean dwarf virus (SbDV) belonging to the genus *Luteovirus* and family *Luteoviridae*. It is a persistent pathogen transmitted only by aphids and was first reported from Japan. Several strains of SbDV have been identified based on symptomology, aphids-vector relationships, physiochemical properties, and molecular characteristics. Terauchi et al. (2001) reported the determination of the complete nucleotide sequences of four SbDV isolates. ELISA using specific monoclonal antibodies has been utilized to discriminate the different strains of SbDV (Mikoshiba et al. 1994). Yamagishi et al. (2006) further reported the use of the dot-blot hybridization technique for the detection of different strains of SbDV. A one-step RT-PCR was employed for the detection of SbDV-infected fields in Wisconsin, USA (Harrison et al. 2005). A generic RT-PCR method was used for the identification and detection of 13 *Luteoviridae* species using three sets of degenerate primers (Chomic et al. 2010). Schneider et al. (2011) reported the identification of SbDV isolates infecting clover using a multiplex RT-PCR detection method that showed co-infections of Y and D components of the virus.

Peanut Clump Disease (PCD): PCD is caused by peanut clump virus (PCV) also known as the Indian peanut clump virus (IPCV) in the Indian subcontinent and is one of the major pathogens causing infection in Poaceae crops and pigeon peas. It is transmitted by the plasmodiophorid obligate parasite, *Polymyxa graminis*, and belongs to the genus *Pecluvirus* and family *Virgaviridae* (Reddy et al. 2008). Several IPCV and PCV isolates have been identified and differentiated based on serological and genomic diversity. While IPCV isolates have been organized into three distinct serotypes viz. IPCV-H (Hyderabad), IPCV-D (Durgapura), and IPCV-L (Ludhiana), the PCV isolates have been grouped into five distinct serotypes based on monoclonal antibodies in triple-antibody sandwich-ELISA (TAS-ELISA) (Nolt et al. 1988; Huguenot et al. 1989; Manohar et al. 1995). Dieryck et al. (2010) reported the development of a broad-spectrum reverse transcription-polymerase chain reaction (RT-PCR) technique to detect IPCV and PCV from West Africa and India. The primers used for the analysis were designed from the conserved 3′ untranslated regions of the genome. Comparative detection of IPCV and PCV using ELISA and RT-PCR showed the RT-PCR to be a better candidate for viral detection. RT-PCR technique also led to the identification and detection of ICV and PCV in two new natural hosts, i.e., *Vigna subterranean* and *Oldenlandia aspera* (Dieryck et al. 2010).

3.4 Conclusions

Viral diseases in plants cause a considerable amount of revenue loss worldwide. These diseases can be managed if timely intervention is applied for effective control. Therefore, it is pertinent that the disease is identified at the initial stage even before symptom expression. An array of detection techniques has been developed for the identification of the viral agents using immunological, nucleic acid–based to a more advanced technique utilizing high-throughput sequencing to detection using biosensors or on-site detection devices. These diagnostic platforms help in the accurate and rapid diagnosis of viral pathogens enabling timely intervention through management. Nucleic acid–based detection techniques have been very successful in detecting and identifying viruses and viroids to their exact taxonomic identity in a vast number of species. Recent variations of the technique like the isothermal nucleic acid amplification have shown greater sensitivity and utility due to less time taken for diagnosis. In the past few years, NGS has come up as a robust tool for the identification of multiple viruses and viroids as it can detect even novel agents whose sequence information is still unavailable. Developments in the field of nanotechnology have enabled point-of-care detection techniques like strip-based lateral flow assays which are gaining popularity due to their fast and reliable detection of symptomless plants. As there is no ideal detection method, therefore, it is important to develop effective techniques for the management of viral diseases.

References

Abd El-Aziz MH, Behiry SI, Younes HA, Hamza KA (2019) The relationship and relativity between three isolates of Potato virus Y Potyvirus infecting potato (Solanum tuberosum L.) at Alexandria and El-Beheira governorates in northern Egypt. Novel Res Microbiol J 3:440–452

Abudayyeh OO, Gootenberg JS, Konermann S, Joung J, Slaymaker IM, Cox DB, Shmakov S, Makarova KS, Semenova E, Minakhin L, Severinov K, Regev A, Lander ES, Koonin EV, Zhang F (2016) C2c2 is a single-component programmable RNA-guided RNA-targeting CRISPR effector. Science 353:aaf5573. https://doi.org/10.1126/science.aaf5573

Adams IP, Glover RH, Monger WA, Mumford R, Jackeviciene E, Navalin-skiene M, Samuitiene M, Boonham N (2009) Next-generation sequencing and metagenomic analysis: a universal diagnostic tool in plant virology. Mol Plant Pathol 10(4):537–545

Adkar-Purushothama CJ, Maheshwar PK, Sano T, Janardhana GR (2011) A sensitive and reliable RT-nested PCR assay for detection of Citrus tristeza virus from naturally infected citrus plants. Curr Microbiol 62:1455–1459

Ahangaran A, Mohammadi GM, Habibi MK, Khezri S, Shahraeen N (2009) Use of rapid serological and nucleic acid- based methods for detecting the soybean mosaic virus. J Agric Sci Technol 11:91–97

Ahmadi S, Almasi MA, Fatehi F, Struik PC, Maradi A (2013) Visual detection of potato leafroll virus by one- step reverse transcription loop-mediated isothermal amplification of DNA with hydroxynaphthol blue dye. J Phytopathol 161:120–124

Algar WR, Krull UJ (2008) Quantum dots as donors in fluorescence resonance energy transfer for the bioanalysis of nucleic acids, proteins, and other biological molecules. Anal Bioanal Chem 391:1609–1618

Al-Khalaf M, Kumari SG, Haj Kasem AA, Makkouk KM, Al-Chaabi S (2009) Use of the two faces of nitrocellulose membrane in tissue blot immunoassay for the detection of bean yellow mosaic virus and the possibility of its mechanically transmitted from the printed membrane to the host plant. Arab J Plant Protect 27:91–94

Anderson TF, Stanley WM (1941) A study by means of the electron microscope of the reaction between tobacco mosaic virus and its antiserum. J Biol Chem 139(1):339–344

Bajet NB, Daquioag RD, Hibino H (1985) Enzyme-linked immunosorbent assay to diagnose rice tungro. J Plant Protect Tropics 2:125–129

Balaji V, Vanitharani R, Karthikeyan AS, Anbalagan S (2004) Veluthambi K infectivity analysis of two variable DNA B components of Mungbean yellow mosaic virus-Vigna in *Vigna mungo* and *Vigna radiata*. J Biosci 29:297–308

Balogh Z, Lautner G, Bardóczy V, Komorowska B, Gyurcsanyi RE, Meszáros T (2010) Selection and versatile application of virus-specific aptamers. FASEB J 24(11):4187–4195

Baranwal VK, Sharma SK, Khurana D, Varma R (2014) Sequence analysis of shorter than genome length episomal banana streak OL virus like sequences isolated from banana in India. Virus Genes 48(1):120–127. https://doi.org/10.1007/s11262-013-0984-1

Becker B, Cooper MA (2011) A survey of the 2006-2009 quartz crystal microbalance biosensor literature. J Mol Recognit 24:754–787

Berger PH, Thornbury DW, Pirone TP (1985) Detection of pictogram quantities of potyviruses using a dot immunobinding assay. J Virol Methods 12:31–39

Bergervoet JH, Peters J, van Beckhoven JR, van den Bovenkamp GW, Jacobson JW, van der Wolf JM (2008) Multiplex microsphere immuno-detection of potato virus Y, X and PLRV. J Virol Methods 149(1):63–68

Bernd Z, Günther Z (2009) Rapid diagnosis of plant virus diseases by transmission electron microscopy. J Virol Methods 162:163–169

Bertolini E, Olmos A, Lopez MM, Cambra M (2003) Multiplex nested reverse-transcription polymerase chain reaction in a single tube for sensitive and simultaneous detection of four RNA viruses and *Pseudomonas savastanoi pv. savastanoi* in olive trees. Phytopathology 93:286–292

Boonham N, Perez LG, Mendez MS, Peralta EL, Blockley A, Walsh K, Barker I, Mumford RA (2004) Development of a real-time RT-PCR assay for the detection of potato spindle tuber viroid. J Virol Methods 116:139–146

Boonham N, Kreuzeb J, Winterc S, Vlugtd R, Bergervoetd J, Tomlinsona J, Mumforda R (2014) Methods in virus diagnostics: from ELISA to next generation sequencing. Virus Res 186:20–31

Bove JM, Vogel R, Albertini D, Bove JM (1988) Discovery of a strain of Tristeza virus (K) inducing no symptoms in Mexican lime. Proceedings of the 10th conference of IOCV. Spain 1988. International Organization of Citrus Virologists Riverside, CA, pp 14–16

Bystricka D, Lenza O, Mraza I, Piherovad L, Kmochd S, Sipc M (2005) Oligonucleotide-based microarray: a new improvement in microarray detection of plant viruses. J Virol Methods 128:176–182

Byzova NA, Safenkova IV, Chirkov SN, Avdienko VG, Guseva AN, Mitrofanova IV, Zherdev AV, Dzantiev BB, Atabekov JG (2010) Interaction of plum pox virus with specific colloidal gold labeled antibodies and development of immunochromatographic assay of the virus. Biokhimiya 75(11):1583–1595

Candresse T, Lot H, German-Retana S, Krause-Sakate R, Thomas J, Souche S, Delaunay T, Lanneau M, Le Gall O (2007) Analysis of the serological variability of *Lettuce mosaic* virus using monoclonal antibodies and surface plasmon resonance technology. J Gen Virol 88:2605–2610

Capote N, Bertolini E, Olmos A, Vidal E, Martinez MC, Cambra M (2009) Direct sample preparation methods for the detection of *Plum pox virus* by real-time RT-PCR. Int Microbiol J 12:1–6

Charlermroj R, Himananto O, Seepiban C, Kumpoosiri M, Warin N, Oplatowska M, Gajanandana O, Grant IR, Karoonuthaisiri N, Elliott CT (2013) Multiplex detection of plant pathogens using a microsphere immunoassay technology. PLoS One 8(4):e62344

Chen ZG, Tang DY (2007) Antigen–antibody interaction from quartz crystal microbalance immunosensors based on magnetic CoFe2O4/SiO2 composite nanoparticle-functionalized biomimetic interface. Bioprocess Biosyst Eng 30:243–249

Chinnaraja C, Viswanathan R, Karuppaiah R, Bagyalakshmi K, Malathi P, Parameswari B (2013) Complete genome characterization of *Sugarcane yellow leaf virus* from India: evidence for RNA recombination. Eur J Plant Pathol 135:335–349

Cho SY, Jeong RD, Yoon YN, Lee SH, Shin DB, Kang HW, Lee BC (2013) One-step multiplex reverse transcription-polymerase chain reaction for the simultaneous detection of three rice viruses. J Virol Methods 193(2):674–678

Chomic A, Pearsonb MN, Clover GRG, Farreyrol K, Sauld D, Hamptona JG, Armstrong KF (2010) A generic RT-PCR assay for the detection of *Luteoviridae*. Plant Pathol 59:429–442. https://doi.org/10.1111/j.1365-3059.2010.02282.x

Clark MF, Adams AN (1977) Characteristics of the microplate method of enzyme-linked immunosorbent assay for the detection of plant viruses. J Gen Virol I977(34):475–483

Colinet D, Nguyen M, Kummert J, Lepoivre P (1998) Differentiation among potyviruses infecting sweet potato based on genus- and virus-specific reverse transcription polymerase chain reaction. Plant Dis 82:223–229

Cui X, Chen X, Wang A (2011) Detection, understanding and control of soybean mosaic virus. In: Sudaric A (ed) Soybean-molecular aspects of breeding. InTech, Rijeka, pp 335–354. isbn:978-953-307-240-1

Cui S, Ling P, Zhu H, Keener HM (2018) Plant Pest detection using an artificial nose system: a review. Sensors (Basel) 18(2):378

Dasgupta I, Das BK, Nath PS, Mukhopadhyay S, Niazi FR, Varma A (1996) Detection of rice tungro bacilliform virus in field and glasshouse samples from India using the polymerase chain reaction. J Virol Methods 58:53–58

Deb M, Anderson JM (2008) Development of a multiplexed PCR detection method for barley and cereal yellow dwarf viruses, *wheat spindle streak virus, wheat streak mosaic virus* and *soil-borne wheat mosaic virus*. J Virol Methods 148:17–24

Derrick KS (1973) Quantitative assay for plant viruses using serologically specific electron microscopy. Virology 56:652–653

Dickert FL, Hayden O, Bindeus R, Mann KJ, Blaas D, Waigmann E (2004) Bioimprinted QCM sensors for virus detection—screening of plant sap. Anal Bioanal Chem 378:1929–1934

Dieryck B, Delfosse P, Reddy AS, Bragard C (2010) Targeting highly conserved 3′-untranslated region of pecluviruses for sensitive broad-spectrum detection and quantitation by RT-PCR and assessment of phylogenetic relationships. J Virol Methods 169:385–390

Drygin YF, Blintsov AN, Osipov AP, Grigorenko VG, Andreeva IP, Uskov AI, Varitsev YA, Anisimov BV, Novikov VK, Atabekov JG (2009) High-sensitivity express immunochromatographic method for detection of plant infection by tobacco mosaic virus. Biochemistry (Mosc) 74:986–993

Drygin YF, Blintsov AN, Osipov AP, Grigorenko VG, Andreeva IP, Osipov AP, Varitsev YA, Uskov AI, Kravchenko DV, Atabekov JG (2012) Highly sensitive field test lateral flow immunodiagnostics of PVX infection. Appl Microbiol Biotechnol 93:179–189

Dugat-Bony E, Peyretaillade E, Parisot N, Biderre-Petit C, Jaziri F, Hill D, Rimour S, Peyret P (2012) Detecting unknown sequences with DNA microarrays: explorative probe design strategies. Environ Microbiol 14:356–371

Eid S, Atamian HS, Abou-Jawdah Y, Havey MJ (2008) Assessing the movement of *Cucurbit yellow stunting disorder virus* in susceptible and tolerant cucumber germplasms using serological and nucleic acid-based methods. J Phytopathol 156:438–445

Elbeaino T, Whitfield A, Sharma M, Digiaro M (2013) Emaravirus-specific degenerate PCR primers allowed the identification of partial RNA-dependent RNA polymerase sequences of Maize red stripe virus and Pigeonpea sterility mosaic virus. J Virol Methods 188:37–40

Elbeaino T, Digiaro M, Uppala M, Sudini H (2014) Deep sequencing of pigeonpea sterility mosaic virus discloses five RNA segments related to emaraviruses. Virus Res 188:27–31

Eun AJC, Huang L, Chew FT, Li SFY, Wong SM (2002) Detection of two orchid viruses using quartz crystal microbalance (QCM) immunosensors. J Virol Methods 99:71–79

Fang Y, Ramasamy RP (2015) Current and prospective methods for plant disease detection. Biosensors 4:537–561

Fellers J, Webb C, Fellers M, Shoup Rupp J, De Wolf E (2019) Wheat virus identification within infected tissue using nanopore sequencing technology. Plant Dis 103:2199. https://doi.org/10.1094/pdis-09-18-1700-re

French R, Robertson NL (1994) Simplified sample preparation for detection of wheat streak mosaic virus and barley yellow dwarf virus by PCR. J Virol Methods 49(1):93–99. https://doi.org/10.1016/0166-0934(94)90059-0

Fukuta S, Kato S, Yoshida K, Mizukami Y, Ishida A, Ueda J, Kanbe M, Ishimoto Y (2003) Detection of tomato yellow leaf curl virus by loop-mediated isothermal amplification reaction. J Virol Methods 112:35–40

García-Arenal F, Fraile A, Malpica JM (2001) Variability and genetic structure of plant viruspopulations. Annu Rev Phytopathol 39:157–186

Gardner M, Kendrick JB (1921) Soybean mosaic. J Agric Res 22:111–114

Gaur RK, Singh M, Rao GP (2003) Molecular characterization of sugarcane mosaic virus isolate from north eastern region of India. Sugar Technol 5:149–153

Girish KR, Usha R (2005) Molecular characterization of two soybean-infecting begomoviruses from India and evidence for recombination among legume-infecting begomoviruses from South-East Asia. Virus Res 108:167–176

Gootenberg JS, Abudayyeh OO, Lee JW, Essletzbichler P, Dy AJ, Joung J, Verdine V, Donghia N, Daringer NM, Freije CA, Myhrvold C, Bhattacharyya RP, Livny J, Regev A, Koonin EV, Hung DT, Sabeti PC, Collins JJ, Zhang F (2017) Nucleic acid detection with CRISPR-Cas13a/C2c2. Science 356:438–442. https://doi.org/10.1126/science.aam9321

Gootenberg JS, Abudayyeh OO, Kellner MJ, Joung J, Collins JJ, Zhang F (2018) Multiplexed and portable nucleic acid detection platform with Cas13, Cas12a, and Csm6. Science 360:439–444. https://doi.org/10.1126/science.aaq0179

Ha C, Coombs S, Revill PA, Harding RM, Vu M, Dale JL (2008) Design and application of two novel degenerate primer pairs for the detection and complete genomic characterization of potyviruses. Arch Virol 153:25–36

Hančević K, Černi S, Radić T, Škorić D (2012) Comparison of different methods for *Citrus tristeza virus* detection in Satsuma mandarins. J Plant Dis Protect 119:2–7

Haq QMI, Rouhibakhsh A, Ali A, Malathi VG (2011) Infectivity analysis of a blackgram isolate of Mungbean yellow mosaic virus and genetic assortment with MYMIV in selective hosts. Virus Genes 42:429–439

Hararika S, Choudhury R, Saikia S, Sarma U (2020) Pathogen detection in Khasi mandarin Orange using serological and electronic nose diagnostic technique. Int J Innov Technol Explor Eng 9(4). https://doi.org/10.35940/ijitee.d2077.029420

Harris JR (2007) Negative staining of thinly spread biological samples. In: Kuo J (ed) Methods in molecular biology, electron microscopy, methods and protocols, vol 369. Humana Press, Totowa, pp 107–142

Harrison B, Steinlage TA, Domier LL, D'Arcy CJ (2005) Incidence of soybean dwarf virus and identification of potential vectors in Illinois. Plant Dis 89:28–32

Helguera PR, Taborda R, Docampo DM, Ducasse DA (2001) Immunocapture reverse transcription-polymerase chain reaction combined with nested PCR greatly increases the detection of *Prunus necrotic ring spot virus* in the peach. J Virol Methods 95:93–100

Helguera PR, Docampo DM, Nome SF, Ducasse DA (2002) Enhanced detection of *Prune dwarf virus* in peach leaves by immunocapture-reverse transcription-polymerase chain reaction with nested polymerase chain reaction (IC-RT-PCR nested PCR). J Phytopathol 150:94–96

Hema M, Savithri HS, Sreenivasulu P (2003) Comparison of direct binding polymerase chain reaction with recombinant coat protein antibody-based dot-immunobinding assay and immunocapture-reverse transcription-polymerase chain reaction for the detection of sugarcane streak mosaic disease in India. Curr Sci 85:1774–1777

Hoang AT, Zhang HM, Yang J, Chen JP, Hébrard E, Zhou GH, Vinh VN, Cheng JA (2011) Identification, characterization, and distribution of southern rice black-streaked dwarf virus in Vietnam. Plant Dis 95:1063–1069

Huguenot C, Givord L, Sommermeyer G, VanRegenmortel MH (1989) Differentiation of Peanut clump virus serotypes by monoclonal antibodies. Virus Res 140:87–102

Hull R (2002) Matthews' plant virology, 4th edn. Academic, San Diego

Hwang TY, Jeong SC, Kim O, Park HM, Lee SK, Seo MJ, Choi MS, Lee YY, Yun HT, Kwon YU, Kim WH (2011) Intra-host competition and interactions between 'Soybean mosaic virus' (SMV) strains in mixed-infected soybean. Aust J Crop Sci 5(11):1379–1387

Islam W, Zhang J, Adnan M, Noman A, Zaynab M, Wu Z (2017) Plant virus ecology: a glimpse of recent accomplishments. Appl Ecol Environ Res 15(1):691–705

Isogai M, Uyeda I, Choi JK (2001) Molecular diagnosis of Rice black-streaked dwarf virus in Japan and Korea. Plant Pathol J 17(3):164–168

James AP, Geijskes RJ, Dale JL, Harding RM (2011) Development of a novel rolling-circle amplification technique to detect *Banana streak virus* that also discriminates between integrated and episomal virus sequences. Plant Dis 95:57–62

Jeong JJ, Ju HJ, Noh J (2014) A review of detection methods for the plant viruses. Res Plant Dis 20(3):173–181

Jones S, Baizan-Edge A, Macfarlane S, Torrance L (2017) Viral diagnostics in plants using next generation sequencing: computational analysis in practice. Front Plant Sci 8:1770. https://doi.org/10.3389/fpls.2017.01770

Joshi R, Dasgupta I (2001) Cloning and molecular analysis of DNA of rice tungro bacilliform viruses from different rice growing regions of India. Indian Phytopathol 54:469–475

Joshi R, Kumar V, Dasgupta I (2003) Detection of molecular variability in rice tungro bacilliform viruses from India using polymerase chain reaction–restriction fragment length polymorphism. J Virol Methods 109:89–93

Julich S, Riedel M, Kielpinski M, Urban M, Kretschmer R, Wagner S, Werres S (2011) Development of a lab-on-a-chip device for diagnosis of plant pathogens. Biosens Bioelectron 26(10):4070–4075. https://doi.org/10.1016/j.bios.2011.03.035

Kamaal N, Mohammad A, Gupta S (2011) Identification of Mungbean yellow mosaic India virus infecting Vigna mungo var. silvestris L. Phytopathol Mediterr 50:94–100

Kanakala S, Sakhare A, Verma HN, Malathi VG (2013a) Infectivity and the phylogenetic relationship of a mastrevirus causing chickpea stunt disease in India. Eur J Plant Pathol 135:429–438

Kanakala S, Verma HN, Vijay P, Saxena DR, Malathi VG (2013b) Response of chickpea genotypes to Agrobacterium-mediated delivery of Chickpea chlorotic dwarf virus (CpCDV) genome and identification of resistance source. Appl Microbiol Biotechnol 97:9491–9501

Kaneko H, Kawana T, Fukushima E, Suzutani T (2007) Tolerance of loop-mediated isothermal amplification to a culture medium and biological substances. J Biochem Biophys Methods 70:499–501

Kannan M, Ismail I, Bunawan H (2018) Maize dwarf mosaic virus: from genome to disease management. Viruses 10:492. https://doi.org/10.3390/v10090492

Kapoor R, Srivastava N, Kumar S, Baranwal VK (2017) Development of recombinase polymerase amplification assay for the diagnosis of Banana bunchy top virus in different banana cultivars. Arch Virol 162(9):2791–2796

Karuppaiah R, Viswanathan R, Ganesh KV (2013) Genetic diversity of sugarcane bacilliform virus isolates infecting Saccharum spp. in India. Virus Genes 46:505–516

Katul L (1992) Characterization by serology and molecular biology of bean leaf roll virus and Faba bean necrotic yellows virus. PhD thesis, University of Gottingen, Gottingen, 115 pp

Kfir R, Genthe B (1993) Advantages and disadvantages of the use of immunodetection techniques for the enumeration of microorganisms and toxins in water. Water Sci Technol 27:243–252

Khedri M, Ramezani M, Rafatpanah H, Abnous K (2018) Detection of food-born allergens with aptamer-based biosensors. TrAC Trends Anal Chem 103:126–136

Khurana SMP (1990) Modern approaches for detection and management of the potato viruses and viroid. In: Grewal JS et al (eds) Current facets in potato research. IPA, CPRI, Simla, pp 98–108

Kohler G, Milsten C (1975) Continuous culture of fused cells secreting antibody of predefined specificity. Nature 256:495–497

Kuan CP, Deng TC, Huang HC, Chi HH, Lu YL (2014) Use of reverse transcription loop-mediated isothermal amplification for the detection of *Zucchini yellow mosaic virus*. J Phytopathol 162:238–244

Kumar PL, Jones AT, Reddy DVR (2003) A novel mite-transmitted virus with a divided RNA genome closely associated with pigeonpea sterility mosaic disease. Phytopathology 93:71–81

Kumar J, Singh SP, Kumar J, Tuli R (2012) A novel mastrevirus infecting wheat in India. Arch Virol 157:2031–2034

Kumar J, Kumar J, Singh S, Shukla V, Singh SP, Tuli R (2015a) Prevalence of wheat dwarf India virus in wheat in India. Curr Sci 180(2):260–264

Kumar S, Rai R, Baranwal VK (2015b) Development of an immunocapture–reverse transcription–polymerase chain reaction (IC-RT-PCR) using modified viral RNA release protocol for the detection of Grapevine leafroll-associated virus 3 (GLRaV-3). Phytoparasitica 43:311–316

Kumar J, Ahmad J, Imtiaz M, Kianian SF (2020) Wheat dwarf India virus and associated betasatellite infecting wheat in Pakistan. Australas Plant Dis Notes 15:16. https://doi.org/10.1007/s13314-020-00383-y

Kumari SG, Makkouk KM, Attar N (2006) An improved antiserum for sensitive serologic detection of chickpea chlorotic dwarf virus. J Phytopathol 154:129–133

Kumari SG, Makkouk KM, Loh MH, Negassi K, Tsegay S, Kidane R, Kibret A, Tesfatsion Y (2008) Viral diseases affecting chickpea crops in Eritrea. Phytopathol Mediterr 47:42–49

Kurosawa S, Park JW, Aizawa H, Wakida SI, Tao H, Ishihara K (2006) Quartz crystal microbalance immunosensors for environmental monitoring. Biosens Bioelectron 22:473–481

Kusano N, Hirashima K, Narahara MKK, Imamura T, Nakahira TMK, Torii K (2007) Immunochromatographic assay for simple and rapid detection of Satsuma dwarf virus and related viruses using monoclonal antibodies. J Gen Plant Pathol 73:66–71

Lau HY, Wang Y, Wee EJH, Botella JR, Trau M (2016) Field demonstration of a multiplexed point-of-care diagnostic platform for plant pathogens. Anal Chem 88:8074–8081

Le DT, Netsu O, Uehara-Ichiki T, Shimizu T, Choi IR, Omura T, Sasaya T (2010) Molecular detection of nine rice viruses by a reverse-transcription loop-mediated isothermal amplification assay. J Virol Methods 170(1–2):90–93

Lee GP, Min BE, Kim CS, Choi SH, Harn HH, Kim SU, Ryu KH (2003) Plant virus cDNA chip hybridization for detection and differentiation of four cucurbit-infecting Tobamoviruses. J Virol Methods 110:19–24

Lee BC, Hong YK, Hong SJ, Park ST, Lee KW (2005) Occurrence and detection of Rice black-streaked dwarf virus in Korea. Plant Pathol J 21(2):172–173

Lievens B, Grauwet TJMA, Cammue BPA, Thomma BPHJ (2005) Recent developments in diagnostics of plant pathogens: a review. Recent Res Dev Microbiol 9:57–79

Lin NS, Hsu YH, Hsu HT (1990) Immunological detection of plant viruses and a mycoplasma like organism by direct tissue blotting on nitrocellulose membranes. Phytopathology 80:824–828

Lister RM (1978) Application of the enzyme-linked immunosorbent assay for detecting viruses in soybean seed and plants. Phytopathology 68:1393–1400

Liu W, Zhao X, Zhang P, Mar TT, Liu Y, Zhang Z, Han C, Wang X (2013) A one step real-time RT-PCR assay for the quantitation of wheat yellow mosaic virus (WYMV). Virol J 10:173

Loconsole G, Onelge N, Potere O, Giampetruzzi A, Bozan O, Satar S, de Stradis A, Savino V, Yokomi RK, Saponari M (2012) Identification and characterization of Citrus Yellow vein clearing virus, a putative new member of the genus Mandarivirus. Phytopathology 102:1168–1175

Lopez MM, Llop P, Olmos A, Marco-Noales E, Cambra M, Bertolini E (2008) Are molecular tools solving the challenges posed by detection of plant pathogenic bacteria and viruses. Curr Issues Mol Biol 11:13–45

Luka G, Ahmadi A, Najjaran H, Alocilja E, DeRosa M, Wolthers K, Malki A, Aziz H, Althani A, Hoorfar M (2015) Microfluidic integrated biosensors: a leading technology towards lab-on-chip and sensing applications. Sensors 15:30011–30031

Maheshwari R, Panigrahi G, Angappan K (2014) Molecular characterization of distinct YMV(yellow mosaic virus) isolates affecting pulses in India with the aid of coat protein gene as a marker for identification. Mol Biol Rep 41:2635–2644

Majumder S, Baranwal VK (2014) Simultaneous detection of four garlic viruses by multiplex reverse transcription PCR and their distribution in Indian garlic accessions. J Virol Methods 202:34–38

Makkouk KM, Kumari SG (1996) Detection of ten viruses by the tissue-blot immunoassay (TBIA). Arab J Plant Protect 14:3–9

Makkouk KM, Kumari SG (2006) Molecular diagnosis of plant viruses. Arab J Plant Protect 24:135–138

Mandal B, Varma A, Malathi VG (1997) Systemic infection of *Vigna mungo* using the cloned DNAs of the blackgram isolate of mungbean yellow mosaic geminivirus through agroinoculation and transmission of the progeny virus by whiteflies. J Phytopathol 145:505–510

Mangrauthia SK, Malathi P, Agarwal S, Ramkumar G, KrishnaveniD NCN, Madhav MS, Ladhalakshmi D, Balachandran SM, Viraktamath BC (2012) Genetic variation of coat protein gene among the isolates of Ricetungro spherical virus from tungro-endemic states of the India. Virus Genes 44:482–487

Manohar SK, Dollet M, Dubern J, Gargani D (1995) Studies on variability of Peanut clump virus: symptomatology and serology. J Phytopathol 143:233–238

Mar TB, Lau D, Schons J, da Silva Pereira PRV, Carminatti AJ (2013) Identification and characterization of wheat streak mosaic virus isolates in wheat-growing areas in Brazil. Int J Agron 1–6. https://doi.org/10.1155/2013/983414

Mark D, Haeberle S, Roth G, von Stetten F, Zengerle R (2010) Microfluidic lab-on-a-chip platforms: requirements, characteristics and applications. Chem Soc Rev 39(3):1153–1182

Martos S, Torres E, El Bakali MA, Raposo R, Gramaje D, Armengol J, Luque J (2011) Co-operational PCR coupled with dot blot hybridization for the detection of *Phaeomoniella chlamydospora* on infected grapevine wood. J Phytopathol 159:247–254

Massart S, Candresse T, Gil J, Lacomme C, Predajna L, Ravnikar M, Reynard JS, Rumbou A, Saldarelli P, Škoric D, Vainio EJ, Valkonen JPT, Vanderschuren H, Varveri C, Wetzel T (2017) A framework for the evaluation of biosecurity, commercial, regulatory, and scientific impacts of plant viruses and viroids identified by NGS technologies. Front Plant Sci 8:45. https://doi.org/10.3389/fmicb.2017.00045

Matsukura K, Towata T, Sakai JI, Onuki M, Okuda M, Matsumura M (2013) Dynamics of southern rice black-streaked dwarf virus in rice and implication for virus acquisition. Phytopathology 103:509–512. https://doi.org/10.1094/PHYTO10-12-0261-R

Meena RP, Baranwal VK (2015) Development of multiplex polymerase chain reaction assay for simultaneous detection of clostero-, badna- and mandari-viruses along with huanglongbing bacterium in citrus trees. J Virol Methods 235:58–64

Mikoshiba Y, Honda K, Kanematsu S, Fujisawa I (1994) Differentiation of soybean dwarf virus strains by monoclonal antibodies. In: 10th international congress of virology, Jerusalem, Israel (abstracts, 210)

Mishra R, Verma RK, Gaur RK (2015) Analysis of genome comparison of two Indian isolates of cowpea aphid-borne mosaic virus from India. Virus Genes 51:306–309

Morales K, Zambrano JL, Stewart LR (2014) Co-infection and disease severity of Ohio maize dwarf mosaic virus and maize chlorotic dwarf virus strains. Plant Dis 98:1661–1665

Nabi SU, Mir JI, Sharma OC, Singh DB, Zaffer S, Sheikh MA (2018) Optimization of tissue and time for rapid serological and molecular detection of apple stem pitting virus and apple stem grooving virus in apple. Phytoparasitica 46:705–713

Naghavi A, Habibi MK, Firouzabadi FN (2008) Detection and identification of some soybean viral mosaic viruses, using molecular techniques in Lorestan Province, south west of Iran. Asian J Plant Sci 7:557–562

Naidu RA, Hughes JDA (2001) Methods for the detection of plant virus diseases. In: Hughes JDA, Odu BO (eds) Plant virology in Sub-Saharan Africa. Proceedings of a conference organized by IITA. International Institute of Tropical Agriculture, Ibadan, pp 233–260

Naidu RA, Miller JS, Mayo MA, Wesley SV, Reddy AS (2000) The nucleotide sequence of Indian peanut clump virus RNA 2: sequence comparisons among pecluviruses. Arch Virol 145:1857–1866

Naimuddin AM, Pratap A (2011) First report of natural infection of Mungbean yellow mosaic India virus in two wild species of Vigna. New Dis Rep 23:21

Naito FY, Melo FL, Fonseca MEN, Santos CA, Chanes CR, Ribeiro BM, Gilbertson RL, Boiteux LS, de Cássia Pereira-Carvalho R (2019) Nanopore sequencing of a novel bipartite new world begomovirus infecting cowpea. Arch Virol 164(7):1907–1910

Nandakishor HV, Kumaraswamy B, Mane SS, Veena GA (2017) RT-PCR based diagnosis of soybean mosaic virus. Int J Pure Appl Biosci 5(4):698–701. https://doi.org/10.18782/2320-7051.3021

Nath PD, Kenyon L, Bartolome VI, Mclaren G, Azzam O (2000) Simple serological assays for detecting rice tungro viruses. Food Agric Immunol 12(2):139–151

Nault LR, Gordon DT, Gingery RE, Bradfute E, Castillo Loayza J (1979) Identification of maize viruses and mollicutes and their potential insect vectors in Pertl. Phytopathology 69:824–828

Nolt BL, Rajeshwari R, Reddy DVR, Barathan N, Manohar SK (1988) Indian Peanut clump virus isolates: host range, symptomatology, serological relationships and some physical properties. Phytopathology 78:310–313

Notomi T, Okayama H, Masubuchi H, Yonekawa T, Watanabe K, Amino N, Hase T (2000) Loop-mediated isothermal amplification of DNA. Nucleic Acids Res 28(12):e63

Ojuederie OB, Odu BO, Ilori CO (2009) Serological detection of seed borne viruses in cowpea regenerated germplasm using protein a sandwich enzyme linked immunorsorbent assay. Afr Crop Sci J 17(3):125–132

Olmos A, Cambra M, Esteban O, Gorris MT, Terrada E (1999) New device and method for capture, reverse transcription, and nested PCR in a single closed-tube. Nucleic Acid Res 27:1564–1565

Olmos A, Bertolini E, Cambra M (2002) Simultaneous and cooperational amplification (Co-PCR): a new concept for detection of plant viruses. J Virol Methods 106:51–59

Owen TW, Al-Kaysi RO, Bardeen CJ, Cheng Q (2007) Microgravimetric immunosensor for direct detection of aerosolized influenza a virus particle. Sens Actuators B Chem 126:691–699

Palanga E, Filloux D, Martin DP, Fernandez E, Gargani D, Ferdinand R, Zabré J, Bouda Z, Neya JB, Sawadogo M, Traore O (2016) Metagenomic-based screening and molecular characterization of cowpea-infecting viruses in Burkina Faso. PLoS One 11(10):e0165188

Pantaleo V, Saponari M, Gallitelli D (2001) Development of a nested PCR protocol for detection of olive-infecting viruses in crude extracts. J Plant Pathol 83:143–146

Parameswari B, Bagyalakshmi K, Viswanathan R, Chinnaraja C (2013) Molecular characterization of Indian sugarcane streak mosaic virus isolate. Virus Genes 46:186–189

Patil BL, Kumar PL (2015) Pigeonpea sterility mosaic virus: a legume-infecting Emaravirus from South Asia. Mol Plant Pathol 16:775–786

Peng J, Zhang J, Xiaa Z, Li Y, Huang J, Fan Z (2012) Rapid and sensitive detection of *banana bunchy top virus* by loop-mediated isothermal amplification. J Virol Methods 185:254–258

Periasamy M, Niazi FR, Malathi VG (2006) Multiplex RT-PCR, a novel technique for the simultaneous detection of the DNA and RNA viruses causing rice tungro disease. J Virol Methods 134:230–236

Powell CA (1987) Detection of three plant viruses by dot—immunobinding assay. Phytopathology 77:306–309

Prabha K, Baranwal VK, Jain RK (2013) Applications of next generation high throughput sequencing technologies in characterization, discovery and molecular interaction of plant viruses. Indian J Virol 24:157–165

Prak Chang KP, Zakaria A, Abdul Nasir AS, Yusuf N, Thriumani R, Shakaff AYM, Adom AH (2014) Analysis and feasibility study of plant disease using e-nose. In: 2014 IEEE international conference on control system, computing and engineering (ICCSCE 2014). https://doi.org/10.1109/iccsce.2014.7072689

Price CP, Kricka LJ (2007) Improving healthcare accessibility through point-of-care technologies. Clin Chem 53:1665–1675

Rani A, Nerida D, Nitin M (2019) The future of plant pathogen diagnostics in a nursery production system. Biosens Bioelectron 145:111–631. https://doi.org/10.1016/j.bios.2019.111631

Rao GP, Chatenet M, Girard JC, Rott P (2006) Distribution of *sugarcane mosaic* and *sugarcane streak mosaic virus* in India. Sugar Technol 8:79–81

Rao GP, Sharma SK, Singh D, Arya M, Singh P, Baranwal VK (2014) Genetically diverse variants of sugarcane bacilliform virus infecting sugarcane in India and evidence of a novel recombinant Badnavirus variant. J Phytopathol 162:779–787

Reddy DVR, Bragard C, Sreenivasulu P, Delfosee P (2008) Pecluviruses. In: Mahy BWJ, Van Regenmortel M (eds) Desk encylopedia of plant and fungal virology. Academic, New York, pp 257–263

Reddy BVB, Obaiah S, Prasanthi L, Sivaprasad Y, Sujitha A, Krishna TG (2015) *Mungbean yellow mosaic India virus* is associated with yellow mosaic disease of blackgram (*Vigna mungo* L.) in Andhra Pradesh, India. Arch Phytopathol Plant Protect 48:345–353

Roberts IM, Milne RG, Van Regenmortel MHV (1982) Suggested terminology for virus antibody interactions observed by electron microscopy. Intervirology 18:147–149

Robinson TS, Scherm H, Brannen PM, Allen R, Deom CM (2016) Blueberry necrotic ring blotch virus in southern high bush blueberry: insights into in planta and in-field movement. Plant Dis 100:575–1579. https://doi.org/10.1094/PDIS-09-15-1035-RE

Roy A, Fayad A, Barthe G, Brlansky RH (2005) A multiplex polymerase chain reaction method for reliable, sensitive and simultaneous detection of multiple viruses in citrus trees. J Virol Methods 129:47–55

Safarpour H, Safarnejad MR, Tabatabaei M, Mohsenifar A, Rad F, Basirat M, Shahryari F, Hasanzadeh F (2012) Development of a quantum dots FRET-based biosensor for efficient detection of Polymyxa betae. Can J Plant Pathol 34(4):507–515

Schneider WL, Damsteegt VD, Stone AL, Kuhlmann M, Bunyard BA, Sherman DJ, Graves MV, Smythers G, Smith OP, Hatziloukas E (2011) Molecular analysis of soybean dwarf virus isolates in the eastern United States confirms the presence of both D and Y strains and provides evidence of mixed infections and recombination. Virology 412:46–54

Selvarajan R, Balasubramanian V (2016) Cutting-edge technologies for detection of plant viruses in vegetatively propagated crop plants. In: Gaur RK et al (eds) Plant viruses: evolution and management. Springer Science Business Media, Singapore, pp 53–72. https://doi.org/10.1007/978-981-10-1406-2_5

Seok KY, Raston ANH, Bock GM (2016) Aptamer-based nano biosensors. Biosens Bioelectron 76:2–19

Shang H, Xie Y, Zhou X, Qian Y, Wu J (2011) Monoclonal antibody-based serological methods for detection of *Cucumber green mottle mosaic virus*. Virol J 8:228–234

Sharma S, Dasgupta I (2012) Development of SYBR green I based real-time PCR assays for quantitative detection of Rice tungro bacilliform virus and Rice tungro spherical virus. J Virol Methods 181:86–92. https://doi.org/10.1016/j.jviromet.2012.01.018

Sharma P, Verma RK, Mishra RD, Choudhary K, Gaur RK (2013) First report of Turnip yellow virus (TuYV) in Brassica juncea (Indian mustard) in India. New Dis Rep 27:21

Sharma SK, Vignesh Kumar P, Geetanjali AS, Pun KB, Baranwal VK (2015) Subpopulation level variation of banana streak viruses in India and common evolution of banana and sugarcane badnaviruses. Virus Genes 50:450–465

Shojaei TR, Mohd Salleh MA, Sijam K, Rahim RA, Mohsenifar A, Safarnejad R, Tabatabaei M (2016) Detection of *Citrus tristeza virus* by using fluorescence resonance energy transfer-based biosensor. Spectrochim Acta A Mol Biomol Spectrosc 169:216–222

Sidharthan VK, Sevanthi AM, Jaiswal S, Baranwal VK (2020) Robust Virome profiling and whole genome reconstruction of viruses and Viroids enabled by use of available mRNA and sRNA-Seq datasets in grapevine (Vitis vinifera L.). Front Microbiol 11:1232. https://doi.org/10.3389/fmicb.2020.01232

Singh D, Tewari AK, Rao GP, Karuppaiah R, Viswanathan R, Arya M, Baranwal VK (2009) RT-PCR/PCR analysis detected mixed infection of DNA and RNA viruses infecting sugarcane crops in different states of India. Sugar Technol 11:373–380

Soumya K, Yogita M, Prasanthi Y, Anitha K, Kavi Kishor PB, Jain RK, Mandal B (2014) Molecular characterization of Indian isolate of peanut mottle virus and immuno diagnosis using bacterial expressed core capsid protein. Virus Dis 25:331–337

Strange RN (2005) Plant disease: a threat to global food security. Annu Rev Phytopathol 43:83–116

Subba Reddy CV, Sreenivasulu P, Sekhar G (2011) Duplex-immunocapture-RT-PCR for detection and discrimination of two distinct potyviruses naturally infecting sugarcane (*Saccharum* spp. hybrids). Indian J Exp Biol 49:68–73

Sullivan TJ, Dhar AK, Cruz-Flores R, Bodnar AG (2019) Rapid, CRISPR-based, field deployable detection of white spot syndrome virus in shrimp. Sci Rep 9:19702. https://doi.org/10.1038/s41598-019-56170-y

Sun W, Jiao K, Zhang S, Zhang C, Zhang Z (2001) Electrochemical detection for horseradish peroxidase- based enzyme immunoassay using p-aminophenol as substrate and its application to detection of plant virus. Anal Chim Acta 434:43–50

Terauchi H, Kanematsu S, Honda K, Mikoshiba Y, Ishiguro K, Hidaka S (2001) Comparison of complete nucleotide sequences of genomic RNAs of four soybean dwarf virus strains that differ in their vector specificity and symptom production. Arch Virol 146:1885–1898

Tiberini A, Tomassoli L, Barba M, Hadidi A (2010) Oligonucleotide microarray-based detection and identification of 10 major tomato viruses. J Virol Methods 168(1–2):133–140

Tomlinson JA, Ostoja-Starzewska S, Adams IP, Miano DW, Abidrabo P, Kinyua Z, Alicai T, Dickinson MJ, Peters D, Boonham N, Smith J (2013) Loop-mediated isothermal amplification for rapid detection of the causal agents of cassava brown streak disease. J Virol Methods 191:148–154

Trzmiel K, Jeżewska M (2008) Identification of maize dwarf mosaic virus in maize in Poland. Dis Notes 92(6):981. https://doi.org/10.1094/PDIS-92-6-0981A

Umamaheswaran K, Jain RK, Bhat AI, Ahlawat YS (2003) Biological and molecular characterization of a Tospovirus isolate from tomato and its relationship with other Tospoviruses. Indian Phytopathol 56:168–173

Urbanavičienė L, Šneideris D, Žižytė M (2015) Wheat streak mosaic virus detected in winter wheat in Lithuania. Zemdirbyste 102(1):111–114. https://doi.org/10.13080/z-a.2015.102.014

Usharani KS, Surendranath B, Haq QMR, Malathi VG (2005) Infectivity analysis of a soybean isolate of Mungbean yellow mosaic India virus by agroinoculation. J Gen Plant Pathol 71:230–237

Varga A, James D (2006) Use of reverse transcription loop-mediated isothermal amplification for the detection of *Plum pox virus*. J Virol Methods 138:184–190

Vashist SK, Luppa PB, Yeo LY, Ozcan A, Luong JHT (2015) Emerging technologies for next-generation point-of-car testing. Trends Biotechnol 33(11):692–705

Vemulapati B, Druffel KL, Husebye D, Eigenbrode SD, Pappu HR (2014) Development and application of ELISA assays for the detection of two members of the family Luteoviridae infecting legumes: pea enation mosaic virus (genus Enamovirus) and bean leafroll virus (genus Luteovirus). Annu Appl Biol 165:130–136

Vidaver AK, Lambrecht PA (2004) Bacteria as plant pathogens. The Plant Health Instructor. www.apsnet.org

Villamor DEV, Mekuria TA, Pillai SS, Eastwell KC (2016) High throughput sequencing identifies novel viruses in nectarine: insights to the etiology of stem pitting disease. Phytopathology 106:519–527. https://doi.org/10.1094/PHYTO-07-15-0168-R

Villamor DEV, Ho T, Al-Rwahnih M, Martin RR, Tzanetakis IE (2019) High throughput sequencing for plant virus detection and discovery. Phytopathology 109:716–725

Viswanathan R, Balamuralikrishnan M, Karuppaiah R (2008a) Identification of three genotypes of sugarcane yellow leaf virus causing yellow leaf disease from India and their molecular characterization. Virus Genes 37:368–379

Viswanathan R, Balamuralikrishnan M, Karuppaiah R (2008b) Characterization and genetic diversity of sugarcane streak mosaic virus causing mosaic in sugarcane. Virus Genes 36:553–564

Viswanathan R, Karuppaiah R, Malathi P, Ganesh Kumar V, Chinnaraja C (2009) Diagnosis of sugarcane yellow leaf virus in asymptomatic sugarcane by RT-PCR. Sugar Technol 11:368–372

Viswanathan R, Karuppaiah R, Balamuralikrishnan M (2010) Detection of three major RNA viruses infecting sugarcane by multiplex reverse transcription-polymerase chain reaction (multiplex-RT-PCR). Australas Plant Pathol 39(1):79–84

Von Lode P (2005) Point-of-care immune-testing: approaching the analytical performance of central laboratory methods. Clin Biochem 38:591–606

Wang J, Wang X, Huang LL (2005) Detection of soybean mosaic virus in soybean dry seed using RT-PCR. Acta Phytophylacica Sin 35(3):214–220

Wang ZH, Fang SG, Zhang ZY, Han CG, Li DW, Yu JL (2006) Development of an ID-ELISA for the detection of Rice black-streaked dwarf virus in plants. J Virol Methods 134:61–65. https://doi.org/10.1016/j.jviromet.2005.11.019

Wang Q, Yang J, Zhou GH, Zhang HM, Chen JP, Adams MJ (2010) The complete genome sequence of two isolates of southern rice black-streaked dwarf virus, a new member of the genus Fijivirus. J Phytopathol 158(11–12):733–737

Webster CG, Wylie JS, Jones MGK (2004) Diagnosis of plant viral pathogens. Curr Sci 86:1604–1607

Wild P (2008) Electron microscopy of viruses and virus–cell interactions. Methods Cell Biol 88:497–524

Wolter F, Puchta H (2018) The CRISPR/Cas revolution reaches the RNA world: Cas13, a new Swiss Army knife for plant biologists. Plant J 94:767–775. https://doi.org/10.1111/tpj.13899

Wu WQ, Guo XG, Zhang HM, Yang J, Lv MF, Chen JP (2013) Simultaneous detection and survey of three rice viruses in China. Plant Dis 97:1181–1186

Wyant PS, Strohmeier S, Schäfer B, Krenz B, Assunção IP, Lima GSDA, Jeske H (2012) Circular DNA genomics (circomics) exemplified for geminiviruses in bean crops and weeds of northeastern Brazil. Virology 427(2):151–157

Yadav N, Khurana SMP (2016) Plant virus detection and diagnosis: progress and challenges. In: Shukla P (ed) Frontier discoveries and innovations in interdisciplinary microbiology. Springer India, pp 97–132. https://doi.org/10.1007/978-81-322-2610-9_7

Yamagishi N, Terauchi H, Honda K, Kanematsu S, Hidaka S (2006) Discrimination of four soybean dwarf virus strains by dot-blot hybridization with specific probes. J Virol Methods 133:219–222

Yardımcı N, Kılıç HÇ (2011) Identification of beet necrotic yellow vein virus in lakes district: a major beet growing area in Turkey. Indian J Virol 22(2):127

Yoon JY, Choi GS, Cho IS, Choi SK (2014) Development of rapid immune-gold strip kit for on-site diagnosis of tomato spotted wilt virus. Res Plant Dis 20(1):15–20

Zan X, Sitasuwana P, Powell J, Dreherb TW, Wang Q (2012) Polyvalent display of RGD motifs on *Turnip yellow mosaic virus* for enhanced stem cell adhesion and spreading. Acta Biomater 8:2978–2985

Zhang P, Mar TT, Liu W, Li L, Wang X (2013) Simultaneous detection and differentiation of Rice black streaked dwarf virus (RBSDV) and Southern rice black streaked dwarf virus (SRBSDV) by duplex real time RT-PCR. Virol J 10(1):24. https://doi.org/10.1186/1743-422X10-24

Zhang S, Ravelonandro M, Russell P, McOwen N, Briard P, Bohannon S, Vrient A (2014) Rapid diagnostic detection of plum pox virus in Prunus plants by iso- thermal AmplifyRP(®) using reverse transcription- recombinase polymerase amplification. J Virol Methods 207:114–120

Zheng GY, Kositratant WC, Gumpf DJ (1988) Detection of plant viruses using Dot - immunobinding assay. Chinese J Virol 4:150–156

Zhou T, Du L, Fan Y, Zhou Y (2012) Reverse transcription loop mediated isothermal amplification of RNA for sensitive and rapid detection of southern rice black streaked dwarf virus. J Virol Methods 180:91–95

Multilocus Gene Characterization of Phytoplasmas

4

Madem Gurivi Reddy and Govind Pratap Rao

Abstract

Using 16S rRNA genes based primers for the identification of phytoplasmas associated with plants and insects has become the universally accepted method worldwide. The highly conserved ribosomal gene sequence (16Sr RNA) is being utilized as the key molecular tool for the identification and taxonomic classification of phytoplasma strains into "*Candidatus* Phytoplasma" species and ribosomal groups and subgroups. However, due to its highly conserved nature, the 16S rRNA gene is considered insufficient for finer differentiation of closely related but distinct phytoplasma strains. In recent years, several other multilocus moderately conserved genes have been employed for the classification and differentiation of phytoplasma strains. These additional loci are able to identify and differentiate between closely related strains of phytoplasmas that may appear similar based on the basis of 16S rRNA gene sequence analysis alone. In the present chapter, the utilization of several multilocus genes for the successful identification of phytoplasma strains within 16S ribosomal subgroups is discussed. These housekeeping genes have more phylogenetically informative characters than the 16S rRNA genes, thus increasing the resolving power for distinguishing genetically closely related phytoplasma strains.

M. G. Reddy
Department of Plant Pathology, S. V. Agricultural College, Tirupati, Acharya N. G. Ranga Agricultural University, Tirupati, Andhra Pradesh, India

G. P. Rao (✉)
Division of Plant Pathology, ICAR—Indian Agriculture Research Institute, Pusa Campus, New Delhi, India

V. K. Singh et al. (eds.), *Diseases of Field Crops: Diagnostics and Management*, https://doi.org/10.1007/978-981-97-6160-9_4

Keywords

Phytoplasmas · Diagnosis · Housekeeping genes · secA gene · secYgene · rp genes · tuf · groEL · *leuS* · *imp* gene · Other multilocus genes

4.1 Introduction

Until the early 1980s, phytoplasma diseases are detected by observation of phytoplasma bodies in ultrathin sections of the infected tissues under electron microscopic (EM) because of their small size (200–800 nm) and the difficulty of culturing them in vitro. However, the EM is very costly and requires special skills and extra time. Simple diagnostic techniques such as direct fluorescence detection (DFD) and DAPI staining were developed, both of which utilize fluorescent microscopy for phytoplasma detection (Hiruki and da Rocha 1986). The DAPI detects phytoplasma DNA, and the DFD method detects autofluorescence of necrotic phloem cells. ELISA was rarely used to detect phytoplasma diseases except in a few cases because of difficult to purify phytoplasma membrane protein and produce a phytoplasma-specific antibody (Lin and Chen 1985). Recent advances in molecular biology allowed the quick detection of phytoplasma DNA by polymerase chain reaction (PCR) and DNA–DNA hybridization (Deng and Hiruki 1991; Kirkpartrick et al. 1987). PCR amplification of the 16S rRNA genes of phytoplasmas has now become the standard method for the diagnosis of phytoplasma diseases (Lee et al. 1993; Namba et al. 1993; Schneider et al. 1995). PCR amplification of the 16S rRNA genes of phytoplasmas and nested PCR together with restriction fragment length polymorphism (RFLP) analyses allowed the broad taxonomic group and subgroup specific detection of phytoplasmas (Ahrens and Seemüller 1992; Schneider et al. 1993; Namba et al. 1993; Lee et al. 1993; 1994, 1995, 1998).

The highly conserved 16S rRNA gene sequence comparison analysis has been utilized generally for the classification of different phytoplasmas strains into "*Candidatus* Phytoplasma species" and groups/subgroups. However, because of its highly conserved nature, the 16S rRNA gene is inadequate for finer differentiation of closely related phytoplasmas strains. It was evident that some subgroups contained more than one biologically significant strain. Thus, in recent years, several moderately conserved multilocus genes have been employed for differentiation and taxonomic classification of closely related phytoplasmas strains, viz. *leu*S, *sec*A, *tuf*, *sec*Y, *gro*EL, *gyr*A, *gyr*B, *imp*, *dna*B, *dna*K, and *rp* genes. Multilocus sequence analyses of the phytoplasma genome provided additional useful information to the 16S rRNA gene–based phylogenetic backbone and enhanced the resolving power in delineating distinct phytoplasma lineages and closely related strains (Martini et al. 2019).

4.2 Ribosomal Protein (*rp*) Genes

This gene represents the first non-ribosomal marker used for phytoplasma strain differentiation which comprises at least 21 genes in phytoplasma genomes (Hodgetts and Dickinson 2010). The genes encoding the proteins rpL2 and rpS19 and were the first phytoplasma *rp* genes to be cloned. Subsequently, the genes encoding proteins rpL22 and rpS3 were cloned by using a heterologous probe derived from the *Mycoplasma capricolum* (Lim and Sears 1991). Successively, several semi-universal primers have been designed to amplify ribosomal genes from phytoplasma genomes belonging to different phylogenetic groups (Martini et al. 2007).

The average *rp* gene sequence similarity index between two 16Sr phytoplasma groups ranged from 50.4 to 83.5%; however, it has been noticed 85–96.9% in the case of the 16S *rRNA* gene. This character of sequence variation and diversity makes *rp* genes a better molecular indexing tool for phytoplasma taxonomic classification. For example, different "*Ca.* Phytoplasma" species ("*Ca.* P. mali," "*Ca.* P. prunorum," and "*Ca.* P. pyri"), which shared 98.9–99.1% 16S rRNA gene sequence similarity, shared only 94.3–94.6% *rp* gene sequence similarity and were readily demarcated by utilizing analysis of *rp* gene sequence comparison (Martini et al. 2004; 2019).

The rp gene–specific primers are used as molecular markers especially for finer taxonomic classification of phytoplasma strains in the 16SrI group (Lee et al. 2004a, 2006), 16SrIII (Davis et al. 2013), 16SrV (Lee et al. 2004b; Arnaud et al. 2007), and 16SrIX (Lee et al. 2012). These results suggested that the analysis of the *rp* gene sequences not only helped in the delineation of subgroups that belong to similar 16S ribosomal subgroups but also identified additional distinct strains within some subgroups that could not be resolved by the use of 16S rRNA gene sequence analyses.

4.3 *SecY* Gene

This gene is similar to *sec*A and *sec*E genes which encodes for an essential component of the Sec protein translocation system (Kakizawa et al. 2001). The secY gene was first identified from 16SrV "flavescence dorée" DNA, and three FD9 primers were designed (Daire et al. 1997). These FD9 primers were employed to identify several 16SrV phytoplasma strains infecting grapevine and alders in Europe (Arnaud et al. 2007). Phylogenetic analysis from the sec Y gene sequences clearly demarcated the existence of three distinct "flavescence dorée" clusters (FD1, FD2, and FD3), consistent with the phylogenetic trees constructed on the map gene and the uvrB-degV region. Interestingly, based on the 16S rDNA gene, the same phytoplasma strains were previously differentiated into two ribosomal subgroups (16SrV-C and 16SrV-D) (Martini et al. 1999).

Based on the published AY and OY genome sequences, secY gene primers were designed to amplify a 1.4 kb secY gene and used for the differentiation of the AY group phytoplasmas strains (Oshima et al. 2004; Lee et al. 2006; Bai et al. 2006). Twenty representative 16SrI strains from ten 16SrI subgroups were used, and the sequencing and RFLP analysis of the resultant PCR products grouped the phytoplasmas into ten SecY subgroups. Overall, the identified subgroups using secY gene sequences in 16SrI and 16SrV groups by RFLP analyses corresponded to those delineated with the rp gene sequences (Lee et al. 2004a, b, 2006).

The SecY gene has also been employed successfully as a genetic marker for finer differentiation of phytoplasma strains from other groups as well such as 16SrIII, 16SrIX, 16SrX, and 16SrXII (Fialová et al. 2009; Danet et al. 2011; Lee et al. 2012; Davis et al. 2013). The *sec*Y and *rp* gene sequence variability is found to be similar. The *sec*Y gene nucleic acid sequence similarity between members of two given 16S ribosomal phytoplasma groups ranged from 53.5 to 77.9%, compared with 85.1–96.9% for 16S rRNA gene sequences. However, the resolving power of *sec*Y was found to be more effective than other gene markers and slightly better than *rp* gene sequences particularly for resolving closely related strains within the same RFLP group of 16S rRNA gene (Hodgetts et al. 2008).

4.4 *Sec*A Gene

It codes for other protein translocase subunits and was employed successfully for phytoplasma strain differentiation (Hodgetts et al. 2008). *Sec*A is a highly conserved protein because of its involvement in cellular processes consuming ATP for protein translocation outside the cell (Kakizawa et al. 2001). Eight phytoplasma strains from four distinct 16S ribosomal groups (16SrXI, 16SrV, 16SrIII, and 16SrI) were detected using *sec*A membrane protein on immunoblots (Wei et al. 2004).

The *sec*A was first sequenced from the published OY genome sequence, and three sets of degenerate primers (SecAFor1, SecAFor2, and SecARev3) were designed to amplify a partial 480 bp of the gene sequence (Kakizawa et al. 2001). Later, conserved sequences of the protein from OY, AY witches' broom, and coconut lethal yellowing phytoplasma were aligned, and degenerate primers were designed.

From 34 phytoplasma DNA samples examined from diverse phytoplasma strains representing 12 16S ribosomal groups, products of predicted size were produced when SecAFor1/SecARev3 and SecAFor2/SecARev3 primer pairs were employed in a semi-nested PCR assay. However, these secA gene–specific primers sometimes yield additional PCR products of varying sizes, making detection more challenging; as a result, new primers SecAFor5 and SecARev2 were developed to amplify a clear and unambiguous product in nested PCR assay (Dickinson and Hodgetts 2013). Sequencing of these PCR results revealed that the region for phytoplasmas in the 16SrI group is two amino acids longer than the region for phytoplasmas in the other reported ribosomal groups. When compared to the 16S rRNA gene, phylogenetic analysis based on *sec*A gene sequences improved the resolution of ribosomal groups and subgroups and also supported the results found with other multilocus genes.

The distinction between strains within the coconut lethal yellowing-type disease group was clearly demarcated (Hodgetts et al. 2008). The phytoplasmas strains affecting coconut palms were separated into three distinct groups on the *sec*A sequence analysis (Bila et al. 2015), an American group typified by LY, an East African group typified by Tanzanian LD, and a West African group typified by lethal decline (LD) from Nigeria and Cape St. Paul wilt disease from Ghana. The sequence similarity among these strains ranged from 69.7 to 84.4% between two 16S ribosomal groups.

4.5 *Tuf* Gene

Tuf gene is a relatively well-conserved gene that has been frequently employed for the differentiation and classification of different phytoplasma strains. It is often found as a single-copy gene in the sequenced whole genome of phytoplasma and is often used in phylogenetic analysis of other bacteria. The *tuf* gene coding the elongation factor (EF-Tu) is a key protein involved in the process of translation.

Schneider et al. (1997) designed a primer pair (fTuf1 and rTuf1) for amplifications of the *tuf* gene sequences from most of the identified phytoplasma ribosomal groups. These gene-specific primers yielded products belonging to 1000 bp from 16SrI, 16SrIII, and 16SrXII phytoplasma groups but failed to amplify from phytoplasmas to 16SrII, 16SrIV, 16SrX, and 16SrXXII groups.

New primers (for direct PCR Tuf340/Tuf890 and for nested PCR Tuf400/Tuf835) have been designed on a conserved region within the *tuf* gene to amplify a 420–444 bp fragment of the *tuf* gene, and the potential of this fragment as a DNA barcode was examined on 16SrI–VII, 16SrIX–XII, 16SrXV, and 16SrXX phytoplasma strains and was found useful (Makarova et al. 2012, 2013; Contaldo et al. 2015).

A comparison of phylogenetic trees based on *tuf* gene sequences confirmed that the tuf tree is highly congruent in comparison with the 16S rRNA tree and had higher inter- and intra-group sequence divergence. The use of the tuf barcode allowed the finer classification of the main ribosomal groups and subgroups and thus provided a better molecular tool for phytoplasma identification (Marcone et al. 2000; Langer and Maixner 2004; Streten and Gibb 2005).

The 16SrI group–specific primers fTufAy and rTufAy (Schneider et al. 1997) were tested on 70 phytoplasma strains and compared with 16S rRNA gene primers (Marcone et al. 2000). RFLP analyses of the 16S rRNA gene and *tuf* gene PCR products divided the 70 strains into 10 16S ribosomal subgroups; however, the *tuf* gene RFLP profiles resulted into seven subgroups. These results suggested that the resolving efficacy for the separation of distinct phytoplasmas in group 16SrI was lower than that of the 16S rRNA gene (Schneider et al. 1997; Marcone et al. 2000).

However, in some cases, the *tuf* gene was found better in the differentiation of various strains within the 16XII-A and 16XII-B subgroups (Langer and Maixner 2004; Contaldo et al. 2011; Makarova et al. 2012; Iriti et al. 2008).

4.6 *Imp* Gene

Immunodominant membrane proteins are the most abundant proteins of the cell membrane located at the outer surface of phytoplasma cells (Shen and Lin 1993). Based on the structure and the position of the protein of the corresponding gene, phytoplasma immunodominant membrane proteins have been classified as (1) immunodominant membrane protein (Imp), (2) antigenic membrane protein (Amp), and (3) immunodominant protein A (IdPA) and (Kakizawa et al. 2006).

They are not orthologues of each other (Kakizawa et al. 2006), and the nonhomologous proteins play the role of "immunodominance" which includes the major portion of the total cellular membrane proteins, in different phytoplasmas. According to a study performed in the Czech Republic, 12 *imp* genotypes from diverse strains of "*Candidatus* Phytoplasma pyri" could be clustered into nine groups, and in total, 17 genotypes were identified based on the diversity of the *imp* gene sequences.

Siampour et al. (2013) used nucleotide sequences of *imp* genes for phylogenetic analysis and showed the separation of the several "*Ca.* Phytoplasma aurantifolia" strains into three distinct sub-clades with very strong bootstrap values. Single amino acid–altering mutation in the *imp* gene sequences divided 11 16SrII-D subgroup phytoplasmas infecting various vegetable, and field crops were divided into two distinct sub-clusters in Oman (Al-Subhi et al. 2018).

4.7 *Sap*11 Gene

The SAP11 gene codes for the SAP11 effector which has an important role in modulating plant development and defense responses (Sugio et al. 2011a, b). SAP11 effector molecules secreted by the phytoplasma alter plant development and induce phytoplasma symptoms like crinkled leaves and proliferation of stems (Sugio et al. 2011b).

The *Sap11* gene after sequence analysis along with other genes like 16S rRNA, *secA*, *imp*, and *tuf* has been used to identify and characterize the phytoplasma associated with the *Crotalaria* witches' broom disease from Oman in a new lineage 16SrII-W subgroup (Al-Subhi et al. 2017). *SAP11* gene sequences were utilized to study the phytoplasma relationship among 11 strains of 16SrII-D group phytoplasmas infecting various horticultural and field crops in Oman (Al-Subhi et al. 2018).

4.8 *Gro*El Gene

It is also known as the *cpn60* or *hsp60* gene and is a conserved gene used for prokaryotic strain characterization (Desai et al. 2009; Vermette et al. 2010). It belongs to the chaperonin family of molecular chaperones and is found in phytoplasma as a supplementary tool for selecting useful and reliable taxonomic information in addition to 16Sr DNA (Mitrovic et al. 2011a, b). The function of the gene is suggested that of a virulence factor (Clark and Tillier 2010). The expected length of amplicons (about 1.4 kb) of the partial *groEL* gene was obtained with aster yellow group (16SrI) which confirmed the specificity of the nested PCR system employed to detect the 16SrI phytoplasma group (Mitrovic et al. 2011a). Based on the RFLP analysis of *cpn60* gene sequences, the classification of different 16SrI, 16SrII, 16SrV, 16SrVII, 16SrIX, 16SrX, 16SrXII, 16SrXIII, and 16SrXIV groups of phytoplasma was proposed (Perez-Lopez and Dumonceaux 2016). This gene allows a finer differentiation of closely related phytoplasma strains and is thus suggested as an additional molecular tool to the existing phytoplasma classification scheme based on the 16S rRNA gene (Martini et al. 2019).

4.9 Leucyl Transfer RNA Synthetase (*leu*S) Gene

It has been validated and used for amplification of a broad range of phytoplasma taxonomic groups. The 16SrRNA gene, however, has relatively low resolution, and therefore, leucyl transfer RNA synthetase (*leu*S) was used to verify and better define the identity of the closely related strains associated with phytoplasmas in group 16SrXI (Abeysinghe et al. 2016; Rao et al. 2017b).

4.10 Other Genes

Other than the aforementioned genes, several other multilocus genes, namely, *nusA* (Shao et al. 2006), dnaB gene (Madhupriya et al. 2015), and folate genes (*folP* and *folk*), have been employed, and were proved useful for differentiation among strains in the 16SrI and 16SrXII groups (Davis et al. 2013). Several constitutive candidate genes, including *dnaA* (encoding chromosomal replication initiator protein), *dnaE* (DNA polymerase III alpha subunit), and *polC* (encoding DNA polymerase III alpha subunit), could potentially be useful for the finer classification of phytoplasmas as well. The sequence variability of these genes is similar or slightly greater than that of the *secY* or *rp* genes (Lee et al. 2010; Hodgetts and Dickinson 2010).

Multilocus characterization on *dna*K, *dpp*A, *tuf*, and *sec*Y genes amplified from several horticultural infected species from Egypt has been achieved, and no variation between the phytoplasma strains of the 16SrII D subgroups has been observed (El-Sisi et al. 2018). Multilocus genes such as *gyr*A, *gyr*B, *sec*Y, *sec*A, *dna*B, and *po*C were also employed for the characterization of the "*Ca.* P. oryzae"-related strains in India (16SrXI-B) (Rao et al. 2017a). Multilocus genes such as *sec*A, *tuf*, and *imp* were also employed for the characterization of 16SrII-D subgroup phytoplasma infecting vegetable crops in Oman (Al-Subhi et al. 2018). In a study in addition to the 16S rRNA gene, *sec*A, *rp*, *sec*Y, *imp*, and *tuf* confirmed the validity and utility of all the multilocus genes as additional molecular markers for authentic characterization of phytoplasma strains belonging to the 16SrII-C and 16SrII-D subgroups in chickpea, sesame, and insect vectors (Reddy et al. 2021). Association of 16SrII-C and 16SrI-B subgroup phytoplasma was confirmed and validated using the various multilocus genes like secA, rp, secY, and tuf in Mexican and French marigold, several weed species, and insect vectors (Panda et al. 2021). According to a study, multilocus gene analysis with secA, rp, secY, and tuf genes confirmed the presence of the 16SrI-B subgroup strain identified and characterized from Tripura state of India and the 16SrII-D subgroup from Uttar Pradesh strain in *Impatiens balsamina* showing flat stem and phyllody in India (Panda et al. 2021).

4.11 Conclusion

The utility of several multilocus genes in the identification and differentiation of phytoplasma strains allowed not only the delineation of subgroups that are recognized and identified with 16S ribosomal subgroups but also capable of identifying, within some subgroups, additional distinct strains that could not be resolved by 16S rRNA gene sequences. These housekeeping genes have better phylogenetically informative characters as compared to the 16S rRNA gene, which significantly increases the resolving power for genetic diversity among closely related phytoplasma strains within the 16S ribosomal group and/or "*Candidatus* Phytoplasma" species. It would be important to exploit other multilocus phytoplasma genes for their relevance in molecular epidemiology and evolution. The characterization of alternative multilocus genes must be implemented and applied in the authentic finer taxonomic classification of phytoplasma strains of economic significance worldwide.

References

Abeysinghe S, Abeysinghe PD, Kanatiwela-de Silva C et al (2016) Refinement of the taxonomic structure of 16SrXI and 16SrXIV phytoplasmas of gramineous plants using multilocus sequence typing. Plant Dis 100(10):2001–2010

Ahrens U, Seemüller E (1992) Detection of DNA of plant pathogenic mycoplasma like organisms by a polymerase chain reaction that amplifies a sequence of the 16S rRNA gene. Phytopathology 82(8):828–832

Al-Subhi A, Hogenhout SA, Al-Yahyai RA, Al-Sadi AM (2017) Classification of a new phytoplasmas subgroup 16SrII-W associated with *Crotalaria* witches' broom diseases in Oman based on multigene sequence analysis. BMC Microbiol 17:221

Al-Subhi AM, Hogenhout SA, Al-Yahyai RA et al (2018) Detection, identification, and molecular characterization of the 16SrII-D phytoplasmas infecting vegetable and field crops in Oman. Plant Dis 102(3):576–588

Arnaud G, Malembic-Maher S, Salar P et al (2007) Multilocus sequence typing confirms the close genetic inter-relatedness between three distinct "flavescence dorée" phytoplasma strain clusters and group 16SrV phytoplasmas infecting grapevine and alder in Europe. Appl Environ Microbiol 73:4001–4010

Bai X, Zhang J, Ewing A et al (2006) Living with genome instability: the adaptation of phytoplasmas to diverse environments of their insect and plant hosts. J Bacteriol 188(10):3682–3696

Bila J, Mondjana A, Samils B et al (2015) High diversity, expanding populations and purifying selection in phytoplasmas causing coconut lethal yellowing in Mozambique. Plant Pathol 64:597–604

Clark GW, Tillier ER (2010) Loss and gain of *GroEL* in the Mollicutes. Biochem Cell Biol 88(2):185–194

Contaldo N, Canel A, Makarova O et al (2011) Use of a fragment of the *tuf* gene for phytoplasma 16Sr group/subgroup differentiation. Bull Insectol 64(Supplement):S45–S46

Contaldo N, Paltrinieri S, Makarova O et al (2015) Q-bank phytoplasma: a DNA barcoding tool for phytoplasma identification. In: Plant pathology. Humana Press, New York, pp 123–135

Daire X, Clair D, Larrue J et al (1997) Detection and differentiation of grapevine yellows phytoplasmas belonging to elm yellows group and to the "stolbur" subgroup by PCR amplification of non-ribosomal DNA. Eur J Plant Pathol 103:507–514

Danet JL, Balakishiyeva G, Cimerman A et al (2011) Multilocus sequence analysis reveals the genetic diversity of European fruit tree phytoplasmas and supports the existence of interspecies recombination. Microbiology 157(2):438–450

Davis RE, Zhao Y, Dally EL et al (2013) '*Candidatus* Phytoplasma pruni', a novel taxon associated with X-disease of stone fruits, *Prunus* spp.: multilocus characterization based on 16S rRNA, secY, and ribosomal protein genes. Int J Syst Evol Microbiol 63(2):766–776

Deng S, Hiruki C (1991) Amplification of 16S rRNA genes from culturable and non-culturable mollicutes. J Microbiol Methods 14(1):53–61

Desai AR, Musil KM, Carr AP et al (2009) Characterization and quantification of feline fecal microbiota using cpn60 sequence-based methods and investigation of animal-to-animal variation in microbial population structure. Vet Microbiol 137(1–2):120–128

Dickinson M, Hodgetts J (2013) PCR analysis of phytoplasmas based on the *secA* gene. In: Phytoplasma. Humana Press, Totowa, 205–215 pp

El-Sisi Y, Omar AF, Sidaros SA et al (2018) Multilocus sequence analysis supports a low genetic diversity among '*Candidatus* Phytoplasma australasia' related strains infecting vegetable crops and periwinkle in Egypt. Eur J Plant Pathol 150(3):779–784

Fialová R, Válová P, Balakishiyeva G et al (2009) Genetic variability of stolbur phytoplasma in annual crop and wild plant species in South Moravia. J Plant Pathol 91:411–416

Hiruki C, da Rocha A (1986) Histochemical diagnosis of mycoplasma infections in *Catharanthusroseus* by means of a fluorescent DNA-binding agent, 4, 6-diamidino-2-phenylindole-2HCl (DAPI). Can J Plant Pathol 8(2):185–188

Hodgetts J, Dickinson M (2010) Phytoplasma phylogeny and detection based on genes other than 16S rRNA. In: Phytoplasmas: genomes, plant hosts, and vectors. CABI, Wallingford, pp 93–113

Hodgetts J, Boonham N, Mumford R et al (2008) Phytoplasma phylogenetics based on analysis of secA and 23S rRNA gene sequences for improved resolution of candidate species of '*Candidatus* Phytoplasma'. Int J Syst Evol Microbiol 58(8):1826–1837

Iriti M, Quaglino F, Maffi D, Casati P, Bianco PA, Faoro F et al (2008) *Solanum malacoxylon*, a new natural host of stolbur phytoplasma. J Phytopathol 156(1):8–14

Kakizawa S, Oshima K, Kuboyama T et al (2001) Cloning and expression analysis of phytoplasma protein translocation genes. Mol Plant-Microbe Interact 14(9):1043–1050

Kakizawa S, Oshima K, Namba S (2006) Diversity and functional importance of phytoplasma membrane proteins. Trends Microbiol 14:254–256

Kirkpartrick BC, Strenger DC, Morris TJ (1987) Cloning and detection of DNA from a nonculturable plant pathogenic mycoplasma like organism. Science 238:197–200

Langer M, Maixner M (2004) Molecular characterisation of grapevine yellows associated phytoplasmas of the stolbur group based on RFLP analysis of non-ribosomal DNA. Vitis 43:191–199

Lee IM, Hammond RW, Davis RE et al (1993) Universal amplification and analysis of pathogen 16S rDNA for classification and identification of mycoplasma like organisms. Phytopathology 83(8):834–842

Lee IM, Gundersen DE, Hammond RW et al (1994) Use of mycoplasma like organism (MLO) group-specific oligonucleotide primers for nested-PCR assays to detect mixed-MLO infections in a single host plant. Phytopathology 84(6):559–566

Lee IM, Bertaccini A, Vibio M et al (1995) Detection of multiple phytoplasmas in perennial fruit trees with decline symptoms in Italy. Phytopathology 85(6):728–735

Lee IM, Gundersen-Rindal DE, Davis RE et al (1998) Revised classification scheme of phytoplasmas based on RFLP analyses of 16S rRNA and ribosomal protein gene sequences. Int J Syst Evol Microbiol 48(4):1153–1169

Lee IM, Gundersen-Rindal DE, Davis RE et al (2004a) 'Candidatus Phytoplasma asteris', a novel phytoplasma taxon associated with aster yellows and related diseases. Int J Syst Evol Microbiol 54(4):1037–1048

Lee M, Martini M, Marcone C et al (2004b) Classification of phytoplasma strains in the elm yellows group (16SrV) and proposal of '*Candidatus* Phytoplasma ulmi' for the phytoplasma associated with elm yellows. Int J Syst Evol Microbiol 54(2):337–347

Lee M, Bottner KD, Secor G et al (2006) 'Candidatus Phytoplasma americanum', a phytoplasma associated with a potato purple top wilt disease complex. Int J Syst Evol Microbiol 56(7):1593–1597

Lee IM, Bottner-Parker KD, Zhao Y et al (2010) Phylogenetic analysis and delineation of phytoplasmas based on secY gene sequences. Int J Syst Evol Microbiol 60(12):2887–2897

Lee IM, Bottner-Parker KD, Zhao Y et al (2012) Differentiation and classification of phytoplasmas in the pigeon pea witches'-broom group (16SrIX): an update based on multiple gene sequence analysis. Int J Syst Evol Microbiol 62(9):2279–2285

Lim PO, Sears BB (1991) DNA sequence of the ribosomal protein genes *rp12* and *rps19* from a plant-pathogenic mycoplasma-like organism. FEMS Microbiol Lett 84:71–74

Lin CP, Chen TA (1985) Monoclonal antibodies against the aster yellows agent. Science 227(4691):1233–1235

Madhupriya, Tiwari AK, Rao GP (2015) Utilization of dnaB gene for characterization of phytoplasmas associated with toria, brinjal and Phlox in India. Phytopathogenic Mollicutes 5(1):11–12

Makarova O, Contaldo N, Paltrinieri S (2012) DNA barcoding for identification of 'Candidatus Phytoplasmas' using a fragment of the elongation factor Tu gene. PLoS One 7(12):e52092

Makarova O, Contaldo N, Paltrinieri S et al (2013) DNA bar-coding for phytoplasma identification. In: Phytoplasma. Humana Press, Totowa, 301–317 pp

Marcone C, Lee I-M, Davis RE et al (2000) Classification of aster yellows-group phytoplasmas based on combined analyses of rRNA and *tuf* gene sequences. Int J Syst Evol Microbiol 50:1703–1713

Martini M, Murari E, Mori N et al (1999) Identification and epidemic distribution of two "flavescence dorée"-related phytoplasmas in Veneto (Italy). Plant Dis 83:925–930

Martini M, Lee I-M, Bottner KD, Zhao Y, Botti S, Bertaccini A et al (2007) Ribosomal protein gene-based phylogeny for finer differentiation and classification of phytoplasmas. Int J Syst Evol Microbiol 57:2037–2051

Martini M, Lee I, Zhao Y et al (2004) Ribosomal protein gene-based phylogeny: a basis for phytoplasma classification. International Organization for Mycoplasmology, 11:156

Martini M, Quaglino F, Bertaccini A (2019) Multilocus genetic characterization of phytoplasmas. In: Bertaccini A, Oshima K, Kube M, Rao GP (eds) Phytoplasmas: plant pathogenic bacteria-III, genomics, host pathogen interactions and diagnosis. Springer, Singapore, pp 161–200

Mitrovic J, Contaldo N, Paltrinieri S et al (2011a) The use of groEL gene for characterisation of aster yellows phytoplasmas in field collected samples. Bull Insectol 64:S17–S18

Mitrovic J, Kakizawa S, Duduk B et al (2011b) The *cpn60* gene as an additional marker for finer differentiation of '*Candidatus* Phytoplasma asteris'-related strains. Ann Appl Biol 159:41–48

Namba S, Oyaizu H, Kato S et al (1993) Phylogenetic diversity of phytopathogenic mycoplasma like organisms. Int J Syst Evol Microbiol 43(3):461–467

Oshima K, Kakizawa S, Nishigawa H et al (2004) Reductive evolution suggested from the complete genome sequence of a plant-pathogenic phytoplasma. Nat Genet 36(1):27–29

Panda P, Nigam A, Rao GP (2021) Multilocus gene analysis reveals the presence of two phytoplasma groups in Impatiens balsamina showing flat stem and phyllody. 3 Biotech 11(3):1–12

Perez-Lopez E, Dumonceaux TJ (2016) Detection and identification of the heterogeneous novel subgroup 16SrXIII-(A/I) I phytoplasma associated with strawberry green petal disease and Mexican periwinkle virescence. Int J Syst Evol Microbiol 66:4406–4415

Rao GP, Thorat V, Manimekalai R et al (2017a) A century progress of research on phytoplasma diseases in India. Phytopathogenic Mollicutes 7(1):1–38

Rao GP, Prakasha TL, Madhupriya TV, Kumar M, Baranwal VK (2017b) First report of association of *Candidatus* Phytoplasma cynodontis (16SrXI-B group) with streak, yellowing, and stunting disease in Durum and Bread wheat genotypes. Plant Dis 101(7):1314

Reddy MG, Baranwal VK, Sagar D et al (2021) Molecular characterization of chickpea chlorotic dwarf virus and peanut witches' broom phytoplasma associated with chickpea stunt disease and identification of new host crops and leafhopper vectors in India. 3 Biotech 11(3):1–23

Schneider B, Ahrens U, Kirkpatrick BC et al (1993) Classification of plant-pathogenic mycoplasma-like organisms using restriction-site analysis of PCR-amplified 16S rDNA. Microbiology 139(3):519–527

Schneider B, Seemüeller E, Smart CD et al (1995) Phylogenetic classification of plant pathogenic mycoplasma-like organisms or phytoplasmas. In: Razin S, Tully JG (eds) Molecular and diagnostic procedures in mycoplasmology. Academic, San Diego, pp 369–380

Schneider B, Gibb KS, Seemüller E (1997) Sequence and RFLP analysis of the elongation factorTu gene used in differentiation and classification of phytoplasmas. Microbiology 143:3381–3389

Shao J, Jomantiene R, Dally EL et al (2006) Phylogeny and characterization of phytoplasmal NusA and use of the *nusA* gene in detection of group 16SrI strains. J Plant Pathol 88:193–201

Shen WC, Lin CP (1993) Production of monoclonal antibodies against a mycoplasma like organism associated with sweet potato witches' broom. Phytopathology 83:671

Siampour M, Izadpanah K, Galetto L et al (2013) Molecular characterization, phylogenetic comparison and serological relationship of the *Imp* protein of several '*Candidatus* Phytoplasma aurantifolia' strains. Plant Pathol 62(2):452–459

Streten C, Gibb KS (2005) Genetic variation in '*Candidatus* Phytoplasma australiense'. Plant Pathol 54:8–14

Sugio A, MacLean AM, Kingdom HN et al (2011a) Diverse targets of phytoplasma effectors: from plant development to defense against insects. Annu Rev Phytopathol 49:175–195

Sugio A, MacLean AM, Grieve VM et al (2011b) Phytoplasma protein effector SAP11 enhances insect vector reproduction by manipulating plant development and defense hormone biosynthesis. Proc Natl Acad Sci U S A 108(48):254–263

Vermette CJ, Russell AH, Desai AR et al (2010) Resolution of phenotypically distinct strains of Enterococcus spp. in a complex microbial community using cpn60 universal target sequencing. Microb Ecol 59(1):14–24

Wei W, Kakizawa S, Jung HY et al (2004) An antibody against the SecA membrane protein of one phytoplasma reacts with those of phylogenetically different phytoplasmas. Phytopathology 94(7):683–686

Diagnosis of Plant Parasitic Nematodes

5

Amit Ahuja, Snehalatha Pasupuleti, Pankaj,
and Vishal Singh Somvanshi

Abstract

Plant parasitic nematodes (PPNs) are the unseen enemies of crop plants. The symptoms caused by PPNs are more often overlooked or confused with symptoms caused by abiotic stress on crops. The management of PPNs is achieved by using resistant crops, a limited number of chemical nematicides, biological agents, or other cultural methods. The success of all of these management practices, however, depends on early detection and diagnosis of PPNs. The currently prevalent methods of nematode diagnosis heavily depend on the examination of plant symptomatology along with morphological identification of the associated nematodes by expert nematologists. In recent times, PCR-based molecular methods such as linear PCR, RT-qPCR, and isothermal PCR-based assays have gained popularity because they are specific and easily doable by non-experts. Several types of isothermal PCR-based assays have the potential to be converted into a point-of-care diagnostic method, which is desperately needed for PPNs.

Keywords

Plant parasitic nematodes · Diagnostics · Morphological · Molecular · LAMP · qPCR · RPA

A. Ahuja · S. Pasupuleti · Pankaj · V. S. Somvanshi (✉)
Division of Nematology, ICAR-Indian Agricultural Research Institute, Pusa Campus,
New Delhi, India
e-mail: vishal.somvanshi@icar.gov.in

V. K. Singh et al. (eds.), *Diseases of Field Crops: Diagnostics and Management*,
https://doi.org/10.1007/978-981-97-6160-9_5

5.1 Introduction

Nematodes or roundworms are among the most numerous invertebrate animals on the earth (Cobb 1914) and are found in all conceivable habitats. Nematodes are broadly categorized into two groups on the basis of their lifestyle—parasitic and free-living (Yeates et al. 1993). The free-living roundworms feed upon soil microorganisms and play an essential role in nutrient recycling and mineralization in the ecosystem. The parasitic nematodes include plant, human, and animal parasites. Approximately 30,000 species of plant parasitic nematodes (PPNs) have been described. On average, PPNs cause approximately 10% loss of yield in agricultural and horticultural crops, amounting to 100–173 billion USD (Kumar et al. 2020). As per a global survey, the top 10 most damaging PPN genera included root-knot nematodes, cyst nematodes, root-lesion nematodes, the burrowing nematode, stem and bulb nematode, the pine wilt nematode, the reniform nematode, *Xiphinema index*, *Nacobbus aberrans*, and leaf and bud nematode (Jones et al. 2013). In India, a yield loss estimate by PPNs for 30 different crops amounted to Rs. 102,039.79 million annually (Jain et al. 2007; Kumar et al. 2020), of which Rs. 51,814.81 million was for 11 field crops. Rice crop alone was damaged to the tune of Rs. 23,272.32 million (Kumar et al. 2020). The most damaging PPNs in India are the root-knot nematodes *M. graminicola*, *M. javanica*, *M. incognita*, *M. hapla,* cyst nematodes *Heterodera* and *Globodera*, *Rotylenchulus reniformis*, *Radopholus similis*, *Pratylenchus coffeae*, *Ditylenchus angustus*, *Aphelenchoides besseyi*, *Hirschmanniella oryzae*, and *Tylenchulus semipenetrans* (Khan 2015). PPNs also worsen plant diseases by acting synergistically with other pathogens. The nematode diseases of crops are difficult to diagnose as their symptoms look more like abiotic stresses (Singh et al. 2015). In addition, PPNs are tremendously hard to control, especially in standing crops and orchards. Chemical pesticide–based options for managing PPNs are scant; therefore, biological control, genetic resistance, sanitation, use of clean planting material, and certain cultural methods are used to manage PPNs (Hafez et al. 2020).

Nematode diagnosis and detection at an early stage is the most critical step for the management of the majority of PPNs. In a field/orchard/greenhouse, early recognition of nematodes' presence in the soil is effective for its' management as once the nematodes are established, it is tough to remove them. Stopping the entry of exotic nematodes into the country through the plant material through effective monitoring and quarantine and thereby preventing its spread to a newer location is also crucial for PPN management. PPN diagnosis relies heavily on manual inspection of fields, crop symptoms, plant tissues, and nematode specimens. In recent years molecular tools have been widely used. Here, various methods presently used to diagnose and detect various PPNs are discussed.

5.2 Methods of Nematode Diagnosis

The diagnosis of PPNs in the field is a cumbersome and challenging task, as in many cases, these nematodes do not induce characteristic symptoms in their hosts. Most of the PPNs induce non-specific symptoms like yellowing, wilting, chlorosis, and stunting (Singh et al. 2015). These unspecific symptoms may overlap with symptoms induced by nutrient deficiency or by the attack of other pests and pathogens. Few nematodes produce certain specific characteristic symptoms which may lead to their identification. The precise identification of PPNs is crucial for adopting suitable management practices in real time and for scientific research purposes. The traditional identification of nematode species wholly based on morphological traits needs a skilled taxonomist. However, several molecular techniques like isozyme profiling, antigen-antibody-based diagnosis, cytogenetics, and DNA markers (RFLP, AFLP, RAPD, SCAR, SSR) were developed and used. Molecular methods were rapid and helped in understanding the evolutionary history and systematics of the nematodes. However, the morphological and molecular methods both have their own sets of benefits and lacunas (Abebe et al. 2011). This chapter is intended to present information regarding available identification methods for plant parasitic nematodes and will provide insight into recently used rapid identification protocols for the detection of PPNs.

5.2.1 Symptomatology and Morphological Identification

To reveal the identity of nematodes infesting the field, the correct procedure of soil and plant sample collection must be performed in the field, followed by microscopic studies in the laboratory to identify the exact cause of symptoms or disease. The PPNs can infect all parts of plants from roots to foliar parts including the stem, leaves, buds, flowers, and seeds. The phytonematodes vary in parasitism on their host plants; therefore, the collection of soil samples and infested plant parts is necessary for nematode diagnosis (Carneiro et al. 2017).

5.2.1.1 Symptom-Specific Diagnosis in the Field (Stage 1)

The specific symptoms produced by nematodes can be categorized into two groups: above-ground symptoms and below-ground symptoms (Fig. 5.1). The preliminary diagnosis of associated nematode problems in the field can be carried out through examining characteristic symptoms. Once the suspected nematode is identified, further examination of associated nematode species in the field should be done in the laboratory by using a microscope. The above-ground symptoms are induced predominantly by foliar nematodes, which produce characteristic symptoms on leaves, stems, flowers, panicles, developing grains, etc. Few of the above-ground symptoms are induced by nematodes feeding on roots. The symptoms on the below-ground plant parts are induced by nematodes which disrupt the normal physiology of the host-plant. In Tables 5.1 and 5.2, characteristic above-ground and below-ground symptoms induced by nematodes are listed.

5.2.1.2 Morphological Identification of Nematodes in the Laboratory Under the Microscope (Stage 2)

The plant parasitic nematodes are not visible with unaided eyes due to their microscopic size and transparent body. The length of most of the PPNs ranges from 0.3 mm to 1.2 cm, which makes it difficult to diagnose them using the naked eye.

Stage-1
Field level diagnosis (Based on symptoms)

Collection of samples:
Collect soil and affected plant parts,
In the case of field crops, collect 15 to 20 soil samples of 250 cc each randomly

Observation of symptoms

- Yellowing, wilting, stunting and patchiness condition in the field
- Observation of specific characteristic symptoms on above and below-ground plant parts
- **The above-ground symptoms** include leaf spots, leaf galls, leaf curling and crinkling, twisting of stem and leaves, degenerate panicles, seed galls, seeds malformation, stem galls, stem discolouration, devitalized buds and floral malady etc.
- **The below-ground symptoms** include root galls, root branching, swelling on roots, stubby roots, coarse roots, curly tips, root forking, reddening of roots, the bushy appearance of roots, rotting of rhizomes, tubers and corms, root lesion and dirty roots etc.

Fig. 5.1 Nematode diagnosis at the field level based on the symptoms

The smallest and longest body length reported is 0.3 mm for the pin nematode, *Paratylenchus*, and 1.2 cm for the stubby root nematode, *Paralongidorus*, respectively. Although the plant parasitic nematodes look very similar, certain key morphological features help in the differentiation between various genera of phylum Nematoda. The study of morphological traits is the basis of traditional taxonomy, which helps in the identification of the nematodes. The most studied morphological traits are the shape and size of the body; the ornamentation on the cuticle; variation in lip region; types of stylet and esophagus; variation in the structure of stylet and esophagus; overlapping of esophageal glands; variation in the reproductive and digestive system; number of ovary and testis; location of vulva, anus, and cloaca; location and type of excretory pore; presence and location of a sensory organ like amphids, phasmids, caudlids, deirids, hemizonids, and hemizonions; variation in the shape of the tail tip, etc. Genus-level identification is possible by using a stereo-binocular microscope and a compound microscope to examine the aforementioned morphological traits. The differentiation of nematodes at the species level requires extensive microscopic studies and might require transmission and scanning electron microscopy. The biological traits like the structure of feeding cells also help in the identification of respective nematodes, e.g., giant cell by the root-knot nematode, syncytium by cyst nematodes, nurse cell by citrus nematodes, and prosyncate by reniform nematodes.

The identification exercise is a step-by-step process (Fig. 5.2), which includes the collection of the required number of soil and plant samples, storage of the samples, washing of the samples in the laboratory with different techniques like Cobb's decanting and sieving, Fenwick's can for cyst, maceration filtration for migratory endoparasitic nematodes, sugar floatation for sluggish nematodes, and bleaching for egg collection, followed by preparation of temporary slides by using lactophenol or

Table 5.1 Characteristic above-ground symptoms, causal nematode, and infested crop

Causal organism	Symptoms	Crop
Strawberry foliar nematode, *Aphelenchoides fragariae*	Dead/devitalised bud or blind plant	Strawberry
Chrysanthemum foliar nematode, *Aphelenchoides ritzemabosi*	Predominant brown angular spots on leaves	Chrysanthemum, Dahlia
White tip nematode, *Aphelenchoides besseyi*	White tip, top three to five centimeters of leaf tip became white	Rice
Tuberose foliar nematode, *Aphelenchoides besseyi*	Distort floral parts and cause floral malady disease	Tuberose
Wheat seed gall nematode, *Anguina tritici*	Seed galls, curling of leaves, crinkled or twisted stem	Wheat
Leaf gall nematode, *Anguina balsamophila*	Galls on leaves	Balsamroot plant
Leaf gall nematode, *Anguina millefolii*	Galls on leaves	Common yarrow, a flowering plant in the family Asteraceae
Stem gall nematode, *Anguina picridis*	Galls on stem	Orchids
Pinewood nematode, *Bursaphelenchus xylophilus*	Wilting in trees, because of their movement in resin canals	Pinewood
Rice stem nematode, *Ditylenchus angustus*	Twisting of basal leaves swollen and ripe Ufra disease, panicle remain enclosed and if emerge then bears a few grains on the tip	Rice
Stem and bulb nematode, *Ditylenchus dipsaci*	Spikkles (pimple-like spots of leaves)	Narcissus
Stem and bulb nematode, *Ditylenchus dipsaci*	Twisting of basal leaves	Onion
Burrowing nematode, *Radopholus similis*, *Radopholus citrophilus*	Spreading decline in citrus orchards, spread to the other citrus trees	Citrus
Burrowing nematode, *Radopholus similis*	Gradual yellowing of leaves	Peppers
Burrowing nematode, *Radopholus similis*	Toppling of fruits head at the bearing stage due to extensive cavities formation by the migration of the nematodes	Banana
Coconut red ring nematode, *Rhadinaphelenchus xylophilus*	Red ring of 3 cm in diameter can be seen inside the stem	Coconut
Stem gall nematode, *Subanguina agropyri*	Galls on stem	Wheat, rye, and barley
Citrus nematode, *Tylenchulus semipenetrans*	Slow decline of citrus plantations, the soil remains stuck to the roots, and a cortical portion of the roots can be torn off easily	Citrus

Table 5.2 Characteristic below-ground symptoms, causal nematode, and infested crop

Causal organism	Symptoms	Crop
Sting nematode, *Belonolaimus longicaudatus*	Coarse appearance of roots	Maize
Stem and bulb nematode, *Ditylenchus dipsaci*	Black-colored streak formation on the underground bulb	Iris
Stem and bulb nematode, *Ditylenchus dipsaci*	Underground bloating on bulbs	Onion and garlic
Potato rot nematode, *Ditylenchus destructor*	Rotting of roots	Potato
Stem and bulb nematode, *Ditylenchus dipsaci*	Brown-colored ring formation on tubers	Daffodils and sweet potato
Stem and bulb nematode, *Ditylenchus dipsaci*	Bushy appearances of roots, appear like a witch's broom	Phlox
Awl nematode, *Dolichodorus heterocephalus*	Reddening discoloration of roots	Celery
Potato cyst nematodes, *Globodera rostochiensis*, *G. pallida*	Appearance of pale or white colored cysts on roots and tubers	Potato
Cereal cyst nematodes, *Heterodera avenae*	Swelling on roots, thickness in roots, and bushy appearance of rootlets (Molya disease)	Wheat, oat, barley
Sheath nematodes, *Hemicycliophora arenaria*	Galls on roots	Lemon
Spiral nematodes, *Helicotylenchus multicinctus*	Superficial lesions formation on roots	Banana
Root-knot nematodes, *Meloidogyne incoginita*	Root forking and galls on roots	Carrots and turnips
Root-knot nematodes, *Meloidogyne incoginita*, *M. javanica*	Galls on the root, appearance of melon shape female on roots	In vegetables like brinjal, tomato
Root-knot nematodes, *Meloidogyne incoginita*, *M. javanica*	Compound galls, small galls coalesce to form a bigger compound gall	Cucurbits
Temperate root-knot nematode, *Meloidogyne hapla*	Excessive branching of roots	Tomato
False root-knot nematode, *Nacobbus batatiformis*	Formation of spindle-shaped galls on roots	Sugar beet and tomato
Stunny root nematodes, *Paratrichodorus* & *Trichodorus*	Root decay from their origin point and appear like a stub, an overall symptom is known as the formation of stubby roots	Maise
Lesion nematode, *Pratylenchus* spp.	Elliptical shape lesions on roots	Various crops
Burrowing nematode, *Radopholus similis*	Formation of lesions on roots and extensive cavities due to migration of nematodes inside root cortex region	Banana

(continued)

Table 5.2 (continued)

Causal organism	Symptoms	Crop
Yam rot nematode, *Scutellonema bradys*	Rotting of roots	Yam
Root gall nematode, *Subanguina radicicola*	Galls on roots	Cereals and grasses
Dagger nematode, *Xiphinema* spp.	Curly tip or fish hook formation on roots	Various woody plants and crops
Dagger nematode, *Xiphinema diversicaudatum*	Galls on roots, spread Arabic mosaic virus disease	Roses

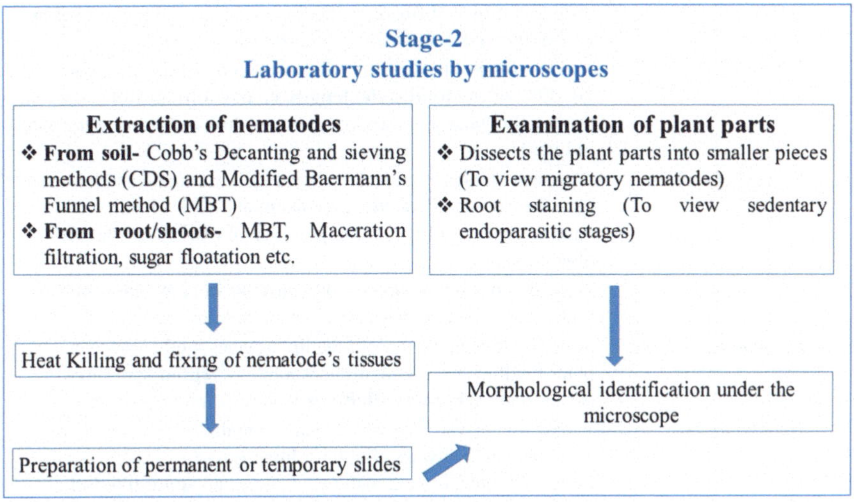

Fig. 5.2 Morphological diagnosis of nematodes in laboratory settings

permanent slides by using glycerol. Once the sample slide is prepared, one should spend time on the microscope to identify the genus of nematode (Southey 1986). Key diagnostic characteristics for the important nematodes' genera are listed in Table 5.3.

Although morphological identification is cost-effective, certain disadvantages are associated with it. The major disadvantages are the lack of technical experts for sample processing and skilled taxonomists to study morphological characters, lengthy and time-consuming procedures, and dependence on the acquisition of several life stages of nematodes.

5.2.2 Nematode Molecular Diagnosis

The molecular methods have become popular for nematode diagnosis and taxonomy in present times. Molecular approaches offer many advantages over the traditional methods, such as faster species identification and the study of evolutionary

Table 5.3 Morphological characteristics important for genus-level identification of various plant parasitic nematodes (http://nemaplex.ucdavis.edu/)

S. no.	Genus	Important morphological characteristics
1.	*Anguina*	Adult nematode's length is around 2.5 mm with a very short stylet of 10 μm, a mature female exhibits sexual dimorphism, where the body of the female becomes **swollen from the middle** and takes an **obese shape**, heat-relaxed **body curved ventrally**, females are monodelphic, and oocytes are arranged in multiple rows around a rachis in ovary
2.	*Aphelenchoides*	Large size **square-shaped metacorpus** is seen in the esophagus, esophageal glands **exhibit dorsal overlapping**, species-level differentiation is difficult, and certain keys are defined by **Allen** to differentiate *A. besseyi, A. fragariae, A. ritzemabosi*
3.	*Belonolaimus*	Stylet is much longer, where the **conus part of the stylet is 70–80% of the total stylet length**, **coarse annulation** is seen on the cuticle under a microscope, two outstretched ovaries are seen in female
4.	*Bursaphelenchus*	Long and slender nematode with set-off lip, rounded stylet knobs, **spicules** are with **disk-like expansion** at the distal end, known as **rostra**, gubernaculum absent, presence of **oval shape caudal alae** in male
5.	*Criconema*	Typical **ornamentation on body cuticles in ring-like structures** with spines, anchor shape stylet knobs are present
6.	*Ditylenchus*	Stylet is short in mushroom-feeding species while longer in plant-feeding species, abutting esophagus, **leptoderan bursa** in males, mature **females slightly swollen** with a single ovary
7.	*Dolichodorus*	Amalgamated esophagus and coarse annulation on cuticle are seen, presence of **wing-like tri-lobbed bursa** and spicules with flanges in male, female with two ovaries and **spike-like tail terminus**
8.	*Globodera*	Vermiform juveniles, 400–500 mm in length, cephalic framework is prominent, females exhibit sexual dimorphism with the **globular body in the adult stage**, and two-ovary condition is seen, curved spicules are seen in male and bursa absent
9.	*Heterodera*	Vermiform juveniles, 400–500 mm in length, cephalic framework is prominent, females exhibit sexual dimorphism with a **lemon-shaped body in adult stage** with **prominent vulval cone**, curved spicules are seen in male and bursa absent
10.	*Hemicycliophora*	Degenerate mouth parts in male; stylet appears in juvenile fourth stage; **spicules are hook, sickle, or U shaped**; female with extra loose cuticular layer and median vulva with a transverse slit
11.	*Helicotylenchus*	Female body coils in a **spiral shape**, paired ovaries are found; males are rare, but if present, the body does not form a spiral shape
12.	*Hirschmanniella*	Body length is near 4 mm, strong labial sclerotization, and the most important key feature is the presence of a **pin-like structure called mucro** at the tail tip
13.	*Hoplolaimus*	Labial framework is heavy, and **tulip shape stylet knobs** are present

(continued)

Table 5.3 (continued)

S. no.	Genus	Important morphological characteristics
14.	*Longidorus*	Body length is around 4–8 mm, at the basal end of the odontophore, **flanges are not present**, presence of a **guiding ring at the anterior end of the stylet**, females have two reflexed ovary
15.	*Meloidogyne*	Vermiform juveniles; the cephalic framework is less prominent than cyst nematodes; females exhibit sexual dimorphism with **melon-shaped bodies in the adult stage**, and two-ovary condition is seen; bursa absent in male
16.	*Nacobbus*	Females exhibit sexual dimorphism with **spindle-shaped to globose-shaped bodies in the adult stage,** phasmids are punctiform, and a small bursa overlapping tail is present
17.	*Paratrichodorus* & *Trichodorus*	**Cigar-shaped body** with round configuration at both ends, bursa present in *Paratrichodorus* and absent in *Trichodorus*, cuticle swells in water for Paratrichodorus and does not swell in *Trichodorus*
18.	*Pratylenchus*	**Ventral esophageal glands overlapping** seen; in females, a single-ovary condition is seen; and males have overlapping bursa
19.	*Paratylenchus*	**Smallest in length among all plant parasitic nematodes** (body length less than 0.5 mm), stylet is very strong and prominent, degenerate mouthparts in males, females have single outstretched ovary
20.	*Radopholus*	**Head is offset** by a slight constriction; **dorsal overlapping** is seen in the esophagus; females have two ovaries and spherical spermatheca, where rod-shaped sperm are stored; presence of bursa in male
21.	*Scutellonema*	Body shape is typical, either C shape or spiral, dorsal overlapping is seen in the esophagus, females have two ovaries, and epiptygma is present in the vulval region, **phasmid is larger in size** and attains a shield-like shape, known as ***Scutella***, hence the name of the genus is *Scutellonema*
22.	*Subanguina*	Female body swollen, morphological features are similar to *Anguina*, but **oocytes in female ovary are arranged in two rows** around a rachis
23.	*Tylenchorhynchus*	**Abutting esophagus** in female with two outstretched ovaries, males have well-developed spicules with vellum
24.	*Rhadinaphelenchus*	Morphological features are **almost similar** to *Bursaphelenchus*, **body length** is greater than the *Bursaphelenchus* and gives a slender appearance to the body
25.	*Rotylenchulus*	Mature female is **reniform or kidney shape**, dorsal esophageal glands overlap the intestine ventrally, disintegrated mouth parts are seen in male, and bursa present
26.	*Xiphinema*	Body length is around 4–8 mm, at the basal end of the odontophore, **flanges are present**, presence of **guiding ring at the base of the odontostyle**, females have one ovary

relationships among nematodes. Besides, molecular characters are less prone to observational biases and outnumber morphological ones, thus providing a larger set of characters for nematode systematics studies. An evaluation of morphological and molecular approaches for nematode identification is presented in Table 5.4.

Molecular methods include techniques like polymerase chain reaction (PCR) and its variants like quantitative polymerase chain reaction (qPCR), DNA sequencing, restriction fragment length polymorphism (RFLP), random amplified polymorphic DNA (RAPD), and isoenzymes (Blok 2005). Isozyme phenotyping can separate proteins based on weight, charge, or conformation, which may be a characteristic of various groups of nematodes. For example, isozymes such as carboxylesterase/esterase were used for discriminating species of root-knot nematodes. In addition, superoxide dismutase, malate dehydrogenase, and glutamate oxaloacetate transaminase have also been utilized for the identification of nematode species

Table 5.4 A comparison between traditional and molecular approaches for nematode diagnosis and taxonomy

Parameters	Traditional approach	Molecular approach
Laboratory setup	Laboratory setup requires less space and cost	Laboratory setup requires significant investment for buying instruments and needs larger space for the lab
Need of technical experts	A skilled person with knowledge of nematode taxonomy is required	Initial training is required to execute protocols
Processing of samples	Needs immediate processing of samples, as nematodes are obligate parasites, they cannot survive without their host, and on the slides, their body degrades, and morphological features may be lost	Once the DNA, RNA, and proteins are isolated, they can be stored in freezers for a later use, or processed immediately
Time required to process a sample	Needs ample time, as it involves numerous steps from the collection of samples to slide preparation	Fast and time-saving
Dependency on the life stages of samples	Morphological studies are mostly based on the key characteristics found in males and females; juveniles are a bit difficult to identify	Any stage-egg, juvenile, or adult-can be used to isolate nucleic acid or protein; thus, the molecular method is not dependent on life stages
Species-level identification	Tough	Easy
Study of gamma taxonomy (evolutionary history)	Tough and lengthy procedure	Possible
Study of phylogeographic history (spread of nematodes)	Tough	Possible
Reproducibility of the results	Often varies from lab to lab	Once the protocol is standardized, the results are reproducible across labs

(Blok and Powers 2009). Serological/immunodiagnostic techniques use polyclonal antibody (PCA) and monoclonal antibody (MCA) raised against specific nematode antigens. Examples of applications of PCA and MCA include the identification of cyst and root-knot nematode species (Siddiqi 1997). Cytogenetic techniques depend on chromosome number and are used for higher taxa classification or identification of pathotypes/species. For example, Heteroderidae and Meloidogynidae have a basic chromosome number of $n = 9$ and $n = 18$ and may have polyploid variants of $n = 18, 36, 54$, whereas Criconematoidea has $n = 5$ and Dolichodoroidea $n = 8$, respectively (Siddiqi 1997). Two species of *Radopholus*, i.e., *R. similis citrophilus* ($n = 5$), could be distinguished from *R. similis similis* ($n = 4$) by having different chromosome numbers (Siddiqi 1997). Each of these methods have inherent pros and cons and must be used depending on the purpose of the study and available resources (Abebe et al. 2011).

The advances in DNA sequencing of diagnostic rDNA region–based methods have made the earlier DNA-based techniques such as RFLP, RAPD, and AFLP redundant (Powers, 2004). Scientists have often used genes/markers of their choice to diagnose nematodes and study the taxonomy of their favorite nematode group by outlining molecular operational taxonomic units (MOTUs) (Powers et al. 2011). The most used genetic locus for nematode diagnosis is the ribosomal gene and its various sub-components, such as 18S, 28S, internal transcribed spacer regions ITS1 and ITS2, and intergenic spacer regions IGS1 and IGS2. In addition, mitochondrial marker genes such as cytochrome oxidase C subunit 1 (*cox-1*) are also heavily used as DNA barcodes for efficient, speedy, and viable alternatives for accurate identification of nematodes in surveys and environmental assessments (Abebe et al. 2011). A high-throughput modification of barcoding is DNA metabarcoding, which is amplicon-based high-throughput sequencing (HTS), i.e., metagenomic analysis of soil samples, the results of which may be directly used for DNA taxonomy (Hebert et al. 2003; Fonseca et al. 2010). The reference databases used for metabarcoding are the same as barcoding (Taberlet et al. 2012; Cristescu 2014).

The surest method to diagnose a nematode is the use of sequence-characterized amplified region (SCAR) markers and other species-specific markers. SCARs are DNA markers designed from specific DNA sequences for a particular nematode species. Several species-specific and other markers are available for various PPNs, a few of which are listed in Table 5.5.

For unknown specimens, sequencing of ITS, 28S rDNA (D2-D3 expansion region), or 18S rDNA is the best approach. Quantitative real-time PCR is also extensively used for nematode species identification (Braun-Kiewnick and Kiewnick 2018). Clear Detections, a Dutch company, offers numerous diagnostic kits for 34 nematode species (https://www.cleardetections.com/index.php/service/nematode-real-time-pcr-diagnostic-kits/). Most recently, loop-mediated isothermal amplification (LAMP) has come up as a technology for nematode detection and diagnosis. LAMP assays are used for the detection of animal and human parasitic helminths (Deng et al. 2019). To run a LAMP reaction, a uniform temperature is required (isothermal condition). Between 4 and 6 oligonucleotide primers are used per reaction, and within 30 min to 1 h, huge amounts of amplified DNA are generated using

Table 5.5 Species-specific and other markers used for the diagnosis of important genera of PPNs

S. no.	Nematode	Host crops	Type of markers	Marker sequences (5′-3′)	References
1.	*Aphelenchoides besseyi*	Rice	KASP markers	1. **PC2 marker sets** **Forward 1: 5′**GACGATCGATTTGCACGTCAGAC**3′** **Forward 2: 5′**AGACGATCGATTTGCACGTCAGA**3′** **Common reverse: 5′**AGGAAACTCTGGTGGAAGT CCGA**3′** 2. **Ab1b marker sets** **Forward 1: 5′**GCGTATCGCACGATCTTGTGTC**3′** **Forward 2: 5′**GCGTATCGCACGATCTTGTGTT**3′** **Common reverse: 5′**TTGATAGGGCAGATACTCGA ATGAAACTT**3′**	Devran and Goknur (2020)
2.	*Meloidogyne graminicola, M. oryzae, M. salasi*	Rice	SCAR markers	**Mg F: 5′** TTCGACTCTGTAGGAAGCC **3′** **Mg R: 5′** CAAAAGTAACCGGACACTCTTTT **3′** **Mo F: 5′** CCAGCATCCGTGTTGTAT **3′** **Mo R: 5′** GCTCCAGGTGAAAAG **3′** **Ms F: 5′** CAAAGAACGGGGTTTATTCG **3′** **MS R: 5′** GGTTATCCGAACTCCCCAAT **3′**	da Silva Mattos et al. (2019)
3.	*Anguina tritici*	Wheat	Species-specific markers	**TW81: 5′**GTTTCCGTAGGTGAACCTGC **3′** **AB28: 5′**ATATGCTTAAGTTCAGCGGGT**3′**	Ami et al. (2019)
4.	*Globodera rostochiensis, G. pallida*	Potato	Species-specific primers	**UNI: 5′**GCAGTTGGCTAGGGATCTTC **3′** **GPA1 5′** GGTGACTCGACGATTGCTGT **3′** **GRO1 5**-ATGTTGTACGTGCCGTACCTT **3′**	Trayanov et al. (2020)
5.	*Heterodera avanae, H. filipjevi*	Wheat	Species-specific primers	**AVEN-COI-forward: 5′** GGGTTTTCGGTTATTTGG **3′** **AVEN-COI-reverse 5′**-CGCCTATCTAAATCT ATA CCA **3′**	Toumi et al. (2013)
6.	*Ditylenchus dipsaci, D. gigas, D. destructor*	Various crops including onion, garlic, oat, rye, etc.	Species-specific primers	**DITuniF: 5′** CTGTAGGTGAACCTGC **3′** **DITdipR: 5′** GACATCACCAGTGAGCATCG **3′** **DITgigR: 5′** GACCACCTGTCGATTC **3′** **DITdesR: 5′** GTTTTTCGCCCACAAATTAGC **3′**	Jeszke et al. (2015)

7.	*Pratylenchus thornei*	Wheat and chickpea	Species-specific primers	**Pthf: 5′** TTC GGAAGACAATAAATC **3′** **Pthr: 5′** TCC AAAATGAAATAATAAA **3′** **Pthsat f1: 5′** CCCGATTCGGATTGAATGCG **3′** **Pthsat r1: 5′** CCCGATTCG GAAAGG GACGA **3′**	Carrasco-Ballesteros et al. (2007)
8.	*Rotylenchulus* spp.	Various crops like pulses, oil crops, field crops, fruit crops, and plantation crops	RT-PCR specific markers	**Forward primer 5′**AAAAACCGGCTTAATTGCAAT GGT **3′** **Reverse primer 5′**TTGTTGACGAGTTGTTGCGTC **3′**	Sayler et al. (2012)

Bst polymerase (Notomi et al. 2000; Tomita et al. 2008). The LAMP-based assays have been shown to be 10–1000 times more sensitive than the conventional PCR. LAMP is cost-effective as compared to PCR as it does not require expensive equipment, and assay readout can be done by using many easy methods combinable with lateral flow–based methods. This makes LAMP a method of choice deployable at point-of-care. Eighteen LAMP assays are available for various PPNs such as *Bursaphelenchus xylophilus*, *B. cocophilus*, and various species of root-knot nematodes (Niu et al. 2011, 2012; Peng et al. 2017; Zhou et al. 2017; Zhang and Gleason 2019), *Aphelenchoides besseyi*, *A. ritzemabosi*, *A. wevelli*, *A. agrostis*, *Radopholus similis*, *Ditylenchus destructor*, and *Tylenchulus semipenetrans*. The details are provided in Table 5.6.

Table 5.6 A list of LAMP assays available for the diagnosis of various PPNs

S. no.	Nematode	Target gene for detection	Reference
1.	*Bursaphelenchus xylophilus*	Internal transcribed spacer of ribosomal DNA	Kikuchi et al. (2009)
2.	*Meloidogyne javanica*, *M. incognita*, *M. arenaria*, *M. hapla*	18srDNA-ITS1 regions and Minc08407 gene sequence	Niu et al. (2011)
3.	*Meloidogyne enterolobii*	5S rDNA-IGS2 region	Niu et al. (2012)
4.	*Radopholus similis*	Internal transcribed spacer of ribosomal DNA	Peng et al. (2012)
5.	*Bursaphelenchus xylophilus*	Pectate lyase-3 (PEL3) genes sequence	Kang et al. (2015)
6.	*Bursaphelenchus xylophilus*	Expansin cDNA/RT-LAMP	Leal et al. (2015)
7.	*Tylenchulus semipenetrans*	Internal transcribed spacer of ribosomal DNA	Lin et al. (2016)
8.	*Tylenchulus semipenetrans*	18S-ITS1-5.8S-ITS2-28S rDNA regions	Song et al. (2017)
9.	*Bursaphelenchus cocophilus*	D2-D3 LSU region	Ide et al. (2017)
10.	*Meloidogyne hapla*	Internal transcribed spacer of ribosomal DNA	Peng et al. (2017)
11.	*Meloidogyne mali*	5.8S ribosomal DNA	Zhou et al. (2017)
12.	*Anguina wevelli*	Internal transcribed spacer of ribosomal DNA	Yu et al. (2018)
13.	*Bursaphelenchus xylophilus*	*syg*-2, synaptogenes	Meng et al. (2018)
14.	*Meloidogyne chitwoodi*, *M. fallax*	(IGS2)-18S region ribosomal rDNA	Zhang and Gleason (2019)
15.	*Aphelenchoides besseyi*	Mitochondrial cytochrome oxidase subunit I (COI) gene	Yang and Yu (2019)
16.	*Ditylenchus destructor*	28S rRNA	Ding et al. (2019)
17.	*Aphelenchoides ritzemabosi*	18S ribosomal RNA gene	Wang et al. (2019)
18.	*Anguina agrostis*	Internal transcribed spacer of ribosomal DNA	Yu et al. (2020)

In the last few years, another new isothermal PCR-based method, the recombinase polymerase amplification method, has emerged for sensitive, reliable, and rapid detection of different PPNs in point-of-care settings with several benefits over PCR and other isothermal techniques like LAMP. RPA is able to detect the target species from the crude DNA of nematodes from infested soil or plant samples at ambient/room temperature (37–42 °C) within 15–20 min, thereby eliminating the use of complex and expensive temperature-controlling equipment. The flexibility of RPA in association with diverse detection methods, simplicity in operation, non-specificity in the heat source, and its accuracy and sensitivity over other molecular techniques make it a revolutionizing point-of-care diagnostic technique for PPNs. Real-time RPA for *Meloidogyne enterolobii* was published (Subbotin 2019), targeting the IGS rRNA gene, allowing its detection from the gall and crude nematode extracts. This assay was very sensitive and detected the presence of 1/10 of a second-stage juvenile after 15–20 min. RPA has since been developed for several other nematode genera, listed in Table 5.7.

Table 5.7 RPA-based diagnostic assays available for various PPNs

Nematode species	RPA method	References
Meloidogyne enterolobii	Basic RPA, real-time RPA	Subbotin (2019)
M. incognita, M. javanica, M. arenaria, M. enterolobii	Basic RPA	Ju et al. (2019)
M. javanica Tropical root-knot nematode complex (*M. incognita, M. javanica, M. arenaria*) and *M. javanica* specific	Basic RPA, LFD-RPA LFD-RPA, Real-time RPA	Chi et al. (2020) Subbotin and Burbridge (2021b)
Bursaphelenchus xylophilus	RPA- POID	Cha et al. (2020)
Ditylenchus destructor	Real-time RPA	Gao et al. (2022)
Heterodera schachtii	LFD-RPA CRISPR/Cas12-RPA	Yao et al. (2021)
Meloidogyne hapla	LFD-RPA Basic RPA, LFD-RPA, real-time RPA	Song et al. (2021) Subbotin and Burbridge (2021a)
Aphelenchoides besseyi	CRISPR/Cas12a RPA	Zhang et al. (2022)
Globodera rostochiensis	RPA-LFA	Wang et al. (2022)
Heterodera avenae	CRISPR/Cas12-RPA	Shao et al. (2022)

5.3 Conclusion, Challenges, and Future Prospects

The count of classical nematode taxonomists has reached an all-time low in the scientific institutions of the world, and the identification of nematodes by morphological methods is a dying art (Abebe et al. 2011; Floyd et al. 2002). Molecular methods are quicker than morphological methods for nematode diagnosis. At present, in a field situation, the nematodes are first diagnosed locally by examination of plant tissue and symptomatology and then sent for confirmation of identity to the nearest molecular laboratory. However, most of the available methods cannot be used for rapid field diagnosis or for point-of-service detection and diagnosis of PPNs. The isothermal methods such as LAMP and RPA are quick, robust, and sensitive, and could be customized as kits for use in the field and point-of-care labs such as quarantine laboratories. Newer technologies, such as CRISPR-based assays, are now developed for the detection of other pathogens but are yet to be used for PPN diagnosis.

References

Abebe E, Mekete T, Thomas WK (2011) A critique of current methods in nematode taxonomy. Afr J Biotechnol 10(3):312–323

Ami SN, Taher IE, Hussen FS, Ahmed AI (2019) First molecular identification of wheat seed gall nematode *Anguina tritici* races parasitised on wheat in Iraq. Acta Univ Sapientiae Agric Environ 11(1):5–15

Blok VC (2005) Achievements in and future prospects for molecular diagnostics of plant-parasitic nematodes. Can J Plant Pathol 27(2):176–185

Blok VC, Powers TO (2009) Biochemical and molecular identification. In: Perry RN, Moens M, Starr J (eds) Root-knot nematodes. CAB International, Wallingford, pp 98–118

Braun-Kiewnick A, Kiewnick S (2018) Real-time PCR, a great tool for fast identification, sensitive detection and quantification of important plant-parasitic nematodes. Eur J Plant Pathol 152(2):271–283

Carneiro RMDG, de Oliveira Lima FS, Correia VR (2017) Methods and tools currently used for the identification of plant parasitic nematodes. In: Shah MM, Mahamood M (eds) Nematology—Concepts, diagnosis and control, pp 19–35

Carrasco-Ballesteros S, Castillo P, Adams BJ, Perez-Artes E (2007) Identification of *Pratylenchus thornei*, the cereal and legume root-lesion nematode, based on SCAR-PCR and satellite DNA. Eur J Plant Pathol 118(2):115–125

Cha D, Kim D, Choi W, Park S, Han H (2020) Point-of-care diagnostic (POCD) method for detecting *Bursaphelenchus xylophilus* in pinewood using recombinase polymerase amplification (RPA) with the portable optical isothermal device (POID). PLoS One 15(1):e0227476

Chi YK, Zhao W, Ye MD, Ali F, Wang T, Qi RD (2020) Evaluation of recombinase polymerase amplification assay for detecting *Meloidogyne javanica*. Plant Dis 104(3):801–807

Cobb NA (1914) Nematodes and their relationships. US Government Printing Office, pp 457–490

Cristescu ME (2014) From barcoding single individuals to metabarcoding biological communities: towards an integrative approach to the study of global biodiversity. Trends Ecol Evol 29(10):566–571

da Silva Mattos V, Mulet K, Cares JE, Gomes CB, Fernandez D, de Sa MFG, Castagnone-Sereno P (2019) Development of diagnostic SCAR markers for *Meloidogyne graminicola, M. oryzae*, and *M. salasi* associated with irrigated rice fields in Americas. Plant Dis 103(1):83–88

Deng MH, Zhong LY, Kamolnetr O, Limpanont Y, Lv ZY (2019) Detection of helminths by loop-mediated isothermal amplification assay: a review of updated technology and future outlook. Infect Dis Poverty 8(1):20

Devran Z, Goknur A (2020) Development and validation of a SNP-based KASP assay for rapid identification of *Aphelenchoides besseyi* Christie, 1942. Crop Prot 136:105235

Ding SW, Yang SH, Wu WJ, Xie H, Xu CL (2019) Rapid diagnosis of *Ditylenchus destructor* by loop-mediated isothermal amplification assay based on 28S rRNA sequences. Eur J Plant Pathol 153(4):1165–1175

Floyd R, Abebe E, Papert A, Blaxter M (2002) Molecular barcodes for soil nematode identification. Mol Ecol 11(4):839–850

Fonseca VG, Carvalho GR, Sung W, Johnson HF, Power DM, Neill SP, Creer S (2010) Second-generation environmental sequencing unmasks marine metazoan biodiversity. Nat Commun 1(1):1–8

Gao B, Ma J, Li X, Wang R, Chen S (2022) A real-time recombinase polymerase amplification assay for fast and accurate detection of *Ditylenchus destructor*. Mol Cell Probes 61:101788. https://doi.org/10.1016/j.mcp.2021.101788. Epub Dec 22. PMID: 34954062.

Hafez SL, Palanisamy S, MacGuidwin AE (2020) Nematode management. In: Stark JC, Thornton M, Nolte P (eds) Potato production systems. Springer International Publishing, pp 259–282

Hebert PD, Cywinska A, Ball SL, Dewaard JR (2003) Biological identifications through DNA barcodes. Proc Roy Soc B Biol Sci 270(1512):313–321

Ide T, Kanzaki N, Giraldo PPP, Giblin-Davis RM (2017) Loop-mediated isothermal amplification (LAMP) for detection of the red ring nematode, Bursaphelenchus cocophilus. Nematology 19(5):559–565

Jain RK, Mathur KN, Singh RV (2007) Estimation of losses due to plant parasitic nematodes on different crops in India. Indian J Nematol 37(2):219–221

Jeszke A, Dobosz R, Obrępalska-Stęplowska A (2015) A fast and sensitive method for the simultaneous identification of three important nematode species of the genus *Ditylenchus*. Pest Manag Sci 71(2):243–249

Jones JT, Haegeman A, Danchin EG, Gaur HS, Helder J, Jones MG, Perry RN (2013) Top 10 plant-parasitic nematodes in molecular plant pathology. Mol Plant Pathol 14(9):946–961

Ju Y, Lin Y, Yang G, Wu H, Pan Y (2019) Development of recombinase polymerase amplification assay for rapid detection of *Meloidogyne incognita, M. javanica, M. arenaria*, and *M. enterolobii*. Eur J Plant Pathol 155(4):1155–1163

Kang JS, Kim AY, Han HR, Moon YS, Koh YH (2015) Development of two alternative loop-mediated isothermal amplification tools for detecting pathogenic pine wood nematodes. For Pathol 45(2):127–133

Khan MR (2015) Nematode diseases of crops in India. In: Recent advances in the diagnosis and management of plant diseases. Springer, New Delhi, pp 183–224

Kikuchi T, Aikawa T, Oeda Y, Karim N, Kanzaki N (2009) A rapid and precise diagnostic method for detecting the pinewood nematode *Bursaphelenchus xylophilus* by loop-mediated isothermal amplification. J Phytopathol 99(12):1365–1369

Kumar V, Khan MR, Walia RK (2020) Crop loss estimations due to plant-parasitic nematodes in major crops in India. Natl Acad Sci Lett 43:409–412

Leal I, Allen E, Foord B, Anema J, Reisle C, Uzunovic A, James D (2015) Detection of living *Bursaphelenchus xylophilus* in wood, using reverse transcriptase loop-mediated isothermal amplification (RT-LAMP). For Pathol 45(2):134–148

Lin B, Wang H, Zhuo K, Liao J (2016) Loop-mediated isothermal amplification for the detection of *Tylenchulus semipenetrans* in soil. Plant Dis 100(5):877–883

Meng F, Wang X, Wang L, Gou D, Liu H, Wang Y, Piao C (2018) A loop-mediated isothermal amplification-based method for detecting *Bursaphelenchus xylophilus* from *Monochamus alternatus*. For Pathol 48(2):e12404

Niu JH, Guo QX, Jian H, Chen CL, Yang D, Liu Q, Guo YD (2011) Rapid detection of *Meloidogyne* spp. by LAMP assay in soil and roots. J Crop Prot 30:1063–1069

Niu J, Jian H, Guo Q, Chen C, Wang X, Liu Q, Guo Y (2012) Evaluation of loop-mediated isothermal amplification (LAMP) assays based on 5S rDNA- IGS2 regions for detecting *Meloidogyne enterolobii*. Plant Pathol 61:809–819

Notomi T, Okayama H, Masubuchi H, Yonekawa T, Watanabe K, Amino N, Hase T (2000) Loop-mediated isothermal amplification of DNA. Nucleic Acids Res 28(12):e63

Peng H, Peng D, Hu X, He X, Wang Q, Huang W, He W (2012) Loop-mediated isothermal amplification for rapid and precise detection of the burrowing nematode, *Radopholus similis*, directly from diseased plant tissues. Nematology 14(8):977–986

Peng H, Long H, Huang W, Liu J, Cui J, Kong L, Peng D (2017) Rapid, simple and direct detection of *Meloidogyne hapla* from infected root galls using loop-mediated isothermal amplification combined with FTA technology. Sci Rep 7:44853

Powers T (2004) Nematode molecular diagnostics: from bands to barcodes. Annu Rev Phytopathol 42:367–383

Powers T, Harris T, Higgins R, Mullin P, Sutton L, Powers K (2011) MOTUs, morphology, and biodiversity estimation: a case study using nematodes of the suborder *Criconematina* and a conserved 18S DNA barcode. J Nematol 43(1):35

Sayler RJ, Walker C, Goggin F, Agudelo P, Kirkpatrick T (2012) Conventional PCR detection and real-time PCR quantification of reniform nematodes. Plant Dis 96(12):1757–1762

Shao H, Jian J, Peng D, Yao K, Abdulsalam S, Huang W, Kong L, Li C, Peng H (2022) Recombinase polymerase amplification coupled with CRISPR-Cas12a technology for rapid and highly sensitive detection of *Heterodera avenae* and *Heterodera filipjevi*. Plant Dis 10:1094

Siddiqi MR (1997) Techniques and methodologies for nematode disease diagnosis and nematode identification. FAO Plant Production and Protection Paper, 144, pp 21–44

Singh S, Singh B, Singh AP (2015) Nematodes: a threat to sustainability of agriculture. Procedia Environ Sci 29:215–216

Song ZQ, Cheng JE, Cheng FX, Zhang DY, Liu Y (2017) Development and evaluation of loop-mediated isothermal amplification assay for rapid detection of *Tylenchulus semipenetrans* using DNA extracted from soil. J Plant Pathol 33(2):184

Song Z, Yang X, Zhang X, Luan M, Guo B, Liu C, Pan J, Mei S (2021) Rapid and visual detection of *Meloidogyne hapla* using recombinase polymerase amplification combined with a lateral flow dipstick assay. Plant Dis 105(9):2697–2703

Southey JF (1986) Laboratory methods for work with plant and soil nematodes. Her Majesty Stationary Office, Norwich, pp 206

Subbotin SA (2019) Recombinase polymerase amplification assay for rapid detection of the root-knot nematode, *Meloidogyne enterolobii*. Nematology 21:243–251

Subbotin SA, Burbridge J (2021a) Sensitive, accurate and rapid detection of the northern root-knot nematode, *Meloidogyne hapla*, using recombinase polymerase amplification assays. Plants 10(2):336

Subbotin SA, Burbridge J (2021b) Recombinase polymerase amplification assays for detection of the major tropical root-knot nematodes. J Nematol 53:1

Taberlet P, Coissac E, Pompanon F, Brochmann C, Willerslev E (2012) Towards next-generation biodiversity assessment using DNA metabarcoding. Mol Ecol 21(8):2045–2050

Tomita N, Mori Y, Kanda H, Notomi T (2008) Loop-mediated isothermal amplification (LAMP) of gene sequences and simple visual detection of products. Nat Protoc 3(5):877–882

Toumi F, Waeyenberge L, Viaene N, Dababat A, Nicol JM, Ogbonnaya F, Moens M (2013) Development of two species-specific primer sets to detect the cereal cyst nematodes *Heterodera avenae* and *Heterodera filipjevi*. Eur J Plant Pathol 136(3):613–624

Trayanov K, Samaliev H, Kostova M, Bojinov B, Besheva A (2020) Morphological and molecular identification of potato cyst nematodes *Globodera rostochiensis* and *Globodera pallida* in Bulgaria. Bulg J Agric Sci 26(2):416–422

Wang DW, Xu CL, Bai ZS, Li JY, Han YC, Zhao LR, Xie H (2019) Development of a loop-mediated isothermal amplification for rapid diagnosis of *Aphelenchoides ritzemabosi*. Eur J Plant Pathol 155(1):173–179

Wang X, Lei R, Peng H, Jiang R, Shao H, Ge J, Peng D (2022) Rapid diagnosis and visual detection of potato cyst nematode (*Globodera rostochiensis*) using recombinase polymerase amplification combination with lateral flow assay method (RPA-LFA). Agronomy 12(10):2580

Yang JI, Yu GY (2019) A loop-mediated isothermal amplification assay for the plant-parasitic nematode *Aphelenchoides besseyi* in rice seedlings. J Nematol 51:1–11

Yao K, Peng D, Jiang C, Zhao W, Li G, Huang W, Kong L, Gao H, Zheng J, Peng H (2021) Rapid and visual detection of *Heterodera schachtii* using recombinase polymerase amplification combined with Cas12a-mediated technology. Int J Mol Sci 22(22):12577

Yeates GW, Bongers TD, De Goede RGM, Freckman DW, Georgieva SS (1993) Feeding habits in soil nematode families and genera—an outline for soil ecologists. J Nematol 25(3):315

Yu LZ, Song SY, Yu C, Qi LJ, Yu ZX, Jiao BB, Yang J (2018) A loop mediated isothermal amplification (LAMP) assay for rapid and reliable detection of *Anguina wevelli*, a grass parasitic nematode. Eur J Plant Pathol 150(3):725–734

Yu L, Song S, Yu C, Jiao BB, Tian YM, Li YJ (2020) Rapid detection of *Anguina agrostis* by loop-mediated isothermal amplification. Eur J Plant Pathol 156(3):819–825

Zhang L, Gleason C (2019) Loop-mediated isothermal amplification for the diagnostic detection of *Meloidogyne chitwoodi* and *M. fallax*. Plant Dis 103(1):12–18

Zhang A, Sun B, Zhang J, Cheng C, Zhou J, Niu F, Luo Z, Yu L, Yu C, Dai Y, Xie K, Hu Q, Qiu Y, Cao L, Chu H (2022) CRISPR/Cas12a coupled with recombinase polymerase amplification for sensitive and specific detection of *Aphelenchoides besseyi*. Front Bioeng Biotechnol 10:912959

Zhou QJ, Cai Y, Gu JF, Wang X, Chen J (2017) Rapid and sensitive detection of *Meloidogyne mali* by loop-mediated isothermal amplification combined with a lateral flow dipstick. Eur J Plant Pathol 148:755–769

Diagnosis and Detection of Seed-Borne Pathogens in Field Crops

6

Seweta Srivastava, Ravindra Kumar, Chhattar Pal, Raghavendra Reddy Manda, Vaibhav Kumar Singh, and H. Prashanth Babu

Abstract

The befitting detection and identification of casual agents responsible for the seed-borne diseases of crop plants are considered to be the most important issue in formulating management strategies. Since diseased seeds frequently show no symptoms, it is critical to diagnose them early in order to prevent pathogen spread through long-distance germplasm exchange. In order to manage diseases of crops, it is crucial to conduct seed health testing in order to identify seed-borne pathogens. By doing so, financial losses and unnecessary fungicide use can be avoided, which lowers costs and reduces the environmental impact of such hazardous chemicals. The primary criteria for choosing seed health test procedures are specificity, sensitivity, speed, ease of use, affordability, and dependability. Conventional methods for identifying fungus transmitted by seeds

S. Srivastava (✉)
School of Agriculture, Lovely Professional University, Phagwara, Punjab, India
e-mail: seweta.21896@lpu.co.in

R. Kumar
ICAR—Indian Institute of Wheat and Barley Research, Karnal, Haryana, India

C. Pal
Regional Sericultural Research Station, Central Silk Board, Miran Sahib, Jammu and Kashmir, India

R. R. Manda
Wageningen University and Research, Wageningen, The Netherlands

V. K. Singh
Wheat Pathology Laboratory, Division of Plant Pathology, ICAR—Indian Agricultural Research Institute, Pusa, New Delhi, India

H. Prashanth Babu
Division of Genetics, ICAR—Indian Agricultural Research Institute, Pusa, New Delhi, India

V. K. Singh et al. (eds.), *Diseases of Field Crops: Diagnostics and Management*,
https://doi.org/10.1007/978-981-97-6160-9_6

rely on the processes of incubation and grow-out. These are lengthy, entail microbiological abilities, and are occasionally not sensitive enough to low levels of seed infection while being widely utilized due to their ease of application. Because of their high levels of specificity and sensitivity, novel identification approaches based on DNA analysis have been used frequently and have proven to be quite efficient. During the identification of fungi, bacteria, viruses, and other seed-borne diseases, a variety of diagnostic techniques—from eye inspection to PCR-based molecular methods—are employed in accordance with ISTA criteria. Conventional PCR is the simplest and most widely used method; more modern approaches include nested PCR, which eliminates low concentrations of specific pathogens; multiplex PCR, which detects many pathogens at once; real-time PCR, which measures fungus on seeds; and magnetic capture hybridization-PCR. The incapacity to differentiate between essential and non-essential inocula and the challenge of getting high-quality DNA templates because of PCR inhibitors in seeds are the primary limitations of molecular techniques. Numerous modified PCR processes, including loop-mediated isothermal amplification, and non-destructive testing techniques have been designed to lessen the inhibitory effect. Considering the many benefits they offer, loop-mediated isothermal amplification and next-generation sequencing have found extensive use in nucleic acid analysis. In the future, their use for the identification of fungal infections in seeds may be greatly expanded.

Keywords

Diagnosis · Detection methods · Field crops · Seed-borne microbes · Management strategies

6.1 Introduction

Since seed is the defender of a cultivar's genetic potential, its level of quality dictates the maximum productivity that may be achieved in a particular cropping system. Growers, dealers, and consumers have been more conscious of seed health in the past few decades, particularly in the aftermath of the GATT era. With the establishment of the WTO, seed health has become a worldwide concern (Upadhyay and Singh 2019; Srivastava et al. 2021). There is an enormous risk of infectious diseases and pest dissemination associated with the transportation of plant germplasm and any other raw plant material (Gergerich et al. 2015). Specifically, seeds are an especially effective means of spreading diseases that are transmitted through seeds (Gupta and Kumar 2020). In addition to the diseases brought on by seed-borne infections, their connection frequently causes a decrease in seed germination and a reduction in the robustness of seedlings (Kumar et al. 2014; Kumar et al. 2023a, b), ultimately resulting in an unhealthy field stand and significant financial losses (Singh et al. 2017). Diseased seeds are typically asymptomatic, in contrast to diseased vegetative plant tissues. Low levels of seed infection can still cause significant crop losses (Pellegrino et al. 2010), especially if the pathogens are quarantine bacteria. One of the main ways to lessen the entrance of diseases into fields is to use seeds that are without disease inoculum. This also helps to lower the creation of inoculum, the effectiveness of secondary spread to the inoculum threshold, and the impact of the disease shortly after it has been established. It is recognized that a minimum of 10–15% rise in output can be attributed to the excellence of the seeds alone. All production-related parameters that impact seed viability, disease-free status, and genetic purity must be considered in order to reach this exceptional level of quality (Gaur et al. 2020).

Pathogens have a snug relationship with seeds, which helps them survive over time and spread widely (Kumar and Gupta 2020). Legislation designed to restrict or stop the spread of exotic infections into unfamiliar regions has been developed in many nations, and these laws are typically backed by detection methods. However, it is challenging to screen seeds for diseases due to low inoculum levels and the frequently uneven distribution within seed batches (Mancini et al. 2016). Numerous harmful microorganisms, including viruses and phytoplasma, are transmitted by seeds. By identifying these pathogens during the tissue culture stage, significant crop loss can be avoided (Patel 2021).

6.2 The Seed: A Microcosm of Microbes

According to Sinclair (1979), seeds are a microcosm of microorganisms, capable of containing a diverse range of fungi, bacteria, viruses, and occasionally nematodes, many of which can infect seedlings or plants and cause illnesses. The process of harvesting, threshing, and processing might lead to infestation or contamination of the seed (Sawicka 2019). The main reasons for threshing losses include seed splitting, insufficient separation from the chaff, and cracked seeds from excessive impact force (Baloch 2010; Shah 2013). There are three possible routes in which the disease can travel with the seeds.

1. ***Admixture***: Although they are not part of seeds, pathogens do accompany them. For instance, during the threshing process, ergot sclerotia are combined with healthy seeds.
2. ***External***: Pathogens can appear on seed surfaces as spores, oospores, or chlamydospores, as in the case of downy mildew in pearl millet, covered smut in barley, and Karnal bunt in wheat.
3. ***Internal***: Pathogens that have a specific link to certain components of the seed, such as fungal incitant that causes loose smut in wheat, infect the wheat embryo.

6.3 Detection Methods for Seed-Borne Pathogens

The following techniques, which combine traditional and contemporary approaches, are employed to identify diseases transmitted by seeds. Due to the rapid spread of plant diseases caused by huge populations of sensitive plants, high relative humidity, high temperatures, and overhead irrigation, there is a significant danger of disease-related economic losses in greenhouse environments. Exclusion is a highly successful disease management technique in these circumstances, and it is achieved by screening and removing contaminated seed lots prior to planting through the use of seed detection assays (Walcott 2003). The technology for detecting seeds is examined in the table below, which also includes new developments. A summary of the features of each assay is presented (Table 6.1).

6.3.1 Conventional Seed Detection Assays

It might be challenging to test seeds for plant diseases. Infested seeds may not show any symptoms, unlike infected vegetative plant tissues, which makes visual detection difficult (Kumar et al. 2020). Furthermore, there could not be many pathogens on the seeds, and the infestation of the seeds might not be spread evenly throughout the lot. Few detection assays meet the minimal requirements for sufficient seed testing, despite the fact that many are available for various seed-borne diseases (Mancini et al. 2016; Kumar and Gupta 2020). A seed assay ought to be easy to use and understand, fast, robust, specific, and affordable (Tsedaley 2015). Numerous technologies, such as eye inspection, selective media, seedling grow-out tests, and serological procedures, have been used to construct seed assays. Even though these exams have been around for a while, there are a few flaws that make them less than perfect. Table 6.2 presents a tabular discussion on the brief descriptions of seed detection assays, along with an analysis of their benefits and drawbacks.

Table 6.1 General characteristics of the different seed detection assays and their effectiveness in detecting seed-borne pathogens

Diagnostic method	Time required	Sensitivity	Specificity	Examples of pathogens detected	References
Visual examination	5–10 min	Low	Low	Phomopsis spp., Cercospora kikuchii, Peronospora manshurica/soybean seed; Cylindrocladium parasiticum/peanut seed	Walcott (2003)
Seed washing technique	10–20 min	Low	Low	*Tilletia* spp./wheat seeds, *Peronospora manshurica*/soybean seed	Walcott (2003)
Freeze blotter incubation	5–7 days	Low/ moderate	Moderate	Leptosphaeria maculans/ Brassicaceae seed	Walcott (2003)
Seedling grow-out assay	2–3 weeks	Low	Low	Seed-borne fungal pathogens of groundnut viz., Macrophomina phaseolina, Aspergillus niger and Aspergillus flavus	Walcott (2003)
Semi-selective media (agar medium incubation)	2–14 days	Moderate	Low–moderate	Alternaria dauci, Alternaria radicina, Alternaria carotiincultae/carrot seed; Verticillium dahliae, Fusarium spp./Cucurbitaceae seed; *Botrytis* spp./ onion seed	Walcott (2003); Mancini et al. (2016)
Serology-based detection	2–4 h	Moderate–high	Moderate–high	Macrophomina phaseolina/cowpea seed	Zhang et al. (2023)
Conventional DNA extraction and polymerase chain reaction (PCR)	5–6 h	High	Very high	Alternaria brassicae, Leptosphaeria *maculans*/ Brassicaceae seed; *Ascochyta lentis*/ lentil seed; *Fusarium oxysporum* f. sp. *basilici*/basil seed	Mancini et al. (2016)

(continued)

Table 6.1 (continued)

Diagnostic method	Time required	Sensitivity	Specificity	Examples of pathogens detected	References
BIO-PCR (selective target colony enrichment followed by PCR)	3–4 days	Very high	Very high	Alternaria brassicae, Leptosphaeria maculans/ Brassicaceae seed; *Ascochyta rabiei*/ chickpea seed	Walcott (2003); Mancini et al. (2016)
Nested PCR	5–6 h	Very high	High	Colletotrichum lindemuthianum/ bean seeds	Danesh et al. (2014)
Real-time PCR	40–60 min	Very high	Very high	Alternaria brassicae, Plasmodiophora *brassicae*/ Brassicaceae seed	Schaad and Frederick (2002)
IMS-PCR (immunomagnetic separation and PCR)	2–5 h	Very high	Very high	Colletotrichum/bean leaves	Walcott and Gitaitis (2000)
MCH-PCR (magnetic capture hybridization and PCR)	2–5 h	Very high	Very high	*Didymella bryoniae*/ Cucurbitaceae seed; *Botrytis* spp./onion seed	Munkvold (2009)
DNA microarray	6 h	Very high	Very high	*Pantoea ananatis*/ bacterial leaf spot of maize, maize dwarf mosaic virus (MDMV), sugarcane mosaic virus (SCMV)	Krawczyk et al. (2017)
Laser biospeckle technique	High	High	High	Fusarium oxysporum, Aspergillus flavus, *Sclerotinia* spp./ bean seed	Singh et al. (2021)
Videometer	High	High	High	Seed-borne pathogens of soybean namely *Phomopsis sojae* and *Colletotrichum dematium* var. *truncata*	Rego et al. (2020)

Table 6.2 Brief descriptions of conventional seed detection assays including their advantages and disadvantages

Conventional seed detection assay	Discovered by	Procedure	Advantage	Disadvantage
Detection of seed-borne fungal pathogens				
Direct inspection of seeds without incubation				
Visual examination	de Tempe and Binnerts (1979)	In order to document the observations on the mixture of other crop seeds, weed seeds, plant parts, inert matter, discolorations, malformations, and fungal structures like encrusted mycelium, sclerotia, galls, bunt balls, acervuli, pycnidia, perithecia, and spore masses, the seed sample first has to be examined with the naked eye before being viewed under a stereoscopic binocular microscope (Srivastava et al. 2020)	In this assay, reducing seedlot infestation can be done by utilizing automatic seed sorting equipment that classifies seeds according to their physical or visual attributes (Paulsen 2002; Walcott et al. 1998)	The assay often exhibits a low sensitivity of detection. Furthermore, seeds contaminated by bacteria, viruses, or fungi may not show any macroscopic signs, rendering visual or physical examination of the seeds ineffective as a detection method (Walcott 2003)
Washing test	Keitreiber (1984)	The inoculum (oospores of downy mildew fungi, teliospores of smuts and bunts, etc.) occurring on the seed surface can be identified by the washing test (Upadhyay and Singh 2019)	Beneficial for detecting the presence of seed-borne fungus present on the seed surface or for testing surface-borne pathogenic fungi (Agarwal et al. 2006; Rao et al. 2006)	The ISTA has not yet authorized a standard working sample for the washing test, which is a qualitative test. As previously indicated, the washing test is typically performed to identify whether fungal propagules are present or absent (International Seed Testing Association 2001)

(continued)

Table 6.2 (continued)

Conventional seed detection assay	Discovered by	Procedure	Advantage	Disadvantage
Incubation methods				
Blotter method	Doyer (1938) and de Tempe (1953)	It is a commonly used technique for determining the health of seeds. It is frequently used in labs as the initial screening test for the process of testing for seed health. According to Srivastava et al. (2020), fungi linked to seeds are closely inspected and detected using "Habit characters."	When looking for fungi that can develop mycelial development and fruiting structures under the test's incubation conditions, the blotter method is frequently employed. This test is used to identify seed infection, and under some circumstances, it can also identify seedling infection (Kumar et al. 2020). This approach is used to evaluate all types of seeds, including grains, vegetables, legumes, ornamentals, and forest seeds	The early growth of some fungi over the slower-growing ones may make examination difficult. One cannot identify pathogenic microbes. That takes a lot of time
Deep-freezing blotter method (DFB)	Limonard (1968)	Following the blotter method's instructions for plating seeds, the plates are incubated for 24 h at 20 °C, then moved to a freezer set at −20 °C for another 24 h. Finally, they are incubated for 5 days at 20 ± 2 °C under cold white, fluorescent light, with 12 h of light and 12 h of darkness alternated Elwakil and Ghoneem (2002)	Deep-freezing does not affect the fungi associated with the seeds as the imbibed seeds on moist blotters are killed (Srivastava et al. 2020)	To destroy the embryo, an extra step in the deep-freeze blotter test involves soaking the seeds for 24 h and then keeping them at −20 °C for another 24 h (Du Toit et al. 2005; Elwakil et al. 2009). On seeds, deep-freezing blotter promotes the growth and development of yeasts and saprophytic bacteria (Neergaard 1979)

Standard moist blotter (SMB) method	Hashmi (1988)	Using sterile tap water (pH 7.2), 25 seeds were plated in a 9-cm-diameter Petri dish with three layers of damp blotter paper. Elwakil and Ghoneem (2002) reported that the plates were incubated at 20 ± 2 °C for 7 days under cold white fluorescent light, with alternate cycles of 12 h of light and 12 h of darkness	The majority of seed pathogens can be detected using this method (Khare et al. 1977)	According to Neergaard (1979), SMB produces saprophytes, which frequently significantly impede the growth of parasitic fungi
2,4-D method	Neergaard (1956a, b)	Neergaard (1956a, b) initially employed 2,4-D in the blotter test to screen for *Phoma lingam* in cabbage seeds, but Hagborg et al. (1950) had previously used it in an agar medium to screen for the anthracnose-causing fungus *Colletotrichum lindemuthianum* on bean seeds	Because 2,4-D slows down seed germination and seedling growth, seeds remain intact, and disease detection is made simpler. International Seed Testing Association (1966) has approved this procedure as a routine test for several diseases that are transmitted through seeds	By partially suppressing the target fungi, external decontamination can lower the occurrence of the infection. In order to get around these problems, excessive amounts of seed infection can be found without compromising the presence of pathogens by applying heat treatment procedures for one hour at 100 °C. This also lessens the seed microflora's masking effects (Soteros 1979)

(continued)

Table 6.2 (continued)

Conventional seed detection assay	Discovered by	Procedure	Advantage	Disadvantage
Rolled paper towel method	Warham (1990)	The rolled paper towel method can be utilized to examine the germination and vigor of both infected and seemingly healthy seed batches in order to determine the impact of a seed-borne inoculum on seed quality characteristics (Srivastava et al. 2020)	This technique is mostly utilized to identify *Ascochyta* disease in peas and *Fusarium* spp. in cereals (Sharma et al. 1987). It is believed that the seed lot with the higher seed vigor index is healthier (Abdul Baki and Anderson 1973)	The main disadvantage to the paper towel germination method is the need for immediate transplant. The root extending from the seed is highly sensitive; even slight touching of the root can cause seed death
Agar plate method	Muskette and Malone (1941)	Although Czeapak's dox agar and potato dextrose agar are the most often used media, selective media provide a more dependable method for pathogen identification. For example, Mangan (1971) modified the agar plate method to isolate *Phoma betae* and named the method "*Hold fast method*." The method helps to observe hold-fast-like structures or hyphal clumps of *P. betae*	The agar plate method provides an efficient tool for the quick identification of specific seed infections ones the analyst has become familiar with the colony characteristics of the fungi in question (Srivastava et al. 2020)	Colony characteristics of the test pathogen are likely to change if another agar medium is employed or if the incubation conditions are changed

Seedling symptom tests or growing-on tests				
Hiltner's brick stone method	Hiltner (1917)	Filling plastic containers up to one-fourth of their volume with sterile crushed brick stone with a maximum grain size of 3–4 mm is the recommended method. Soak the crushed brick stone in water and then plant the seeds one centimeter deep in the pot. After 2 weeks, the seedlings are removed from the containers, and they are left in the darkness at room temperature with observations made for disease symptoms. It was primarily created to provide information on seedling symptoms for field performance	It is also used for testing of treated seeds	The main disadvantage of all types of seedling symptom tests is that these methods are particularly useful when seedlings show symptoms in a short time otherwise are not able to give proper results (Upadhyay and Singh 2019). Sometimes, it may also take longer time than other conventional methods (Vannacci et al. 2014)
Sand method	Endo (1961)	Rhizoctonia solani isolates, both pathogenic and non-pathogenic, can be identified using this technique, which involves seeding sterile quartz sand on culture plates and irrigating them with sterile water	When compared to "roll towel" testing, the results from sand media are more precise and repeatable, particularly when applied to seed lots that have undergone extensive chemical treatment or are older	

(continued)

Table 6.2 (continued)

Conventional seed detection assay	Discovered by	Procedure	Advantage	Disadvantage
Standard soil method	Karlberg (1974)	Plastic multipot trays are filled with a pre-sterilized, consistent soil mixture that includes four parts clay, six parts peat, and the necessary amount of fertilizer. To keep the moisture in after sowing, they are covered with a polyethylene bag. Depending on a number of factors, including the type of seed and temperature, the symptoms might be seen anywhere from 2–4 weeks after incubation	Used soil can be sterilized by heating it to 95 °C for 20–30 min. This allows it to be used repeatedly for seed planting	
Test tube agar method	Khare et al. (1977)	This technique was primarily created to find *Septoria nodorum* in wheat grains	Additionally, it is useful for identifying different seed-borne fungal genera of *Colletotrichum* spp., *Fusarium* spp., and *Drechslera* spp., particularly of barley and legumes	
Embryo count method	Hewett (1970)	For microscopic studies, the embryo must be separated from the remainder of the seed when the inoculum of a fungus, such as *Ustilago nuda* or *U. tritici*, is present (Rennie 1982)	Hewett (1970) developed this unique technique for separating the embryo in order to find loose smut of wheat and barley	By eliminating the need for phenol and lowering the quantity of embryos analyzed for each seed sample, this procedure has been refined (Khanzada et al. 1989; Cappelli et al. 1993)

Staining tests	Welty and Rennie (1984)	When inoculum is present in the internal tissues of the seed and it is not feasible to detect the fungi in a reasonable amount of time using the previously mentioned conventional methods (Upadhyay and Singh 2019), some staining compounds specific for fungal hyphae, such as trypan blue, aniline blue, and rose bengal, are used (Srivastava et al. 2020)	These techniques are used to identify seed-borne infections that are biotrophs, or that may grow on artificial substrate, but only very slowly. These pathogens typically lack the ability to form fruiting structures (like Ustilaginales) or take a long time to develop (like *Septoria*), often taking over 10 days (Marcinkowska 2002)	Since different species have varying capacities for sporulation, it may take several days to obtain results (for example, for *Colletotrichum lindemuthianum* and *Alternaria* spp.) or even longer than 2 weeks if additional mycelium structure characteristics (such as chlamydospores from *Phoma* spp. and microsclerotia from *Verticillium* spp.) are required for identification (Marcinkowska 2002)
Serological technique				
Monoclonal antibody technique	Irwin (1987)	Fungi can only be detected by serological procedures with the use of monoclonal antibody techniques	Monoclonal assays have been shown to identify *Pyrenophora graminea* (Burns et al. 1994) in barley as well as seed-borne fungus of rice (Dewey 1992) and spruce (Mitchell 1988)	Considering the kind of immunogen most likely to produce a highly specific polyclonal antiserum, opinions differ widely (Chu et al. 1989)

(continued)

Table 6.2 (continued)

Conventional seed detection assay	Discovered by	Procedure	Advantage	Disadvantage
Detection of seed-borne bacterial phytopathogens				
Visual separation of infected seed	Wharton (1967)	Many seeds that do not exhibit significant encrustation will go unnoticed when using the visual separation method. When inspected under near ultraviolet light, seeds that are not easily identifiable as infected under standard lighting conditions exhibit small, diffused luminous patches (Srivastava et al. 2020)	By concentrating contaminated seeds into small subsamples and increasing the likelihood of their detection, selecting all fluorescent seeds from large samples for bacteriological testing can significantly boost the sensitivity of the testing method (Wharton 1967)	According to Parker and Dean (1968), many seeds that do not exhibit significant encrustation will go unnoticed when using the visual separation method
Phage method	Katznelson and Sutton (1951)	The method relies on the growth and multiplication of particular bacteriophage particles when a vulnerable bacterium is present. With this technique, it was possible to find and identify seed-borne *Pseudomonas phaseolicola* or *Xanthomonas campestris* pv. *phaseoli* (Katznelson and Sutton 1951)	Outstanding specificity, a long track record of performance, and an assay that is well-established and well-characterized. Cheap and simple instrumentation is not required (Smartt and Ripp 2011)	It is time-consuming and arduous, necessitates maintaining big phage sets, can be difficult to interpret results, and needs a culturable phytobacterial host strain (Smartt and Ripp 2011)

Enzyme-linked immunosorbent assay (ELISA)	Candlish et al. (1988)	In order to make a distinction between strains of *Pseudomonas syringae* pv. *pisi*, the causal organism of bacterial blight on pea seeds, extremely sensitive monoclonal antibodies were first created and used in both indirect and competitive ELISA assays	Incredibly specialized and labor-saving because of its automation potential. permits the identification of bacterial poisons as well (Law Jodi et al. 2015)	The primary disadvantages include their limited sensitivity, false-negative results, and occasionally the potential for cross-reactivity with closely similar antigens (Law Jodi et al. 2015)
	Ball and Reeves (1991)	The commercial test for the detection of bacterial seed-borne illness was primarily created and marketed.		
	Rajeshwari et al. (1998)	Created an ELISA method that is quick, sensitive, and specific for identifying isolates of *Ralstonia solanacearum* in tomato seeds		

(continued)

Table 6.2 (continued)

Conventional seed detection assay	Discovered by	Procedure	Advantage	Disadvantage
Immunofluorescence microscopy	Malin et al. (1983) (indirect immunofluorescence microscopy) Franken and van Vuurde (1990) (Direct immunofluorescence microscopy)	It is another crucial test and an extensively advised immunodiagnostic technique for identifying bacterial infestations in seeds (van Vuurde 1987). Through the use of homologous antibodies conjugated with fluorescent dyes (such as fluorescein isothiocyanate), antigens can be detected under ultraviolet light via both indirect and direct immunofluorescence cell staining techniques. The amount of bacteria present in the seed preparation is correlated with fluorescence. For instance, *Pseudomonas syringae* pv. *phaseolicola* was found by direct immunofluorescence, while *Xanthomonas campestris* pv. *phaseoli* was discovered in bean seeds by an indirect approach (Srivastava et al. 2020)	For routine application, IF has the benefit of being inexpensive and able to be completed in a matter of days by technicians with little background in microbiology	Antiserum-to-saprophytic microbe cross- or non-specific responses pose the most hazards in immune function failure (IF)

Immunofluorescence colony-staining (IFC) method	van Vuurde (1997)	The routine indexing and quantitative assessment of pathogenic seed-borne bacterial field thresholds are accomplished through the use of the immunofluorescence colony-staining (IFC) method	Compared to conventional isolation and the previously mentioned serological tests, the IFC assay is more sensitive and specific. The assay is simple to run, and visual scoring or an image analyzer makes it easy to evaluate the data. 4–5 days are needed for results, whereas 30–45 days are needed for conventional procedures. The resultant bacterial culture can be preserved for later use and analyzed via PCR or host infectivity. It can be employed as an extremely sensitive detection method and has potential use in localization studies when combined with highly specific antibodies (commercially available monoclonal and recombinant antibodies) (Veena and van Vuurde 2002)	However, selecting the appropriate secondary conjugate is essential to achieving the best assay findings (Veena and van Vuurde 2002)

(continued)

Table 6.2 (continued)

Conventional seed detection assay	Discovered by	Procedure	Advantage	Disadvantage
Growing-on test	Srinivasan et al. (1973)	The seedlings are cultivated from infected seeds in environments that are best suited for the development of disease symptoms in the developing seedlings (Srivastava et al. 2020)	Seeds are sown in greenhouses using appropriate substrates (such as agar tubes, sand, or other sterile substrate) and various ambient conditions (Upadhyay and Singh 2019)	The primary drawback of all seedling symptom tests is their limited ability to yield accurate results. These approaches are most helpful when seedlings exhibit symptoms quickly thereafter (Upadhyay and Singh 2019). Sometimes, it may also take longer time as compared to other methods (Vannacci et al. 2014)
Detection of seed-borne viral pathogens				
Dry examination	Phatak (1974)	A thorough visual examination of the seed may reveal whether a virus is present (Srivastava et al. 2020)	Nonetheless, it has occasionally proved to be a useful diagnostic tool. For instance, there is a positive link between the morphology of healthy and infected barley stripe mosaic virus-infected seeds, with the former showing pronounced variations in texture, shape, and size (Phatak and Summanwar 1967)	Visual inspection of dry seed is likely the least effective method for determining the presence of virus in seed, since the relationship between seed abnormalities and seed infection is inconsistent (Bryant 1981)

Biological methods				
Growing-on tests	Srinivasan et al. (1973)	The seedlings are cultivated from infected seeds in environments that are best suited for the development of disease symptoms in the developing seedlings (Srivastava et al. 2020)	Bean common mosaic virus (BCMV) in beans, bean yellow mosaic virus (BYMV) in peas, soybean mosaic virus (SMV) in soybeans, etc. can all show symptoms when a seed grows in moderately heated and light environments. However, certain viruses—like barley stripe mosaic virus (BSMV)—need particular light and temperature conditions to grow (Srivastava et al. 2020)	These tests are not practicable for routinely screening large batches of commercial seed due to the significant time and space requirements. Additionally, the strong effects of host cultivar, virus strain, and climate on symptom development in these bioassays frequently make detection more difficult (Bryant 1981)
Infectivity test	Pelet and Gagnebin (1963) and Pelet (1965)	Test plants are injected with a test seedling extract, which may include ungerminated or germinated seeds (Srivastava et al. 2020)	This biological method can be used to detect the presence of the bean common mosaic virus (BCMV) in bean seeds (Srivastava et al. 2020)	
Biochemical/ histochemical methods	Phatak (1974)	This approach is predicated on the idea that viral infections alter host tissues' biochemistry and physiology. Phatak (1974) used 1% aqueous phloxine solution to stain infected tissues in order to show viral inclusions	The detection of pinwheel and bundle inclusions in the cytoplasm of tissues infected with viruses belonging to the Poty virus group serves as an example of how such techniques identify the pathogen solely to its group (Edwardson 1966)	Usually sufficient just to measure the degree of field stock deterioration, not to precisely screen for seed infection (Rollings 1965)

(continued)

Table 6.2 (continued)

Conventional seed detection assay	Discovered by	Procedure	Advantage	Disadvantage
Serological test				
Enzyme-linked immunosorbent assay (ELISA)	Clark and Adams (1977)	The broad applicability of ELISA, an acronym that was first used by Engvall and Perlmann (1971) to characterize the fundamental characteristics of the assay, has led to its use in the serological diagnosis of various viral diseases, including pea seed-borne mosaic virus (Hamilton and Nichols 1978), barley stripe mosaic virus (Lister et al. 1978), and soybean mosaic virus (Lister 1978)	Incredibly specialized and labor-saving because of its automation potential. Permits the identification of bacterial toxins as well (Law Jodi et al. 2015)	Their low sensitivity, false-negative results, and occasionally the possibility of cross-reactivity with closely similar antigens are their key disadvantages (Law Jodi et al. 2015)
Dot-ELISA (dot-blot, dot immunobinding, or spot immunodetection assay)	Lange (1986)	An updated version of the ELISA method is this one (Lange 1986; Stead 1992). The modified dot assay was designed to identify proteins by identifying the antigen on a nitrocellulose membrane, developing the membrane in 4-chloro-1-naphthol, and then treating the membrane with a test antibody, a second antibody coupled with peroxidase, and the first antibody (Abd El-Aziz 2019)	Compared to previous ELISA techniques, this indirect method proved to be marginally more sensitive for pea seed-borne mosaic virus (Courtioux et al. 2005). Due to its extremely tiny sample volume (2 μL compared with 250 μL for ELISA), as noted by Powell (1987) and Abd El-Aziz and Younes (2019) (4 μL compared with 250 μL for ELISA), DBIA has the benefit of being able to detect significantly smaller quantities of virus	The dot-blot analysis is only useful if the antibody's specificity—that is, if it reacts only with the target antigen—is known. An additional possible disadvantage of the DBIA is the requirement for a high amount of 50 mL of moderately concentrated (1 mg/mL) viral antiserum (Abd El-Aziz 2019)

Tissue blotting immunoassay (TBIA)	Makkouk and Kumari (1996)	Razor blades were used to cut the tissue of both healthy and infected seeds. The exposed cut edges had been followed by pressing onto nitrocellulose membrane, and the entire process was completed using a dot blot after blocking buffer (2% bovine serum albumin and 2% Triton x-100 solution in carbonate buffer, pH 9.6) and was left for an hour at room temperature. The blots' development of purple hue indicated that the assay had produced a positive result. No color or a green color formed in the negative reaction (Mostafa and Abou-Ela 2011)	It is proposed that this test is more practical than standard ELISA and can be used to detect viruses that are transmitted through seeds once they have germinated. It is manageable in less than 4 h without compromising on delicacy. It is less expensive and does not need specialized infrastructure (Abd El-Aziz 2019)	However, due to the somewhat laborious sample preparations required in comparison to ELISA and PCR, its limitations are mostly related to automation challenges and limited application to large-scale diagnostic procedures. To turn TBIA into a high-throughput screening technique for plant virus identification, sample preparation needs to be improved (Han et al. 2007)

(continued)

Table 6.2 (continued)

Conventional seed detection assay	Discovered by	Procedure	Advantage	Disadvantage
Gel diffusion techniques				
Double diffusion test	Scott (1961) followed by Hamilton (1965)	The seed that has to be examined is often steeped in tap water. Subsequently, the entire seed or a portion of it (like the embryo) is subjected to a trituration process, and the resultant mixture is moved to a well segmented within a diffusion medium. Agar gel is a typical diffusion medium. Next, a different well is filled with an antiserum that is particular to a virus that may have infected the seed. The antibody molecules and the virus particles (antigen) gradually converge on one another. Double diffusion is the name given to this diffusion because it is occurring in two different directions (Srivastava et al. 2020)	In broad terms, the double diffusion approach is fairly sensitive and quite specific. Concentrations of 10–25 μg/mL of virus can be detected by it (Carroll 1979)	Gel diffusion, for all its popularity, might not always provide an acceptable resolution. To further minimize erroneous reactions, tissue extracts must frequently be clarified before testing. Since many viruses only sometimes produce excellent results (Garnsey et al. 1979) or do not reach sufficiently high concentrations in tissues to yield meaningful results (Landon 1977), the relatively low sensitivity level reached with gel diffusion testing remains a severe restriction. Another restriction states that before testing in immunodiffusion, non-isometric viruses must be physically or chemically broken to reduce the particles to diffusible protein subunits (Langenberg and Ball 1972; Purcifull and Batchelor 1977; Carroll et al. 1979; Simmonds and Gumming 1979)
Radial or single diffusion test	(Shepherd 1972)	The single or radial diffusion method is similar to the double diffusion technique except that only the virus antigen diffuses from a well out into the agar diffusion medium. Normally, a subunit antiserum is used with the radial diffusion method and, therefore, only degraded viruses can be detected (Shepherd 1972)	This method is also very sensitive and rapid. It can detect as little as 1 μg/mL of degraded virus (approx. a ten-fold greater sensitivity than conventional double-diffusion test). In a matter of minutes or a few hours, results can be obtained with pre-made agar diffusion plates. This method has been used for the detection of seed-borne barley stripe mosaic virus (Slack and Shepherd 1975)	

Agglutination technique				
Chloroplast test	Jermoljev and Chod (1966)	The first application of this method was to find BCMV in bean seeds that had germinated; the existence of the virus is indicated by the clumping of chloroplasts (Jermoljev and Chod 1966)	Take less time, are simple to execute, and have a high level of analytical sensitivity. It is capable of simultaneously detecting a huge diversity of antibodies	When the quantity of antibodies exceeds that of antigens, agglutination responses may be prevented. Prozone phenomenon is the term used to describe this kind of inhibition (Singer and Plotz 1956). At most, the responses are semi-quantitative
Latex test	Phatak (1974)	For everyday tasks, this strategy is among the most promising. In order to conduct this test, antibodies or antigens are bound to the latex beads—an artificial carrier made of polystyrene—surface. Latex beads range in size from 0.8 to 1 μm. In the past, BSMV in barley, soybean mosaic virus in soybean, bean yellow mosaic virus in broad bean, and BCMV in soybean were all found using this latex agglutination test (Srivastava et al. 2020)		

(continued)

Table 6.2 (continued)

Conventional seed detection assay	Discovered by	Procedure	Advantage	Disadvantage
Biophysical methods				
Electron microscopy (EM)	Gold et al. (1954)	Detection of viruses in seed health can be done with an electron microscope. Barley stripe mosaic virus (BSMV) particles had previously been observed by Gold et al. (1954) using an electron microscope on barley seeds	The fact that EM does not require organism-specific reagents to recognize the pathogenic agent constitutes one of the primary benefits of employing EM for viral diagnosis (Goldsmith and Miller 2009)	As multiple viruses have the same morphological class, it is challenging to identify the virus with precision using electron microscopy (Wolff and Bárcena 2021)
Immunosorbent electron microscopy (ISEM)	Derrick (1973)	According to Roenhorst et al. 2018), ISEM combines electron microscopy and serology. Using this technique, the grid is coated with a diluted antiserum and then rinsed. Next comes the trapping phase, where the grid is incubated with the virus solution and the virus particles are bound. The layer of antibody and other serum components, which may be even more significant (Lesemann et al. 1980), prevent non-specific binding of structures other than the viral antigen that may be present in the sample. The virus on the grid is selectively immunopurified as a result	The ability to see and count the virions on the grid was likely the advantage of ISEM, as opposed to the ELISA technique, which had a relatively high background from healthy seeds and thus performed poorly in differentiating between healthy and diseased samples, particularly in cases where the virus's concentration in the seeds was extremely low (Baker et al. 1985)	Unquestionably, one drawback of ISEM for mass screening is that it requires more effort than ELISA (Milne 1991)

Contact radiography with "Soft" X-rays	Phatak and Summanwar (1967)	Soft X-rays are electromagnetic waves that have wavelengths between 1 and 100 nm and energy between 0.12 and 12 keV. Soft X-rays are a good tool for assessing agricultural products because of their low penetration power and capacity to show variations in internal density (Rahman and Cho 2016). Barley seeds infected with BSMV and runner bean (Phaseolus multiflorus) seeds infected with runner bean mosaic virus (RBMV) could be distinguished from one another based on lower opacity to X-ray penetration (Srivastava et al. 2020)	The widely used method known as "soft X-ray imaging" generates an X-ray image in 3–5 s. In the seed industry, soft X-ray imaging has started to be utilized to find interior holes and flaws (Neethirajan et al. 2006; Mathanker et al. 2013). Because X-rays are short electromagnetic waves (range from 0.01 to 10 nm) with high penetrating strength (Kotwaliwale et al. 2014), they have shown tremendous potential for data collection from interior structures (e.g., damages in embryo and endosperm) (Arkhipov et al. 2019). The seed samples receive less X-ray radiation because of the shorter exposure period. In summary, quick image capture and the shorter exposure duration of X-ray projection imaging are helpful for the prompt examination of seed samples (Ahmed et al. 2020)	Prior to image feature extraction, the obtained X-ray pictures were processed using image processing techniques such as noise reduction, contrast, and intensity augmentation (Ahmed et al. 2020)

(continued)

6.4 From Conventional to Molecular

Conventional procedures such as direct inspection of dry seeds, filter and centrifuge extraction technique, washing, soaking, incubation, blotter, and embryo count tests are the main means of detecting seed-borne diseases (Castro et al. 1994). According to Majumder et al. (2013), the aforementioned methods were regularly used until the last few decades of the twentieth century. In most circumstances, a visual examination might not prove sufficient to reliably identify a pathogen invasion in seeds. By cracking open the seed, one can inspect the interior contents for signs of illness (Michelle and Stephen 2010). Seed-borne infections can also exist on seeds even in the absence of overt signs of illness (Agarwal and Sinclair 1997). As a result, numerous detection techniques for different seed-borne diseases have been developed over time. However, the identification of morphological traits, microscopical observation, biochemical characterization, and cultural assets continue to be crucial in the detection of seed-borne diseases (López et al. 2015).

In order to circumvent and enhance these restricted accesses, a nucleic acid–based detection technique has recently been developed. For the precise and quick identification of seed-borne diseases, nucleic acid–based molecular techniques—specific, quick, and dependable—were developed (Chahal and Pannu 1997; Lopez et al. 2003). Molecular diagnostics gained a significant pace following the mid-1980s PCR introduction. For the most common seed-borne diseases affecting field crops, PCR-based diagnostic techniques have been successfully developed. Over the past few decades, there has been a significant improvement and simplification of pathogen detection and identification due to the swift advancement of genomic approaches for plant pathogen characterization (Schilling et al. 1996).

Although phytopathogens can exist at smaller DNA amounts, the higher level of confidence, accuracy, specificity, and sensitivity associated with DNA-based molecular techniques (Capote et al. 2012; Midorikawa et al. 2018) allow for the detection of phytopathogens at the initial phases of disease progression (Luchi et al. 2013; Rollins et al. 2016). These days, polymerase chain reaction–based assays, restriction fragment length polymorphism analysis, real-time polymerase chain reaction, and nucleic acid hybridization techniques are some of the most widely used nucleic acid–based methods. Particularly for pathogens identified to be spread by seeds as well as by other planting materials, the growth of diagnostic DNA probes peculiar to the pathogen(s) in question has aided in the quick detection, differentiation, and quantification of the population of pathogenic inoculum inside the seed (Hariharan and Prasannath 2021). Because nucleic acid–based approaches are faster, more precise, highly sensitive, and more comprehensive compared to conventional procedures, they are widely accepted. Real-time PCR, multiplex PCR, and bio-PCR are a few detection techniques that offer quick and accurate data analysis. Compared to culture plate testing and conventional PCR, real-time PCR has been used successfully to identify and quantify a wide range of plant pathogens with enhanced specificity (Belete 2018). Thus, in order to improve efficiency, specificity, and sensitivity in regular seed health monitoring, real-time PCR tests must be developed and implemented (Narayanasamy 2011; López et al. 2015).

6.5 Molecular Detection of Seed-Borne Microflora

6.5.1 Nucleic Acid–Based Techniques

Nucleic acid–based approaches have been recognized as one of the most practical and effective approaches to identifying pathogens among the current instruments. The majority of nucleic acid–based detection methods, especially those that use PCR, are quick, accurate, and extremely sensitive. Nevertheless, there are obstacles to the correct application of these techniques, which range from the high cost of the technology to the requirement for a sufficient understanding of the procedures to enable interpretation (Majumder et al. 2013). A number of attempts have been made to apply this technique to viruses (Lange 1986), fungi (Reeves 1995), and bacteria (Schaad et al. 1989). However, probes are not employed in routine seed health testing because contaminating DNA may interfere with accurate organism detection when seed extracts are probed directly. In this instance, it is preferable to remove organisms from seeds and refine them prior to applying the probes (Maude 1996).

6.5.1.1 Polymerase Chain Reaction (PCR)

Because of its accuracy, rapidity, and sensitivity, the polymerase chain reaction (PCR) (Mullis and Faloona 1987) is a technique that allows DNA to be amplified and multiplied up to a million times (Saiki et al. 1988; Oliver 1993). The biology and identification of fungi have primarily been accomplished by PCR techniques (Reeves 1995). A few earliest instances of the detection and diagnosis of fungi in seeds include the *Phomopsis* species found in soybean seeds (Jaccoud-Filho and Reeves 1993). Some bacterial species were also identified by using PCR techniques, such as *Pantoea* (*Erwinia*) *stewartii* in maize seeds (Blakemore and Reeves 1993) and *Pseudomonas syringae* pv. *phaseolicola* and *Pseudomonas syringae* pv. *pisi* in bean and pea seeds, respectively (Prosen et al. 1993; Rasmussen and Wulff 1991; Reeves et al. 1994).

6.5.1.2 Multiplex PCR

Due to the simultaneous amplification of several gene targets, multiplex PCR provides a faster detection than standard PCR (Law Jodi et al. 2015). Since RAPD was not appropriate for immediate identification, Stevens et al. (1997) used it in conjunction with the nuclear ribosomal unit's internal transcribed spacer (ITS) region to create a multiplex PCR seed health test that could identify and distinguish *Pyrenophora* spp. pathogenic to barley. Conventional PCR and mPCR have a similar basic concept. In contrast to the standard PCR assay, which uses just one set of specific primers, the mPCR assay uses multiple sets of specific primers. In order to create an effective mPCR assay, primer sets must have comparable annealing temperatures, which makes primer design crucial to the development of mPCR (Zhao et al. 2014). An effective mPCR assay also depends on the quantity of the PCR buffer, the proportion of deoxynucleotide to magnesium chloride, the amount of DNA template, the cycling temperatures, and the Taq DNA polymerase (Markoulatos et al. 2002; Cheah et al. 2008; Khoo et al. 2009).

6.5.1.3 Real-Time PCR

Real-time PCR has a number of significant improvements over conventional PCR (Table 6.3), which may increase its acceptability for application in routine seed testing (Walcott 2003). Combining real-time PCR with the other PCR modifications (MCH-PCR, IMS-PCR, BIO-PCR) described previously can yield substantial advantages. For the detection of *Microdochium nivale* in wheat (*Triticum* spp.) seeds and *Acidovorax avenae* subsp. *citrulli* in watermelon seeds, real-time PCR seed detection tests have been published (Taylor et al. 2002). As the technology grows more accessible, more real-time PCR seed assays are probably going to be created.

6.5.1.4 Bio-PCR

For the purpose of detecting *Pseudomonas syringae* pv. *phaseolicola* in bean seed extracts, a polymerase chain reaction (PCR) technique known as "Bio-PCR" is described. This technique combines biological and enzymatic amplification of PCR

Table 6.3 The summary of advantages and limitations of nucleic acid–based rapid detection methods for seed-borne microbes

Detection method	Advantages	Limitations	References
Conventional PCR	Highly sensitive and specific. Gives rapid and precise results when using the primer of specific species	DNA must be purified since it is impacted by PCR inhibitors. It is challenging to discern between living and dead cells	Aslam et al. (2017); Srivastava et al. (2020)
Nested PCR	Use of two sets of primers increases the yield and specificity of the amplification of the target DNA	Risk of contamination, because of two cycles of amplification	Aslam et al. (2017)
Multiplex PCR	Likewise, this method is extremely specific, automated, and sensitive. It may identify various diseases, yielding accurate findings	PCR inhibitors damage these cells, making it harder to tell viable from non-viable cells. Primer design has a critical role	Mandal et al. (2011); Zhang (2013)
Real-time PCR	High sensitivity and specificity, rapid cycling, and reproducible. Processing of post-amplification products is not necessary with this method	Extremely expensive and challenging for a multiplex real-time PCR test. Affected by PCR inhibitors as well. It can be challenging to discern between viable and non-viable cells, and cross-contamination can occasionally happen. Requires employees with training	Mandal et al. (2011); Park et al. (2014)

(continued)

Table 6.3 (continued)

Detection method	Advantages	Limitations	References
Bio-PCR	The advantages of bio-PCR have also been demonstrated in a comparison between direct PCR, double-antibody sandwich ELISA, and semi-selective media for the detection of *Xanthomonas albilineans*, the causal agent of leaf scald of sugarcane (*Saccharum* spp.). Lastly, in order for bio-PCR to identify the target organism, it must grow on the selective medium. Therefore, DNA from non-viable cells is not discovered, only viable colonies are	One of the drawbacks of bio-PCR is that every pathogen requires a semi-selective medium. As was previously said, semi-selective media typically take time to design and necessitate specialized expertise in the nutritional needs and chemical tolerances of the target organism. The inability of bio-PCR to be applied to obligatory parasites (such as viruses) is an additional significant disadvantage. It is therefore mostly restricted to bacteria and fungi that are easily cultivable	Wang et al. (1999); Walcott (2003)
Immunomagnetic separation-PCR (IMS-PCR)	Compared to traditional PCR, IMS-PCR greatly increases detection efficiency and accuracy. For the purpose of detecting seed-borne microorganisms, this method turned out to be more precise as well as sensitive than extracting DNA from hexacetyldimethylethyl ammonium bromide (CTAB), which was followed by ELISA and direct PCR	IMS-PCR has numerous benefits; however, its ability to selectively capture organisms using polyclonal antibodies limits its applicability. It is necessary to create polyclonal antibodies for every disease since they are not easily accessible. It is also doubtful that IMS-PCR would be successful in recovering the filamentous mycelia of plant pathogenic fungus because it depends on microscopic beads for pathogen capture	Ausubel et al. (1987); Walcott and Gitaitis (2000)

(continued)

Table 6.3 (continued)

Detection method	Advantages	Limitations	References
Magnetic capture hybridization-PCR (MCH-PCR)	This test has the advantages of being quick (it can be finished in a day), applicable to all pathogens, and capable of overcoming the inhibitory effects of seed chemicals. DNA capture probes can be easily synthesized and may be more accessible than polyclonal antibodies, in comparison with IMS-PCR. Thus, a wider use of this technology may be possible	MCH-PCR captures and detects DNA as opposed to viable pathogen propagules, and it would be impossible to determine whether amplified DNA originated from viable or non-viable organisms. Nevertheless, by focusing on specific mRNAs for detection, this possible issue might be resolved. The target DNA might then be amplified and detected via reverse transcription PCR. Since mRNA is only created by living cells and breaks down quickly, this would support the finding of only living cells and organisms. This method would be more expensive and technically difficult, though	Walcott (2003)
Luminex PCR	This method is employed for the multiplex detection of bacteria and viruses that are spread by seeds. It is an interesting technology for laboratory-based sample testing because of its huge multiplexing capacity, which allows up to 100 pathogens to be analyzed simultaneously in a single run, and its quick analysis	The interpretation and reporting of analyte data at very low concentrations have a constraint. The assay's apparent benefit is that, compared to traditional ELISAs, its dynamic range appears to be far wider. Numerous sensitivity problems, particularly for a number of cytokines, are still unclear at the very low concentration range	Liu et al. (2005); Peters et al. (2007); Bergervoet et al. (2008); Leng et al. (2008); De Boer and López (2012)
Oligonucleotide DNA microarray (DNA chip technology)	The lack of a need for extensive DNA sequencing makes it simple to employ	Extremely expensive, needs a lot of mRNA, oligonucleotide probes, and target gene labeling	Lauri and Mariani (2009); Mandal et al. (2011); Park et al. (2014); Aslam et al. (2017)

(continued)

Table 6.3 (continued)

Detection method	Advantages	Limitations	References
Nucleic acid sequence–based amplification (NASBA)	Costly apparatus is not required. Superior to RT-PCR	Enzymes that are thermolabile are necessary for the reactions to be specific. It is impossible to raise the reaction temperature above 42 °C without compromising it	Lauri and Mariani (2009); Zhao et al. (2014)
Loop-mediated isothermal amplification (LAMP)	Because the LAMP assay offers a greater degree of specificity and sensitivity than other traditional PCR techniques, it is frequently employed to identify diseases that are transmitted by seeds. Less costly, simpler to use, and does not need a heat cycling system	The intricacy of primer design renders it inadequate for identifying unfamiliar or unsequenced targets	Prasannakumar et al. (2020); Prasannakumar et al. (2021)
Laser biospeckle technique	Researchers are paying more attention to laser biospeckle analysis because of its many built-in benefits, such as its non-invasive imaging, affordability, and quick and easy procedures	A few of the limitations discovered include the instability of the solid-state lasers, which were utilized to minimize the setup's size, as well as the impact of outside light and vibration. A further obstacle related to the portability and easy use of the devices is the standardization of the procedures to maintain repeatability	Braga Jr (2017); Singh et al. (2021)
Videometer	A videometer was used to examine the ability to distinguish between non-infected and infected lots of cereal seeds that were infected with various pathogens (such as different species of *Fusarium* or *Tilletia* sp.). The methodology of the videometer is based on light reflection. To determine whether this technique could distinguish between infected and non-infected seed lots, a variety of seed lots were examined	Since it is an untargeted approach, any fungi present in a sample should be detected in theory	Olesen et al. (2011); Mancini et al. (2016); Srivastava et al. (2020)

targets with streamlined sample processing methods (Schaad et al. 1995). In order to remove seed-borne bacteria, seed samples are either smashed or cleaned in a suitable buffer. Semi-selective media is covered with aliquots of the seed wash, and the mixture is then incubated for two to three days. PCR is then carried out using particular primers following the collection of colonies and DNA extraction. When it comes to fungi that spread through seeds, the first step in the process is to introduce the desired fungus mycelial mass into the seeds followed by high relative humidity incubation. This is then followed by DNA extraction and PCR (Pryor and Gilbertson 2001).

6.5.1.5 IMS-PCR

This technique involves the selective binding of target cells from solutions having a mixture of cells employing small magnetic beads coated with antibodies specific to a given pathogen (Olsvik et al. 1994; Walcott and Gitaitis 2000). After that, the cells can either be utilized immediately for DNA extraction and PCR testing, or else they can be grown on specific media to produce more of the target pathogen (Upadhyay and Singh 2019). Following immunocapture, the IMBs are completely cleaned to remove inhibitors and nontarget microorganisms, and they are immobilized using a magnet. After that, IMBs can be placed on a selective agar medium and left to incubate until visible bacterial colonies form. Alternatively, DNA suitable for PCR can be extracted from collected bacteria by boiling them (Walcott 2003).

6.5.1.6 MCH-PCR

The procedures of polymerase chain reaction and magnetic capture hybridization are identical to those of IMS-PCR. But the main difference between the two methods is that MCH-PCR concentrates and captures particular DNA fragments that can be utilized as PCR templates by using single-stranded DNA probes (Walcott 2003). This technique involves capturing DNA fragments for PCR using magnetic beads covered with single-stranded DNA probes. Using materials containing PCR-inhibiting chemicals, this approach has been utilized to detect viruses, bacteria, and fungi (Jacobsen 1995). Although this method is quick and can get past the inhibitory effects of seed chemicals, it is unable to differentiate between cells that are viable and those that are not (Upadhyay and Singh 2019).

6.5.1.7 Luminex PCR

Since 2008, Luminex has been using their innovative technique to diagnose plant pathogens (Ishii et al. 2008). The Luminex PCR technology works by allowing an oligonucleotide probe attached to separate beads to hybridize with a biotin-labeled PCR product to detect variations in nucleotide sequences. A substitute to ELISA is the Luminex xMAP technique, which uses immunobeads, which are paramagnetic microspheres coated in antibodies and internally fluorescently labeled. As tiny ELISA wells, these beads function (Mancini et al. 2016).

6.5.1.8 DNA Microarray

Many oligo nucleotides are arranged on a tiny glass or silica surface (chip) in DNA chip technology (Lemieux et al. 1998; Vernet 2002). The DNA of the target pathogen is broken down into tiny pieces, which are then hybridized with oligonucleotides on the DNA chip and marked with fluorescent markers. Fluorescence is a sign that the target pathogen is present in the chip. According to Upadhyay and Singh (2019), DNA chip technology offers the potential to identify many diseases simultaneously and within a short period of time. The ability to link hundreds or thousands of oligonucleotides to precise spots on every single chip is what gives this approach its power. These oligonucleotides can be used to identify pathogens in seed samples because they complement DNA sequences specific to some microbes (Walcott 2003).

6.5.1.9 Nucleic Acid Sequence–Based Amplification (NASBA)

Compton (1991) created a unique nucleic acid amplification technique called nucleic acid sequence–based amplification (NASBA), which is capable of amplifying RNA. It was created to amplify nucleic acids without the use of a heat cycler at the beginning of the 1990s (Compton 1991; Fakruddin et al. 2012). A battery of three enzymes (avian myeloblastosis virus reverse transcriptase, RNase H, and T7 RNA polymerase) is used in this sensitive transcription-based amplification system, which produces single-stranded RNA as the primary amplification product. The system has been designed for the specific detection of RNA (Rodriguez-Lazaro et al. 2006). The NASBA approach is a potent, cutting-edge gene amplification technology that is becoming a popular, quick, and easy diagnostic tool for the early identification and detection of microbial illnesses. A pathogen detection kit that is inexpensive, portable, and simple to use—even in laboratories with limited resources—might be developed with the help of NASBA (Fakruddin et al. 2012).

6.5.1.10 LAMP Assay

Field-based usage of loop-mediated isothermal diagnostic (LAMP) assays may be made possible by a real-time surveillance approach (King et al. 2018). Because of its high level of specificity and sensitivity in comparison to other traditional PCR methods, the LAMP test has been widely used to detect viruses (Lopez-Jimena et al. 2018), bacteria (Yan et al. 2017; Prasannakumar et al. 2020), and fungal pathogens (King et al. 2018). These assays were initially reported by Notomi et al. (2000) and have a number of potential benefits over earlier reported PCR diagnostics. They target numerous unique sequences of target DNA using various sets of primers. These benefits consist of the potential to use lyophilized (freeze-dried) reaction reagents; rapid reaction times (less than 30 min); the ability to operate at a single, steady temperature, which lowers power requirements; and portable equipment (Beissner et al. 2015; Dickinson 2015).

6.5.1.11 "Biospeckle" Laser Technique

Laser biospeckle analysis (Goodman 1975) is an optical imaging technology that has drawn interest from researchers in the past few decades because of its quick and easy procedures. When a coherent light source illuminates an object with chemical or biological activity, the phenomenon known as "biospeckle" is seen (Zdunek et al. 2014). The optical phenomenon of interference, which is produced when a laser light interacts with the seed coat, is the basis for this approach. By examining seeds under a laser beam, distinct activity zones can be identified (Braga Jr et al. 2005; Rabelo et al. 2011). This approach can identify the existence of fungi on seeds since they exhibit biological activity (Mancini et al. 2016). Biospeckle analysis has been widely applied in agriculture during the last few years for a variety of purposes (Singh et al. 2020).

6.5.1.12 Videometer

Olesen et al. (2011) developed the videometer methodology, which can be used as a guide for capturing multispectral images. This method, which combines characteristics from photos taken at visible and near-infrared wavelengths, was useful in differentiating between healthy and contaminated spinach seeds. A multispectral vision system, such as the VideometerLab device, can differentiate between healthy and sick seeds. It can also be used to assess the color, texture, and chemical composition of seed surfaces (Mancini et al. 2016).

6.6 Conclusion

All pathogen detection techniques, from traditional to the most advanced real-time PCR, have benefits and drawbacks. However, choosing the best technique based on its multifaceted significance is essential right now, and integrated approaches involving multiple tests are advised. The methodical implementation of these methodologies guarantees the provision of high-quality seed that is both disease-free and in good health. Molecular techniques are currently being used by the majority of research stations and diagnostic labs for the detection and identification of seed-borne diseases. In the upcoming years, more durable, adaptable, and affordable technologies that enable higher throughput, enhanced sensitivity as well as specificity, and the detection of numerous microorganisms will be developed. Because of a novel nucleic acid–based assay's versatility and dependability, old procedures may have to be abandoned. In the near future, seeds free from a variety of pathogens will be available thanks to the precision of newly developed detection protocols based on molecular methods.

References

Abd El-Aziz MH (2019) Three modern serological methods to detect plant viruses. J Plant Sci Phytopathol 3:101–106

Abd El-Aziz MH, Younes HA (2019) Detection of Cucumber mosaic cucumovirus in infected cowpea plants (*Vigna unguiculata* L.) from northern Egypt. Novel Res Microbiol J 3:326–340

Abdul Baki AA, Anderson JP (1973) Vigour determination in soybean seeds by multiple criteria. Crop Sci 13:630–633

Agarwal VK, Sinclair JB (1997) Principles of seed pathology, 2nd edn. CRC Press, Boca Raton, p 539

Agarwal P, Dev U, Singh B, Indra R, Khetarpal R (2006) Seed-borne fungi detected in consignments of soybean seeds (*Glycine max*) imported into India. EPPO Bull 36:53–58

Ahmed MR, Yasmin J, Park E, Kim G, Kim MS, Wakholi C, Mo C, Cho BK (2020) Classification of watermelon seeds using morphological patterns of x-ray imaging: a comparison of conventional machine learning and deep learning. Sensors 20(23):6753

Arkhipov MV, Priyatkin NS, Gusakova LP, Potrakhov NN, Gryaznov AY, Bessonov VB, Obodovskii AV, Staroverov NE (2019) X-ray computer methods for studying the structural integrity of seeds and their importance in modern seed science. Tech Phys 64:582–592

Aslam S, Tahir A, Aslam MF, Alam MW, Shedayi AA, Sadia S (2017) Recent advances in molecular techniques for the identification of phytopathogenic fungi—A mini review. J Plant Interact 12(1):493–504

Ausubel FM, Bent R, Kingston RE, Moore DJ, Smith JA, Silverman G, Struhl K (1987) Current protocols in molecular biology. Wiley, New York

Baker KK, Ramsdell DC, Gillett JM (1985) Electron microscopy: current applications to plant virology. Plant Dis 69(1):85–90

Ball SFL, Reeves JC (1991) The application of new techniques in the rapid testing for seed-borne pathogens. Plant Var Seeds 4:169–176

Baloch UK (2010) In: Lewis B, Mejia D (eds) Wheat: post-harvest operations. Islamabad, Pakistan Agricultural Research Council, pp 1–21

Beissner M, Phillips RO, Battke F, Bauer M, Badziklou K, Sarfo FS, Maman I, Rhomberg A, Piten E, Frimpong M, Huber KL, Symank D, Jansson M, Wiedemann FX, Kere KB, Herbinger KH, Löscher T, Bretzel G (2015) Loop-mediated isothermal amplification for laboratory confirmation of buruli ulcer disease—towards a point-of-care test. PLoS Neglect Trop Dis 9:e0004219

Belete T (2018) An overview of conventional and molecular detection methods of seed-borne bacterial pathogens. Res Rev J Bot Sci 7(2):33–35

Bergervoet JHW, Peters J, van Beckhoven JRCM, van den Bovenkamp GW, Jacobson JW, van der Wolf JM (2008) Multiplex microsphere immuno-detection of potato virus Y, X and PLRV. J Virol Methods 149:63–68

Blakemore EJA, Reeves JC (1993) PCA used in the development in a new seed health test to identify *Erwinia stewartii*, a bacterial pathogen of maize. Proc first ISTA plant disease committee symp on seed health testing agric Canada central seed laboratory, Ottawa, Ontario, pp 19–22

Braga RA Jr (2017) Challenges to apply the biospeckle laser technique in the field. Chem Eng Trans 58:577–582

Braga RA Jr, Rabelo GF, Granato LR, Santos EF, Machado JC, Arizaga R, Rabal HJ, Trivi M (2005) Detection of fungi in beans by the laser biospeckle technique. Biosyst Eng 91(4):465–469

Bryant GR (1981) The development of serodiagnostic procedures for the detection and quantitation of soybean mosaic virus in seeds. Retrospective Theses and Dissertations 6894

Burns R, Vernon ML, George EL (1994) Monoclonal antibodies for the detection of *Pyrenophora graminea*. In: Schots A, Dewey FM, Oliver R (eds) Modern assays for plant pathogenic fungi-identification, detection and quantification. CAB International, Wallingford, pp 199–203

Candlish AAG, Taylor JD, Cameron J (1988) Immunological methods as applied to bacteria pea blight. Proc Brighton Crop Prot Conf—Pests Dis 2, pp 787–794

Capote N, Mara A, Aguado A, Snchez-Torres P (2012) Molecular tools for detection of plant pathogenic fungi and fungicide resistance. In: Cumagun CJR (ed) Plant pathology. InTech, Rhode Island, pp 151–202

Cappelli C, Buonaurio R, Pezzotti M et al (1993) A simplification of the embryo test method (ETM) to detect Ustilago nuda (Jens.) Rostr. in barley seeds. Phytopathol Mediterr 32:143–144

Carroll TW (1979) Methods of detecting seedborne plant viruses. J Seed Tech 4(2):82–95

Carroll TW, Gossel PL, Batchelor DL (1979) Use of sodium dodecyl sulfate in serodiagnosis of barley stripe mosaic virus in embryos and leaves. Phytopathology 69:12–14

Castro C, Schaad NW, Bonde MR (1994) A technique for extracting *Tilletia indica* teliospores from contaminated wheat seeds. Seed Sci Technol 22:91–98

Chahal SS, Pannu PPS (1997) Detection of *Tilletia indica* in Wheat and *T. barclayana* in Rice samples and its implication for seed certification. In: Hutchins JD, Reeves JC (eds) Seed health testing. CAB International, Wallingford, pp 153–158

Cheah YK, Noorzaleha AS, Lee LH, Radu S, Sukardi S, Sim JH (2008) Comparison of PCR fingerprinting techniques for the discrimination of *Salmonella enterica* subsp. *enterica* serovar Weltevreden isolated from indigenous vegetables in Malaysia. World J Microbiol Biotechnol 24:327–335

Chu PWG, Waterhouse PM, Martin RR, Gerlach WL (1989) New approaches to the detection of microbial plant pathogens. Biotechnol Genet Eng Rev 7:45–111

Clark MF, Adams AN (1977) Characteristics of the microplate method of enzyme-linked immunosorbent assay. J Gen Virol 34:475–483

Compton J (1991) Nucleic acid sequence-based amplification. Nature 350:91–92

Courtioux B, Bisser S, M'Belesso P, Ngoungou E, Girard M, Nangouma A, Josenando T, Jauberteau-Marchan MO, Bouteille B (2005) Dot enzyme-linked immunosorbent assay for more reliable staging of patients with human African trypanosomiasis. J Clin Microbiol 43(9):4789–4795

Danesh YR, Tajbakhsh M, Najafi S, Amirnia R, Ghiyas M (2014) Using nested-PCR for detection of seed-borne fungi *Ustilaginoidea virens* in rice fields of Iran. In: Turkey 5th seed congress with international participation, pp 670–672

De Boer SH, López MM (2012) New grower-friendly methods for plant pathogen monitoring. Annu Rev Phytopathol 50:197–218

de Tempe J (1953) The blotter method of seed health testing programme. ISTA 28:133–151

de Tempe J, Binnerts J (1979) Introduction to methods of seed health testing. Seed Sci Technol 7:601–636

Derrick KS (1973) Quantitative assay for plant viruses using serologically specific electron microscopy. Virology 56:652–653

Dewey FM (1992) Detection of plant-invading fungi by monoclonal antibodies. In: Duncan JM, Torrance L (eds) Techniques for the rapid detection of plant pathogens. Blackwell Scientific Publications, Oxford, pp 47–63

Dickinson M (2015) Loop-mediated isothermal amplification (LAMP) for detection of phytoplasmas in the field. In: Lacomme C (ed) Plant pathology: techniques and protocols. Springer, New York, pp 99–111

Doyer LC (1938) In: Veenman H, Zonen (eds) Manual for the determination of seed-borne diseases. International Seed Testing Association, Wageningen

Du Toit LJ, Derie ML, Hernandez-Perez P (2005) *Verticillium* wilt in spinach seed production. Plant Dis 89:4–11

Edwardson JR (1966) Electron microscopy of cytoplasmic inclusions in cells infected with rod-shaped viruses. Am J Bot 53:359–364

Elwakil M, Ghoneem KM (2002) An improved method of seed health testing for detecting the lurked seed-borne fungi of fenugreek. Plant Pathol J 1:11–13

Elwakil MA, El-Refai IM, Awadallah OA, El-Metwally MA, Mohammed MS (2009) Seed-borne pathogens of faba bean in Egypt: detection and pathogenicity. Plant Pathol J 8:90–97

Endo RM (1961) Turf grass diseases in Southern California. Plant Dis Rep 45:869–873

Engvall E, Perlmann P (1971) Enzyme-linked immunosorbent assay (ELISA) quantitative assay of immunoglobulin G. Immunochemistry 8:871–874

Fakruddin M, Mazumdar RM, Chowdhury A, Mannan KSB (2012) Nucleic acid sequence-based amplification (NASBA)—Prospects and applications. Int J Life Sci Pharma Res 2(1):L106–L121

Franken AAJM, van Vuurde JWL (1990) Problems and new approaches in the use of serology for seedborne bacteria. Seed Sci Technol 18:415–426

Garnsey SM, Gonsalves D, Purcifull DE (1979) Rapid diagnosis of citrus tristeza virus infections by sodium dodecyl sulfate-immuno-diffusion procedures. Phytopathology 69:88–95

Gaur A, Kumar A, Kiran R, Kumari P (2020) Importance of seed-borne diseases of agricultural crops: economic losses and impact on society. In: Kumar R, Gupta A (eds) Seed-borne diseases of agricultural crops: detection, diagnosis & management. Springer, Singapore, p 23. https://doi.org/10.1007/978-981-32-9046-4_1

Gergerich RC, Welliver RA, Gettys S et al (2015) Safeguarding fruit crops in the age of agricultural globalization. Plant Dis 99:176–187

Gold AH, Suneson CA, Houston BR, Oswald JW (1954) Electron microscopy and seed and pollen transmission of rod-shaped particles associated with the false stripe virus disease of barley. Phytopathology 44:115–117

Goldsmith CS, Miller SE (2009) Modern uses of electron microscopy for detection of viruses. Clin Microbiol Rev 22(4):552–563

Goodman JW (1975) Chapter 1. Statistical properties of laser speckle patterns. In: Laser speckle and related phenomena, 1st edn. Springer, Berlin, pp 9–75

Gupta A, Kumar R (2020) Management of seed-borne diseases: an integrated approach. In: Kumar R, Gupta A (eds) Seed-borne diseases of agricultural crops: detection, diagnosis & management. Springer, Singapore, pp 717–745. https://doi.org/10.1007/978-981-32-9046-4_25

Hagborg WAF, Warner GM, Philips NA (1950) Use of 2,4-D as an inhibitor of germination in routine examinations of beans for seed-borne infection. Science 111:91

Hamilton RI (1965) An embryo test for detecting seed-borne barley stripe mosaic virus in barley. Phytopathology 55:789–799

Hamilton RI, Nichols C (1978) Serological methods for detection of pea seed-borne mosaic virus in leaves and seeds of *Pisum sativum*. Phytopathology 68:539–543

Han JH, Shin JS, Kim YH, Kim BD (2007) Improvement of antigen blotting in a tissue blot immunobinding assay for the detection of two chili pepper viruses. J Microbiol Biotechnol 17(11):1885–1889

Hariharan G, Prasannath K (2021) Recent advances in molecular diagnostics of fungal plant pathogens: a mini review. Front Cell Infect Microbiol 10:600234

Hashmi MH (1988) Seed-borne mycoflora of *Trigonella foenum-graecum* L. Pak J Bot 20:233–237

Hewett PD (1970) A note on extraction rate in the embryo method for loose smut of barley, *Ustilago nuda* (Jens.) Rostp. Proc Int Seed Test Assoc 35:181–183

Hiltner L (1917) Technische Vorschriften fur die prufung von saatgut, gulting vom1. Jub. 1916 an. B. Besonderer Teil. 1. Getreide. Landwirtsch versuchs-Stationen 89:379–383

International Seed Testing Association (1966) International rules for seed testing. Proc Int Seed Test Assoc 31, pp 1–152

International Seed Testing Association (2001) International rules for seed testing. Rules Amendments Seed Sci Technol 29, pp 1–127

Irwin JAG (1987) Recent advances in the detection of seedborne pathogens. Seed Sci Technol 15:755–763

Ishii H, Tanoue J, Oshima M et al (2008) First application of PCR-Luminex system for molecular diagnosis of fungicide resistance and species identification of fungal pathogens. J Gen Plant Pathol 74:409–416

Jaccoud-Filho DS, Reeves JC (1993) Detection and identification of *Phomopsis* species in soya bean seeds using PCR. In: Proceedings of first ISTA plant disease committee symp. on seed health testing. Agric Canada Central Seed Laboratory, Ottawa, Ontario, pp 34–37

Jacobsen CS (1995) Microscale detection of specific bacterial DNA in soil with a magnetic capture hybridization and PCR amplification assay. Appl Environ Microbiol 61:3347–3352

Jermoljev E, Chod J (1966) Pouziti serologicke metodyk testovani obesne mozaiky Fazoluna lkicich semen. Fazolu Ochr Rost 2:145–150

Karlberg S (1974) Gronings-och sundhet sunderso-kningar i enhets jord enlight multipotmetoden. Medd Statens Centrala fronkontrollanstalt 49:45–51

Katznelson H, Sutton MD (1951) A rapid phage plaque count method for the detection of bacteria as applied to the demonstration of internally borne bacterial infections of seed. J Bacteriol 61:689–701

Keitreiber M (1984) ISTA handbook on seed health testing, working sheet no. 53. Bundesanstalt für Pflanzenbau, Wien

Khanzada AK, Shetty HS, Mathur SB et al (1989) Avoidance of phenol in the embryo count procedure. In: 22nd international seed testing association congress, seed symposium, Edinburgh, Scotland, June 21–30, 1989. Abstracts of papers. ISTA, Zürich: no. 38

Khare MN, Mathur SB, Neergaard P (1977) A seedling symptom test for detection of *Septoria nodorum* in wheat seed. Seed Sci Technol 5:613–617

Khoo CH, Cheah YK, Lee LH, Sim JH, Noorzaleha AS, Sidik MS, Radu S, Sukardi S (2009) Virulotyping of *Salmonella enterica* subsp. *enterica* isolated from indigenous vegetables and poultry meat in Malaysia using multiplex-PCR. Antonie Van Leeuwenhoek 96:441–457

King KM, Krivova V, Canning GGM, Hawkins NJ, Kaczmarek AM, Perryman SAM, Dyer PS, Fraaije BA, West JS (2018) Loop-mediated isothermal amplification (LAMP) assays for rapid detection of *Pyrenopeziza brassicae* (light leaf spot of brassicas). Plant Pathol 67:167–174

Kotwaliwale N, Singh K, Kalne A, Jha SN, Seth N, Kar A (2014) X-ray imaging methods for internal quality evaluation of agricultural produce. J Food Sci Technol 51:1–5

Krawczyk K, Uszczyńska-Ratajczak B, Majewska A, Borodynko-Filas N (2017) DNA microarray-based detection and identification of bacterial and viral pathogens of maize. J Plant Dis Protect 124:577–583

Kumar R, Gupta A, Maheshwari VK, Atwal SS (2014) Health status of farmers' saved seed of various paddy varieties in Haryana, India. Plant Pathol J 13(3):186–192

Kumar R, Gupta A, Srivastava S, Devi G, Singh VK, Goswami SK, Gurjar MS, Aggarwal R (2020) Diagnosis and detection of seed-borne fungal phytopathogens. In: Kumar R, Gupta A (eds) Seed-borne diseases of agricultural crops: detection, diagnosis and management. Springer, Singapore, pp 107–142. https://doi.org/10.1007/978-981-32-9046-4_5

Kumar R, Gupta A (eds) (2020) Seed-borne diseases of agricultural crops: Detection, diagnosis and management, Springer, Singapore. p 871. https://doi.org/10.1007/978-981-32-9046-4

Kumar R, Gupta A, Verma K (2023a) Studies on seed longevity of pigeonpea (*Cajanus cajan* L.) and chickpea (*Cicer arietinum* L.) under ambient storage condition. Legume Res. 46(7):869–875. https://doi.org/10.18805/LR-4597

Kumar R, Gupta A, Verma K, Singh A (2023b) Effect of seed treatments and storage period on seed health parameters of pea (*Pisum sativum* L.) under ambient storage conditions. Legume Res. 46(9):1233–1239. https://doi.org/10.18805/LR-4634

Landon J (1977) Enzymoimmunoassay; techniques and uses. Nature 268:483–484

Lange L (1986) The practical application of new developments in test procedures for the detection of viruses in seeds. Dev Appl Biol 1:269–281

Langenberg WG, Ball EM (1972) High pH-ammonia agar immunodiffusion for plant viruses. Phytopathology 62:1214–1217

Lauri A, Mariani PO (2009) Potentials and limitations of molecular diagnostic methods in food safety. Genes Nutr 4:1–12

Law Jodi WF, Ab Mutalib NS, Chan KG, Lee LH (2015) Rapid methods for the detection of foodborne bacterial pathogens: principles, applications, advantages and limitations. Front Microbiol 5:770

Lemieux B, Aharomi A, Schena M (1998) Overview of DNA chip technology. Mol Breed 4:277–289

Leng SX, McElhaney JE, Walston JD, Xie D, Fedarko NS, Kuchel GA (2008) ELISA and multiplex technologies for cytokine measurement in inflammation and aging research. J Gerontol A Biol Sci Med Sci 63(8):879–884

Lesemann DE, Bozarth RF, Koenig R (1980) The trapping of tymovirus particles on electron microscope grids by adsorption and serological binding. J Gen Virol 48:257–264

Limonard T (1968) Ecological aspects of seed-health testing. Proc Int Seed Test Assoc 33, pp 17–73

Lister RM (1978) Application of the enzyme-linked immunosorbent assay for detecting viruses in soybean seed and plants. Phytopathology 68:1393–1400

Lister RM, Wright SE, Kloots JM (1978) Sensitive detection of barley stripe mosaic virus in barley seeds and embryos by ELISA. Proc Amer Phytopathol Soc 12, pp 198

Liu MY, Xydakis AM, Hoogeveen RC, Jones PH, Smith EO, Nelson KW, Ballantyne CM (2005) Multiplexed analysis of biomarkers related to obesity and the metabolic syndrome in human plasma, using the Luminex-100 system. Clin Chem 51(7):1102–1109

Lopez MM, Bertolini E, Olmos A, Caruso P, Gorris MT, Llop P, Penyalver R, Cambra M (2003) Innovative tools for detection of plant pathogenic viruses and bacteria. Int Microbiol 6:233–243

López MM, Llop P, Olmos A, Marco-Noales E, Cambra M, Bertolini E (2015) Are molecular tools solving the challenges posed by detection of plant pathogenic bacteria and viruses? Curr Issues Mol Biol 11:13–46

Lopez-Jimena B, Wehner S, Harold G, Bakheit M, Frischmann S, Bekaert M, Faye O, Sall AA, Weidmann M (2018) Development of a single-tube one-step RT-LAMP assay to detect the Chikungunya virus genome. PLoS Negl Trop Dis 12:1–14

Luchi N, Ghelardini L, Belbahri L, Quartier M, Santini A (2013) Rapid detection of *Ceratocystis platani* inoculum by quantitative real-time PCR assay. Appl Environ Microbiol 79:5394–5404

Majumder D, Rajesh T, Suting EG, Debbarma A (2013) Detection of seed borne pathogens in wheat: recent trends. Aust J Crop Sci 7(4):500–507

Makkouk KM, Kumari SG (1996) Detection of ten viruses by the tissue-blot immunoassay (TBIA). Arab J Plant Protect 14:3–9

Malin EM, Roth DA, Belden EL (1983) Indirect immunofluorescent staining for detection and identification of *Xanthomonas campestris* pv. *phaseoli* in naturally infected bean seed. Plant Dis 67:645–647

Mancini V, Murolo S, Romanazzi G (2016) Diagnostic methods for detecting fungal pathogens on vegetable seeds. Plant Pathol 65:691–703

Mandal PK, Biswas AK, Choi K, Pal UK (2011) Methods for rapid detection of foodborne pathogens: an overview. Am J Food Technol 6:87–102

Mangan A (1971) A new method of the detection of *Pleospora bjoerlingii* infection of sugar-beet seed. Trans Br Mycol Soc 57:169–172

Marcinkowska JZ (2002) Methods of finding and identification of pathogens in seeds. Plant Breed Seed Sci 46(1):31–48

Markoulatos P, Siafakas N, Moncany M (2002) Multiplex polymerase chain reaction: a practical approach. J Clin Lab Anal 16:47–51

Mathanker SK, Weckler PR, Bowser TJ (2013) X-ray applications in food and agriculture: a review. Trans ASABE 56:1227–1239

Maude RB (1996) Seedborne diseases and their control. Principles and practice. CAB International, Wallingford, p 280

Michelle C, Stephen F (2010) Seed diseases and seed-borne pathogens of North America. Plant Pathol 53(2):35–44

Midorikawa GEO, Miller RNG, Bittencourt DM (2018) Molecular identification and detection of foodborne and feedborne mycotoxigenic fungi. In: El Sheikha AF, Levin R, Xu J (eds) Molecular techniques in food biology. Wiley, Hoboken, pp 385–407

Milne RG (1991) Chapter 7: Immunoelectron microscopy for virus identification. In: Mendgen K, Lesemann DE et al (eds) Electron microscopy of plant pathogens. Springer, Berlin, pp 87–102

Mitchell LA (1988) A sensitive dot immunoassay employing monoclonal antibodies for detection of *Sirococcus strobilinus* in spruce seed. Plant Dis 72:644–667

Mostafa FAM, Abou-Ela AA (2011) Sensitive detection of watermelon mosaic and zucchini yellow mosaic viruses from infected squash plants using serological methods and polymerase chain reaction. Egypt J Exp Biol (Bot) 7(2):179–185

Mullis KB, Faloona FA (1987) Specific synthesis of DNA *in vitro* via a polymerase-catalyzed chain reaction. Methods Enzymol 155:335–350

Munkvold GP (2009) Seed pathology progress in academia and industry. Annu Rev Phytopathol 47:285–311

Muskette AE, Malone JP (1941) The ulster method for the examination of flax seed for the presence of seed borne parasites. Ann Appl Biol 28:8–13

Narayanasamy P (2011) Microbial plant pathogens-detection and disease diagnosis, vol 2. Springer, Dordrecht

Neergaard P (1956a) Arsberetning vedrende frpatologisk Kontrol. St. Plantetilsyn, 1 April 1953–31 May, 1954. Statens Plantetilsyn, Kbenhavn
Neergaard P (1956b) Nye angreb. 6 Arsberetning vedrprende frppatologisk kontrol, pp 15–16
Neergaard P (1979) Seed pathology, vol 1–2. MacMillan Press, London, p 1191
Neethirajan S, Karunakaran C, Symons S, Jayas DS (2006) Classification of vitreousness in durum wheat using soft X-rays and transmitted light images. Comput Electron Agric 53:71–78
Notomi T, Okayama H, Masubuchi H, Yonekawa T, Watanabe K, Amino N, Hase T (2000) Loop-mediated isothermal amplification of DNA. Nucleic Acids Res 28:e63
Olesen MH, Carstensen JM, Boelt B (2011) Multispectral imaging as a potential tool for seed health testing of spinach (*Spinacia oleracea* L.). Seed Sci Technol 39:140–150
Oliver R (1993) Nucleic acid-based methods for detection and identification. In: Fox RTV (ed) Principles of diagnostic techniques in plant pathology. CAB International, Wallingford, pp 153–169
Olsvik O, Popovic T, Skjerve E, Cudjoe KS, Hornes E, Ugelstand J, Uhlen M (1994) Magnetic separation techniques in diagnostic microbiology. Clin Microbiol Rev 7:43–54
Park SH, Aydin M, Khatiwara A, Dolan MC, Gilmore DF, Bouldin JL, Ahn S, Ricke SC (2014) Current and emerging technologies for rapid detection and characterization of Salmonella in poultry and poultry products. Food Microbiol 38:250–262
Parker MC, Dean LL (1968) Ultraviolet as sampling aid for detection of bean seed infected with *Pseudomonas phaseolicola*. Plant Dis Rep 52:534–538
Patel P (2021) A review on plant disease diagnosis through biosensor. Int J Biosens Bioelectron 7(2):50–62
Paulsen MR (2002) Using machine vision to inspect oil seeds. Inform 1:50–55
Pelet E (1965) Dosage due virus de la Mosaique de la laitue par indexage de la graine sur *Chenopodium quinoa*. Revue Hort Suisse 38:7–10
Pelet E, Gagnebin F (1963) Production de graines de laitue sans virus. Revue Hort Suisse 36:22–38
Pellegrino C, Gilardi G, Gullino ML, Garibaldi A (2010) Detection of *Phoma valerianellae* in lamb's lettuce seeds. Phytoparasitica 38:159–165
Peters J, Sledz W, Bergervoet JHW, van der Wolf JM (2007) An enrichment microsphere immuno-assay for the detection of *Pectobacterium atrosepticum* and *Dickeya dianthicola* in potato tuber extracts. Eur J Plant Pathol 117:97–107
Phatak HC (1974) Seed borne plant viruses—identification and diagnosis in seed health testing. Seed Sci Technol 2:3–146
Phatak HC, Summanwar AS (1967) Detection of plant viruses in seeds and seed stocks. Proc Int Seed Test Assoc 32, pp 625–631
Powell CA (1987) Detection of three plant viruses by dot - immunobinding assay. Phytopathology 77:306–309
Prasannakumar MK, Buela Parivallal P, Manjunatha C, Mahesh HB, Pramesh D, Narayan KS, Gopal VB, Priyanka K, Puneeth ME, Rangaswamy KT (2020) Loop-mediated isothermal amplification assay for pre-symptomatic stage detection of *Xanthomonas axonopodis* pv. *punicae* infection in pomegranate. Australas Plant Pathol 49:467–473
Prasannakumar MK, Parivallal PB, Pramesh D, Mahesh HB, Raj E (2021) LAMP-based foldable microdevice platform for the rapid detection of *Magnaporthe oryzae* and *Sarocladium oryzae* in rice seed. Sci Rep 11:178
Prosen D, Hatziloukas E, Schaad NW, Panopoulos NJ (1993) Specific detection of *Pseudomonas syringae* pv. *phaseolicola* DNA in bean seed by polymerase chain reaction-based amplification of a phaseolotoxin gene region. Phytopathology 83:965–970
Pryor BM, Gilbertson RL (2001) A PCR-based assay for detection of *Alternaria radicina* on carrot seed. Plant Dis 85:18–23
Purcifull DE, Batchelor DL (1977) Immuno-diffusion tests with sodium dodecyl sulfate (SDS)-treated plant viruses and plant viral inclusions. Florida Agric Exp Sta Bull, pp 788
Rabelo GF, Enes AM, Braga RA Jr, Fabbro IMD (2011) Frequency response of biospeckle laser images of bean seeds contaminated by fungi. Biosyst Eng 110(3):297–301
Rahman A, Cho B (2016) Assessment of seed quality using non-destructive measurement techniques: a review. Seed Sci Res 26(4):285–305

Rajeshwari N, Shylaja MD, Krishnappa M et al (1998) Development of ELISA for the detection of *Ralstonia solanacearum* in tomato: its application in seed health testing. World J Microb Biotechnol 14(5):697–704
Rao NK, Hanson J, Dulloo ME, Ghosh K, Nowell D, Larinde M (2006) Manual of seed handling in Genebanks. Bioversity International, Rome
Rasmussen OF, Wulff BS (1991) Detection of *Ps*. pv. *pisi* using PCR. In: Proceedings of 4th international working group on *Pseudomonas syringae* pathovars, pp 369–376
Reeves JC (1995) Nucleic acid techniques in testing for seedborne diseases. In: Skerritt JH, Appels R (eds) New diagnostics in crop sciences. CAB International, Wallingford, pp 127–149
Reeves JC, Rasmussen OF, Simpkins SA (1994) The use of DNA probe and PCR for the detection of *Pseudomonas syringae* pv. *pisi* in pea seed. Plant pathogenic bacteria, 8th international conference, pp 383–389
Rego CHQ, Silva FF, Gomes-Junior FG, de Moraes MHD, de Medeiros AD, da Silva CB (2020) Using multispectral imaging for detecting seed-borne fungi in cowpea. Agriculture 10:361
Rennie WJ (1982) Barley: loose smut. In: ISTA handbook on seed health testing, working sheet no. 25. ISTA, Zürich
Rodriguez-Lazaro D, Hernández M, D'Agostino M, Cook N (2006) Application of nucleic acid sequence-based amplification for the detection of viable foodborne pathogens: progress and challenges. J Rapid Methods Automat Microbiol 14(3):218–236
Roenhorst JW, Krom C, Fox A, Mehle N, Ravnikar M, Werkman AW (2018) Ensuring validation in diagnostic testing is fit for purpose: a view from the plant virology laboratory. EPPO Bull 48(1):105–115
Rollings M (1965) Disease control through virus-free stock. Annu Rev Phytopathol 3:367–396
Rollins L, Coats K, Elliott M, Chastagner G (2016) Comparison of five detection and quantification methods for *Phytophthora ramorum* in stream and irrigation water. Plant Dis 100:1202–1211
Saiki RK, Gelfand DH, Stoffel S, Scharf SJ, Higuchi R, Horn GT, Mullis KB, Erlich HA (1988) Primer-directed enzymatic amplification of DNA with a thermostable DNA polymerase. Science 239:487–491
Sawicka B (2019) Post-harvest losses of agricultural produce. Sustain Dev 1(1):1–16
Schaad NW, Frederick RD (2002) Real-time PCR and its application for rapid plant disease diagnostics. Can J Plant Pathol 24:250–258
Schaad NW, Azad H, Peet RC et al (1989) Identification of *Pseudomonas syringae* pv. *phaseolicola* by a DNA hybridization probe. Phytopathology 79:903–907
Schaad NW, Cheong SS, Tamaki S, Hatziloukas E, Panopoulos N (1995) A combined biological and enzymatic amplification (BIO-PCR) technique to detect *Pseudomonas syringae* pv. *phaseolicola* in bean seed extracts. Phytopathology 85(2):243–248
Schilling AG, Moller EM, Geiger HH (1996) Polymerase chain reaction-based assays for species specific detection of *Fusarium culmorum*, *F. graminearum* and *F. avenaceum*. Mol Plant Pathol 86:515–522
Scott HA (1961) Serological detection of Barley stripe mosaic virus in single seeds and dehydrated leaf tissues. Phytopathology 51:200–201
Shah D (2013) Assessment of pre and post-harvest losses in tur and soyabean crops in Maharashtra. Agro-Economic Research Centre Gokhale Institute of Politics and Economics, Pune
Sharma HL, Randhawa HS, Kapur A et al (1987) Seed discoloration in Rice. Oryza 24(1):37–41
Shepherd RJ (1972) Transmission of viruses through seed and pollen. In: Kado CI, Agrawal HA (eds) Techniques in plant virology. Van Nostrand Reinhold, New York, pp 267–292
Simmonds DH, Gumming BG (1979) Detection of lily symptomless virus by immunodiffusion. Phytopathology 69:1212–1215
Sinclair J (1979) The seed: a microcosm of microbes. J Seed Technol 4(2):68–73
Singer JM, Plotz CM (1956) The latex fixation test. I. Application to the serological diagnosis of rheumatoid arthritis. Am J Med 21:888
Singh S, Sinha A, Yadav SM, Singh BK, Singh H, Srivastava S, Kumar R (2017) Antagonistic behavior of different bioagents against dominant seed borne fungi of mungbean seeds under *in*

vitro condition. Proc Natl Acad Sci India Sect B Biol Sci 87:599–602. https://doi.org/10.1007/s40011-015-0604-z
Singh P, Chatterjee A, Bhatia V, Prakash S (2020) Application of laser biospeckle analysis for assessment of seed priming treatments. Comput Electron Agric 169:1–13
Singh P, Chatterjee A, Rajput LS, Kumar S, Nataraj V, Bhatia V, Prakash S (2021) Biospeckle-based sensor for characterization of charcoal rot (*Macrophomina phaseolina* (Tassi) Goid) disease in soybean (*Glycine max* (L.) Merr.) crop. IEEE Access 9:31562–31574. https://doi.org/10.1109/ACCESS.2021.30598689
Slack SA, Shepherd RJ (1975) Serological detection of seed-borne *Barley stripe mosaic virus* by a simplified radial-diffusion technique. Phytopathology 65:948–955
Smartt AE, Ripp S (2011) Bacteriophage reporter technology for sensing and detecting microbial targets. Anal Bioanal Chem 400:991–1007
Soteros JJ (1979) Detection of *Alternaria radicina* and *A. dauci* from imported carrot seed in New Zealand. N Z J Agric Res 22:185–190
Srinivasan MC, Neergaard P, Mathur SB (1973) A technique for detection of *Xanthomonas campestris* in routine seed health testing of crucifers. Seed Sci Technol 1:853–859
Srivastava S, Kumar R, Bindal S, Singh VP, Rana M, Singh JP, Sinha A (2020) Ancient, mid-time, and recent history of seed pathology. In: Kumar R, Gupta A (eds) Seed-borne diseases of agricultural crops: detection, diagnosis & management. Springer, Singapore. https://doi.org/10.1007/978-981-32-9046-4_4
Srivastava S, Bindal S, Pavithra G, Kumar R, Rana M (2021) Seed-borne mycoflora of edible oilseed crops of India. In: Tiwari AK (ed) Advances in seed production and management. Springer Science + Business Media, Singapore, pp 307–324
Stead DE (1992) Techniques for detecting and identifying plant pathogenic bacteria. In: Duncan JM, Torrance L (eds) Techniques for the rapid detection of plant pathogens. Blackwell Scientific Publications, Oxford
Stevens EA, Blakemore EJA, Reeves JC (1997) Development of a PCR-based test to detect and identify *Pyrenophora* spp. In: Hutchins JD, Reeves JC (eds) Seed health testing. CAB International, Wallingford, pp 139–145
Taylor E, Bates J, Kenyon D, Maccaferri M, Thomas J (2002) Modern molecular methods for characterisation and diagnosis of seed-borne fungal pathogens. J Plant Pathol 83:75–81
Tsedaley B (2015) Review on seed health tests and detection methods of seedborne diseases. J Biol Agric Healthc 5(5):176–184
Upadhyay P, Singh SP (2019) Detection methods for seed borne pathogens. Int J Curr Microbiol App Sci 8(3):318–323
van Vuurde JWL (1987) Detecting seedborne bacteria by immunofluorescence. In: Civerolo EL, Collmer A, Davis RE, Gillaspie AG (eds) Plant pathogenic bacteria. Current plant science and biotechnology in agriculture, vol 4. Springer, Dordrecht
van Vuurde JWL (1997) Immunofluorescence colony-staining as a tool for sample indexing and for the determination of field thresholds for pathogenic seedborne bacteria. In: Hutchins JD, Reeves JC (eds) Seed health testing. CAB International, Wallingford, pp 165–173
Vannacci G, Sarrocco S, Porta-Puglia A (2014) Chapter 6. Improved detection and monitoring of seed-borne fungal plant pathogens in Europe. In: Gullino ML, Munkvold G (eds) Global perspectives on the health of seeds and plant propagation material, plant pathology in the 21st century, vol 6. Springer Science + Business Media, Dordrecht, pp 67–85
Veena MS, van Vuurde JWL (2002) Indirect immunofluorescence colony staining method for detecting bacterial pathogens of tomato. J Microbiol Methods 49(1):11–17
Vernet G (2002) DNA-Chip technology and infectious diseases. Virus Res 82:65–71
Walcott RR (2003) Detection of seedborne pathogens. HortTechnology 13(1):40–47
Walcott RR, Gitaitis RD (2000) Detection of *Acidovorax avenae* subsp. *citrulli* in watermelon seed using immunomagnetic separation and the polymerase chain reaction. Plant Dis 84:470–474
Walcott RR, McGee DC, Misra MK (1998) Detection of asymptomatic fungal infections of soybean seeds by ultrasound analysis. Plant Dis 82:584–589

Wang ZK, Comstock JC, Hatziloukas E, Schaad NW (1999) Comparison of PCR, BIO-PCR, DIA, ELISA and isolation on semi-selective medium for detection of *Xanthomonas albilineans*, the causal agent of leaf scald of sugarcane. Plant Pathol 48:245–252

Warham EJ (1990) Effect of *Tilletia indica* infection on viability, germination and vigour of wheat seed. Plant Dis 74:130–135

Welty RE, Rennie WJ (1984) Grasses, endophyte. In: ISTA handbook on seed health testing, working sheet n. 55. ISTA, Zürich

Wharton AL (1967) Detection of infection by *Pseudomonas phaseolicola* (Burkh.) Dowson in while white seeded dwarf bean stocks. Ann Appl Biol 60:305–312

Wolff G, Bárcena M (2021) Multiscale electron microscopy for the study of viral replication organelles. Viruses 13(2):197

Yan M, Li W, Zhou Z, Peng H, Luo Z, Xu L (2017) Direct detection of various pathogens by loop-mediated isothermal amplification assays on bacterial culture and bacterial colony. Microb Pathog 102:1–7

Zdunek A, Adamiak A, Pieczywek PM, Kurenda A (2014) The biospeckle method for the investigation of agricultural crops: a review. Opt Lasers Eng 52:276–285

Zhang G (2013) Foodborne pathogenic bacteria detection: an evaluation of current and developing methods. Meducator 1:15

Zhang M, Shi Z, Chen G, Cao A, Wang Q, Yan D, Fang W, Li Y (2023) Detection and identification methods and control techniques for crop seed diseases. Agriculture 13:1786

Zhao X, Lin CW, Wang J, Oh DH (2014) Advances in rapid detection methods for foodborne pathogens. J Microbiol Biotechnol 24:297–312

Application of Machine Learning in Plant Disease Detection and Classification

7

R. N. Singh, P. Krishnan, Sonam Sah,
and Vaibhav Kumar Singh

Abstract

Plant disease detection and monitoring is a crucial issue that needs to be addressed to increase agricultural productivity and ensure global food security. Conventional visual detection and monitoring methods of plant diseases are tedious and costly and require expert knowledge for accurate decisions. Machine learning and deep learning methods are gaining popularity among researchers as automatic disease detection methods which can provide quick and accurate disease detection and overcome several limitations of the traditional methods. This chapter covers the basic concepts of machine learning and deep learning, followed by the general method for detecting and monitoring plant diseases. The chapter also covers the basics of some recent and popular machine learning and deep learning classification methods, including support vector machine, K-nearest neighbours, random forest, artificial neural network, and convolutional neural network. The methods for assessing the machine learning models and their performances for agricultural crop disease detection were also covered. At the end, the major challenges in developing machine learning models for plant disease detection are presented.

Keywords

Plant disease detection · Image analysis · Machine learning · Deep learning

R. N. Singh (✉) · S. Sah
ICAR—National Institute of Abiotic Stress Management, Baramati, Maharashtra, India

P. Krishnan
Division of Agricultural Physics, ICAR—Indian Agricultural Research Institute, New Delhi, India

V. K. Singh
Wheat Pathology Laboratory, Division of Plant Pathology, ICAR—Indian Agricultural Research Institute, New Delhi, India

V. K. Singh et al. (eds.), *Diseases of Field Crops: Diagnostics and Management*,
https://doi.org/10.1007/978-981-97-6160-9_7

7.1 Introduction

Plant diseases severely impact both the quantity and quality of food crops, which in turn impacts the socio-economic and food security around the globe. Therefore, timely and quick detection of plant diseases is paramount for plant disease management to reduce agricultural losses and for ensuring food security for the growing population. India, an agriculture-based country, contributes significantly to world agricultural production. At the same time, agriculture contributes up to 17% of India's GDP, with around 70% of the population dependent on agriculture or agri-businesses for their living (Agarwal et al. 2019a, b). Any abiotic or biotic factor affecting crop production inevitably affects the socio-economic condition of the farmers and the country. Among biotic factors, pests are reported to cause an annual loss of around 10.8% in the world and around 15% in India, while up to 16% global yield loss is estimated due to plant pathogens like bacteria, fungi, and viruses (Ficke et al. 2018). Because of these factors, farmers face major challenges in managing pests and diseases due to the increasing frequency of incidences with changing climate. Apart from this, most farmers still use the orthodox methods of manual field visits and visual observation–based disease identification and its management, which is less accurate and resource-consuming. It is highly subjective to the expertise of the farmer to identify diseases in different crops accurately, as the farmers may be aware of a few crop diseases, but there is a high probability of incorrect disease identification and management practices along with missing several other diseases in the field which will lead to more damage to the whole crop. Hence, using advanced tools and techniques like machine learning and deep learning for accurate and timely detection becomes obligatory. Artificial intelligence, machine learning, and deep learning are trending research topics worldwide with applications in the majority of fields, including marketing, business, medical, and agriculture. However, these three terms are quite confusing and often used interchangeably, so it is necessary to define and understand the fundamental differences between these three terms properly.

The term artificial intelligence was coined by John McCarthy in 1955, who defined it as "the science and engineering of making intelligent machines" (McCarthy 1955). In simplest terms, artificial intelligence (AI) combines computer science with robust datasets for problem-solving. AI simulates human intelligence in machines, especially computer systems, to mimic human intelligence for solving problems or doing certain work. The main objective of AI is to create expert decision systems and act on behalf of humans. AI is a broad umbrella under which machine learning (ML) and deep learning (DL) come. ML is a subset of AI, while the DL is a subset of ML (Fig. 7.1). Machine learning (ML) is the process in which the computer extracts consistencies from training data. In ML, computers are trained using data to act without complex programming, allowing them to improve their knowledge and actions based on training datasets. The ML model develops

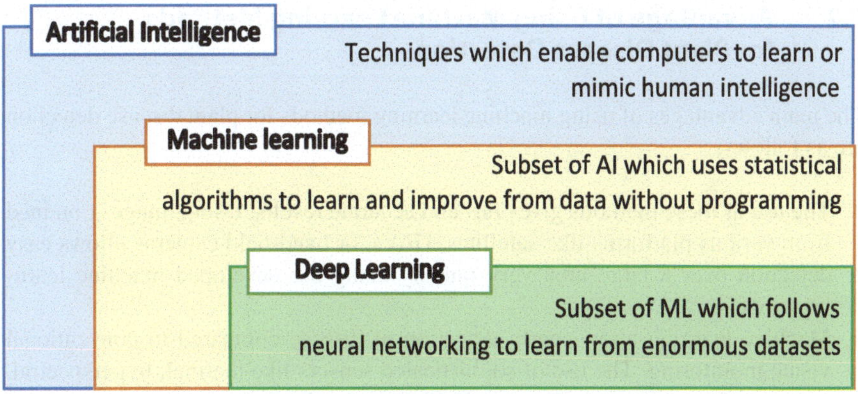

Fig. 7.1 Venn diagram showing relationships between artificial intelligence, machine learning, and deep learning

algorithms to learn the patterns in the input dataset and uses them to perform that task. ML aims to automate analytical model building to perform tasks like object detection and language translation, which is achieved by using algorithms that iteratively learn from task-specific training data to find hidden or complex patterns without programming (Bishop 2006). ML has good applicability in tasks related to high-dimensional data such as classification, clustering, and regression. By extracting consistencies from training datasets, ML can produce accurate and reliable decisions. Deep learning (DL) is an extension of ML, which adds more "depths" in algorithms and data transformations. The basic fundamental of DL is inspired by the artificial neural network, a ML method which uses multiple layers of processing to learn from the given dataset. DL has a comparatively higher ability to quickly solve more complex problems because of more complex algorithms. Based on the given task and input dataset, the machine learning techniques are classified into three major groups:

1. Supervised learning: It requires users to give output values (y) against each input value (x) of the training dataset. These pairs of input and output data are used to calibrate the model parameters, and once developed, the model takes new or unseen data x and predicts corresponding y.
2. Unsupervised learning: In this system, the model is not trained by the user, and the machine learning system itself detects the pattern in the input dataset. Problems of clustering and dimension reductions are a few examples where unsupervised machine learning is employed.
3. Reinforced learning: In reinforcement learning, the state of the system, goal, allowable action, and constraints are described to the system, and the model uses trial and error to learn to generate the best output based on the goal.

7.2 Advantage of Using Machine Learning Methods for Plant Disease Detection

The main advantages of using machine learning methods for plant disease detection are as follows:

1. The use of these methods gives fast and accurate results. Using images obtained from various platforms like satellites, UAVs, or hand-held cameras allows easy detection over a large area very quickly using the developed machine learning models.
2. Machine learning allows early detection of disease compared to conventional visual monitoring. The use of sophisticated sensors like thermal, hyperspectral, and even a combination of simple RGB indices allows the detection of changes in plants that are not visible to the naked eye.
3. Do not require any expert knowledge to identify the disease in the field. A farmer may know a few diseases and their symptoms; however, due to emerging pests and diseases, it becomes difficult to identify diseases in every crop. However, the established model using machine learning is capable of easily identifying various diseases in different crops.
4. Saves time and other resources. Conventional disease detection and management are highly subjective to the expertise of the observer. Incorrect identification of a disease or its severity may lead to excessive use of chemicals, which increases pesticide residue in the soil-plant-atmosphere continuum along with the increased cost of cultivation. The use of machine learning gives accurate and early information about disease which saves time of manual monitoring, labor, and other resources. Machine learning models can be integrated with sensor-based pesticide application methods like variable rate technology of precision agriculture or delivery of pesticides using drones etc. for effective use of resources.

7.3 Major Steps Involved in Plant Disease Detection and Classification Using Machine Learning

7.3.1 Image Acquisition

The first step to developing a plant disease detection system is to capture or acquire images of both healthy and diseased plants. It is one of the crucial steps and plays a vital role in the success of the final detection and classification system developed. The images can be captured using optical, infrared, or hyperspectral sensors using various platforms like satellites, UAVs, or hand-held sensors and even smartphones. These high-quality images are used to train models after pre-processing. Apart from collecting images manually, there are also some plant disease datasets available which can also be used for training machine learning models depending on the user's requirement.

7.3.1.1 Available Image Datasets of Plant Disease

The available image datasets for plant disease have descriptions and images of several diseases of crops in both healthy and diseased conditions. There are several datasets, some of which are free, and some are paid. The PlantVillage dataset is a famous and freely available image dataset that consists of more than 54,000 healthy and disease images divided into more than 30 categories based on plant species and disease. However, the images in PlantVillage datasets are taken under laboratory conditions, which is the major limitation for its applications in real conditions. The PlantDoc dataset is another plant disease dataset that consists of more than 20,000 images of non-lab data of more than 30 species of plants and their diseases (Singh et al. 2020). Some important publicly available plant disease datasets are given in Table 7.1. Apart from this, several institutions like IRRI and BRRI give access to their datasets on reasonable requests. Many researchers have also collected their own images under various environmental and lab conditions, which can be accessed on personal requests.

7.3.2 Pre-processing

The high-quality images after acquisition are used at this stage. The images obtained were subjected to many noises, distortions, and atmospheric and sensor-specific aberrations, which need to be corrected and improved before extraction of disease data from raw images. Pre-processing strategies aim to remove these issues and bring the data to a stage in which machine learning can be applied. Noise removal, distortion removal, color space conversion, image cropping, smoothening, filtering, and enhancement are major techniques used for pre-processing the machine learning input data.

7.3.3 Image Segmentation

Image segmentation is the process of breaking down the image into various subgroups to reduce the complexity and improve the accuracy of disease recognition. For example, field crop image segmentation involves dividing the image pixels into

Table 7.1 Some important publicly available datasets of plant diseases

S. no.	Dataset	Important crops
1.	PlantVillage	Pepper bell, tomato, potato
2.	PlantDoc	Apple, tomato, potato, pepper bell, grapes, corn
3.	IPM Images	Tomato, potato, pepper, pea, soya beans, turnip rice, carrot, onion
4.	New Plant Diseases Dataset	Apple, grape, orange, blueberry, tomato, potato, pepper bell, soybean, strawberry
5.	APS Images	Tomato, potato, pepper pea

different classes or regions of interest like disease-affected and healthy areas. Segmentation techniques are mainly classified under four categories:

(a) Pixel-based segmentation
   1. Thresholding: Otsu, global, adaptive, etc.
   2. Clustering: K-means clustering, fuzzy C-mean
(b) Edge-based: edge detection, active contours, etc.
(c) Region-based: region growing segmentation, split merge, etc.
(d) ANN-based segmentation

7.3.4 Feature Extraction

In this process, the segmented data is transformed (preserving the original information of the dataset) into numerical features that ML models can process to yield better results than raw data. Shape, color, and texture are major information extracted from input images during feature extraction. Feature extraction reduces the data dimension to create manageable and meaningful groups. Principal component analysis, linear discriminant analysis, scale-invariant feature transform, etc. are some of the methods used for feature extraction in machine learning (Roy et al. 2021).

7.3.5 Classification

It is the process of identifying and categorizing input data into different possible known classes. In this step, models separate input data into known categories, and with the help of these pre-categorized datasets, the models classify future datasets into relevant categories. At this stage, the actual detection and separation into different classes is performed based on the selected method of classification. There are several classification techniques available for plant disease classification like support vector machine (SVM), decision tree, random forest (RF), K-means, K-nearest neighbors (KNN), genetic algorithm (GA), fuzzy logic, fuzzy C-means, artificial neural network (ANN), and back propagation neural network (BPNN), convolutional neural network (CNN). We will discuss a few of them in detail along with their application for plant disease classification.

7.3.5.1 Support Vector Machine (SVM)

SVM is one of the most popular machine learning algorithms capable of doing both classifications as well as regression. The aim of SVM is to find the best decision boundary, known as hyperplane, with the highest margin (distance between hyperplane and support vectors) between different classes, which distinctly classifies data points (Fig. 7.2). SVM is of two types: linear and non-linear. The linear SVM is used to separate linearly separable data into two groups. The hyperplane created in this case is a straight line (Fig. 7.3). Similarly, non-linear SVM is used for non-linear problems where the data cannot be separated using a single straight line;

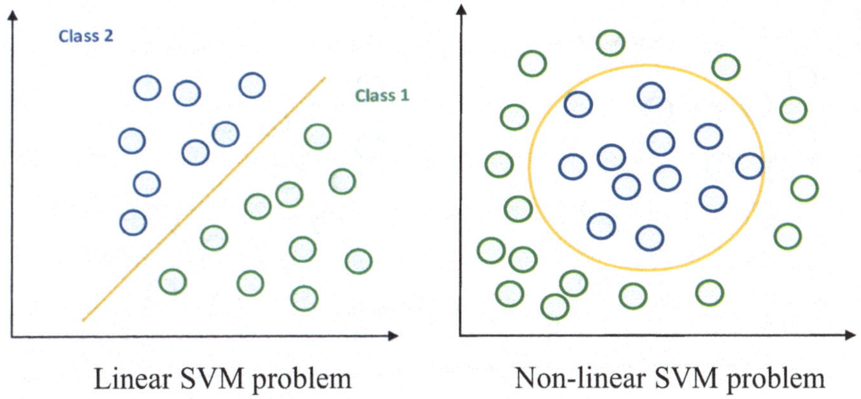

Fig. 7.2 Linear and non-linearly separable SVM problems

Fig. 7.3 SVM in two-dimensional space

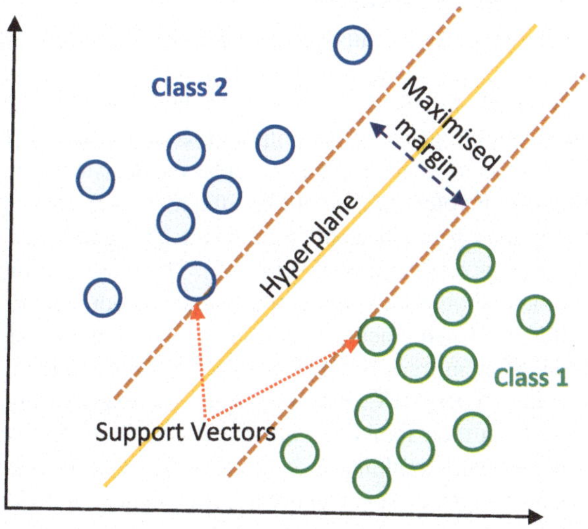

however, it can be converted to linearly separable data of higher dimensions (Fig. 7.3). Several researchers use SVM in many crop disease detection studies to detect diseases of wheat, rice, corn, apple, and many other fruits and vegetables.

7.3.5.2 K-Nearest Neighbors (KNN)

KNN is one of the simplest and oldest supervised machine learning algorithms. KNN is non-parametric in nature and can be used in both classification and regression problems. However, it is mainly used for classification. KNN is also known as a lazy learner because it does not learn immediately from the training data. Instead, it stores and learns at the time of classification (Wang et al. 2022). KNN calculates the Euclidean distance of the data point from k number of neighbors and assigns the point to a class for which the number of nearest neighbors is maximum (Fig. 7.4).

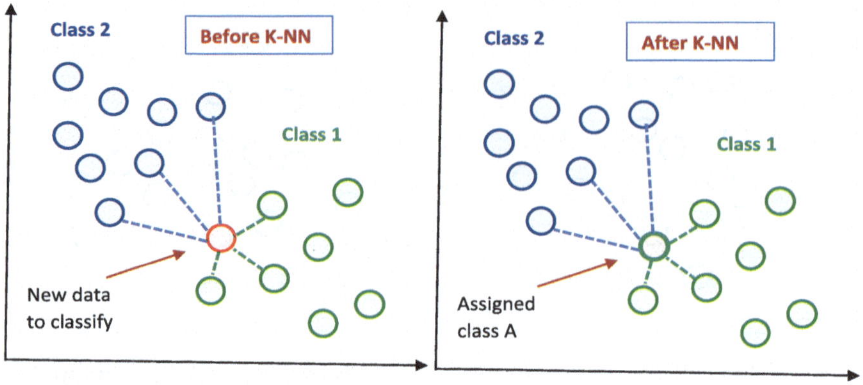

Fig. 7.4 Concept of KNN algorithm

7.3.5.3 Random Forest

Random forest is a multiple decision tree–based robust algorithm capable of handling both classification and regression problems of high-dimensional data. RF is a supervised classification algorithm which builds multiple decision trees and combines them to generate the output class or prediction. The "forest" is the ensemble of decision trees, the models consider the decision from each tree, and based on the majority of votes (in case of classification) or the average of predictions (in case of regression), it produces the final result. The greater the number of trees, the higher the accuracy of the model (Fig. 7.5). There are two important assumptions of RF: (1) there should be some actual values in our feature variable so that the model builds itself using actual data rather than random guesses and (2) the predictions from different trees should have very low correlations among themselves (Fratello and Tagliaferri 2018). When the trees are relatively uncorrelated, they protect each other from their individual errors. In a nutshell, RF uses bagging and feature randomness while building individual trees, trying to create an uncorrelated forest of trees whose prediction is more accurate than an individual tree.

7.3.5.4 Artificial Neural Network (ANN)

ANN is inspired from the working of the human brain. ANN has a set of neural network algorithms that try to learn from experiences like te functioning of the human brain's neurons. The ANN is made up of three layers which are known as the input layer, hidden layer, and output layer, as shown in Fig. 7.6 (Nguyen et al. 2021). The input layer takes data or information from the outside world and transfers it to the next layers. The next layer(s), known as hidden layers, consists of multiple interconnected nodes (neurons) which act as processors that communicate with each other and do all the computations. There can be any number of hidden layers in

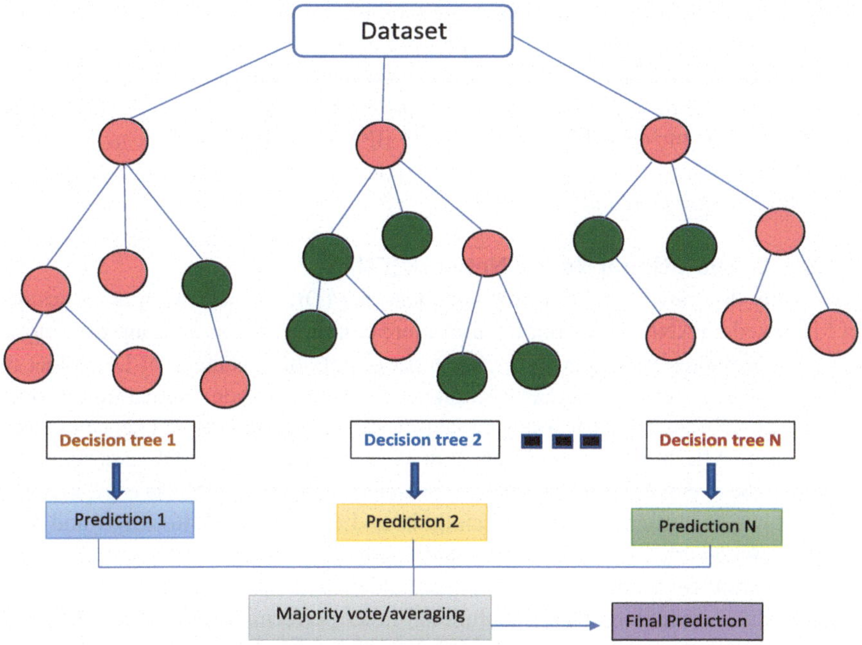

Fig. 7.5 Diagram showing the concept of random forest algorithm

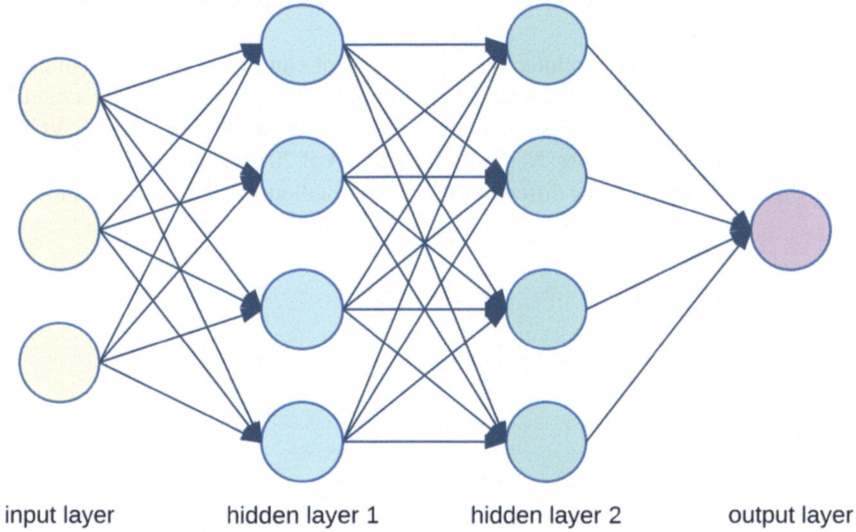

Fig. 7.6 Diagram showing the concept of ANN algorithm

ANN. After processing, the hidden layers send the results or predictions to the output layer, which may have single or multiple nodes, depending on the task. ANN generally uses feed-forward methodology and an activation function to yield a desired output with minimum error. Back propagation neural network (BPNN) is similar to ANN; the only difference is that in BPNN, the information can be propagated back to adjust the weight of nodes to reduce errors and improve the accuracy of the final output (Liu and Zhou 2009).

7.3.5.5 Convolutional Neural Network (CNN)

As already discussed, we know that deep learning (DL) is the branch or extension of ML based on ANN. However, DL uses more complex models and can solve more complex problems faster than ML. The most important advantage of DL is that it automatically performs feature extraction from raw data without human supervision or training. DL has higher learning capabilities than ML models and hence outperforms most ML models, giving accurate and precise results. DL considerably improves the accuracy of disease detection and classification. CNN is a DL method that follows feed-forward ANN (Kamilaris and Prenafeta-Boldú 2018) and has three hidden layers, namely, convolutional, rectified linear unit, and pooling layers, assembled to form a single network. These hidden layers perform the work of automatic feature extraction. The convolutional layer performs feature extraction, the rectified linear unit introduces non-linearity, and the pooling layer acts as the dimension reduction layer. The output layers in CNN are fully connected, which performs the classification. The main advantages of CNN are the removal of an expensive feature extraction task and faster testing. However, the training of CNN takes more time than other methods.

There are several DL architectures available that can be used to start building new DL models, and this saves researchers from the extra effort of starting a model from scratch. Some of the popular architectures are AlexNet, CaffeNet, VGG, GoogleNet, and Inception-ResNet50 (Saleem and Arif 2019). Some recent work on plant disease detection using different ML and DL methods is presented in Table 7.2.

Table 7.2 Recent advances in plant disease detection using machine learning and deep learning methods

S. no.	Year	Crop	Data	Method	Accuracy	References
1.	2016	Apple	Field image	RF	80%	Samajpati and Degadwala (2016)
2.	2017	Potato	PlantVillage and university data	RF	79%	Patil et al. (2017)
				SVM	84%	
				ANN	92%	
3.	2017	Apple	PlantVillage	SVM	94%	Chuanlei et al. (2017)
4.	2017	Rice	Field image	KNN	76%	Suresha et al. (2017)
5.	2019	Potato	PlantVillage	CNN	90%	Arya and Singh (2019)
				AlexNet	98%	
				DCNN	98%	Al-Amin et al. (2019)
6.	2019	Apple	PlantVillage	CNN	96%	Militante et al. (2019)
7.	2019	Rice	IoT	CNN	89%	Chen et al. (2019)
8.	2019	Tomato	PlantVillage	RF	95%	Govardhan and Veena (2019)
9.	2020	Rice	Field data	SVM	98%	Sethy et al. (2020)
				ResNet50	98%	
10.	2020	Rice	Field data	DCNN	90%	Li et al. (2020)
11.	2020	Tomato	PlantVillage	CNN-based ResNet	99%	Chakravarthy and Raman (2020)
12.	2021	Citrus	Field image	SVM	87%	Sujatha et al. (2021)
				RF	77%	
				CNN (VGG-16)	89%	
13.	2021	Wheat	Field image	VGG-16	90%	Goyal et al. (2021)
				RESNET50	80%	
14.	2021	Tomato	PlantVillage	CNN	97%	Sembiring et al. (2021)
15.	2022	Soyabean	Field image	CNN	84%	Rastogi and Swaminarayan (2022)

7.4 Evaluation Measures

Statistical evaluation of the ML model results is essential to understand the accuracy of the results. For regression problems, well-established standard statistical measures like bias, mean bias error, root-mean-square error, and coefficient of determination were used (Singh et al. 2022a, b, 2023a, b), whereas classification models were assessed using classification accuracy, sensitivity, specificity, accuracy, precision, F1 score, etc. (Kamilaris and Prenafeta-Boldú 2018).

7.5 Challenges of Machine Learning Methods in Plant Disease Detection

1. Big data requirements are the first and foremost challenge in using machine learning models for crop disease detection. There is a lack of freely available datasets for plant diseases and collecting massive data to develop the ML model becomes very costly.
2. The second major challenge in using ML models is selecting the model and its parameters (hyperparameters) for a specific disease. As there are no set rules or guidelines on how models should be built for a specific task, the researchers have nearly endless options to choose from.
3. The explainability of machine learning models is another major issue when it comes to why the model is performing well or badly. Most ML and DL models work as black box models, and there are no explanations for model results or predictions.
4. Machine learning models need the user to be aware of model bias and drift. The models will exhibit similar tendencies as the input data, so any (human-induced) bias in the input data will also reflect in the model predictions. Similarly, the relations between predictors and the target variables may change over time, affecting the model performance and the model developed using old data.
5. Ideally, images used for developing models should be captured from different angles, at different crop stages under varying environmental conditions. However, this is practically impossible to capture all variations. Moreover, adding more variables also increases the size of the training dataset. The field conditions always have certain deviations from those which were used for model development which affects the accuracy of the model.
6. Variations of disease symptoms with stages of crop or environmental conditions make it difficult for machine learning to classify the disease.
7. ML models are developed usually for a single disease. However, in the field, the appearance of multiple diseases simultaneously or diseases with some nutritional deficiency makes it difficult for the ML model to detect and classify the disease accurately.
8. Various crop diseases have similar symptoms; such cases also make it very difficult to identify or classify plant diseases.

7.6 Conclusion

Accurate detection and classification of plant diseases are essential for minimizing field losses and increasing farmer's income. The conventional methods of manual identification of plant disease need considerable effort, cost, labor, and expertise and yet suffer from major limitations of the subjectivity of the expert and may not be a practical approach for large areas. The machine learning approach provides efficient and robust options for disease detection, monitoring, and classification. Providing quality input data is the key to getting accurate results from machine learning. However, good accuracy of disease detection has been achieved in identifying many crop diseases using machine learning algorithms like SVM, ANN, and CNN, proving the potential of machine learning as an important tool for sustainable disease management.

References

Agarwal A, Sarkar A, Dubey AK (2019a) Computer vision-based fruit disease detection and classification. In: Smart innovations in communication and computational sciences. Springer, pp 105–115. https://doi.org/10.1007/978-981-13-2414-7_11

Agarwal M, Kaliyar RK, Singal G et al (2019b) FCNN-LDA: a faster convolution neural network model for leaf disease identification on apple's leaf dataset. In: 2019 12th international conference on information & communication technology and system (ICTS), pp 246–251. https://doi.org/10.1109/ICTS.2019.8850964

Al-Amin M, Karim DZ, Bushra TA (2019) Prediction of rice disease from leaves using deep convolution neural network towards a digital agricultural system. In: 2019 22nd International conference on computer and information technology (ICCIT), pp 1–5. https://doi.org/10.1109/ICCIT48885.2019.9038229

Arya S, Singh R (2019) A comparative study of CNN and AlexNet for detection of disease in potato and mango leaf. In: 2019 international conference on issues and challenges in intelligent computing techniques (ICICT) 1, pp 1–6. https://doi.org/10.1109/ICICT46931.2019.8977648

Bishop CM (2006) Information science and statistics. Pattern Recognition and Machine Learning Springer

Chakravarthy AS, Raman S (2020) Early blight identification in tomato leaves using deep learning. In: 2020 international conference on contemporary computing and applications (IC3A), pp 154–158. https://doi.org/10.1109/IC3A48958.2020.233288

Chen WL, Lin YB, Ng FL et al (2019) RiceTalk: rice blast detection using internet of things and artificial intelligence technologies. IEEE Internet Things J 7(2):1001–1010. https://doi.org/10.1109/JIOT.2019.2947624

Chuanlei Z, Shanwen Z, Jucheng Y et al (2017) Apple leaf disease identification using genetic algorithm and correlation based feature selection method. Int J Agric Biol Eng 10(2):74–83. https://doi.org/10.3965/j.ijabe.20171002.2166

Ficke A, Cowger C, Bergstrom G et al (2018) Understanding yield loss and pathogen biology to improve disease management: Septoria Nodorum blotch—A case study in wheat. Plant Dis 102(4):696–707. https://doi.org/10.1094/PDIS-09-17-1375-FE

Fratello M, Tagliaferri R (2018) Decision trees and random forests. Encyclopedia of bioinformatics and computational biology: ABC of bioinformatics, vol 1, pp 3

Govardhan M, Veena MB. (2019) Diagnosis of tomato plant diseases using random forest. In: 2019 global conference for advancement in technology (GCAT), pp 1–5. https://doi.org/10.1109/GCAT47503.2019.8978431

Goyal L, Sharma CM, Singh A et al (2021) Leaf and spike wheat disease detection & classification using an improved deep convolutional architecture. Inform Med Unlocked 25:100642. https://doi.org/10.1016/j.imu.2021.100642

Kamilaris A, Prenafeta-Boldú FX (2018) Deep learning in agriculture: a survey. Comput Electron Agric 147:70–90. https://doi.org/10.1016/j.compag.2018.02.016

Li D, Wang R, Xie C et al (2020) A recognition method for rice plant diseases and pests video detection based on deep convolutional neural network. Sensors (Basel) 20(3):578. https://doi.org/10.3390/s20030578

Liu L, Zhou G (2009) Extraction of the rice leaf disease image based on BP neural network. In: 2009 international conference on computational intelligence and software engineering, pp 1–3. https://doi.org/10.1109/CISE.2009.5363225

McCarthy J (1955) Human-level AI is harder than it seemed

Militante SV, Gerardo BD, Dionisio NV (2019) Plant leaf detection and disease recognition using deep learning. In: 2019 IEEE Eurasia conference on IOT, communication and engineering (ECICE), pp 579–582. https://doi.org/10.1109/ECICE47484.2019.8942686

Nguyen TA, Ly HB, Tran VQ (2021) Investigation of ANN architecture for predicting load-carrying capacity of castellated steel beams. Complexity. https://doi.org/10.1155/2021/6697923

Patil P, Yaligar N, Meena SM (2017) Comparison of performance of classifiers-svm, rf and ann in potato blight disease detection using leaf images. In: 2017 IEEE international conference on computational intelligence and computing research (ICCIC), pp 1–5. https://doi.org/10.1109/ICCIC.2017.8524301

Rastogi MN, Swaminarayan P (2022) Data collection and analysis using IOT device and machine learning techniques for disease classification on soybean plant. J Optoelectron Laser 41(6):143–150. 10050086.2022.06.18

Roy S, Ray R, Dash SR et al (2021) Plant disease detection using machine learning tools with an overview on dimensionality reduction. In: Data analytics in bioinformatics: a machine learning perspective, pp 109–144. https://doi.org/10.1002/9781119785620.ch5

Saleem P, Arif M (2019) Plant disease detection and classification by deep learning. Plants 8(11):468. https://doi.org/10.3390/plants8110468

Samajpati BJ, Degadwala SD (2016) Hybrid approach for apple fruit diseases detection and classification using random forest classifier. In: 2016 international conference on communication and signal processing (ICCSP), pp 1015–1019. https://doi.org/10.1109/ICCSP.2016.7754302

Sembiring A, Away Y, Arnia F et al (2021) Development of concise convolutional neural network for tomato plant disease classification based on leaf images. J Phys Conf Ser 1845(1):012009. https://doi.org/10.1088/1742-6596/1845/1/012009

Sethy PK, Barpanda NK, Rath AK et al (2020) Deep feature based rice leaf disease identification using support vector machine. Comput Electron Agric 175:105527. https://doi.org/10.1016/j.compag.2020.105527

Singh D, Jain N, Jain P et al (2020) PlantDoc: a dataset for visual plant disease detection. In: Proceedings of the 7th ACM IKDD CoDS and 25th COMAD, pp 249–253. https://doi.org/10.1145/3371158.3371196

Singh RN, Krishnan P, Bharadwaj C, Das B (2022a) Improving prediction of chickpea wilt severity using machine learning coupled with model combination techniques under field conditions. Ecol Inform 73:101933. https://doi.org/10.1016/j.ecoinf.2022.101933

Singh RN, Krishnan P, Singh VK, Banerjee K (2022b) Application of thermal and visible imaging to estimate stripe rust disease severity in wheat using supervised image classification methods. Ecol Inform 71:101774. https://doi.org/10.1016/j.ecoinf.2022.101774

Singh R, Krishnan P, Singh VK, Das B (2023a) Estimation of yellow rust severity in wheat using visible and thermal imaging coupled with machine learning models. Geocarto Int:38. https://doi.org/10.1080/10106049.2022.2160831

Singh RN, Krishnan P, Singh VK et al (2023b) Combining biophysical parameters with thermal and RGB indices using machine learning models for predicting yield in yellow rust affected wheat crop. Sci Rep 13:18814. https://doi.org/10.1038/s41598-023-45682-3

Sujatha R, Chatterjee JM, Jhanjhi N et al (2021) Performance of deep learning vs machine learning in plant leaf disease detection. Microprocess Microsyst 80:103615. https://doi.org/10.1016/j.micpro.2020.103615

Suresha M, Shreekanth KN, Thirumalesh BV (2017) Recognition of diseases in paddy leaves using KNN classifier. In: 2017 2nd international conference for convergence in technology (I2CT), pp 663–666. https://doi.org/10.1109/I2CT.2017.8226213

Wang K, Zhang Z, Wu X et al (2022) Multi-class object detection in tunnels from 3D point clouds: an auto-optimized lazy learning approach. Adv Eng Inform 52:101543. https://doi.org/10.1016/j.aei.2022.101543

Implementing Artificial Intelligence in Wheat Disease Identification: A Mobile Application Approach

8

Sapna Nigam, Rajni Jain, Vaibhav Kumar Singh, Samyak Jain, Sudeep Marwaha, and Alka Arora

Abstract

Wheat, a crucial crop globally and a staple food for millions, faces numerous disease threats, impacting yields and food security. However, wheat crops are vulnerable to many diseases, which can reduce yields and threaten food security. Therefore, developing accurate and efficient methods for identifying and diagnosing these diseases is essential. Traditional visual inspection methods are laborious and subjective, prompting the exploration of image-based approaches. Image-based identification methods offer a promising alternative, leveraging advances in computer vision and deep learning to detect and diagnose plant diseases from images automatically. In recent years, artificial intelligence (AI) techniques have shown great potential for image-based identification of plant diseases in crops, including wheat. This chapter explores the application of AI techniques for wheat disease identification using deep learning. Additionally, it discusses the creation of an image dataset repository for Indian wheat diseases and the development of a user-friendly mobile application to aid farmers in disease detection and management.

S. Nigam (✉) · S. Marwaha · A. Arora
ICAR—Indian Agricultural Statistics Research Institute, New Delhi, India
e-mail: sapna.nigam@icar.gov.in

R. Jain
ICAR—National Institute of Agricultural Economics and Policy Research, New Delhi, India

V. K. Singh
Wheat Pathology Laboratory, Division of Plant Pathology, ICAR—Indian Agricultural Research Institute, Pusa, New Delhi, India

S. Jain
Shiv Nadar University, New Delhi, India

V. K. Singh et al. (eds.), *Diseases of Field Crops: Diagnostics and Management*,
https://doi.org/10.1007/978-981-97-6160-9_8

Keywords

Artificial intelligence · Image repository · Plant diseases · Wheat rusts · Mobile application development

8.1 Introduction

Artificial intelligence in agriculture has become crucial for handling the various challenges of crop production regarding food security, productivity, sustainability, and the overall environmental impact. In order to meet the food demands of the continuously increasing global population, overall food production must be enhanced faster (FAOSTAT 2023). Food productivity is majorly challenged by plant diseases across the globe. Plant diseases are a significant threat to agriculture and can lead to crop losses, decreased yield, and reduced production quality. Traditionally, crop inspection and plant disorders have been carried out visually by farmers with some training or experience.

Additionally, identifying plant diseases manually is costly and time-consuming, requiring continuous expert oversight which may not be readily available, especially in remote or economically disadvantaged areas. Even experienced agronomists and pathologists can struggle to accurately diagnose specific diseases due to their complexity and variability. Misdiagnoses can lead to ineffective treatments and exacerbate the problem. Moreover, variations in symptoms displayed by diseased plants can further complicate diagnosis. This can result in farmers using inappropriate pesticides, potentially leading to the development of pathogen resistance and worsening the situation.

Moreover, earlier, farmers used to visit their nearest Krishi Vigyan Kendra (KVK) or call the Kisan Call Centre (KCC) to seek advice and guidance related to forewarning of diseases (Jain et al. 2009) and mobile advisories for plant protection and other. However, this conventional method of visiting KVKs is time-consuming and not always feasible for farmers. Also, the response time for specific customized recommendations is longer in both these cases. Kisan Call Centre (KCC) receives a high volume of calls related to plant diseases, indicating a significant need for automated plant disease management in the farming community with the increasing use of digital technologies and online platforms.

The above challenges of automated disease identification and advisories can be easily addressed in this new era of artificial intelligence (AI) powered by strong and robust algorithms and less expensive computational hardware. AI has the potential to revolutionize plant disease identification and management in agriculture by providing accurate, timely, and customized recommendations for disease management, thereby addressing some of the challenges associated with traditional approaches (Mohanty et al. 2016; Nigam and Jain 2020). AI is a branch of computer science discovered in the early 1950s that enables the computer to mimic human problem-solving behavior (Rich et al. 2009). Techniques like machine learning and deep learning, powered by AI, offer a promising avenue to advance and refine precise plant protection practices while also broadening the scope of computer applications in disease detection within agriculture. These algorithms can be trained to identify plant diseases using various data sources such as images, videos, and sensor data. This can help in the accurate and timely diagnosis of plant diseases and provide customized recommendations for disease management based on the specific crop, region, and environmental conditions (Dheeraj and Chand 2024). These AI-powered platforms can enable farmers to access real-time information about plant diseases and their management through their smartphones or other digital devices (Jain et al. 2021). This can reduce the need for farmers to visit extension centers or call KCCs, thereby saving time and improving access to information (Nigam et al. 2021a).

Hence, it is imperative to investigate automated AI methods like machine learning and deep learning for the cost-effective and early detection of disease occurrences in plants.

8.2 AI-Based Image Identification of Plant Diseases in Wheat

AI-based image identification of plant diseases is a field of research involving machine learning, deep learning algorithms, and computer vision techniques to analyze images of plant tissues and identify the presence of disease symptoms (Haque et al. 2022). These technologies can potentially revolutionize how plant diseases are diagnosed and managed in the agricultural domain. Different types of AI models can be used for image-based plant disease identification, including convolutional neural networks (CNNs), support vector machines (SVMs), and decision trees (Kamilaris and Prenafeta-Boldú 2018). CNNs are the most widely used AI models for image analysis, and they have shown excellent performance in identifying plant diseases in crops. AI algorithms can be trained on large datasets of images and associated disease labels, allowing them to learn patterns and features specific to different diseases. Currently, deep learning algorithms stand as the predominant choice for AI-driven computer vision methodologies, empowering computer models to autonomously extract features from data, a departure from the machine learning algorithms prevalent in the preceding decade (LeCun et al. 2015). The PlantVillage dataset for horticultural crops, publicly accessible, has been widely employed by numerous researchers (Hughes and Salathé 2015). Despite the importance of cereal crops for global food security, negligible attempts are available in the literature on cereal crops due to the unavailability of datasets in the public domain. Among cereals, wheat (*Triticum aestivum* L.) stands out as one of the most crucial crops of global significance economically. India, being the second-largest producer and consumer of wheat worldwide, contributes approximately 13.5% to the global wheat production (FAOSTAT 2023). Wheat serves as a vital source of protein (20%) and calorie (19%) intake through consumption. However, wheat production in India faces numerous biotic stressors, posing significant threats to its productivity. Among the worldwide pests and diseases, wheat rust, including leaf rust (brown rust), stripe rust (yellow rust), and stem rust (black rust), is notorious for causing substantial yield losses in wheat (Chen et al. 2014; Singh et al. 2017; Srinivas et al. 2023a). Yield losses due to wheat rust have been estimated to range from 10% to 70%, while during severe epidemics, grain damage could reach up to 100% (Chen 2005; Srinivas et al. 2023b). Therefore, ensuring timely and effective control of wheat rusts is imperative for sustaining global productivity and ensuring food security (Singh et al. 2016; Nigam et al. 2021b).

To the best of our knowledge, only limited work is reported on mobile application–based wheat disease identification specifically for Indian conditions. This gap in research is largely attributed to the absence of suitable image datasets for Indian wheat diseases. Also, the algorithms developed for other crop disease identification will not be suitable for wheat diseases due to different features and symptoms. Therefore, recognizing the importance of automated wheat rust identification, there is a need to develop robust and efficient algorithms for identifying wheat diseases in the Indian scenario along with the disease severity stages for timely and accurate preventive measures. With this backdrop, there is a need to develop AI models and user interfaces to identify the selected wheat diseases at the field level at the earliest.

8.3 Key Components of an Image-Based Wheat Disease Identification System

The key components of an image-based disease identification system involve acquiring high-quality images, pre-processing and extracting relevant features, and classifying the images using deep learning algorithms. The key components of the wheat disease identification system are discussed below:

8.3.1 Image Acquisition

This involves capturing high-quality images of the plant or crop of interest using appropriate cameras, lighting, and environmental conditions. We investigated three types of wheat rust diseases: stem rust (black rust), leaf rust (brown rust), and stripe rust (yellow rust). *Puccinia graminis f.* sp. *tritici* West causes stem rust and has large oval-shaped pustules filled with dark orange to red spores. These pustules convert into black spores at the end of the wheat season. It is prevalent all over the country, specifically in the Central and South Indian regions of the country due to the warmer climate. Brown rust is caused by *Puccinia triticina* Eriks, having circular and orange-brown spore pustules on wheat leaves. On maturing of fungus, the pustules turn from orange to black. It is the most prevalent rust type found in the north-western, north-eastern, central, and peninsular regions of the country. Unlike stem rust, this rust variant does not manifest on the epidermal tissues. Stripe rust, also known as yellow rust, is attributed to *Puccinia striiformis f. tritici* Eriks, which stands as one of the most destructive diseases, significantly constraining wheat yield potential globally (Chen 2005). It manifests as yellow stripes on leaves, leading to substantial yield losses by impairing the photosynthetic system, causing foliar damage, stunting plant growth, and notably diminishing grain weight and quality (Chen 2005). In India, stripe rust has garnered increasing attention in recent years, especially in the North-Western Plain Zone (NWPZ) and Northern Hills Zone (NHZ) due to favorable weather conditions characterized by low temperatures and high humidity. We gathered images of plants affected by various diseases, as well as healthy plants, from the experimental fields of the Division of Plant Pathology at ICAR-Indian Agricultural Research Institute in New Delhi and Madhya Pradesh, from 2019 to 2021 year. These images were taken on sunny days to ensure good lighting conditions. We made sure to capture the images every 7–10 days after the initial occurrence of disease symptoms on the plant leaves. This interval allowed us to capture images at consistent stages of leaf growth and minimized any potential harm to the plants during collection. We collected images using a commonly available handheld mobile camera OnePlus7, OnePlus 6T, and Canon camera, having resolutions ranging from 16–20 megapixels, focus 1.7, and 25 mm wide features. The image dataset is called "WheaRust21," comprising 6556 original images, which were further augmented using different techniques (Nigam et al. 2023). This dataset is used for developing the deep learning model for identifying wheat diseases.

8.3.2 Pre-processing Techniques

Image pre-processing involves the preparation of image data for machine learning models. The goal of pre-processing is to clean, transform, and normalize the data to make it suitable for modeling. Some common steps in image data pre-processing are image resizing, normalization, augmentation, and noise reduction. In image resizing, the images are transformed to a specific size to ensure that they are compatible with the deep learning algorithms being used. The common size of the images was 256 × 256 pixels. Image normalization normalizes the pixel values of the images to a specific range by dividing pixel values by 255 to ensure that all images have the same scale. Image augmentation increases the size and diversity of the dataset by using flipping, rotation, and scaling of the images. Also, the blurred images or images with noise were removed to improve the quality of the data. These steps are important to ensure that the image data is of high quality and suitable for modeling and that the results of the machine learning models are reliable and interpretable (Fig. 8.1).

8.3.3 Creation of Image Repository

A well-organized image repository is developed at a local host using Piwigo software to facilitate the management of large amounts of our image data. The image repository helps in easy access and includes organizing data by experimental conditions, metadata, and other relevant factors. Other advantages of developing an image repository for research image data include the following:

Fig. 8.1 Sample images of wheat rust dataset for disease identification

- Data organization: Image repository helps in organizing large amounts of images in a well-structured manner.
- Centralized storage: Researchers can easily access and share data with others in their organization or domain.
- Data security: Data is protected from unauthorized access, accidental deletion, or any type of data loss.
- Data standardization: It can facilitate the standardization of data formats and metadata across experiments, making it easier to compare and analyze data.
- Data sharing: It enables the easy sharing of data between researchers and collaborators, promoting research integrity and transparency.
- Reproducibility: It provides a means to store and share data, enabling other researchers to verify and build upon the results of published studies.

Therefore, our image repository consists of a main album, named "WheatRustData." The main "WheatRustData" album consists of four sub-albums as per the disease name "Stem Rust," "Yellow Rust," "Brown Rust," and "Healthy" leaves of wheat crop. The screenshot of the main page of the repository is shown in Fig. 8.2.

Figures 8.3, 8.4, 8.5, and 8.6 depict the screenshots of each of the four sub-albums within the image repository. These figures illustrate how the images are viewed within the repository, with the option for users to easily scroll down to see the images. Users can customize the number and size of images displayed using the "custom" option in the left corner of the screen. Additionally, images can be sorted by various options, such as alphabetical order or date of image taken, to provide greater accessibility and ease of use for users.

On clicking on the image, the associated metadata of that image is shown on the right side of the screen. Figures 8.7, 8.8, and 8.9 show the details of the image such

Fig. 8.2 Home page of image repository

Fig. 8.3 Yellow rust–infected leaf images in repository

as creation, the posting date of the image on the repository, the album path, and the device and model name from which it is captured. The date and time of the image on which it is captured are also mentioned along with the aperture and focus.

8.3.4 Feature Extraction and Classification

This step involves identifying and extracting relevant features from the pre-processed images, which can be used to distinguish between healthy and diseased plants. In deep learning, there is automated extraction of features, unlike machine

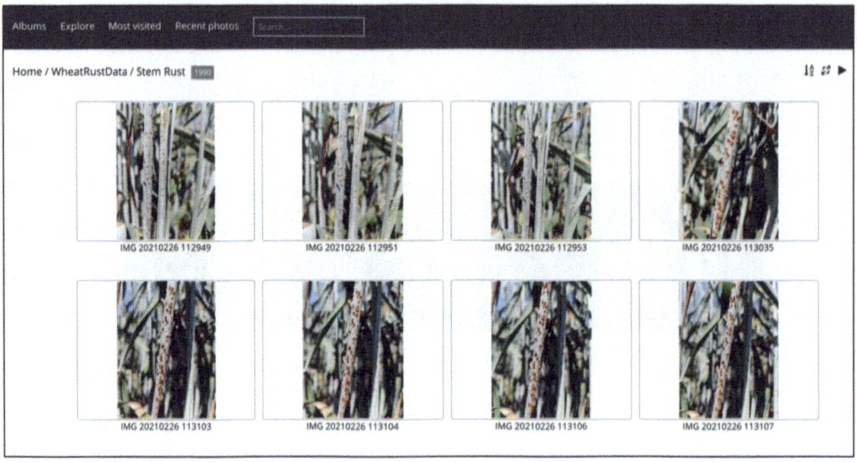

Fig. 8.4 Stem rust–infected leaf images in repository

Fig. 8.5 Healthy leaves of wheat in repository

learning algorithms. Further, in the classification task, the extracted features of input images are assigned to specific classes, such as healthy or diseased, based on predefined criteria. The feature extraction and classification steps are combined in the overall framework developed for wheat disease identification and discussed in a subsequent section.

Fig. 8.6 Brown rust–infected leaves in repository

Fig. 8.7 Screenshot of yellow rust–infected leaf image along with the metadata

8.4 AI Model Framework for a Wheat Disease Identification System

In our wheat disease identification AI model, we utilized a powerful pre-existing architecture called EfficientNet, which was created by Tan and Le in Tan and Le 2020. To customize it for our specific task, we made adjustments to the last layer of this architecture and added some extra layers. This allowed our model to accurately classify four types of damaged leaves in the WheatRust21 dataset. Our approach involved a technique called transfer learning, where we replaced the final layer of

Fig. 8.8 Screenshot of brown rust–infected leaf image along with the metadata

Fig. 8.9 Screenshot of stem rust–infected leaf image along with the metadata

each version of the EfficientNet model (B0-B4) with a few different types of layers at the bottom of the network. These changes help our model make sense of the input images and categorize them into four groups: healthy leaves, leaves with yellow spots, leaves with brown spots, and leaves affected by stem rust. Figure 8.10 provides a visual representation detailing the architectural intricacies of the AI model employed for wheat disease identification.

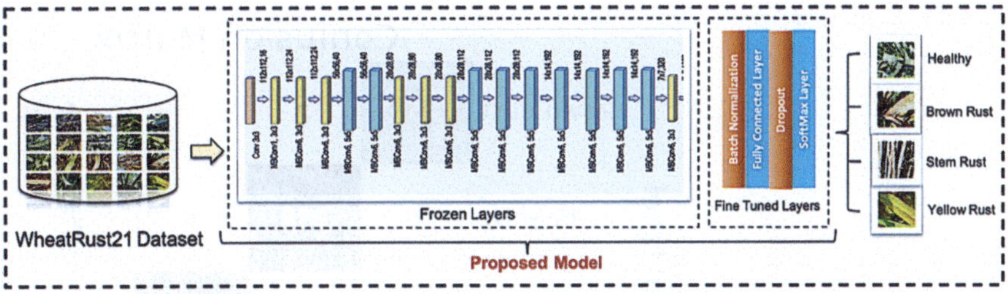

Fig. 8.10 AI model framework for wheat disease identification. (Source: Nigam et al. 2023)

Table 8.1 Comparative analysis of EfficientNet models

EfficientNet model	Size		WheatRust21 dataset accuracy (%)			Training time/epoch (s)	Prediction time/image (s)
	Input size (px)	Batch size	Train	Val	Test		
B0	224	32	97.70	97.79	96.55	32	0.19
B1	240	32	98.84	97.93	96.70	78	0.19
B2	260	32	98.52	97.47	97.09	108	0.20
B3	300	32	99.78	98.32	97.10	122	0.22
B4	380	32	99.83	98.47	98.93	171	0.30

8.4.1 Hyperparameters

The hyperparameters for the disease identification models are the number of epochs from 15 to 50 epochs due to early stopping. The batch sizes for models EfficientNet (B0-B4) were 32. The dropout rate to avoid overfitting was set to 0.2, and the Adamax optimizer was employed for training.

8.4.2 Hardware Implementation

The experimental setup for the above experiments was conducted on a DGX server. The deep learning framework used in the experiments included Keras with TensorFlow as the backend. The coding was done using the Python programming language on Jupyter Notebook within the Conda environment. The graphical processing unit (GPU) utilized in the experiments was the NVIDIA Tesla V100-SXM2.

8.4.3 Performance Evaluation

In Table 8.1, the outcomes of assessing how well each EfficientNet model (B0–B4) performed on the WheatRust21 dataset are presented. To make learning more effective and speed up training, we utilized early stopping and a learning rate scheduler. Total training time refers to the duration when the model's loss values start to rise.

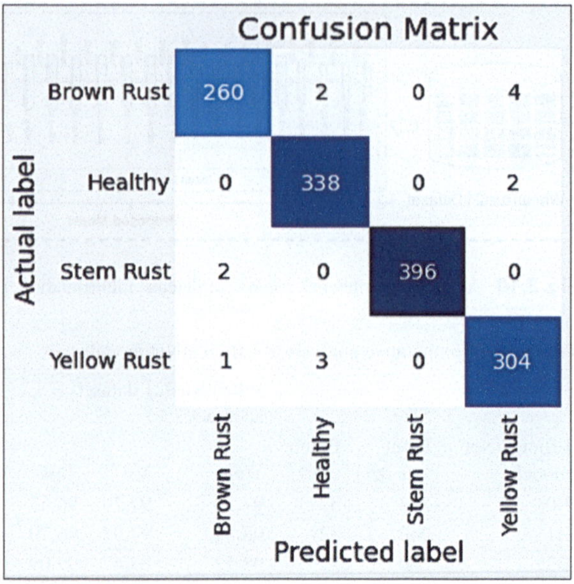

Fig. 8.11 Confusion matrix of the wheat model

We calculated the training time per epoch by dividing the total training time by the number of epochs. The hyperparameters remained unchanged across all experiments, as outlined in Table 8.1.

Table 8.1 illustrates that our suggested model, which is built on the EfficientNet B4 architecture, obtains the highest levels of accuracy in both training and testing phases when applied to our WheatRust21 dataset. Out of all the hardware restrictions, the suggested model demonstrates the highest training and testing accuracy, with scores of 99.83% and 98.47%, respectively.

8.4.4 Evaluation Metrics

Figure 8.11 displays the confusion matrix of the proposed wheat disease identification model. The numbers along the diagonal represent the count of accurately identified photos, while the numbers outside the diagonal reflect instances of misclassification.

Furthermore, the classification report resulting from the confusion matrices includes the F1, precision, and recall metrics for the suggested disease identification model (Table 8.2). A model is deemed appropriate if its F1 score is near to one.

8.4.5 Accuracy and Loss Curves

Figure 8.12 shows the accuracy and loss trends of the wheat disease identification model during training and validation. According to Hossin and Sulaiman (2015), a

Table 8.2 Performance measures of the proposed network

	WheatRust21 dataset		
Disease classes	Precision	Recall	F1
Brown rust	0.99	0.98	0.99
Healthy	0.99	0.99	0.99
Stem rust	1.00	0.99	1.00
Yellow rust	0.98	0.99	0.98

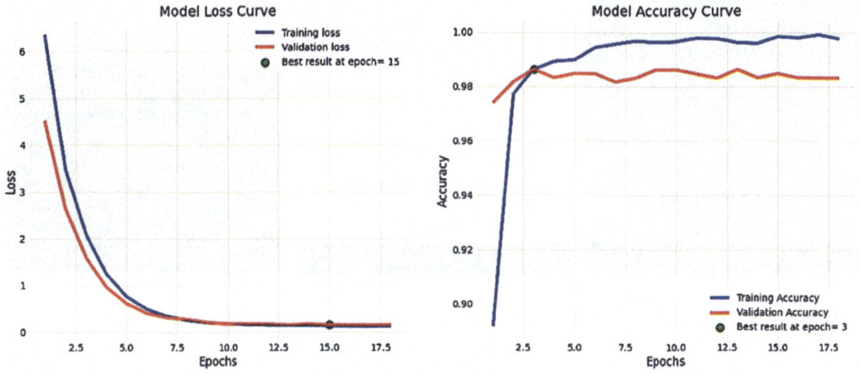

Fig. 8.12 Accuracy curves for the proposed model

model is considered effective when both training and validation accuracy consistently improve without significant fluctuations. The results of our proposed model indicate its suitability for automated wheat disease detection.

8.5 Deployment of Wheat Disease Identification Model in a Mobile Application

A mobile application is also developed for deploying an AI model for wheat disease identification to provide increased accessibility, high-throughput screening, data collection and feedback, and remote deployment capabilities. With a mobile app, users can easily and quickly identify disease symptoms in their wheat crops without the need for expertise. This can save time and resources and enable faster response times during an outbreak. Figure 8.13 illustrates the home screen of the mobile application.

Figure 8.13 shows a splash screen that illustrates the name and description of the mobile app. Once the splash screen disappears, an option for clicking the leaf image from the mobile camera and uploading the image from the device gallery is shown. Upload the image from a mobile device and then click on the "*Identify*" button to predict the input image. When a leaf image is uploaded to the server and if it is in the healthy class, it will display "Image Identified as Healthy," and no additional processing of the image is required (Fig. 8.14).

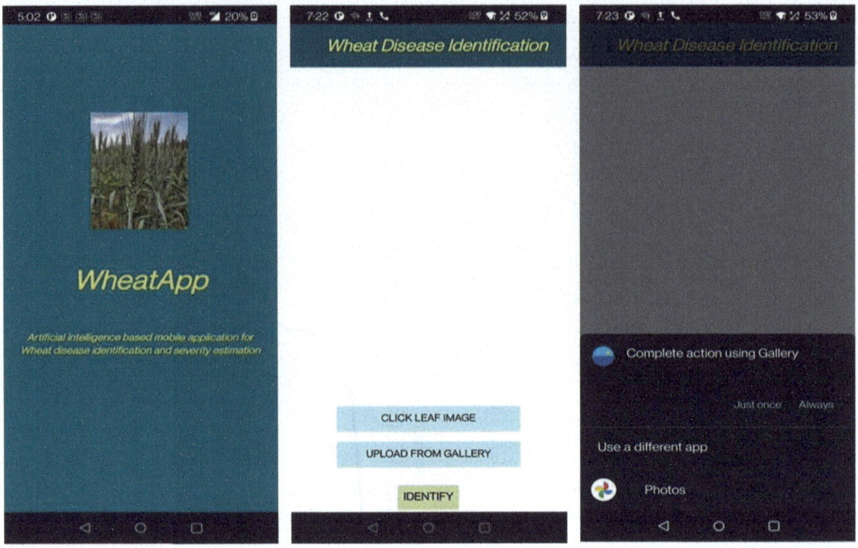

Fig. 8.13 Screenshots of a developed mobile application for wheat disease identification

If the input wheat leaf image is disease-infected, then the "Identify" button predicts the type of disease. In Fig. 8.15, the uploaded image is disease-infected, and by clicking the "*Identify*" button, the following screen shows the result. The disease identified is yellow rust (Fig. 8.15) and stem rust (Fig. 8.16).

8.5.1 Limitations and Future Scope

The study has covered only three significant diseases of wheat. Models including other diseases of wheat and symptoms due to nutritional deficiency in the wheat plant may be incorporated into the model. The study may be integrated with various recommendations regarding applying protection chemicals. The study can also be extended further by including additional severity levels of wheat diseases. Furthermore, combining deep learning with hyperspectral imaging could be a potential approach to early prediction that drastically lowers the use of pesticides on crops. The authors anticipate that the proposed frameworks will be a remarkable contribution to agricultural sciences.

8.6 Conclusion

In recent years, deep learning algorithms have relied heavily on the diagnosis and classification of diseases. These algorithms offer a major advance over the conventional method by automatically extracting the phenotypic features of disease-infected plant leaves. Developing mobile applications that can detect crop disease at

Fig. 8.14 Identification of healthy or infected leaf

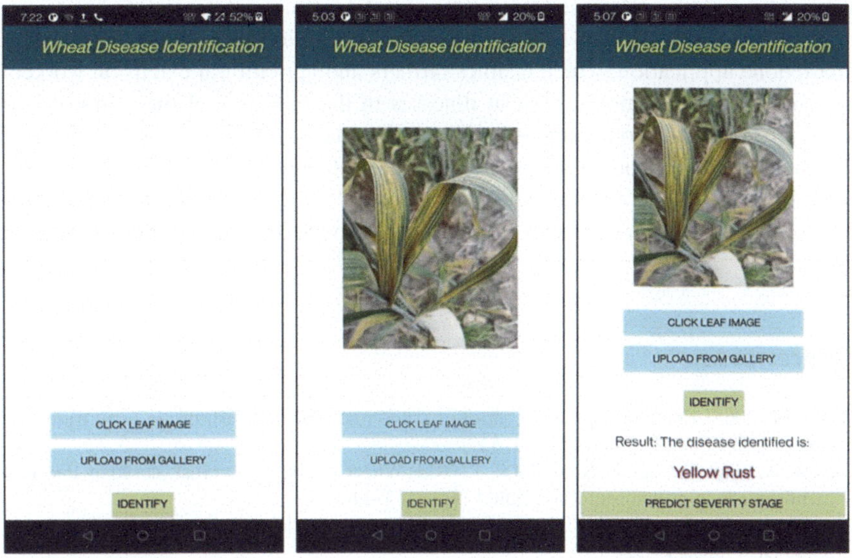

Fig. 8.15 Image-based identification of yellow rust disease

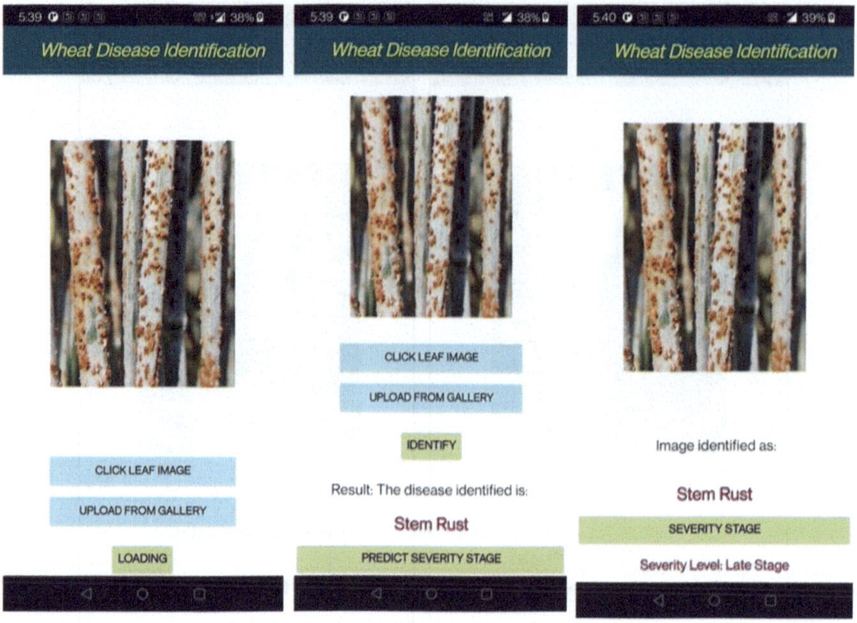

Fig. 8.16 Image-based identification of stem rust disease in a mobile application

an early stage is crucial to maximizing the potential of these algorithms and making them available to farmers. Our proposed model was developed in this study using the WheatRust21 dataset of rust-infected wheat crops. A precision of 98.93% was attained by our wheat disease identification model, which was constructed using the EfficientNet architecture. The concept is further implemented in our WheatApp smartphone application, which enables farmers and agricultural extension workers to easily access and monitor wheat illnesses in the field in real time. In order to receive fast and accurate suggestions based on the severity level of the disease, stakeholders can upload an image of their wheat plant to the app. Through the prevention of crop losses and the reduction of excessive use of plant protection chemicals, our automated system will support environmentally friendly and sustainable agriculture.

References

Chen XM (2005) Epidemiology and control of stripe rust [Puccinia striiformis f. sp. tritici] on wheat. Can J Plant Pathol 27(3):314–337

Chen W, Wellings C, Chen X, Kang Z, Liu T (2014) Wheat stripe (yellow) rust caused by Puccinia striiformis f. sp. tritici. Mol Plant Pathol 15(5):433–446

Dheeraj A, Chand S (2024) LWDN: lightweight DenseNet model for plant disease diagnosis. J Plant Dis Protect 131:1–17

FAOSTAT 2022, Statistical Database of the Food and Agriculture of the United Nations (2023) FAO. http://www.fao.org

Haque M, Marwaha S, Deb CK, Nigam S, Arora A, Hooda KS, Agrawal RC (2022) Deep learning-based approach for identification of diseases of maize crop. Sci Rep 12(1):1–14
Hossin M, Sulaiman MN (2015) A review on evaluation metrics for data classification evaluations. Int J Data Mining Knowledge Manage Process 5(2):1
Hughes D, Salathé M (2015) An open-access repository of images on plant health to enable the development of mobile disease diagnostics. arXiv preprint arXiv:1511.08060
Jain R, Minz R, Ramasubramanian V (2009) Machine learning for forewarning crop diseases. J Indian Soc Agric Stat 63(1):97–107
Jain R, Nigam S, Santrupth S (2021) Artificial intelligence based models for plant protection. Int J Agric Environ Sustain 3(1):1–7
Kamilaris A, Prenafeta-Boldú FX (2018) Deep learning in agriculture: a survey. Comput Electron Agric 147:70–90
LeCun Y, Bengio Y, Hinton G (2015) Deep learning. Nature 521(7553):436–444
Mohanty SP, Hughes DP, Salathé M (2016) Using deep learning for image-based plant disease detection. Front Plant Sci 7:1419
Nigam S, Jain R (2020) Plant disease identification using deep learning: a review. Indian J Agric Sci 90(2):249
Nigam S, Jain R, Marwaha S, Arora A, Singh VK, Singh AK, Paul RK (2021a) Automating yellow rust disease identification in wheat using artificial intelligence. Indian J Agric Sci 91(9):1391–1395
Nigam S, Jain R, Marwaha S, Arora A (2021b) Wheat rust disease identification using deep learning. In: Chatterjee J, Kumar A, Rathore P, Jain V (eds) Internet of things and machine learning in agriculture: technological impacts and challenges. DeGruyter, Berlin, pp 239–250
Nigam S, Jain R, Marwaha S, Arora A, Haque MA, Dheeraj A, Singh VK (2023) Deep transfer learning model for disease identification in wheat crop. Eco Inform 75:102068
Rich E, Knight K, Nair SB (2009) Artificial intelligence, 3rd edn. Tata McGraw-Hill, New Delhi
Singh VK, Mathuria RC, Gogoi R, Aggarwal R (2016) Impact of different fungicides and bio-agents, and fungicidal spray timing on wheat stripe rust development and grain yield. Indian Phytopathol 69(4):357–362
Singh VK, Singh GP, Singh PK, Harikrishna, Mathuria RC, Gogoi R, Aggarwal R (2017) Assessment of slow rusting resistance components to stripe rust pathogen in some exotic wheat germplasm. Indian Phytopathol 70(1):52–57. https://doi.org/10.24838/ip.2017.v70.i1.48994
Srinivas K, Singh Vaibhav K, Srinivas B, Sameriya KK, Kumar U, Gangwar OP, Kumar S, Prasad L, Singh GP (2023a) Multi-environment phenotyping to identify broad-based, stable resistance in wheat germplasms against leaf and stripe rust diseases. Cereal Res Commun 51:931. https://doi.org/10.1007/s42976-022-00315-9
Srinivas K, Singh VK, Srinivas B et al (2023b) Determining the impact of stripe rust and leaf rust on grain yield and yield components' losses in Indian wheat cultivars. Cereal Res Commun 52:733. https://doi.org/10.1007/s42976-023-00435-w
Tan M, Le Q (2020) EfficientNet: rethinking model scaling for convolutional neural networks. In: International conference on machine learning. PMLR, pp 6105–6114

Part II

Management of Field Crop Pathogens/ Diseases

Bakanae Disease of Rice: Current Status and Management Strategies

9

Sunil Kumar Sunani, S. T. Prashantha, and B. M. Bashyal

Abstract

Bakanae disease of rice, caused by *Fusarium fujikuroi*, is emerging as one of the most devastating diseases, resulting in significant yield losses across various rice-producing states in India. Reductions in crop production ranging from 15% to 25% have been documented in different states. This disease not only leads to quantitative losses but also affects the quality of crop production. The incidence of bakanae disease has been steadily increasing, particularly in basmati rice cultivars in north-western India, notably in Punjab, Haryana, and Uttar Pradesh. This chapter focuses on the current status of Bakanae disease and its management strategies.

Keywords

Bakanae · *Fusarium fujikuroi* · Rice · Genomics · Variability · Management

9.1 Introduction and History

The disease, initially described by Hori (1898) and later referenced by Ito and Kimura (1931), has been recognized in Japan since 1828. Hori attributed it to a *Fusarium* species, specifically *F. heterosporum* Nees. Sawada (1917) discovered the ideal form of the causal organism, which he named *Lisea fujikuroi* Saw. Subsequently,

S. K. Sunani
ICAR-Indian Institute of Pulse Research, Regional Station, Khordha, Bhubaneswar, Odisha, India

S. T. Prashantha · B. M. Bashyal (✉)
Division of Plant Pathology, ICAR-Indian Agricultural Research Institute, New Delhi, Delhi, India

V. K. Singh et al. (eds.), *Diseases of Field Crops: Diagnostics and Management*, https://doi.org/10.1007/978-981-97-6160-9_9

it was reclassified by Ito and Kimura (1931) into the genus *Gibberella* as *Gibberella fujikuroi*, with *Fusarium moniliforme* Sheld. serving as its anamorph. The present name of the bakanae pathogen is *F. fujikuroi* (Nirenberg 1976). In 1926, Kurosawa established the pathogenic nature of the fungus and its link to the peculiar hypertrophy seen in the disease. This phenomenon captured the interest of biochemists and plant physiologists, prompting further exploration that resulted in the isolation of gibberellin and other growth regulators by Yabuta and Hayashi (1939). In India, the disease was initially identified as and termed foot rot by Thomas (1931, 1933). Extensive research on bakanae disease has been conducted in Japan since 1898. The discovery of the teleomorph in nature prompted Sun and Snyder (1978) and other researchers to investigate the disease's epidemiology in Taiwan. Their findings revealed that the fungus exhibits heterothallic characteristics. "Bakanae" is a Japanese term meaning "foolish" or "bad" seedlings, referring to the distinctive elongation symptoms observed in the disease. This condition is recognized by various names in different parts of the world. The disease is referred to as "otokenae" (male seedling) in Japan, "white stalk" and "Fusariosis" in China, "palaylalake" (man rice) and "Equatorial foot rot" in British Guiana, and "bakanae" in French-speaking regions. In India, it is known as "foolish plant" or "foot rot" (Singh and Sunder 1997; Singh and Sunder 2012). Additionally, it goes by "white head disease" and "root rot" (Saremi et al. 2008).

9.2 Distribution

The bakanae disease prevails throughout the world in all the rice-growing countries such as Japan, European countries, America, African countries, Bangladesh, Brazil, California, Cameroon, China, European countries, India, Indonesia, Iran, Ivory Coast, Malaysia, Mexico, Nepal, Nigeria, Pakistan, Philippines, Spain, Sri Lanka, Thailand, Trinidad, Turkey, Uganda, Venezuela, and Vietnam (Cumagun et al. 2011; Desjardins et al. 2000; Kanjanasoon 1965; Khokhar 1990; Nelson et al. 1993; Surek 1997). Karov et al. (2009) documented the first occurrence of bakanae disease in Macedonia. This disease is increasingly becoming a significant concern in the regions of South and Southeast Asia, including India, Indonesia, Japan, Nepal, and Thailand. The changing cultivation environments in these areas have been attributed to the emergence of bakanae disease, as noted by Bashyal et al. (2016a), Desjardins et al. (2000), Kini et al. (2002), Saremi (2005), and Webster and Gunnell (1992). The first recorded observation of bakanae disease in India was made by Thomas (1931). Since then, instances of the disease have been reported from various states across the country. These include Andhra Pradesh (Vidyasekaran et al. 1966), Maharashtra (Parate and Lanjewar 1987), Odisha (Kauraw 1981), Punjab (Bedi and Dhaliwal 1970), Tripura (Sarkar 1986), Uttar Pradesh (Pavgi and Singh 1964), and West Bengal (Hajra et al. 1994). In India, bakanae disease is particularly widespread in states known for basmati rice cultivation, such as Haryana, Punjab, and Uttar Pradesh (Bashyal and Aggarwal 2013; Bashyal et al. 2014, 2016a). High incidences of bakanae disease have been noted in specific basmati rice varieties including Pusa Basmati 1121, Pusa Basmati 1509, Pusa 2511, CSR 30, Dehradoon Basmati, and Pakistani Basmati (Bashyal et al. 2016a). In 2011, reports indicated a disease incidence of up to 40% in districts including Kapurthala, Ropar, Patiala, Ludhiana, Amritsar, Gurudaspur, and Hoshiarpur in Punjab, India (Bashyal et al. 2014). Gupta et al. (2014) reported high bakanae disease incidence (1.2–40%) in basmati and other aromatic rice varieties from Uttarakhand, Punjab, Haryana, Rajasthan, and Bihar. Disease incidence of 1.25% to 50% was observed in Odisha where varieties like Pooja were observed to be susceptible under field conditions (Raghu et al. 2018). The disease was found to have a 100% prevalence rate in Pusa Basmati 1121 in districts such as Meerut, Hapur, Ghaziabad, Baghpat, and Shamli in Western Uttar Pradesh (Jain et al. 2019). This high prevalence underscores the significance of addressing bakanae disease in rice cultivation for the foreseeable future.

9.3 Economic Importance and Yield Loss

The decrease in rice yield caused by bakanae disease varies across different locations and even within the same location at different stages of crop growth. This disease can lead to yield reductions of up to 70% under favorable environmental conditions, as observed in various parts of the world (Hajra et al. 1994; Singh et al. 1996). Hossain et al. (2007) have reported a 25% yield loss in Bangladesh. It has been reported that nearly 40% yield reduction in Nepal due to this disease (Desjardins et al. 2000). Bakanae disease is responsible for a 3.7–14.7% reduction in crop production in Thailand, whereas, in Japan, 20–50% yield reduction has been observed (Ito and Kimura 1931; Kanjanasoon 1965). In Pakistan, yield losses ranging from 10% to 50% have been documented (Bhalli et al. 2001; Ghazanfar et al. 2013; Khokhar and Jaffrey 2002). Bakanae disease is notably more severe and widespread in basmati rice varieties in India (Bashyal et al. 2014; Gupta et al. 2014). Crop production losses ranging from 15% to 25% have been reported across various regions of India (Pannu et al. 2012; Pavgi and Singh 1964; Rathaiah et al. 1991; Sunder et al. 2014). The losses inflicted by this disease on crop production encompass both qualitative and quantitative aspects. The bakanae pathogen has been detected in paddy grains of various basmati rice genotypes (Butt et al. 2011; Bashyal and Aggarwal 2013; Bashyal et al. 2016b).

9.4 Symptoms

Fusarium fujikuroi, responsible for bakanae disease in rice, exhibits a spectrum of symptoms ranging from pre-emergence seedling death to grain infection during maturity. These symptoms include abnormal elongation, seedling blight, root rot, crown rot, etiolation, stunted growth, discoloration, grain sterility, and the formation of empty panicles, varying across different climates worldwide (Desjardins et al. 2000; Ou 1985; Sun and Snyder 1981; Webster and Gunnell 1992). The disease occurs both in the main crop and ratoon in Japan (Sasaki 1976) and in the field, as well as in the nursery. In the nursery, either the grains fail to germinate or the seedlings fail to emerge above the soil. Such pre-emergence damage occurs in a restricted manner. Pathogen growth manifests visibly in coleoptiles just above the soil line and also at the joint of palea and lemma in affected grains, as noted by Thomas (1931). In wet nurseries, seedling mortality begins merely a week after establishment. Post-emergence seedling mortality is frequently elevated during the nursery stage. These symptoms become apparent as early as the seventh day after sowing in wet nurseries and may persist for up to 6 weeks. Infected seedlings exhibiting such symptoms may wilt and perish within 3 to 6 days of their appearance. Not all infected seedlings may exhibit overgrowth. However, certain seemingly healthy plants may display symptoms after transplantation. When the nodal tissues of infected plants are split open, brown discoloration becomes apparent. Initially, this discoloration is confined to lower nodes, but as the infection progresses, upper nodes also exhibit the same brown discoloration. Mycelial masses, carrying conidia, are commonly found within the hollow internodes. As the disease incidence peaks, fungal growth becomes evident on the node surfaces, manifesting as a pink or white color on the plant's exterior, as documented by Surek and Gumustekin (1994). Thomas (1931) documented the development of adventitious roots from the lower part of the culm as one of the disease symptoms observed in India. Additionally, the pathogen induces lesions on the surface of rice plant leaves, as reported by Sasaki (1976).

Older diseased plants consistently yield fewer tillers compared to their healthy counterparts. This contrast becomes particularly noticeable during the seventh and eighth weeks, coinciding with the peak tillering period. As the flowering period draws near, affected plants exhibit slender and elongated tillers, which begin to bear spikelets a week to 10 days earlier than the rest of the crop. Upon splitting open the culms of infected plants, brown discoloration of the nodal tissues becomes evident. Initially, this discoloration is localized to the lower nodes. However, in later stages, the upper nodes also display similar symptoms. Within the hollow internodes, mycelial masses containing spores can be found. Subsequently, fungal growth emerges externally, primarily in the basal nodal zones, resulting in a white or pink bloom on the surface.

9.5 Host Range of the Pathogen

The bakanae pathogen has a wide host range. Besides rice, the pathogen infects maize, sorghum, sugarcane, wheat, barley, cotton, mango, pineapple, fig, pine, and asparagus (Singh and Sunder 1997).

9.6 The Pathogen

Fusarium fujikuroi Nirenberg is recognized as the primary causal organism of bakanae disease in rice. However, other species of *Fusarium*, such as *F. proliferatum* (Mats.) Nirenberg and *F. verticillioides* (Sacc.) Nirenberg, have also been observed to be linked with this disease, as documented by Amoah et al. (1995) and Desjardins et al. (2000). *G. fujikuroi* Sawada is recognized as the sexual stage of *F. fujikuroi* Nirenberg, as noted by O'Donnell et al. (1998) and Leslie and Summerell (2006). This pathogen belongs to the Kingdom Fungi, Phylum Ascomycota, Class Sordariomycetes, Order Hypocreales, and Family Nectriaceae (CABI, Index Fungorum, and CAB International). The pathogen produces ascospores within a sac-like structure known as ascus. Asci developed in fruiting bodies are known as ascocarps or perithecia, spherical to oval in shape and somewhat roughened outside, with a size of 250–330 × 220–280 μm. The asci are piston-shaped, cylindrical, with a flattened upper portion, measuring approximately 90–102 × 7–9 μm. Asci are typically 4- or 6-spored, although occasionally they may contain eight spores. The spores themselves are one-septate and measure about 15 × 5.2 μm. The hyphae exhibit branching and are septate in structure. The pathogen produces white color mycelium. Usually, the pathogen does not produce sporodochia, but if produced, they are pale orange in color. The pathogen is capable of producing both macro and micro conidiophores, each carrying macro and micro conidia, respectively. A single microconidiophore typically bears sub-lateral phialides. The microconidia exhibit agglutination, forming chains and often bearing false heads. They are typically 1–2 celled, with a fusiform oval shape. Macroconidiophores feature basal cells and apical phialides, from which macroconidia are generated. These macroconidia are slightly sickle-shaped or occasionally straight, as observed by Leslie and Summerell (2006) and Karov et al. (2009).

The disease is attributed to species within the *G. fujikuroi* species complex, which is further categorized into ten distinct fully fertile mating populations (MP-A–J), as detailed by Kvas et al. (2009), Bashyal et al. (2012), and Bashyal et al. (2020). Mating populations A, C, and D within the *G. fujikuroi* species complex are linked with the bakanae disease of rice. The conidial stages associated with MP-A, MP-C, and MP-D are *F. verticillioides*, *F. fujikuroi*, and *F. proliferatum*, respectively, as documented by Amoah et al. (1995) and Desjardins et al. (2000). In Taiwan, the mating population C (MP-C) was first time identified from the strains of bakanae pathogen–infecting rice (Hseieh et al. 1977). In Italy, this mating population has also been identified in association with the disease (Amatulli et al. 2010). Additionally, two mating populations, A and D, were detected in bakanae-infected rice plants from Australia, the USA, and Africa. Moreover, MP-D has been isolated from rice plants infected with bakanae disease in Asia (Amoah et al. 1995; Desjardins et al. 1997). Most researchers have observed that *F. fujikuroi* is the most virulent and abundant among all the species associated with bakanae disease in rice, as indicated by studies conducted by Nancy (2002), Zainudin and Salleh (2010), Zainudin et al. (2008a, b), Amatulli et al. (2010), and Bashyal et al. (2016a). Nevertheless, various other *Fusarium* species, including *F. moniliforme*, *F. andiyazi*, *F. sacchari*, and

F. subglutinans, have been identified in association with bakanae disease in diverse rice-growing nations. Researchers from countries such as India, Bangladesh, Malaysia, Nepal, Pakistan, Italy, Indonesia, and Iran have documented these findings (Bashyal and Aggarwal 2013; Quazi et al. 2013; Khokhar 1990; Desjardins et al. 2000; Pra et al. 2010; Saremi and Farrokhi 2004; Wulff et al. 2010; Zainudin et al. 2008a, b).

In India, three distinct *Fusarium* species, namely, *F. fujikuroi*, *F. proliferatum*, and *F. verticillioides*, have been identified in relation to bakanae disease in rice (Bashyal and Aggarwal 2013; Bashyal et al. 2016b). In India, 1–24% of rice seeds were reported to be infected by *G. fujikuroi* species complex (Bashyal and Aggarwal 2013). These three *Fusarium* spp. were varied in pathogenicity and symptom production. The crown and stem rot symptoms were produced by *F. verticillioides*, while elongation symptoms were induced by *F. fujikuroi* (Bashyal and Aggarwal 2013). Bashyal et al. (2016b) investigated the individual and combined effects of three pathogenic *Fusarium* species—*F. verticillioides* (Fv), *F. proliferatum* (Fp), and *F. fujikuroi* (Ff)—on the severity of bakanae disease in the susceptible rice genotype Pusa Basmati 1121. In single pathogen inoculations, *F. fujikuroi* demonstrated high virulence with a disease severity exceeding 70%. Conversely, disease severity was 43% for *F. proliferatum* and 62% for *F. verticillioides*. The highest disease severity of 75% was observed when rice plants were co-inoculated with (Fv + Ff), followed by 70% in the case of (Ff + Fv + Fp), 63% with (Ff + Fp), and 48% with (Fv + Fp). Bashyal et al. (2016a) conducted an extensive survey on bakanae disease incidence in various basmati rice-growing states of India. They found high disease incidence, reaching up to 20%, in the Mathura District of Uttar Pradesh and the Karnal District of Haryana. Out of the 126 *Fusarium* spp. isolates collected, a staggering 99.2% were identified as *F. fujikuroi*. These isolates were further categorized based on their virulence, with 28.6% classified as highly virulent, 34.1% as virulent, and 37.3% as moderately virulent.

9.7 Isolation and Pathogenicity

The isolation of bakanae pathogen culture can be accomplished using various mediums including potato dextrose agar (a common medium), peptone pentachloro nitrobenzene agar (a selective medium), and carnation leaf agar and Spezieller Nahrstoffarmer agar (natural mediums) (Pra et al. 2010; Zainudin et al. 2008a). Typically, fungal cultures are incubated at 25 ± 1 °C for a duration of 7 days. However, Burgess et al. (1994) observed good growth and sporulation of cultures under slightly different conditions. They reported successful incubation in a room with fluctuating temperatures, reaching 25 °C during the day and dropping to 22 °C at night, with a 12-h photoperiod. The soil dilution method is indeed a viable approach for isolating pathogens from the soil surrounding the root zone of infected plants (Saremi and Farrokhi 2004; Saremi et al. 2008; Sharma and Singh 1997). Seed inoculation has been noted as the most suitable method for studying the development of bakanae disease compared to the soil inoculation method (Amatulli et al. 2010; Agarwal et al. 1989; Karov et al. 2009; Quazi et al. 2013; Zainudin et al. 2008a).

9.8 Secondary Metabolites

Gibberellic acid is the main secondary metabolite produced by the bakanae pathogen which is isolated from the culture filtrate of *G. fujikuroi* and named gibberellins by Yabuta and Sumiki (1938). The metabolic pathway genes of GA3 were reported in fungi, bacteria, *Arabidopsis*, and also in cereal plants (Sakamoto et al. 2004). GA biosynthetic genes are typically organized in a gene cluster, as elucidated by Linnemanstons et al. (1999). This cluster comprises one GA4 desaturase gene (des) and four cytochrome P450 monooxygenase genes (P450-1 to P450-4) (Rojas et al. 2001; Tudzynski et al. 2003).

The *G. fujikuroi* species complex is known to produce various secondary metabolites, including fumonisin B1 and B2, moniliformin, beauvericin, fusaproliferin, fusaric acid, and gibberellins (Bacon and Hinton 1996: Leslie et al. 1996; Desjardins et al. 2000; Reynoso et al. 2004). Among the secondary metabolites produced by the *G. fujikuroi* species complex, fumonisin, fusaric acid, and gibberellin have been identified as the most potent contributors to disease development and pathogenicity on rice (Amoah et al. 1995; Zainudin et al. 2008b).

9.9 Variability

Considerable variability has been reported in the development of symptoms, as well as in the production of gibberellin and fusaric acid. Factors such as the amount of inoculum, various pathogen isolates, and the levels of gibberellin and fusaric acid production significantly influence the development of symptoms (Amatulli et al. 2010; Kaur et al. 2014; Nyvall 1999; Sharma and Bagga 2007). Nisikado and Matsumoto (1933) noted variation in pathogenicity among 66 strains of *G. fujikuroi* isolated from rice, based on seedling overgrowth. They observed that rice plants exhibit etiolating symptoms when the pathogen inoculum is low, which correlates with reduced gibberellin production upon host infection. Isolates or strains that produce high levels of gibberellin induce rice seedlings and plants to grow tall, thin, and yellow and ultimately perish. Conversely, high levels of fusaric acid production lead to the development of chlorotic seedlings that are stunted, followed by root and crown rots, ultimately resulting in their death (Amoah et al. 1995; Karov et al. 2009). A positive correlation has been observed between bakanae disease and the production of gibberellin in culture, as noted by Sunder and Satyavir (1998b). Additionally, Thakur (1974) found a correlation between the development of foot rot and the production of pectic enzymes. Ma et al. in 2008 recorded a significant correlation between bakanae symptoms and the length of seedlings treated with gibberellic acid. Bashyal and Aggarwal (2013) also observed variations of the *F. fujikuroi* isolates collected from rice seeds in terms of morphology and pathogenicity.

Thakur (1974) classified 48 isolates of *F. moniliforme* into four groups based on their variable growth rates in India. Meanwhile, Sunder and Satyavir (1998b) categorized 28 isolates into five groups according to their gibberellic acid production and virulence levels. Kaur et al. (2014) discovered that all isolates of the bakanae pathogen produced gibberellic acid, while only 45% of the isolates produced fusaric acid. Additionally, Ogawa (1988) reported resistance to (benzimidazole) benomyl and carbendazim against the bakanae pathogen, possibly due to successive applications as a seed disinfectant. According to Wada et al. (1990), isolates resistant to benomyl but sensitive to ergosterol biosynthesis inhibitors like perfurazoate did not produce gibberellins. However, all types of isolates, regardless of their resistance or sensitivity, produced fusaric acid. Additionally, Hamamura et al. (1989) reported that isolates less sensitive to triflumizole produced lower quantities of gibberellins and exhibited reduced pathogenicity. Similarly, in carbendazim-tolerant strains, reduced sporulation, mycelial growth, and lower pathogenicity were observed (Pan et al. 1997).

9.10 Diagnosis of Bakanae Disease

Bakanae disease of rice was initially discovered in 1828 in Japan, as documented by Ito and Kimura (1931). However, the specific *Fusarium* species associated with multiple disease symptoms remained unclear. Early authorities, such as Snyder and Hansen (1945), considered *G. fujikuroi* to be a single species. However, it is now recognized that this taxon comprises several distinct species, collectively known as the *G. fujikuroi* species complex (GFSC).

9.10.1 Conventional Approach to Diagnose Bakanae Disease

Rice plants infected with bakanae disease are visually inspected for a range of symptoms. These include rotting of seedlings before emergence, seedling blight, crown rot, root rot, stunted growth, excessive or abnormal elongation of plants, sterility, and the production of empty panicles. These symptoms can manifest under different climatic conditions across the world (Sun and Snyder 1981; Ou 1985; Webster and Gunnell 1992; Desjardins et al. 2000).

9.10.2 Identification of *F. fujikuroi* Based on Morphological and Cultural Characters

The rice plants infected with bakanae symptoms were cultured on a PDA medium. After obtaining pure culture, they were studied for their morphological and cultural characteristics. *F. fujikuroi* undergoes morphological changes, initially developing white mycelia that later transition to gray, violet, or magenta hues. Sporodochia are typically not present, but if they do occur, they appear pale orange and contain macroconidia. These macroconidia are moderately slender with tapered apical ends and poorly developed basal cells, usually having three to five septa. Microconidia are either club- or oval-shaped with a flattened base and may have zero to one septa, while chlamydospores are not found (Leslie and Summerell 2006).

9.10.3 DNA-Based Detection Assay for Detection of the *Fusarium fujikuroi*

The prompt and precise detection of plant pathogens is crucial for effective disease management. Traditionally, diagnosing a disease involves recognizing visual symptoms, isolating the associated organism using specialized media in a laboratory setting, and confirming its identity through microscopic analysis. These conventional approaches are time-consuming and less sensitive, some of the pathogen species are closely related which are difficult to identify, and high-throughput diagnosis is also problematic (Leslie and Summerell 2006). Therefore, nucleic acid amplification technologies are an alternative for specific and rapid detection of plant pathogens which include polymerase chain reaction, micro array, and isothermal amplification of DNA techniques.

PCR-based detection is a valuable tool for the majority of closely related pathogens, where morphological identification is difficult, or seed-borne pathogens, which can carry the disease in newer areas (Mills et al. 1992; Ward et al. 2004; Tisserat et al. 1991). Amatulli et al. (2010) employed species-specific primer pairs (Fuji1F/TEF1R and Proli1F/TEF1R) to detect *F. fujikuroi* and *F. proliferatum* from seeds or infected rice tissue. Identification of *Fusarium* species was primarily achieved at the species level through sequencing of the translation elongation factor 1-α (TEF) gene (Pra et al. 2010; Petrovic et al. 2013; Wulff et al. 2010; Geiser et al. 2013), internally transcribed spacer regions (ITS1 and ITS2), and β-tubulin (*tub2*) (O'Donnell et al. 1998; Bashyal and Aggarwal 2013). Additional techniques have been employed to differentiate members of the *G. fujikuroi* species complex. Petrovic et al. (2013) utilized AFLPs, Xu et al. (1995) utilized CHEF gel karyotypes, and Amoah et al. (1995) used RAPD. These studies collectively revealed that the *G. fujikuroi* complex comprises three distinct lineages: the African, Asian, and American clades (O'Donnell et al. 1998). Phylogenetic analysis of the ITS region in India, as conducted by Bashyal and Aggarwal (2013), resulted in a dendrogram comprising three distinct clades. Clade I comprised isolates of *F. verticillioides*, while Clade II encompassed *F. fujikuroi*, and Clade III was constituted by *F. moniliforme*. Hsuan et al. (2011) identified *Fusarium* species within the *G. fujikuroi* species complex, isolated from rice, sugarcane, and maize, utilizing both morphological characteristics and DNA sequencing of the TEF-1α gene. Additionally, Amatulli et al. (2012) employed the PCR and qPCR methods for the detection of *F. fujikuroi* and *F. proliferatum* in rice tissue. Lin et al. (2014) used speciesspecific TaqMan PCR to distinguish two *Fusarium* species (*F. verticillioides* and *F. proliferatum*).

PCR and loop-mediated isothermal amplification (LAMP) assays were developed for the detection of *F. fujikuroi*. A specific marker, BMFf1900, was created for amplifying *F. fujikuroi* with a sensitivity of 10 pg. This marker was designed based on unique sequences (NRPS31 gene) of *F. fujikuroi* obtained from GenBank accession KY953210. Real-time primers (NRPS31RT-F and NRPS31RT-R) were subsequently designed, resulting in a further increase in the sensitivity of the assay to 10 fg. This enhanced sensitivity allows for more precise quantification of the inoculum. Later on, the LAMP assay was standardized, for the visualization of results under isothermal conditions. The PCR and LAMP-based assays have undergone validation under field conditions, seed testing, and resistance evaluation of rice genotypes. These diagnostics have proven to be rapid, efficient, and sensitive in detecting *F. fujikuroi*. Their utility extends to testing seed health, validating varietal resistance, and monitoring pathogen inoculum levels, thereby providing valuable tools for disease management strategies (Sunani et al. 2019).

9.11 Disease Epidemiology

Fusarium fujikuroi, the causal agent of bakanae disease in rice, is primarily transmitted through infected seeds, making it predominantly seed-borne. However, the pathogen can also persist in plant debris and soil. Infected seeds serve as the key source of primary inoculum, initiating disease spread in rice fields (Anderson and Webster 2005; Watanabe 1974) as the inoculum present in soil diminishes rapidly following the decomposition of rice plant debris in field conditions (Kanjanasoon 1965; Ou 1985; Sun 1975). During adverse conditions, the pathogen can persist on the seed coat as spores and in host debris within the soil, adopting a survival strategy through thick-walled hyphae or macroconidia (Ou 1987; Saremi and Farrokhi 2004; Watanabe 1974). Sun (1975) reported that the fungus can survive in the soil for approximately 3–4 months, primarily in the form of thick-walled hyphae or macroconidia. Additionally, Karov et al. (2009) observed significantly reduced bakanae disease incidence when clean and healthy seeds were planted in fields with a history of the disease.

Puyam et al. (2017) conducted a study on the survival of *F. fujikuroi* in soil, concluding that the bakanae pathogen's survival under field conditions diminishes over time, indicating its weak soil-inhabiting nature. They further noted that conidia primarily spread through wind and water dispersal. Seeds become infected either by airborne ascospores produced during the flowering stage or through contamination by conidia during seed harvesting. These infectious agents subsequently lead to seedling infection upon germination (Sun 1975). Pathogen presence in seeds has been observed in the husk, embryo, and endosperm, with the highest colonization occurring in the husk region (Sunani et al. 2020). In different cultivars, the incidence of embryo infection varies from 2% to 41%. Manandhar (1999) reported that the isolation frequency of the bakanae pathogen from infected, untreated rice seeds was 15%, 25%, and 75% from the embryo, endosperm, and hulls, respectively. While the fungus is known to be systemic in plants, it is not typically observed to infect the panicles systemically. Nisikado and Kimura (1941) found the pathogen inside the xylem vessels of rice plants. Sunani et al. (2017) observed that the conidial germination of the pathogen was highest in root tissue, followed by stems.

The optimum temperature for pathogen infection ranges from 27–30 °C, while the most conducive temperature for disease development is 35 °C, which coincidentally is also favorable for plant growth. Decreasing the temperature resulted in reduced disease incidence (Burgess et al. 1996; Saremi and Farrokhi 2004). Mandal and Chaudhuri (1988) observed a reduction in the pathogen population in soil with the application of high doses of nitrogenous fertilizers. Additionally, they noted that less disease incidence was observed in directly seeded rice compared to transplanted rice (Saremi and Farrokhi 2004). During the summer crop season, dry nurseries experiencing high temperatures and relative humidity conditions exhibited high disease incidence. Kanjanasoon (1965) observed that bakanae incidence was higher in dry seeds compared to pre-soaked seeds.

9.12 Genomics of *Fusarium fujikuroi*

Fusarium fujikuroi is recognized as a phylogenetically distinct species within the *G. fujikuroi* species complex (O'Donnell et al. 1998). Over the past decade, whole genomes of multiple *Fusarium* species have been sequenced, aiding in the comprehension of host-pathogen interactions (King et al. 2015). Currently, 13 genomes of *F. fujikuroi* have been published from various nations (Bashyal et al. 2017; Niehaus et al. 2017; Chiara et al. 2015; Wiemann et al. 2013; Jeong et al. 2013). Wiemann et al. (2013) identified two genes unique to *F. fujikuroi*, namely, the polyketide synthase gene (PKS19) and the non-ribosomal peptide synthetase gene (NRPS31), based on the genome sequence of the *F. fujikuroi* isolate IMI58289. The genomes of eight *F. fujikuroi* isolates from various geographic locations were sequenced. They observed differences in the types of asexual spores produced (macroconidia and/or microconidia), variations in chromosome sizes, and differences in the expression profiles of secondary metabolite gene clusters among these isolates (Niehaus et al. 2017). In addition to this, the isolates were characterized as two distinct pathotypes based on symptoms produced (rotting and stunting) and secondary metabolite profiling. Bashyal et al. (2017) identified 1194 secretory proteins in the *F. fujikuroi* isolate "F250." Among the 356 carbohydrate-active enzyme (CAZyme) genes identified, glycoside hydrolase (GH) families such as GH3 and GH5, which are involved in the degradation of cellulose and hemicelluloses, were predominant. Additionally, *F. fujikuroi* harbored α-galactosidases encoding CAZymes GH67 and GH36, which are deficient in other plant pathogens. The genome of *F. fujikuroi* also contained an abundance of pectin-degrading enzyme families (GH28, GH78, CE8, PL3, PL1, and PL9), implicated in the colonization of root tissue. Based on whole-genome analyses of five isolates (IMI58289, B14, FGSC8932, KSU3368, and KSU X-10,626) from various geographic locations, phylogenetic analysis revealed that the Indian isolate "F250" is genetically closer to the Taiwan isolate "IMI58289." Furthermore, through comparative analysis, 12,240 common clusters were identified among the different genomes of *F. fujikuroi* (Bashyal 2018).

9.13 Histopathology/Infection Process

Histopathological studies give important information on pathogenesis. The presence of *F. fujikuroi* inside the kernel and its extensive distribution throughout seemingly healthy mature plants indicate that the infection is systemic in nature (Kato et al. 2012; Wiemann et al. 2013). Additionally, it is known to infect both maize and rice (Koehler 1942).

Kushiro et al. (2012) explored *F. proliferatum* pathogenic potential in rice plants concerning browning, fumonisin production, and survival rates in rice seeds. They administered pathogen inoculation by spraying a conidial suspension of *F. proliferatum* onto rice plants during the flowering period. Browning manifested on the ear, leaf, and stalk of rice. Six months later, pathogen recovery revealed that *F. proliferatum* sustained viability in rice grains, indicating its strong association with rice plants.

Sunani et al. (2020) studied the infection process of *F. fujikuroi* in seed, seedlings, and adult tissues. The inoculated seeds were observed under a light microscope and scanning electron microscope at different hours after inoculation. The pathogen was observed in glumes, embryos, and husks of rice seeds, and maximum localization was found in husks followed by embryo and endosperm.

A chain of events of colonization was reported in the rice seed husk. At 6 h post-inoculation, the conidia start germinating, the infectious structures like swollen tip hyphae (STH) were also observed. At 12 h post-inoculation, various infectious structures such as appressoria (AP), infection cushion-like (IC) structures, and infectious hyphae (IH) were observed. After 24 h of, inoculation the intercellular and intracellular colonization of *F. fujikuroi* was also found. The maximum husk surface colonization and browning of husk along with fungal mycelium were visible at 72 h post-inoculation. These infectious structures potentially enable the fungus to directly breach the barrier, which could be accountable for the colonization and localization within husk tissue (Sunani et al. 2020).

The progress of the pathogen was observed on 21-day-old inoculated seedlings at different hours after inoculation. Macroconidia and microconidia of *F. fujikuroi* initiated germination at 6 h post-inoculation, initially on the root surface. Subsequently, they developed swollen tip hyphae (STH) at the end of the hyphal tip, which exerted pressure on the rice tissue. After 12 h of inoculation, various types of infectious structures, including foot structure (FS), infectious hyphae (IH), appressoria (AP), and infection cushion (IC), were observed. Root tissue colonization commenced at 24 h post-inoculation, characterized by the intra- and intercellular movements of mycelium (M). By 48 h post-inoculation, tissue colonization was evident, progressing to complete colonization of rice root tissue by 72 h post-inoculation (Sunani et al. 2020).

9.14 Disease Management

Ma et al. (2008) screened genotypes carrying dwarf and semi-dwarf genes against bakanae disease of rice under field conditions and identified genotypes carrying d1 as susceptible whereas genotypes carrying sd6 of sdq(t) and d2q genes as resistant. Lu (1994) identified that the genotype Qingxi96 exhibited moderate resistance at the seedling stage and was resistant at the adult stage. Conversely, the genotype Longjiao 86,074-6 was resistant at the seedling stage but showed moderate resistance at the adult stage. Some genotypes, including Dongrong 84-21, G-6, Sui 89-17, and Zupei 7, were stable and exhibited moderately resistant reactions at both stages, suggesting that bakanae disease resistance may vary depending on the growth stage. Studies on varietal screening have indicated that non-scented rice cultivars demonstrate higher resistance to bakanae disease compared to aromatic cultivars and germplasm (Sunder and Singh 1998; Bashyal et al. 2017; Fiyaz et al. 2014; Pannu et al. 2012; Ghazanfar et al. 2013; Gupta et al. 2014). Currently, high disease severity has been observed in Pusa Basmati 1509 and Pusa Basmati 1121 cultivars in Northern India. However, other basmati rice varieties such as Pusa Basmati 1401, Pusa 2511, CSR 30, Dehradun Basmati, and Pakistani Basmati have also shown susceptibility, with disease incidence ranging from 0.5% to 15% under field conditions in India (Bashyal et al. 2016a, 2017; Gupta et al. 2014). Sunder and Singh (1998) identified that the resistant rice genotypes were C4-64 (green base), Karjat x 13-21, BR1067-84-1-3-2-1, BR 4363-8-11-4-9, BR 1257-31-1-1, IR 58109-109-1-1-3, HKR07-40, HKR07-53, HKR08-13, HKR08-22, HKR08-21, HKR96-561, HKR96-565, PAU3456-46-6-1-1, MAUB2009-1, PNR 600, and RDN 01-2-10-9. Rice genotypes Athadapunnu, C101A51, IR58025B, Panchami, PAU 201, Pusa 1342, Varun Dhan, and Chan-dana as highly resistant whereas BPT 5204, Himju, Peeli Badam, and Suphala as resistant to bakanae disease were identified by Fiyaz et al. (2014). Ghazanfar et al. (2013) conducted a screening of several rice germplasms, identifying IR-6 and KKS-133 as resistant cultivars with 18.80% and 19.82% plant infection rates, respectively. Additionally, Saremi et al. (2008) and Saremi and Farrokhi (2004) identified cultivar Binam as highly resistant, while cultivars Kadous, Sahel, and Shafagh were classified as moderately resistant against bakanae disease.

Khan et al. (1999) characterized bakanae disease resistance as monogenic, with dominance observed in cv. KS 282 and recessiveness in cv. IR 6. Furthermore, two quantitative trait loci (QTLs), qB1 on chromosome 1 and qB10 on chromosome 10, were identified using a japonica/indica double haploid population Chunjiang06 and TN1 (Yang et al. 2006). Both QTLs exhibited an additive effect, contributing to approximately 13% of the phenotypic variation each. Another major QTL, designated as qBK1, was identified on the long arm of chromosome 1 using near-isogenic lines (NILs) derived from a cross between the highly resistant *indica* variety Shingwang and the susceptible *japonica* variety Ilpum. This QTL accounted for 65% of the phenotypic variation (Hur et al. 2015). Fiyaz et al. (2016) analyzed the recombinant inbred line (RIL) population of Pusa Basmati 1121 and Pusa 1342 to identify QTLs associated with bakanae disease resistance using the QTL IciMapping software. Through interval mapping (IM), they identified four QTLs located on chromosome 1 (qBK1.1, qBK1.2, and qBK1.3) and chromosome 3 (qBK3.1). Notably, two QTLs detected on chromosome 1, namely, qBK1.2 and qBK1.3, were found to be novel. qBK1.2 was identified as a major QTL, mapped within a 0.26 Mb region between RM10153 and RM5336, where approximately 55 annotated genes were identified. Similarly, in another study $qBK1^{WD}$ conferring resistance to bakanae disease of rice was identified in recombinant inbred lines (Lee et al. 2018).

For bakanae disease management, the most common and widely accepted practice involves seed treatment with fungicides (Gupta et al. 2015; Bashyal et al. 2016c). Various fungicides are utilized for seed treatment, including thiram, benomyl, thiram + benomyl, thiram + carbendazim, thiram + carboxin, fludioxonil, iprodione + triticonazole, ipconazole, mancozeb, prochloraz, and thiophanate-methyl. These fungicides, applied at a rate of 1–2% of seed weight, have been found effective in different countries such as Bangladesh, Japan, Korea, India, Iran, Italy, Nepal, Pakistan, Taiwan, and Turkey (Ou 1987; Tateishi et al. 1998; Bagga and Sharma 2006; Bagga et al. 2007; Karov et al. 2009; Ora et al. 2011). In addition to the fungicides mentioned earlier, propiconazole, triflumizole, prochloraz, and ipconazole have also shown effectiveness against benomyl-resistant strains (Tateishi et al. 1998; Karov et al. 2009). Treatment of seedlings with benlate, carbendazim, and topsin has been highly effective against the disease (Javed et al. 1996). Furthermore, soil drenching of the nursery with benzimidazole fungicide at a concentration of 0.2% has been found to be effective against rice bakanae disease (Bhalli et al. 2001). Treating seedlings with benlate or carbendazim at a concentration of 0.1% for 6 to 8 h significantly decreased disease incidence and enhanced grain yield (Bagga and Sharma 2006). Moreover, the hot water immersion technique (Hayasaka et al. 2001) and a combination of thermotherapy with antagonistic yeasts proved effective against bakanae disease (Matić et al. 2014). Ochi et al. (2016) documented that immersing rice seeds in distilled water and subjecting them to plasma irradiation decreased the severity of bakanae disease from 18.1% to 7.8%, with the effectiveness varying based on the duration of plasma irradiation. In their evaluation of 11 fungicides as seed treatments against bakanae disease, Bavistin, Carzeb, Nativo, and Sulphonate, when applied at a concentration of 2.5 g/L, completely inhibited pathogen growth and significantly reduced seedling infection (Hossain et al. 2015). Seed treatment, along with the application of carbendazim-sand mix at a rate of 1 g/m^2 in nursery beds, combined with a 3-h seedling dip in 0.1% carbendazim, proved effective against the disease. Additionally, foliar spraying of carbendazim at the flowering stage significantly reduced bakanae disease incidence from 73.9% to 35.0% (Sunder et al. 2014). Moreover, foliar spraying of tebuconazole 250 EC was also effective in reducing bakanae disease incidence and increasing grain yield (Kumar et al. 2016).

Hossain et al. (2016) reported that the treatment of rice rhizosphere by *Bacillus oryzicola* (YC7007) suspension @ 2.0 × 10^7 cfu/mL reduced bakanae disease severity up to 78%. The surfactin-producing *Bacillus* (SPB) strains NH-100 and NH-217 reduced the incidence of bakanae disease by up to 80% (Sarwar et al. 2018). *Talaromyces flavus* SAY-Y-94-01 (Tough-block, Idemitsu Kosan Co.) and

Trichoderma asperellum SKT-1 (Eco-hope, Kumiai Chemical Industry Co.) were registered biofungicides in Japan for the control of bakanae disease (Kato et al. 2012). Tateishi et al. (2006) found that treating seeds with the biocontrol agent *Talaromyces* sp. isolate (KNB-422) effectively controlled bakanae disease in rice. Application of *Talaromyces flavus* reduced disease severity and incidence by 70–75% in India while also increasing aboveground biomass, grains per panicle, and overall rice yield (Rawat et al. 2016; Bashyal et al. 2016c). An integrated management module against bakanae disease in rice was developed, resulting in a reduction of 95% in disease incidence. The efficacy of nursery drenching with carbendazim against bakanae disease in rice was evaluated. Eighteen-day-old seedlings of the Pusa Basmati 1121 rice variety were drenched with varying concentrations of carbendazim and transplanted 5 days after treatment. The lowest incidence of disease was observed with 0.2% carbendazim treatment. Seeds from each treatment underwent residue analysis, revealing that the use of carbendazim at a concentration of 0.2% as nursery drenching is safe, with no residual effects (Bashyal 2018).

References

Agarwal PC, Mortensen CN, Mathur CB (1989) Bakanae disease and foot-rot Fusarium moniliforme. Seed-Borne Diseases and Seed Health Testing on Rice, Technical Bulletin (3):31–36

Amatulli MT, Spadaro D, Gullino ML, Garibaldi A (2010) Molecular identification of *Fusarium spp.* associated with bakanae disease of rice in Italy and assessment of their pathogenicity. Plant Pathol 59(5):839–844

Amatulli MT, Spadaro D, Gullino ML, Garibaldi A (2012) Conventional and qRT PCR for the identification of *Fusarium fujikuroi* and *Fusarium proliferatum* from diseased rice tissues and seeds. Eur J Plant Pathol 134(2):401–408

Amoah BK, Rezanoor HN, Nicholson P, Mac-Donald MV (1995) Variation in the *Fusarium* section Liseola: pathogenicity and genetic studies of *Fusarium moniliforme* Sheldon from different hosts in Ghana. Plant Pathol 44:563–572

Anderson LL, Webster RK (2005) A comparison of assays for *Gibberella fujikuroi* and their ability to predict resulting bakanae from seed sources in California. Phytopathology 95(S4):6

Bacon CW, Hinton DM (1996) Symptomless endophytic colonization of maize by *Fusarium moniliforme*. Can J Bot 74(8):1195–1202

Bagga PS, Sharma VK (2006) Evaluation of fungicides as seedling treatment for controlling bakanae/foot rot (*Fusarium moniliforme*) disease in Basmati rice. Indian Phytopathol 59:305–308

Bagga PS, Sharma VK, Pannu PPS (2007) Effect of transplanting dates and chemical seed treatments on foot rot disease of basmati rice caused by *Fusarium moniliforme*. Plant Dis Res 22(1):60–62

Bashyal BM (2018) Etiology of an emerging disease: bakanae of rice. Indian Phytopathol 71:485–494

Bashyal BM, Aggarwal R (2013) Molecular identification of *Fusarium* spp. associated with bakanae disease of rice in India. Indian J Agric Sci 83(1):72–77

Bashyal BM, Aggarwal R, Gupta S, Banerjee S (2012) Ecology and genetic diversity of *Fusarium* spp. associated with bakanae disease of rice. In: Sinha A, Sharma BK, Srivastava M (eds) Modern trends in microbial biodiversity of natural ecosystem. Biotech Books Publisher, New Delhi, p 577

Bashyal BM, Aggarwal R, Banerjee S, Gupta S, Sharma S (2014) Pathogenicity, ecology and genetic diversity of the *Fusarium* spp. associated with an emerging bakanae disease of rice

(*Oryza sativa* L.) in India. In: Kharwar RN, Upadhyay RS, Dubey NK, Raghuwanshi R (eds) Microbial diversity and biotechnology in food security. Springer, New York, pp 307–314

Bashyal BM, Aggarwal R, Rawat K, Sharma S, Gupta AK, Choudhary R, Bhat J, Krishnan SG, Singh AK (2020) Genetic diversity and population structure of Fusarium fujikuroi causing Bakanae, an emerging disease of rice in India. Indian J Exp Biol 58(1):45–52

Bashyal BM, Aggarwal R, Sharma S, Gupta S, Rawat K, Singh D, Krishnan SG (2016a) Occurrence, identification and pathogenicity of *Fusarium* species associated with bakanae disease of basmati rice in India. Eur J Plant Pathol 144(2):457–466

Bashyal BM, Aggarwal R, Sharma S, Gupta S, Singh UB (2016b) Single and combined effects of three *Fusarium* species associated with rice seeds on the severity of bakanae disease of rice. J Plant Pathol 98(2):405–412

Bashyal BM, Aggarwal R, Singh VK (2016c) Bakanae: an emerging problem in rice cultivation. Indian Farm 65(11):25–26

Bashyal BM, Rawat K, Sharma S, Kulshrestha D, Singh AK, Dubey H, Solanki AKU, Sharma TR, Aggarwal R (2017) Whole genome sequencing of *Fusarium fujikuroi* provides insight into the role of secretory proteins and cell wall degrading enzymes in causing bakanae disease of rice. Front Plant Sci 8:2013. https://doi.org/10.3389/fpls.2017.02013

Bedi PS, Dhaliwal DS (1970) Spermatosphere of rice variety native I from different states of India. Indian Phytopatholo 23(4):708–710

Bhalli JA, Aurangzeb M, Ilyas MB (2001) Chemical control of bakanae disease of rice caused by *Fusarium moniliforme*. Online J Biol Sci 1:483–484

Burgess LW, Summerell BA, Bullock S, Gott KP, Backhouse D (1994) Laboratory manual for *Fusarium* research, 3rd edn. Department of Crop Science, University of Sydney, Royal Botanic Gardens, Sydney

Burgess LW, Backhouse D, Swan LJ, Esdaile RJ (1996) Control of *Fusarium* crown rot of wheat by late stubble burning and rotation with sorghum. Aust Plant Pathol 25(4):229–233

Butt AR, Yaseen SI, Javaid A (2011) Seed-borne mycoflora of stored rice grains and its chemical control. J Anim Plant Sci 21(2):193–196

Cumagun CJR, Arcillas E, Gergon E (2011) UP-PCR analysis of the seedborne pathogen *Fusarium fujikuroi* causing bakanae disease in rice. Int J Agric Biol 13(6):1029–1032

Chiara M, Fanelli F, Mule G, Logrieco AF, Pesole G, Leslie JF, Horner DS, Toomajian C (2015) Genome sequencing of multiple isolates highlights sub-telomeric genomic diversity within Fusarium fujikuroi. Genome Biol Evol 7:3062–3069. https://doi.org/10.1093/gbe/evv198

Desjardins AE, Plattner RD, Nelson PE (1997) Production of fumonisin B1 and moniliformin by *Gibberella fujikuroi* from rice from various geographic areas. Appl Environ Microbiol 63:838–842

Desjardins AE, Manandhar HK, Plattner RD, Manandhar GG, Poling SM, Maragos CM (2000) *Fusarium spp.* from Nepalese rice and production of mycotoxins and gibberellic acid by selected species. Appl Environ Microbiol 66:1020–1025

Fiyaz RA, Krishnan SG, Rajashekara H, Yadav AK, Bashyal BM, Bhowmick PK, Singh NK, Prabhu KV, Singh AK (2014) Development of high throughput screening protocol and identification of novel sources of resistance against bakanae disease in rice (Oryza sativa L.). Indian J Genet 74(4):414–422

Fiyaz RA, Yadav AK, Krishnan SG, Ellur RK, Bashyal BM, Grover N, Bhowmick PK, Nagarajan M, Vinod KK, Singh NK, Prabhu KV, Singh AK (2016) Mapping quantitative trait loci responsible for resistance to bakanae disease in rice. Rice 9:45. https://doi.org/10.1186/s12284-016-0117-2

Geiser DM, Aoki T, Bacon CW, Baker SE, Bhattacharyya MK, Brandt ME et al (2013) One fungus, one name: defining the genus Fusarium in a scientifically robust way that preserves long-standing use. Phytopathology 103(5):400–408

Ghazanfar MU, Javed N, Wakil W, Iqbal M (2013) Screening of some fine and coarse rice varieties against bakanae disease. J Agric Res 51:41–49

Gupta AK, Singh Y, Jain AK, Singh D (2014) Prevalence and incidence of bakanae disease of rice in Northern India. J Agric Res 1(4):233–237

Gupta AK, Solanki IS, Bashyal BM, Singh Y, Srivastava K (2015) Bakanae of rice-an emerging disease in Asia. J Anim Plant Sci 25:1499–1514

Hajra KK, Ganguly LK, Khatua DC (1994) Bakanae disease of rice in West Bengal. J Mycopathol Res 32(2):95–99

Hamamura H, Kawahara M, Shimoda S (1989) Some characteristics of *Gibberella fujikuroi* (*Fusarium moniliforme*) isolates less sensitive to triflumizole. Ann Phytopathol Soc Jpn 55(3):275–280

Hayasaka T, Ishiguro K, Shibutani K, Namai T (2001) Seed disinfection using hot water immersion to control several seed-borne diseases of rice plants. J Phytopathol 67(1):26–32

Hori S (1898) Some observations on bakanae disease of the rice plant. Mem Agric Res Sta (Tokyo) 12:110–119

Hossain MA, Latif MA, Kabir MS, Kamal MM, Mian MS, Akter S, Sharma NR (2007) Dissemination of integrated disease management practices through farmers' participatory field trial. A report on Agricultural Technology Transfer (ATT) project. Bangladesh Agricultural Research Council, Dhaka

Hossain MS, Ayub Ali M, Mollah MIU, Khan MAI, Sajjadul Islam AKM (2015) Evaluation of fungicides for the control of bakanae disease of rice caused by *Fusarium moniliforme* (Sheldon). Bangladesh Rice J 19(1):49–55

Hossain MT, Khan A, Chung EJ, Rashid MHO, Chung YR (2016) Biological control of rice bakanae by an endophytic *Bacillus oryzicola* YC7007. Plant Pathol J 32(3):228–241

Hseieh WH, Smith SN, Snyder WC (1977) Mating groups in *Fusarium moniliforme*. Phytopathology 67:1041–1043

Hsuan HM, Salleh B, Zakaria L (2011) Molecular identification of Fusarium species in Gibberella fujikuroi species complex from rice, sugarcane and maize from Peninsular Malaysia. Int J Mol Sci 12(10):6722–6732

Hur YJ, Lee SB, Kim TH, Kwon T, Lee JH, Shin DJ, Park SK, Hwang UH, Cho JH, Yoon YN, Yeo US, Song YC, Kwak DY, Nam MH, Park DS (2015) Mapping of qBK1, a major QTL for bakanae disease resistance in rice. Mol Breed 35(2):78. https://doi.org/10.1007/s11032-015-0281-x

Ito S, Kimura J (1931) Studies on the bakanae disease of the rice plant. Rep Hokkaido Agr Exp Stn 27:1–95

Jain SK, Khilari K, Dongre M, Pal S (2019) Occurrence of bakanae disease of rice in Western Uttar Pradesh, India. Int J Curr Microbiol App Sci 8(05):207–212

Javed MS, Ayoub M, Mahmood A, Ahmad M (1996) Chemical control of bakanae disease of rice by nursery dip treatment method. Pak J Sci 48:90–91

Jeong H, Lee S, Choi GJ, Lee T, Yun SH (2013) Draft genome sequence of *Fusarium fujikuroi* B14, the causal agent of the bakanae disease of rice. Genome Announc 1(1):e00035–e00013. https://doi.org/10.1128/genomea.00035-13

Kanjanasoon P (1965) Studies on the bakanae disease of rice in Thailand. Doc Agr Thesis Tokyo University, Tokyo

Karov I, Mitrev S, Kostadinovska E (2009) *Gibberella fujikuroi* (Sawada) Wollenweber, the new parasitical fungus on rice in the Republic of Macedonia. Zbornik Matice Srpske Za Prirodne Nauke 4(116):175–182

Kato A, Miyake T, Nishigata K, Tateishi H, Teraoka T, Arie T (2012) Use of fluorescent proteins to visualize interactions between the bakanae disease pathogen *Gibberella fujikuroi* and the biocontrol agent *Talaromyces* sp. KNB-422. J Gen Plant Pathol 78:54–61

Kaur J, Pannu PPS, Sharma S (2014) Morphological, biochemical and molecular characterization of *Gibberella fujikuroi* isolates causing bakanae disease of basmati rice. J Mycol Plant Pathol 44:78–82

Kauraw LP (1981) Effect of length of treatment and fungicide concentration on seed germination and incidence of foot rot disease of rice. Int Rice Res Newsl 6(6):15

Khan TZ, Gill MA, Khan MG (1999) Inheritance studies for bakanae disease of rice. Pak J Phytopathol 11:181–182

Khokhar LK (1990) Bakanae and foot rot of rice in Punjab. Pakistan National Agricultural Research Centre, Crop Diseases Research Institute, Islamabad (Pakistan). IRRN 15:30

Khokhar LK, Jaffrey AH (2002) Identification of sources of resistance against bakanae and foot rot disease in rice. Pak J Agric Res 17(2):176–177

King R, Urban M, Hammond MCU, Pak KH, Hammond KE (2015) The completed genome sequence of the pathogenic ascomycete fungus Fusarium graminearum. BMC Genomics 16:544. https://doi.org/10.1186/s12864-015-1756-1

Kini KR, Leth V, Mathur SB (2002) Genetic variation in *Fusarium moniliforme* isolated from seeds of different host species from Burkina Faso based on random amplified polymorphic DNA analysis. J Phytopathol 150(4–5):209–212

Koehler B (1942) Natural mode of entrance of fungi into corn ears and some symptoms that indicate infection. J Agric Res 64:421–442

Kumar P, Sunder S, Singh R, Kumar A (2016) Management of foot rot and bakanae of rice through fungicides. Indian Phytopathol 69(2):124–127

Kurosawa E (1926) Experimental studies on the nature of the substance secreted by the "bakanae" fungus. Nat Hist Soc Formosa 16:213–227

Kushiro M, Saitoh H, Sugiura Y, Aoki T, Kawamoto SI, Sato T (2012) Experimental infection of *Fusarium proliferatum* in *Oryza sativa* plants; fumonisin B1 production and survival rate in grains. Int J Food Microbiol 156(3):204–208

Kvas M, Marasas WFO, Wingfield BD, Wingfield MJ, Steenkamp ET (2009) Diversity and evolution of *Fusarium* species in the *Gibberella fujikuroi* complex. Fungal Divers 34:1–21

Lee SB, Hur YJ, Cho JH, Lee JH, Kim TH, Cho SM, Song YC, Seo YS, Lee J, Kim TS, Park YJ, Oh MK, Park DS (2018) Molecular mapping of $qBK1^{WD}$, a major QTL for bakanae disease resistance in rice. Rice 11:3. https://doi.org/10.1186/s12284-017-0197-7

Leslie JF, Summerell BA (2006) The *Fusarium* laboratory manual. Blackwell, London

Leslie JF, Marasas WFO, Shephard GS, Sydenham EW, Stockenstrom S, Thiel PG (1996) Duckling toxicology and the production of fumonisin and moniliformin by isolates in the A and F mating populations of *Gibberella fujikuroi* (*Fusarium moniliforme*). Appl Environ Microbiol 62:1182–1187

Lin Z, Xu S, Que Y, Wang J, Comstock JC, Wei J, McCord PH, Chen B, Chen R, Zhang M (2014) Species-specific detection and identification of Fusarium species complex, the causal agent of sugarcane Pokkah Boeng in China. PLoS One 9(8):e104195. https://doi.org/10.1371/journal.pone.0104195

Linnemanstons P, Vob T, Hedden P, Gaskin P, Tudzynski B (1999) Deletions in the gibberellin biosynthesis gene cluster of *Gibberella fujikuroi* by restriction enzyme-mediated integration and conventional transformation mediated mutagenesis. Appl Environ Microbiol 65:2558–2564

Lu B (1994) Preliminary report on evaluation of resistance to rice bakanae disease in rice germplasms. Plant Prot 20(3):20–21

Ma L, Ji Z, Bao J, Zhu X, Li X, Zhuang J, Xia Y (2008) Responses of rice genotypes carrying different dwarf genes to *Fusarium moniliforme* and gibberellic acid. Plant Prod Sci 11(1):134–138

Manandhar J (1999) *Fusarium moniliforme* in rice seeds: its infection, isolation, and longevity/*Fusarium moniliforme* in Reissamen. J Plant Dis Prot 106(6):598–607

Mandal DN, Chaudhuri S (1988) Survivality of *Fusarium moniliforme* Sheld. Under different moisture regimes and soil conditions. Int J Trop Plant Dis 6:201–206

Matić S, Spadaro D, Garibaldi A, Gullino ML (2014) Antagonistic yeasts and thermotherapy as seed treatments to control Fusarium fujikuroi on rice. Biol Control 73:59–67

Mills PR, Sreenivasaprasad S, Brown AE (1992) Detection and differentiation of Colletotrichum gloeosporioides isolates using PCR. FEMS Microbiol Lett 98(1–3):137–143

Nancy AE (2002) Foolish seedlings and DELLA regulators: the functions of rice SLR1 and Arabidopsis RGL1 in GA signal transduction. Plant Cell 14:1–6

Nelson PE, Desjardins AE, Plattner RD (1993) Fumonisins, mycotoxins produced by *Fusarium species*: biology, chemistry and significance. Annu Rev Phytopathol 31:233–252

Niehaus EM, Kim HK, Munsterkotter M, Janevska S, Arndt B, Kalinina SA, Houterman PM, Ahn PI, Alberti I, Tonti S, Kim DW, Sieber CMK, Humpf HU, Yun SH, Guldener U, Tudzynski B (2017) Comparative genomics of geographically distant *Fusarium fujikuroi* isolates revealed two distinct pathotypes correlating with secondary metabolite profiles. PLoS Pathog 13(10):e1006670. https://doi.org/10.1371/journal.ppat.1006670

Nirenberg HI (1976) Untersuchungen uber die morphologische und biologische Differenzierung in der *Fusarium*-Sektion *Liseola*. Mitt Biol Bundesanst Land-u Forstwirtsch Berlin-Dahlem 169:1–117

Nisikado Y, Kimura K (1941) A contribution to the pathological anatomy of rice plants affected by *Gibberella fujikuroi* (Saw.). Wr I Ber Ohara Inst Landw Forsch 8:421–426

Nisikado Y, Matsumoto H (1933) Studies on the physiological specialization of *Gibberella fujikuroi* the causal fungus of rice bakanae disease. Trans Tottori Soc Agric Sci 4:200–211

Nyvall RF (1999) Field crop diseases. Iowa State University Press, Ames

Ochi A, Konishi H, Ando S, Sato K, Yokoyama K, Tsushima S, Yoshida S, Morikawa T, Kaneko T, Takahashi H (2016) Management of bakanae and bacterial seedling blight diseases in nurseries by irradiating rice seeds with atmospheric plasma. Plant Pathol 66(1):67–76. https://doi.org/10.1111/ppa.12555

O'Donnell K, Cigelnik E, Nirenberg HI (1998) Molecular systematics and phylogeography of the *Gibberella fujikuroi* species complex. Mycologia 90:465–493

Ogawa K (1988) Damage by bakanae disease and its chemical control. Jpn Pestic Inf 52:13–15

Ora N, Faruq AN, Islam MT, Akhtar N, Rahman MM (2011) Detection and identification of seed borne pathogens from some cultivated hybrid rice varieties in Bangladesh. Middle East J Sci Res 10:482–488

Ou SH (1985) Bakanae disease and foot rot. Rice diseases survey. Commonwealth Mycological Institute, Kew, pp 262–272

Ou SH (1987) Rice diseases. CAB International, Commonwealth Mycological Institute, Kew, p 256

Pan Y, Wang ZY, Wu HZ (1997) Mycelial growth, sporulation and pathogenicity of *Fusarium moniliforme* isolates with differential resistance to carbendazim (MBC). Jiangsu J Agric Sci 13:90–93

Pannu PPS, Kaur J, Singh G, Kaur J (2012) Survival of *Fusarium moniliforme* causing foot rot of rice and its virulence on different genotypes of rice and basmati rice. (Abstr.). Indian Phytopathol 65(5):149–209

Parate DK, Lanjewar RD (1987) Studies on seed mycoflora of two rice cultivars grown in rice tract of Vidarbha. PKV Res J 11(1):47–50

Pavgi MS, Singh JL (1964) Bakanae and foot rot of rice in Uttar Pradesh, India. Plant Dis Reptr 48:340–342

Petrovic T, Burgess L, Cowie I, Warren R, Harvey P (2013) Diversity and fertility of *Fusarium sacchari* from wild rice (*Oryza australiensis*) in Northern Australia, and pathogenicity tests with wild rice, sorghum and maize. Eur J Plant Pathol 136:773–788

Pra MD, Tonti MS, Pancaldi D, Nipoti P, Alberti I (2010) First report of *Fusarium andiyazi* associated with rice bakanae in Italy. Plant Dis 94:1070

Puyam A, Pannu PPS, Kaur J, Sethi S (2017) Investigations on survival and host range of *Fusarium moniliforme* Sheld causing foot rot disease of rice. Environ Ecol 35(3B):2071–2075

Quazi SAJ, Meon S, Jaafar H, Ahmad ZABM (2013) Characterization of *Fusarium proliferatum* through species specific primers and its virulence on rice seeds. Int J Agric Biol 15:649–656

Rathaiah Y, Das GR, Singh KHU (1991) Estimation of yield loss and chemical control of bakanae disease of rice. Oryza 28:509–512

Raghu S, Yadav MK, Prabhukarthikeyan SR, Baite MS, Lenka S, Jena M (2018) Occurrence, pathogenicity, characterization of Fusarium fujikuroi causing rice bakanae disease from Odisha and in vitro management. Oryza 55(1):214–223

Rawat K, Bashyal BM, Sharma S, Aggarwal R (2016) *Talaromyces flavus*: a potential biocontrol agent against the bakanae resistance against bakanae disease in rice (*Oryza sativa* L.). Indian J Genet 74:414–422

Reynoso MM, Torres AM, Chulze SN (2004) Fusaproliferin, beauvericin and fumonisin production by different mating populations among the *Gibberella fujikuroi* complex isolated from maize. Mycol Res 108:154–160

Rojas MC, Hedden P, Gaskin P, Tudzynski B (2001) The *P450-1* gene of *Gibberella fujikuroi* encodes a multifunctional enzyme in gibberellin biosynthesis. Proc Natl Acad Sci U S A 98:5838–5843

Sakamoto T, Miura K, Itoh H, Tatsumi T, Tanaka MU, Ishiyama K, Kobayashi M, Agrawal GK, Takeda S, Abe K, Miyao A, Hirochika H, Kitano H, Ashikari M, Matsuoka M (2004) An overview of gibberellin metabolism enzyme genes and their related mutants in rice. Plant Physiol 134(4):1642–1653

Saremi H (2005) *Fusarium*, biology, ecology and taxonomy. Jihad Daneshgahi, Ferdossy Mashhad University, Iran

Saremi H, Farrokhi F (2004) Study on bakanae disease of rice and evaluation of cultivars in Gilan and Zanjan provinces. In: Iran Proceedings of fourth international Iran and Russia conference, pp 358–364

Saremi H, Ammarellou A, Marefat A, Okhovvat SM (2008) Binam a rice cultivar, resistant for root rot disease on rice caused by *Fusarium moniliforme* in Northwest, Iran. Int J Bot 4:383–389

Sarkar BB (1986) Controlling bakanae and foot rot diseases with seed treatments. Int Rice Res Newsl 11(3):18

Sarwar A, Brader G, Corretto E, Aleti G, Ullah MA, Sessitsch A, Hafeez FY (2018) Qualitative analysis of biosurfactants from *Bacillus* species exhibiting antifungal activity. PLoS One 13(7):e0201624. https://doi.org/10.1371/journal.pone.0201624

Sasaki T (1976) Elongation of ratoon in rice plants inoculated with *Fusarium moniliforme* Sheldon. Ibid 42:606–608

Sawada K (1917) Beitrage uber Formosas-Pilze no. 14. Transac Nat Hist Soc Formosa 3:31–133

Sharma VK, Bagga PS (2007) Pathogenic behaviour of *Fusarium moniliforme* isolates causing foot rot disease in basmati rice. Plant Dis Res 22:165–166

Sharma RD, Singh RS (1997) A technique for selective isolation of *Fusarium moniliforme* from soil and plant tissues. Indian J Mycol Plant Pathol 3:67–70

Singh R, Sunder S (1997) Foot rot and Bakanae of rice: retrospect's and prospects. Int J Trop Plant Dis 15:153–176

Singh R, Sunder S (2012) Foot rot and bakanae of rice: an overview. Rev Plant Pathol 5:566–604

Singh NI, Devi RKT, Singh LNK (1996) Withering of growing shoot of rice caused by *Fusarium moniliforme*. Plant Dis Res 11(1):99–100

Snyder WC, Hansen HN (1945) The species concept in *Fusarium* with reference to discolor and other sections. Am J Bot 32:657–666

Sun SK (1975) The disease cycle of rice bakanae disease in Taiwan. Proc Natl Sci Counc Repub China 8(2):245–256

Sun SK, Snyder WC (1981) The bakanae disease of the rice plant. In: Nelson PE, Toussoun TA, Cook RJ (eds) Fusarium diseases, biology and taxonomy. The Pennsylvania State University Press, University Park, pp 104–113

Sun SK, Snyder WD (1978) The bakanae disease of rice plant. Sci Bull Taiwan 10(7):2

Sunani SK, Bashyal BM, Aggarwal R, Prakash G (2017) Conidial germination of *Fusarium fujikuroi* causing bakanae disease of rice. Environ Ecol 35(4):2790–2794

Sunani SK, Bashyal BM, Kharayat BS, Prakash G, Krishnan SG, Aggarwal R (2020) Identification of rice seed infection routes of Fusarium fujikuroi inciting bakanae disease of rice. J Plant Pathol 102(1):113–121

Sunani SK, Bashyal BM, Rawat K, Manjunatha C, Sharma S, Prakash G, Aggarwal R (2019) Development of PCR and loop mediated isothermal amplification assay for the detection of bakanae pathogen Fusarium fujikuroi. Eur J Plant Pathol 154(3):715–725

Sunder S, Satyavir (1998b) Vegetative compatibility, biosynthesis of GA3 and virulence of *Fusarium moniliforme* isolates from bakanae disease of rice. Plant Pathol 47:767–772

Sunder S, Singh A (1998) Screening of rice genotypes for resistance to bakanae disease. Indian Phytopathol 51:299–300

Sunder S, Singh R, Dodan DS (2014) Management of bakanae disease of rice caused by *Fusarium moniliforme*. Indian J Agric Sci 84(2):224–228

Surek H (1997) Rice diseases in Turkey. In: Chataigner J (ed) Maladies durizenrégion méditerran éenneetles possibilitésd'amélioration desarésistance. CIHEAM, Montpellier, pp 45–47

Surek H, Gumustekin H (1994) Research activities on controlling rice bakanae and foot rot disease (*Fusarium moniliforme*) in Turkey. In: Clément G, Cocking EC (eds) FAO MedNet rice: breeding and biotechnology groups: proceedings of the workshops. CIHEAM, Montpellier, pp 27–30

Tateishi H, Saishoh T, Suzuki T, Chida T (1998) Antifungal properties of the seed disinfectant ipconazole and its protection against bakanae and other diseases of rice. Ann Phytopathol Soc Jpn 64(5):443–450

Tateishi H, Sakuma Y, Miyake T (2006) Isolation and screening of antagonistic fungi against rice seedling diseases and their efficacy (abstract in Japanese). Jpn J Phytopathol 72:265–266

Thakur KSS (1974) Role of gibberellic acid, fusaric acid and pectic enzyme in the foot rot disease of rice. II-RISO 23:191–207

Thomas KM (1931) A new paddy disease in Madras. Madras Agric J 19:34–36

Thomas KM (1933) Foot rot of paddy and its control. Madras Agric J 21:263–272

Tisserat NA, Hulbert SH, Nus A (1991) Identification of Leptosphaeria korrae by cloned DNA probes. Phytopathology 81(8):917–921

Tudzynski B, Mihlan M, Rojas MC, Linnemannstons P, Gaskin P, Hedden P (2003) Characterization of the final two genes of the gibberellin biosynthesis gene cluster of *Gibberella fujikuroi*–des and P450-3 encode GA4 desaturase and the 13-hydroxylase, respectively. J Biol Chem 278:28635–28643

Vidyasekaran P, Muthaswamy G, Subramanian C (1966) Role of seed borne microflora in paddy seed spoilage. Indian Phytopathol 19:333–341

Wada T, Kuzuma S, Takenaka M (1990) Sensitivity of *Fusarium moniliforme* isolates to perfurazoate. Ann Phytopathol Soc Jpn 56:449–456

Ward E, Foster SF, Fraaije AB, McCartney HA (2004) Plant pathogen diagnostics: immunological and nucleic acid-based approaches. Ann Appl Biol 145:1–16

Watanabe Y (1974) The possibility of soil transmission in bakanae disease and the contamination of seeds with causal fungus during the hastening process of seed germination. Bull Tokai Kinki Natl Agric Exp Station 27:35–41

Webster RK, Gunnell PS (1992) Compendium of rice diseases. The APS Press, St. Paul

Wiemann P, Sieber CM, Von Bargen KW, Studt L, Niehaus EM, Espino JJ et al (2013) Deciphering the cryptic genome: genome-wide analyses of the rice pathogen *Fusarium fujikuroi* reveal complex regulation of secondary metabolism and novel metabolites. PLoS Pathog 9(6):e1003475. https://doi.org/10.1371/journal.ppat.1003475

Wulff EG, Sorensen JL, Lubeck M, Nielsen KF, Thrane U, Torp J (2010) *Fusarium* spp. associated with rice Bakanae: ecology, genetic diversity, pathogenicity and toxigenicity. Environ Microbiol 12(3):649–657

Xu JR, Yan K, Dickman MB, Leslie JF (1995) Electrophoretic karyotypes distinguish the biological species of Gibberella fujikuroi (Fusarium section Liseola). Mol Plant-Microbe Interact 8(1):74–84

Yabuta T, Hayashi T (1939) Biochemical studies on bakanae fungus of rice. Part III Studies on physiological action of gibberellin on the plant. J Agric Cliem Soc Jpn 15:403–414

Yabuta T, Sumiki Y (1938) The crystallization of gibberellins A and B. J Agr Chem Soc, Japan 14:1526

Yang CD, Guo LB, Li XM, Ji ZJ, Ma LY, Qian Q (2006) Analysis of QTLs for resistance to rice bakanae disease. Chin J Rice Sci 20(6):657–659

Zainudin NAIM, Salleh B (2010) Variability of *Fusarium* species associated with bakanae disease of rice based on virulence, vegetative and biological compatibility. Sydowia 62:89–104

Zainudin NAIM, Razak AA, Salleh B (2008a) Bakanae disease of rice in Malaysia and Indonesia: etiology of the causal agent based on morphological, physiological and pathogenicity characteristics. J Plant Prot Res 48:475–485

Zainudin NAIM, Razak AA, Salleh B (2008b) Secondary metabolite profiles and mating populations of *Fusarium* species in section *Liseola* associated with bakanae disease of rice. Malaysian J Microbiol 4:6–13

CRISPR Technology: Plant Disease Management

10

Sandhya Upadhyay and Gohar Taj

Abstract

Overcoming the drawbacks of numerous tools and techniques, CRISPR has made an immense place in the field of genome editing. It is versatile, conscientious, inexpensive, and easy to accomplish with no bioethical controversies. Due to the described reasons, CRISPR technology is also implicated in plant disease management to enhance the harvest potential for up-growing populations with constricted farming land and dynamic environmental conditions. The management strategy includes early detection of phytopathogens and the production of disease-resistant plant varieties that are eradicated by numerous diseases and bring up the food supply. This chapter also provides the kinds of approaches used to develop resistant varieties of plants with respect to the type of pathogen and disease involved, added with some examples.

Keywords

CRISPR · Nuclease · Gene editing · Cas nuclease · Disease management

10.1 Introduction

For hundreds of years, pathogens (bacteria, fungi, and viruses) have been continually invading plants in order to feed, reproduce, or complete their life cycle and afflict the plant with illness. These pathogens were never confined to any geographical region and disseminated through various means; however, some of them are conditional and host-specific. Despite this, due to the rise in global temperature and

S. Upadhyay · G. Taj (✉)
Department of Molecular Biology and Genetic Engineering, College of Basic Sciences and Humanities, GBPUAT, Pantnagar, Uttarakhand, India

V. K. Singh et al. (eds.), *Diseases of Field Crops: Diagnostics and Management*,
https://doi.org/10.1007/978-981-97-6160-9_10

fluctuation in environmental conditions, some pathogens have shattered the barricades of seasonal dynamics, occupy variation in their physiology, attain resistance toward agrochemicals (fungicides and bactericides), and have increased disease severity (Baker et al. 2000; Luck et al. 2011). The consequences of the rising global temperature arise detrimental effects on the plant's yield, quality, and overall health. It is estimated that the global loss of crop productivity due to pests and pathogens is about 20–40% (Savary et al. 2019). As per the census, the human population is also increasing logarithmically, ensuring the need for food and its products. However, the diseases within edible crops cause a reduction in food and the economy [each year, plant diseases cost the global economy around $220 billion (FAO)]. Tilman et al. (2002) also claim that by 2050, population increase will cause a doubling of the world's grain consumption.

Meanwhile, some plants have evolved multiple strategies to defend themselves from harmful invaders. They had developed structural barriers (thrones, high lignin content), anti-microbial substances (phytoalexins), increased synthesis of stress-related hormones (abscisic acid and ethylene), and secondary metabolites and bioactive compounds (salicylic and jasmonic acid) to mitigate the consequences of attack (Miedes et al. 2014). These molecular defense systems of plants toward varieties of pathogens depend upon the specific recognition events that occur between molecular patterns from pathogens and plant transmembrane receptors followed by receptor-mediated signal transduction (Hammond-Kosack and Jones 1996). However, not all kinds of plants are able to protect themselves from such disease-causing pathogens.

The approaches used to overcome disease induced by detrimental phytopathogens depend on the prevention and reduction of infection such as (i) the implementation of simple agronomic practices which are pre-sowing, crop rotation, intercropping, mix cropping, field monitoring, forecasting models, and use of healthy seeds and fertilizers (Heitefuss and Welch 1989) and (ii) use of chemicals (fungicides), bio-control agents, nanoparticles, soil organic matter, and bioactive plant compounds. Some of these approaches impose risk in the environment and evoke resistance in pathogens (Vannier et al. 2019). However, many plants' genome exhibits tolerance toward pathogens that were used to transfer to other beneficiary crops through breeding. However, this technique requires a lengthy duration and is tedious to perform, and some cause the addition of unrequired traits. Other techniques like RNA interference, molecular breeding, and genetic engineering were employed to develop disease resistance but ended up with some hectic procedures and bioethical controversies as of the source of transgene. Besides the drawbacks of explained approaches and techniques. Many of the disease-resistant plants were developed and approved in various countries. According to ISAAA, there are several approved genetically modified viral-resistant crops including tomato, sweet pepper, squash, potato, plum, papaya, and beans. In addition to it, a late blight disease-resistant potato was also approved in Indonesia. Moreover, to overcome problems related to the former techniques, a new technique was established, known as gene editing.

10.2 Gene Editing

The idea behind gene editing is that if the regulatory and gene DNA sequences are altered, the resulting products and regulation will change, respectively. This gene editing system relies on the programmable sequence-specific nucleases such as transcription activator-like effector nuclease (TALEN), zinc-finger nuclease (ZFN), meganuclease (MN), and clustered regularly interspaced short palindromic repeats (CRISPR). ZFNs and TALENs are artificial bipartite enzymes that consist of a modular DNA-binding domain and the Fok1 nuclease domain required for targeting the nuclease. In both cases, the DNA-binding domain can be engineered to recognize a specific DNA sequence. However, the designing and construction of large modular proteins of ZFN and TALEN are laborious and expensive. In addition, there is a high rate of failure, at least for ZFNs, to recognize and cleave the intended DNA sequence (Voytas 2013). Therefore, in comparison to other nuclease systems, CRISPR is far better due to its simplicity, affordability, and versatility thus the most advanced and potent editing system which was acquired from the prokaryotic immune system (*Streptococcus pyogenes*, *Pseudomonas aeruginosa*, *Pyrococcus furiosus*, and *Sulfolobus solfataricus*).

The CRISPR gene editing system is made up of essential parts including Cas proteins, crRNA (guide RNA), and protospacer adjacent motif (PAM), which act in harmony to provide precise DNA targeting and editing capabilities. The Cas nuclease is required to cause double-stranded breaks in nucleotide sequences; the gRNA aids the Cas in targeting the appropriate location, while PAM is necessary for Cas9 recognition and binding (Jinek et al. 2012). Additionally, by changing the sequence of gRNA with respect to the target site, it is possible to modify or edit the genome throughout, which is one of the reasons for the name CRISPR, an RNA-guided endonuclease system (Belhaj et al. 2015). Furthermore, some CRISPR-associated Cas proteins, each tailored for specific objectives and targets are described in Table 10.1.

The genetic modifications through CRISPR technology rely on the DNA repair system that is induced after the double-stranded break (DSB) generated by the nuclease. Then, the two different repair pathways: homologous direct recombination (HDR) and nonhomologous end joining (NHEJ) come into play. The HDR system is activated only when there is an availability of homologous strands whether it could be from the cell's chromosome or from the donor DNA, though it can be used to produce knock-in, frameshift, and replacements, while the NHEJ has a role when no homologous sequence is available thus the reason it is used to generate knockouts (Fig. 10.1). The knockouts are easy to perform as they only utilize the CRISPR/Cas and NHEJ pathways, while the knock-in editing is a bit typical in which the donor DNA is supplied with the CRISPR/Cas machinery. However, there is a greater chance of repairing through the NHEJ pathway though some NHEJ inhibitors (SCR7) are introduced to promote the HDR pathway (Dong 2020). These DNA templates and CRISPR are transported via various delivery systems in plants such as polyethylene glycol–mediated, agrobacterium-mediated, biolistic, pollen

Table 10.1 Activity of different Cas proteins in genome editing

S. No.	Type	Activity	References
1	Cas9	DNA targeting	Jinek et al. (2012)
2	Cytosine and adenine base editor (made up of nCas + deaminase domain)	Base editing (substitution of nucleotide bases)	Gaudelli et al. (2017)
3	Cas9n (nickase) and dCas9 (dead) (mutated version of Cas9 where RuvC domain is inactivated)	DNA targeting	Xu and Qi (2019)
4	Cas12	DNA targeting	Tian et al. (2022)
5	Cas13a	Collateral cleavage activity RNA	Tian et al. (2022)
6	Cas14	DNA targeting	Aman et al. (2020)

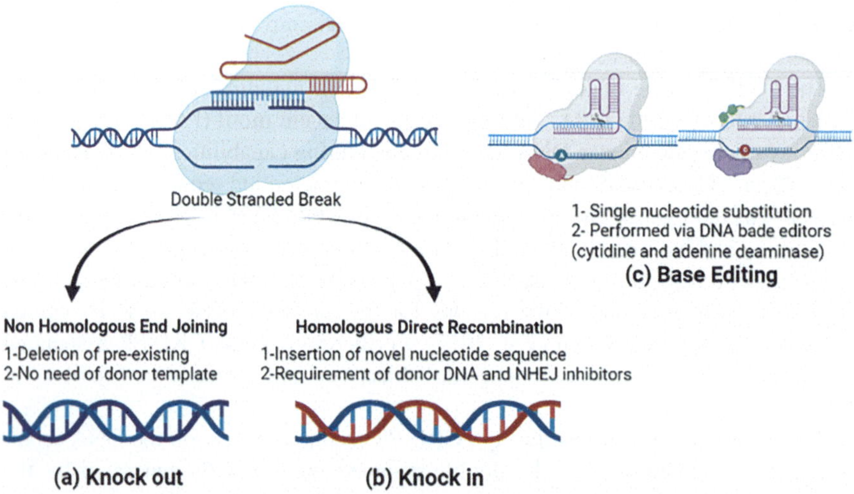

Fig. 10.1 Type of editing can be performed by using the NHEJ and HDR pathways with types of Cas proteins. (**a**) Deletion is achieved through the NHEJ pathway during knockout; (**b**) knock-in utilizes HDR to introduce replacements or insertions; (**c**) base editing is performed by specialized Cas protein which causes substitution to rephrase it

magnetofection–mediated, and nanoparticle-mediated transformation and floral dip methods (Sandhya et al. 2020).

10.3 Designing the Target for CRISPR System

The development of next nucleotide sequencing (NGS) technology and bioinformatics databases provide fast and faultless information about target nucleotide sequences that will be utilized for the synthesis and designing of gRNA which plays a role in providing direction to Cas nuclease within the genome of plants. This gRNA has a complementary sequence with respect to the target region so it anneals toward the target and activates Cas (Alkan et al. 2018). There are various tools for designing gRNA such as CasFinder, CRISPR-P, CRISPR design tool, and E-CRISPR (Manghwar et al. 2020). Designing the gRNA is the key success of CRISPR technology because some homologous sequences in the genome may cause off-target effects. Actually, to reduce the off-target effects, there is a combined strategy depending upon the number of nucleotides by which the gRNA is made and finding the other off-target regions through bioinformatics tools. To reduce the off-target, the designing of gRNA is highly dependent on the 20 nucleotide sequences and protospacer adjacent motif (PAM). Moreover, PAM is also responsible for the binding of Cas to target (Jinek et al. 2012). There are various algorithms, and methods were designed that are used to identify off-targets such as whole genome sequencing, T7E1 assay, CHIP, GUIDE-Seq, CHOPCHOP, CRISPR-PLANTv2, CRISPOR, and CROP.IT (Manghwar et al. 2020).

10.4 Approach to Develop Disease-Resistant Plants via CRISPR

The idea behind developing resistant varieties of plants is based on the plant response system with respect to the type of pathogen involved. The two major types of genes associated with plant response are the susceptibility gene (S gene) required for disease induction thus called the negative regulator of immunity, and the resistant gene (R gene) provides immunity against the pathogen. The modification within the S helps in the interruption of plants and pathogens in spite of the R gene helping in reducing the severity and helping build up tolerance. Therefore, the editing within the regulatory sequences (promoter, effector) of the S and R genes is helpful (Zaidi et al. 2020). This strategy is majorly utilized in bacterial- and fungal-resistant systems, while in viruses, the editing is performed within the viral machinery and gene-assisting viral assembly systems.

10.4.1 Bacterial Disease

The bacterial pathogen may be soil borne, resides internally within the plant concept in the CRISPR system, and relies on the knockout of susceptible genes so it expresses the altered product such as the modification in plant pathogen receptor and causes no signal transduction. Another includes the insertion in between the effector gene required for the pathogen recognition (Schenke and Cai 2020). There are several instances of CRISPR-edited crops bringing bacterial disease resistance, some of which are described in Table 10.2. Additionally, resistance can be prompted by a single base replacement rather than only knock-ins and knockouts.

Table 10.2 CRISPR/Cas-engineered crops resistant to bacterial, fungal, and viral phytopathogens

Pathogens		Disease	Crop name	Gene name	Type of modification	References
Bacteria	*Xanthomonasoryzae*	Bacteria blight	Rice	Xa13	Deletion of 149 nucleotides from promoter region	Li et al. (2020)
	Burkholderiaglumae	Blight blight	Rice	OsMPK5	Knockout	Xie and Yang (2013)
	Pseudomonas syringae, *Pseudomonas capsici*, and *Xanthomonas species*	Bacterial speck	Tomato	S1DMR6-1	Deletion in the DMR6 gene resulting in increased salicylic level	Paula de Toledo Thomazella et al. (2016)
	Xanthomonascamestrispv. musacearum	Banana wilt	Banana	Downy mildew resistance 6 (DMR6)	Deletion	Tripathi et al. (2021)
	Erwiniaamylovora	Fire blight	Apple	DIPM-1, DIPM-2 and DIPM-4	Indel (insertion and deletion)	Malnoy et al. (2016)
Fungi	*Xanthomonascitri*	Citrus cancer	Orange	Lateral organ boundaries 1 (CsLOB1) gene	Deletion of EBEPthA4 sequence present in CsLOB1 gene	Peng et al. (2017)
	Blumeriagraminis	Powdery mildew	Wheat		Mutation	Wang et al. (2014)
	Magnaportheoryzae	Rice blast	Rice	OsERF922	Knockout	Wang et al. (2016)
	Ustilaginoideavirens	False smut	Rice	USTA and UvSLT2	Gene replacement	Liang et al. (2018)
	Oidium species	Powdery mildew	Tomato	Mildew resistance locus (MLO) gene	Knockout	Pramanik et al. (2021)
	Leptosphariamaculans	Blackleg	Mustard	HK	Knockout	Idnurm et al. (2017)
	Ustilagomaydis	Corn smut	Maize	bW2 and bE1	Site-directed mutagenesis	Schuster et al. (2016)
	Botrytis cinerea	Gray mold	Grape	WRKY52	Knockout	Wang et al. (2018)

(continued)

Virus	Potyviridae family	Cucumber vein yellowing	Cucumber	eIF4E (eukaryotic translation initiation factor 4E) gene	Point mutation	Chandrasekaran et al. (2016)
	Yellow mosaic virus	Mosaic	Zucchini	eIF4E	Knockout	Chandrasekaran et al. (2016)
	Tomato yellow leaf curl virus	Yellow leaf curl disease	Tomato	SIPelo gene	Knockout	Pramanik et al. (2021)
	Streak virus	Banana streak virus	Banana	eBSV	Indel	Tripathi et al. (2019)
	Brown streak virus	Leaf streak	Cassava	nCBP-1 & nCBP-2	Knockout	Gomez et al. (2019)
	Cotton leaf curl multan virus	Leaf curl	Benthi	IR and C1 coding region	Destabilizing the viral replication intermediates through linearization of circular genomes	Yin et al. (2019)

10.4.2 Fungal Disease

Majorly, severe diseases in plants are caused by fungi, and about 30% of plant diseases are due to fungi (Giraud et al. 2010) especially the saprophytes. The strategies utilized to develop fungal-resistant plants are based upon various criteria such as inhibiting the interaction of fungus and plant receptors, mutating the susceptible factors (Nekrasov et al. 2017), knocking out the negative regulators genes required for interaction (Fister et al. 2018), and deletion or alteration in cis promoters (Oliva et al. 2019). Some of the CRISPR-edited pathogen-restricted plants are discussed in Table 10.2. In addition to it, another approach is added to reduce fungal disease in plants. Here, a new potential fungal genotype produced through CRISPR that has to compete with destructive fungal strains was deliberated in the field so when it performs parasexual activities with pathogenic strains, the progeny produced receives reduced pathogenicity (Muñoz et al. 2019).

10.4.3 Viral Disease

Plant viruses are obligatory in nature which utilizes plant machinery for their replication and survival. These pathogens may have DNA or RNA as their genome. The resistance developed toward the DNA viruses can be achieved through the destruction of essential viral genes expressing replication-associated protein, coat protein, intergenic region having origin of replication, or targeting the plant factors responsible for assisting the viral life cycle (Ji et al. 2015). For RNA viruses, the editing is performed toward the virus-supportive host plant factors such as eukaryotic translation initiation factor (eIF) and replication initiator protein (Rep) that are required to maintain the replication of viruses (Khatodia et al. 2017). In this case, CRISPR/Cas is used to generate knockout mutation which halts viral replication. The reported strategies were developed for single- and double-stranded DNA and single-stranded RNA viruses, and some examples are given in Table 10.2.

10.5 Role in Disease Diagnosis

Beyond its gene editing capabilities, CRISPR also plays a crucial role in sensing phytopathogens, offering an alternative to laborious hybridization and PCR-based nucleic acid detection methods. By modifying the Cas9 nuclease, precise pathogen detection in plants can be achieved without DNA degradation. To achieve accurate pathogen detection, researchers deactivate the nuclease activity of Cas9 and incorporate reporters such as fluorescence, quenchers, or biotin-FAM. This modification allows Cas9 to bind to the target region without causing DNA cleavage, effectively flagging the presence of specific pathogens (Li et al. 2019). The detection system relies on unique molecular markers, which are specific DNA markers carried by particular pathogens. These markers can be genes associated with pathogen species that express proteins or factors integrated within the pathogen's genome.

Besides Cas9, other Cas proteins, such as Cas12a, Cas13a (Tian et al. 2022), and Cas14, have been utilized in biosensing detection systems based on the phenomenon of collateral cleavage. In these systems, the Cas enzyme is combined with probes made of nucleotides similar to the target sequence and attached to a reporter molecule. When the Cas protein is activated and binds to its target, it initiates cleavage of all single-stranded nucleotide chains carrying the introduced probes in the sample (East-Seletsky et al. 2016).

This innovative approach offers several advantages over traditional detection methods. Firstly, it eliminates the need for time-consuming and complex hybridization or PCR steps, streamlining the detection process. Additionally, it enables highly specific and sensitive detection of phytopathogens, facilitating timely and targeted responses to plant diseases. Furthermore, the versatility of CRISPR allows the system to be tailored to detect various pathogens by simply designing new guide RNAs for different targets.

Overall, CRISPR-based pathogen detection represents a promising tool for agriculture and plant health, providing a rapid, precise, and efficient method to monitor and manage phytopathogens, ultimately contributing to better crop yields and food security. Continued research and advancements in this field hold the potential for transformative applications in agricultural biotechnology and disease control.

Various effective sensing methods have been demonstrated for pathogen detection, exemplified by the case of wheat and rice blast caused by the fungus *Magnaporthe oryzae Triticum* (MoT). For wheat blast, gRNA targets MoT-6098 and MoT-6099, and detection is achieved using recombinase polymerase amplification (RPA), Cas, and nucleic acid rapid lateral flow immunoassay (NALFIA) techniques, leveraging Cas12a collateral cleavage (Kang et al. 2021). In rice, crRNAs are designed for two genes, scytalone dehydratase (SDH1) and translation elongation factor 1 alpha (TEF1), to detect the pathogen (Zhang et al. 2020). Similarly, the grapevine red blotch virus is detected using a CRISPR/Cas12a biosensing system assisted by plasmonic nanoparticles. These methods offer efficient and specific pathogen detection, holding great promise for crop protection and disease management.

10.6 Conclusion

CRISPR technology has successfully addressed limitations seen in traditional methods, offering precise pathogenic disease diagnosis without triggering bioethical or bio-threatening controversies. The technique's ease of use, cost-effectiveness, and reliable maintenance have enabled its evolution from prokaryotes to advanced living systems. This progress has resulted in the development of numerous disease-resistant plants and diagnostic kits, making CRISPR a promising tool for plant disease management. By curbing the co-evolution of phytopathogens, CRISPR aids in safeguarding plant health. Notably, the approval of CRISPR-edited cabbage for human consumption by the Swedish Board of Agriculture in 2016 underscores its potential to revolutionize agriculture. Embracing CRISPR technology offers great promise in managing plant diseases, boosting productivity, and ultimately mitigating food insecurity challenges for the burgeoning global population.

References

Alkan F, Wenzel A, Anthon C, Havgaard JH, Gorodkin J (2018) CRISPR-Cas9 off-targeting assessment with nucleic acid duplex energy parameters. Genome Biol 19(1):1–13

Aman R, Mahas A, Mahfouz M (2020) Nucleic acid detection using CRISPR/Cas biosensing technologies. ACS Synth Biol 9(6):1226–1233. https://doi.org/10.1021/acssynbio.9b00507

Baker RHA, Sansford CE, Jarvis CH, Cannon RJC, MacLeod A, Walters KFA (2000) The role of climatic mapping in predicting the potential geographical distribution of non-indigenous pests under current and future climates. Agric Ecosyst Environ 82(1–3):57–71. https://doi.org/10.1016/S0167-8809(00)00216-4

Belhaj K, Chaparro-Garcia A, Kamoun S, Patron NJ, Nekrasov V (2015) Editing plant genomes with CRISPR/Cas9. Curr Opin Biotechnol 32:76–84. https://doi.org/10.1016/j.copbio.2014.11.007

Chandrasekaran J, Brumin M, Wolf D, Leibman D, Klap C, Pearlsman M et al (2016) Development of broad virus resistance in non-transgenic cucumber using CRISPR/Cas9 technology. Mol Plant Pathol 17(7):1140–1153. https://doi.org/10.1111/mpp.12375

Dong OX (2020) Gene replacement and knockin made easy in plants. Mol Plant 13(8):1104. https://doi.org/10.1016/j.molp.2020.07.015

East-Seletsky A, O'Connell MR, Knight SC, Burstein D, Cate JH, Tjian R, Doudna JA (2016) Two distinct RNase activities of CRISPR-C2c2 enable guide-RNA processing and RNA detection. Nature 538(7624):270–273

Fister AS, Landherr L, Maximova SN, Guiltinan MJ (2018) Transient expression of CRISPR/Cas9 machinery targeting TcNPR3 enhances defense response in Theobroma cacao. Front Plant Sci 9:268. https://doi.org/10.3389/fpls.2018.00268

Gaudelli NM, Komor AC, Rees HA, Packer MS, Badran AH, Bryson DI, Liu DR (2017) Programmable base editing of A•T to G•C in genomic DNA without DNA cleavage. Nature 551(7681):464–471

Giraud T, Gladieux P, Gavrilets S (2010) Linking the emergence of fungal plant diseases with ecological speciation. Trends Ecol Evol 25(7):387–395. https://doi.org/10.1016/j.tree.2010.03.006

Gomez MA, Lin ZD, Moll T, Chauhan RD, Hayden L, Renninger K et al (2019) Simultaneous CRISPR/Cas9-mediated editing of cassava eIF 4E isoforms nCBP-1 and nCBP-2 reduces cassava brown streak disease symptom severity and incidence. Plant Biotechnol J 17(2):421–434

Hammond-Kosack KE, Jones JDG (1996) Resistance gene-dependent plant defense responses. Plant Cell 8(10):1773. https://doi.org/10.1105/tpc.8.10.1773

Heitefuss R, Welch J (1989) Crop and plant protection: the practical foundations. Ellis Horwood, Chichester

Idnurm A, Urquhart AS, Vummadi DR, Chang S, Van de Wouw AP, López-Ruiz FJ (2017) Spontaneous and CRISPR/Cas9-induced mutation of the osmosensor histidine kinase of the canola pathogen Leptosphaeriamaculans. Fungal Biol Biotechnol 4(1):1–12

Ji X, Zhang H, Zhang Y, Wang Y, Gao C (2015) Establishing a CRISPR–Cas-like immune system conferring DNA virus resistance in plants. Nat Plants 1(10):1–4

Jinek M, Chylinski K, Fonfara I, Hauer M, Doudna JA, Charpentier E (2012) A programmable dual-RNA–guided DNA endonuclease in adaptive bacterial immunity. Science 337(6096):816–821. https://doi.org/10.1126/science.1225829

Kang H, Peng Y, Hua K, Deng Y, Bellizzi M, Gupta DR et al (2021) Rapid detection of wheat blast pathogen Magnaportheoryzae Triticumpathotype using genome-specific primers and Cas12a-mediated technology. Engineering 7(9):1326–1335. https://doi.org/10.1016/j.eng.2020.07.016

Khatodia S, Bhatotia K, Tuteja N (2017) Development of CRISPR/Cas9 mediated virus resistance in agriculturally important crops. Bioengineered 8(3):274–279. https://doi.org/10.1080/21655979.2017.1297347

Li Y, Li S, Wang J, Liu G (2019) CRISPR/Cas systems towards next-generation biosensing. Trends Biotechnol 37(7):730–743. https://doi.org/10.1016/j.tibtech.2018.12.005

Li C, Li W, Zhou Z, Chen H, Xie C, Lin Y (2020) A new rice breeding method: CRISPR/Cas9 system editing of the Xa13 promoter to cultivate transgene-free bacterial blight-resistant rice. Plant Biotechnol J 18(2):313. https://doi.org/10.1111/pbi.13217

Liang Y, Han Y, Wang C, Jiang C, Xu JR (2018) Targeted deletion of the USTA and UvSLT2 genes efficiently in Ustilaginoideavirens with the CRISPR-Cas9 system. Front Plant Sci 9:699. https://doi.org/10.3389/fpls.2018.00699

Luck J, Spackman M, Freeman A, Trebicki P, Griffiths W, Finlay K, Chakraborty S (2011) Climate change and diseases of food crops. Plant Pathol 60(1):113–121. https://doi.org/10.1111/j.1365-3059.2010.02414.x

Malnoy M, Viola R, Jung MH, Koo OJ, Kim S, Kim JS et al (2016) DNA-free genetically edited grapevine and apple protoplast using CRISPR/Cas9 ribonucleoproteins. Front Plant Sci 7:1904. https://doi.org/10.3389/fpls.2016.01904

Manghwar H, Li B, Ding X, Hussain A, Lindsey K, Zhang X, Jin S (2020) CRISPR/Cas systems in genome editing: methodologies and tools for sgRNA design, off-target evaluation, and strategies to mitigate off-target effects. Adv Sci 7(6):1902312. https://doi.org/10.1002/advs.201902312

Miedes E, Vanholme R, Boerjan W, Molina A (2014) The role of the secondary cell wall in plant resistance to pathogens. Front Plant Sci 5:358. https://doi.org/10.3389/fpls.2014.00358

Muñoz IV, Sarrocco S, Malfatti L, Baroncelli R, Vannacci G (2019) CRISPR-Cas for fungal genome editing: a new tool for the management of plant diseases. Front Plant Sci 10:135. https://doi.org/10.3389/fpls.2019.00135

Nekrasov V, Wang C, Win J, Lanz C, Weigel D, Kamoun S (2017) Rapid generation of a transgene-free powdery mildew resistant tomato by genome deletion. Sci Rep 7(1):482

Oliva R, Ji C, Atienza-Grande G, Huguet-Tapia JC, Perez-Quintero A, Li T et al (2019) Broad-spectrum resistance to bacterial blight in rice using genome editing. Nat Biotechnol 37(11):1344–1350

Paula de Toledo Thomazella D, Brail Q, Dahlbeck D, Staskawicz B (2016) CRISPR-Cas9 mediated mutagenesis of a DMR6 ortholog in tomato confers broad-spectrum disease resistance. BioRxiv 2016:064824. https://doi.org/10.1101/064824

Peng A, Chen S, Lei T, Xu L, He Y, Wu L et al (2017) Engineering canker-resistant plants through CRISPR/Cas9-targeted editing of the susceptibility gene Cs LOB 1 promoter in citrus. Plant Biotechnol J 15(12):1509–1519. https://doi.org/10.1111/pbi.12733

Pramanik D, Shelake RM, Park J, Kim MJ, Hwang I, Park Y, Kim JY (2021) CRISPR/Cas9-mediated generation of pathogen-resistant tomato against tomato yellow leaf curl virus and powdery mildew. Int J Mol Sci 22(4):1878. https://doi.org/10.3390/ijms22041878

Sandhya D, Jogam P, Allini VR, Abbagani S, Alok A (2020) The present and potential future methods for delivering CRISPR/Cas9 components in plants. J Genet Eng Biotechnol 18:1–11

Savary S, Willocquet L, Pethybridge SJ, Esker P, McRoberts N, Nelson A (2019) The global burden of pathogens and pests on major food crops. Nat Ecol Evol 3(3):430–439

Schenke D, Cai D (2020) Applications of CRISPR/Cas to improve crop disease resistance: beyond inactivation of susceptibility factors. IScience 23(9):101478. https://doi.org/10.1016/j.isci.2020.101478

Schuster M, Schweizer G, Reissmann S, Kahmann R (2016) Genome editing in Ustilagomaydis using the CRISPR–Cas system. Fungal Genet Biol 89:3–9. https://doi.org/10.1016/j.fgb.2015.09.001

Tian T, Qiu Z, Jiang Y, Zhu D, Zhou X (2022) Exploiting the orthogonal CRISPR-Cas12a/Cas13a trans-cleavage for dual-gene virus detection using a handheld device. Biosens Bioelectron 196:113701. https://doi.org/10.1016/j.bios.2021.113701

Tilman D, Cassman KG, Matson PA, Naylor R, Polasky S (2002) Agricultural sustainability and intensive production practices. Nature 418(6898):671–677

Tripathi JN, Ntui VO, Ron M, Muiruri SK, Britt A, Tripathi L (2019) CRISPR/Cas9 editing of endogenous banana streak virus in the B genome of Musa spp. overcomes a major challenge in banana breeding. Commun Biol 2(1):46

Tripathi JN, Ntui VO, Shah T, Tripathi L (2021) CRISPR/Cas9-mediated editing of DMR6 orthologue in banana (Musa spp.) confers enhanced resistance to bacterial disease. Plant Biotechnol J 19(7):1291. https://doi.org/10.1111/pbi.13614

Vannier N, Agler M, Hacquard S (2019) Microbiota-mediated disease resistance in plants. PLoS Pathog 15(6):e1007740. https://doi.org/10.1371/journal.ppat.1007740

Voytas DF (2013) Plant genome engineering with sequence-specific nucleases. Annu Rev Plant Biol 64:327–350. https://doi.org/10.1146/annurev-arplant-042811-105552

Wang Y, Cheng X, Shan Q, Zhang Y, Liu J, Gao C, Qiu JL (2014) Simultaneous editing of three homoeoalleles in hexaploid bread wheat confers heritable resistance to powdery mildew. Nat Biotechnol 32(9):947–951

Wang F, Wang C, Liu P, Lei C, Hao W, Gao Y et al (2016) Enhanced rice blast resistance by CRISPR/Cas9-targeted mutagenesis of the ERF transcription factor gene OsERF922. PLoS One 11(4):e0154027. https://doi.org/10.1371/journal.pone.0154027

Wang X, Tu M, Wang D, Liu J, Li Y, Li Z et al (2018) CRISPR/Cas9-mediated efficient targeted mutagenesis in grape in the first generation. Plant Biotechnol J 16(4):844–855. https://doi.org/10.1111/pbi.12832

Xie K, Yang Y (2013) RNA-guided genome editing in plants using a CRISPR–Cas system. Mol Plant 6(6):1975–1983

Xu X, Qi LS (2019) A CRISPR–dCas toolbox for genetic engineering and synthetic biology. J Mol Biol 431(1):34–47. https://doi.org/10.1016/j.jmb.2018.06.037

Yin K, Han T, Xie K, Zhao J, Song J, Liu Y (2019) Engineer complete resistance to Cotton Leaf Curl Multan virus by the CRISPR/Cas9 system in Nicotianabenthamiana. Phytopathol Res 1(1):1–9

Zaidi SSEA, Mahas A, Vanderschuren H, Mahfouz MM (2020) Engineering crops of the future: CRISPR approaches to develop climate-resilient and disease-resistant plants. Genome Biol 21(1):1–19

Zhang YM, Zhang Y, Xie K (2020) Evaluation of CRISPR/Cas12a-based DNA detection for fast pathogen diagnosis and GMO test in rice. Mol Breed 40:1–12

Diagnosis and Management Strategies for Soybean Diseases

11

Himani Jeena, Krishna Pratap Singh, Gaurav Rakhonde, Manisha Dev, Kumari Surbhi, and T. Aravind

Abstract

Soybean is one of the most important oilseed crops cultivated in tropical and subtropical regions of the world since ancient times. The productivity of soybeans is affected by several abiotic and biotic constraints, which include diseases caused by fungi, bacteria, and viruses. The most important diseases of soybean include anthracnose, pod blight, charcoal rot, Rhizoctonia aerial blight, bacterial blight, damping off, seedling rot and blight, yellow mosaic virus, soybean mosaic, frogeye leaf spot, etc., which can affect yield if environmental conditions favor disease growth. The correct management of diseases depends on accurate identification and early detection. The economic significance, symptomatology, causal organism, disease cycle, epidemiology, and management of a few economically significant diseases of soybean are all covered in this chapter. It takes into account the main diseases of soybean and pathogens affecting agricultural productivity, as well as new emerging threats.

Keywords

Diagnosis · Diseases · Management strategies · Soybean crop

H. Jeena (✉) · K. P. Singh (✉) · M. Dev · K. Surbhi
Department of Plant Pathology, College of Agriculture, Govind Ballabh Pant University of Agriculture and Technology, Pantnagar, Uttarakhand, India

G. Rakhonde
Department of Plant Pathology, University of Agricultural Sciences, GKVK, Bengaluru, Karnataka, India

T. Aravind
Department of Plant Pathology, M. S. Swaminathan School of Agriculture, Centurion University of Technology and Management, Paralakhemundi, Odisha, India

V. K. Singh et al. (eds.), *Diseases of Field Crops: Diagnostics and Management*, https://doi.org/10.1007/978-981-97-6160-9_11

11.1 Introduction

Soybean (*Glycine max* L. Merril) is the most significant seed legume in the world, which produces 25% of the world's edible oil and almost two-thirds of the protein concentrate used in cattle feed. A valuable component of formulated meals for poultry and fish is soybean meal (Wang et al. 2019). Carl Linnaeus first described the soybean, *Glycine max* (L.) Merr, in his first printing of Genera Plantarum. Leguminous and self-pollinating, soybean (*Glycine max*) is a crop that belongs to the *Fabaceae* family and subfamily *Papilionaceae*. Instead of being classified as a pulse, the plant is an oilseed (Yadav et al. 2020). In terms of protein production per hectare among field crops, soybean ranks top, making it one of the most significant crops for global food security. As early as the eleventh century, it is believed that soybeans first appeared in China. Soybean farming has been documented in writing since 2200 BC. Around the world, soybean is grown in regions with ideal agro-climatic conditions in a number of nations, including the United States, Brazil, Argentina, China, India, Paraguay, Canada, and Mexico. It was designated as one of the five sacred grains by Emperor Sheng-Nung. Thus, soybean has been farmed in China for more than 4000 years (Hymowitz 1970). India began growing soybeans in the early 1800s (Rathore et al. 2021). Majority of the literature demonstrates that the nation's commercial soybean farming began in the 1970s.

In India, the cultivation of soybeans has increased more quickly than many other crops. Due to its better returns compared to other crops, the soybean crop has displaced other crops and taken on a significant role in agriculture. Due to the substitution of soybean for other crops like sunflower in Argentina and cotton in the United States, as well as the use of pasture land in Argentina and Brazil, the area of soybean significantly increased due to the shifting of area from cotton, groundnut, cereals, etc. and in different countries around the world (Rathore et al. 2021). With an expected production of 368.47 million metric tonnes and a productivity of 2.89 t/ha, soybean production was estimated to be taking place on 127.28 mha of land worldwide in 2019–20. In India, the total area dedicated to soybean farming in 2020 is projected to be roughly 11.3 mha, with a production estimate of 13.2 million metric tonnes and a productivity rate of 1.17 t/ha. Madhya Pradesh, Maharashtra, Rajasthan, Uttar Pradesh, and Uttarakhand are the principal states in India that cultivate soybeans. Some diseases provide the greatest harm to soybean agriculture among the various production restrictions, thus identifying these diseases is essential. Many diseases severely affect this crop, which results in yield losses of up to 20–100%. In India, 35 different soybean diseases have been found, some of which are significant based on the amount of production loss. Viruses, Sclerotium blight, anthracnose, rust, charcoal rot, and Rhizoctonia aerial blight were the main causes of yield losses in 2006 (Wrather and Koenning 2006). Some diseases were widespread across the nation, while others were confined to particular locations. Other diseases that lowered yield included brown leaf spot, frogeye leaf spot, damping off, and sclerotinia stem rot, though these occurred only occasionally or infrequently. Several diseases have been reported in India, including anthracnose, bacterial diseases, brown spot, charcoal rot, frogeye leaf spot, pod and stem blight, purple seed stain and Cercospora leaf blight, Rhizoctonia aerial blight, seedling diseases, soybean rust, virus diseases, and others. According to a different study, insect pests such as stem flies, girdle beetles, and different defoliators, as well as diseases like the yellow mosaic virus, rust, rhizoctonia, anthracnose, etc., are the main biotic pressures on the soybean crop in India. Asian soybean rust disease was first noted on soybeans in India in 1951 (Sharma and Mehta 1996). Banerjee et al. (2014) confirmed the presence of Soybean Mosaic Virus (SMV) in soybeans grown in Meghalaya, India, under mid-hill conditions. In the North Eastern Hill region, moderate to severe cases of frogeye leaf spot (*Cercospora sojina*), rust (*Phakospora pachyrhizi*), and purple seed stain (*Cercospora kikuchii*) have been reported. In this chapter, some of the significant soybean diseases are discussed, along with features of their epidemiology and management.

11.2 Diseases

11.2.1 Category: Fungal Diseases

11.2.1.1 Anthracnose

Economic Importance

Anthracnose disease is economically significant in humid tropics and tropical climate parts of the world. In warm, humid tropics and subtropics, where soybean production is now increasing, the disease reduces seed germination and seed quality, and can result in substantial yield losses (Sinclair 1989). The top eight soybean producing nations (Argentina, Bolivia, Brazil, Canada, China, India, Paraguay, and the United States) are estimated to have suffered a yield loss of 25.4 million tonnes as a result of *C. truncatum*, with China suffering the greatest losses (16.6 million tonnes), followed by the United States (4.9 million tonnes) and India (1.17 million tonnes) (Wrather et al. 2010). Anthracnose disease causes a loss of soybean yield of between 16% and 25% in India (Sharma et al. 2014). According to Bhadauria et al. (2013) and Nataraj et al. (2020), *Colletotrichum truncatum* is a hemi-biotroph. In Alabama, estimates of the maximum seed production drop due to anthracnose varied from 16% to 26%. Thailand and India both reported yield losses of 30–50% (Sinclair 1989).

Pathogen

The fungus *Colletotrichum truncatum* is responsible for anthracnose in soybean. Although numerous other *Colletotrichum* species have also been found to be implicated, this is the species most frequently linked to the disease. On infected tissues, the fungus grows dense, black acervuli. When magnified by 20X or more, these black entities will resemble "pin cushions" on the surface of the plant. The spores (conidia) that are produced by acervuli are what cause the disease to spread. It has been determined that the fungus is seed-borne and that it overwinters in contaminated agricultural detritus.

Disease Symptoms and Epidemiology

Preemergence and postemergence damping-off may occur when infected seeds are planted. The cotyledons of seedlings acquire sunken, dark brown blemishes. Lesions on seedlings can spread to the stem and harm young plants. Although infections can occur in plants at any stage of growth, in Nebraska, the likelihood of infection seems to rise with maturity. Brown dots with uneven shapes that appear on stems, pods, and petioles are the most typical symptoms (Fig. 11.1). Premature defoliation is caused by extensive lesions girdling the petioles (Boufleur et al. 2021). Mycelium may entirely cover the cavity of infected pods, preventing the production of seeds (a condition known as pod blanking), or fewer and/or smaller seeds may develop. When a seed does form, it may appear normal or discolored, moldy, and shriveled. In lesions in all host tissue locations, dark acervuli form. Rolling of the leaves, laminar vein necrosis, petiole cankers, and early defoliation are all effects of leaf diseases (Fig. 11.1). At maturity, this disease is frequently seen on soybean stems. Warm, humid weather is often when anthracnose infections occur. When rain or dew spells last longer than 12 h/day, plant leaf moisture, or infection often ensues (Nataraj et al. 2020).

Fig. 11.1 Symptoms of anthracnose on leaf (**a**) and pod (**b**, **c**, **d**)

Disease Cycle

C. truncatum and *C. glycines* both have the ability to over-season as mycelium in infected crop residue or infected seeds and can produce microsclerotia that are infectious to soybeans (Yang and Hartman 2016). At every stage of growth, but especially from bloom through pod fill, plants are prone to infection. Infected seeds are the primary source of *C. truncatum and C. plurivorum's* primary inoculum, which aids in their long-distance dispersal and the introduction of novel fungal isolates to a region. Following conidial germination and the development of an appressorium on the plant surface, fungal penetration takes place. The infection and colonization of soybean leaves by *C. truncatum* and *C. sojae* follow similar patterns (Manandhar et al. 1985). In the initial stage of *C. truncatum*'s hemibiotrophic lifestyle, the penetration peg transforms into a typical primary hypha, a biotrophic vesicle situated between the cell wall and the plasma membrane. After the biotrophic phase, the tissue enters the necrotrophic phase, where secondary hyphae are produced and invade the tissue intra- and intercellularly, leading to cell death. According to the possibility of quiescent/endophytic infection that has been described (Bhadauria et al. 2013), plant tissues may become infected even while no symptoms are present (Chen et al. 2006). Anthracnose symptoms begin during the necrotrophic stage, when acervuli carrying conidia are formed. The conidia serve as the disease's secondary inoculum, dispersed by water splashes that break down the mucilage that covers them and promote short-range dispersal (Madden 1997).

Management

The management of soybean anthracnose should begin with the sowing of disease-free seeds and the practice of crop rotation due to the potential for off-season survival of species of *Colletotrichum* that infect soybean and the long-distance dissemination of the pathogen by seeds (Yang and Hartman 2016). Although seeds are typically symptomless, even small percentages of contamination can cause significant crop losses (Pellegrino et al. 2010). Seeds can be treated with systemic fungicides such as carbendazim, benzimidazoles, or triazoles to stop the disease. Additionally, new research has been done on potential biological control utilizing biopriming techniques. *Pseudomonas aeruginosa* and *Trichoderma harzianum* injected into soybean seeds reduced *C. truncatum* field incidence by up to 92%, matching the effectiveness of the fungicide benomyl. Currently, fungicides such as azoxystrobin, captan, mancozeb, carbendazim, and thiophanate methyl are employed as preventatives. However, certain isolates of *C. truncatum* have shown resistance to these chemicals as well as reduced sensitivity to difenoconazole and propiconazole. Mancozeb (63%) and carbendazim (12%) decreased the disease incidence. This was regarded as the best treatment among the evaluated fungicides since it improved the yield of the treated plots when compared to the control. Farmers in India's anthracnose-affected regions use Tebuconazole at a rate of 625 mL/ha or Thiophanate methyl at a rate of 1 kg/ha to chemically suppress the infection. For treating the condition, SAAF and Mancozeb @ 2.5 g/L of water might be utilized (Subedi et al. 2016).

11.2.1.2 Aerial Blight of Soybean

Economic Importance

The disease may result in a crop output drop of up to 35%. Due to local environmental conditions, disease-related damage varies in northern Brazil, where crop losses may be between 31% and 60%. According to reports from the United States, the disease reduced agricultural yields by up to 50%. Stand establishment and crop output both experienced severe losses of up to 50% and 60%, respectively (Chang et al. 2017). *R. solani* infection in soybeans reduces the crop's shoot length and pod formation. Additionally, it is known to decrease the weight and overall output of seeds. The pathogen could reduce crop yields by 35% to 60% in humid weather conditions. In Chhattisgarh, Madhya Pradesh, and Uttarakhand, excessive rainfall seasons were associated with a severe occurrence of the disease (Wrather et al. 2010; Joshi et al. 2018).

Disease Symptoms and Epidemiology

Aerial blight symptoms often occur on soybean plants that are 45–60 days old, and the infection first affects the lower or middle parts of the plant before moving upward to infect the leaves, stem, and pods. The first signs of the disease appear as water-soaked, light green, erratic lesions on the plant's leaflets. The tiny lesions quickly combine to produce larger, irregular patches that cover the entire leaf (Fig. 11.2). The larger irregular spots have reddish-brown borders and browned grayish cores (Williamson et al. 2006). On the diseased plant sections, the fungus' hyphae appear as a brownish spider web. Small, uneven, brown or dark-colored sclerotia are visible on the infected plant surface. Plants that have been extensively affected lose their leaves and die. In the disease's aerial blight phase, the plant's drooping leaves adhere to the stem section and pods directly below them, acting as a source of infection for starting an infection in the plant's pods. At this stage of infection, petioles and stems begin to exhibit the pathogen's distinctive

Fig. 11.2 Symptoms of Rhizoctonia aerial blight or web blight on leaf

brown-colored lesions. Infected seeds are produced as a result of pod infection. In environments with high humidity, the fungus's mycelial proliferation and sclerotia generation are facilitated. The importance of weather elements, including temperature, humidity, rainfall, daylight hours, etc., in the development of the disease over time in the crop season has been underlined by a number of workers (Mathpal 2016; Joshi et al. 2018; Singh et al. 2019; Surbhi and Singh 2020). The ideal temperature for the atmosphere is 28 °C, within the range of 10–40 °C, whereas temperatures below 10 °C hinder the fungus's ability to produce sclerotia. The development of *R. solani* in the field is supported by relative humidity levels of 85–90% (Surbhi and Singh 2020).

Disease Cycle

R. solani is primarily soil-borne, and during heavy rains, soil splashes onto the leaves serve as the initial inoculum. The pathogen is capable of extensive soil growth and relies on crop debris to survive. It can last for a long time in soil. Phases 1 and 2 are soil-borne and leaf-borne, respectively (Yang et al. 1990). In addition to colonizing soil organic materials as saprophytes, *R. solani* also survives as mycelium, often as a result of plant pathogenic activity. No inoculum is discovered below 40 cm of earth, which has the maximum inoculum potential. The fungus can spread quickly and over a larger area in the soil, where it can persist as mycelia or sclerotia. Weed hosts are the first to acquire the infection and promote the formation of basidiospores due to their close proximity to the soil containing the primary inoculum and their distinct microclimate. The disease then spreads through the air, infecting leaves, stems, or pods while producing a new cycle of basidiospores on the primary host.

Disease Management

The disease can be managed in a cost-efficient, simple, and reliable manner by using resistant varieties. Although no commercial soybean cultivars with complete immunity to the pathogen's infection are currently known, moderately resistant varieties can be grown (Thind 2005). Disease incidence can be decreased through crop rotation with nonleguminous crops, crop residue removal or burning, summer ploughing, and the application of fertilizers in the recommended amounts. Maintaining appropriate distances between plant rows in the field is another important management technique. A dense canopy could produce a humid environment that encourages disease growth. Additionally, a dense canopy may increase the risk of infection spreading through contact between healthy plant leaves and infected plant leaves or through the splashing of mycelial fragments caused by rain. Fungicides including Pyraclostrobin (0.1%), Pencycuron (0.1%), and Fluopyram (0.1%) have been reported to be successful in controlling the disease under chemical management (Mathpal 2016). Thiram treatment of seeds is reported to reduce the pathogen's initial infection (Thind 2005). There have been reports that two foliar applications of Tilt (0.1%) spaced 15 days apart can lessen the disease's severity in the field. According to Sharma and Tripathi, some additional fungicides, including Contaf, Bavistin, and Indofil M-45, have also been observed to provide effective disease control (2001). The disease can also be controlled by using biocontrol agents such as *Trichoderma harzianum*, *Pseudomonas fluorescens*, and *B. subtilis* (Mathpal 2016).

11.2.1.3 Charcoal Rot

Economic Importance

Charcoal rot is a soil-borne disease. Generally, soybean losses range from 30% to 50% annually worldwide and can reach 77% in India. Crops at any growth stage can become infected by it, and microsclerotia persist in the field for several years.

Pathogen

The soil-dwelling organism *Macrophomina phaseolina* can infect soybeans at any stage of crop development, but it typically does so during the post-flowering period. Additionally, the fungus is seed-borne in numerous crops, including soybean. In order to persist in soil for 2–15 years and serve as the main source of inoculum, it develops microsclerotia in the root and stem tissues of the host plants. The Macrophomina's pycnidiospores are ellipsoid to obovoid and 20–24/7–9 μm in size (Gupta et al. 2012).

Symptoms and Epidemiology

After emergence, symptoms on seedlings can be seen as brown to dark patches on the cotyledons. The cotyledon edges can occasionally turn brown or black and shed early. From the unifoliate leaf stage onward, symptoms manifest as circular to rectangular, reddish-brown lesions on the growing hypocotyls of infected seedlings. After a few days, these lesions may turn dark brown or black. Such lesions could mount the stem. If hot and dry circumstances continue, infected seedlings could perish. On the roots, stems, pods, and seeds, the pathogen produces lesions. Foliar symptoms develop from the plant's top down. Infected plants have smaller-than-normal leaves that eventually turn yellow before wilting. As the fungus moves up the stem throughout the season, a reddish-brown discoloration of the vascular components of roots and lower stems precedes the premature yellowing. Black bodies, either locally or widely dispersed, cover the infected mature and dry pods (microsclerotia). Numerous tiny, pinhead-sized microsclerotia develop after the plant dies, and they are easily visible when the epidermal tissue of the lower stems and roots is peeled away from the affected areas (Fig. 11.3). In patches, the affected crop in the field displays premature yellowing (Gupta et al. 2012). Despite some infections on soybean seedlings occurring at 20–25 °C, the disease was more prevalent at 30–35 °C. Over a wide temperature range, infected seedlings may act as a latent source of inoculum for the mature plant phase (Gupta and Chauhan 2005). The number of diseased seedlings increased along with the number of *M. phaseolina* propagules that were viable. When soil conditions are abnormally dry and soil temperatures are consistently over 35 °C for 2 to 3 weeks, seedlings can become sick. Dry soils, high soil C: N ratios, and low bulk density are all favorable for the germination of microsclerotia in soybean fields (Gupta et al. 2012).

Fig. 11.3 Symptoms of charcoal rot

Disease Cycle

Microsclerotia, which are tough, black fungal structures, are how *Macrophomina phaseolina* survives in the soil. Between 20 °C and 40 °C, microsclerotia germinate and infect root tissue. Both mature plants and seedlings might be affected by the fungus. Once infected, *Macrophomina phaseolina* develops enzymes and toxins that break down root tissue, and within 2–3 weeks, the root and stem tissue is colonized (Islam et al. 2012). Amylase, cellulose, hemicellulose, lipase, and pectinase are some of the extracellular enzymes (Ahmad et al. 2006; Kaur 2012). The primary function of these extracellular enzymes, their association with fungal development, and their impact on the infection process are still largely unclear despite being linked to cell and membrane wall breakdown (Schinke and Germani 2012). Phaseolina and botryodiplodin, two toxins identified from *M. phaseolina*, are likely to be important in the infection process. The vascular system will become infected with *M. phaseolina* under favorable environmental conditions, interfering with the plant's normal ability to transfer water and nutrients to the leaves and resulting in the usual disease signs of wilting and premature leaf loss (Gupta and Chauhan 2005; Khan 2007). Microsclerotia in the plant waste from harvested soybean plants decompose and return to the soil. Short et al. (1980) reported that *M. phaseolina* may persist on leftovers at or below the soil surface for almost 2 years. Brazilian studies revealed that even while microsclerotia could persist on contaminated agricultural residue for approximately 35 months, pathogenicity declined after that point.

Disease Management

Crop rotation is one of the most important cultural practices. This lessens the chance of soil microsclerotia accumulating. Adhere to suggested plant populations. When there is a drought, soybeans that are densely planted will compete for moisture, which favors the development of diseases. Use varieties that are resistant. Choose plant types that are less prone to charcoal rot. There are some known origins of resistance. Neem, mustard cake, and farmyard manure are a few examples of natural amendments that can treat the disease known as charcoal rot (Rubayet et al. 2018). Sodium methyldithiocarbamate soil fumigation decreased pathogen populations on soybean residue and in the roots of plants cultivated in field plots. Treatment of the seed with 3 g/kg of captan, 3 g/kg of thiram, 3 g/kg of thiram + carbendazim (2: 1), or 2 g/kg of thiram + carboxin has been suggested in India to treat soybean seeds with *T. viride* or *Trichoderma harzianum* (4–5 g/kg of seed) to prevent charcoal rot. *Pseudomonas aeruginosa* seed treatment decreased *M. phaseolina* infection by 14–100%, depending on the bacterial strain and the type of soybean employed (Choudhary 2011). *M. phaseolina* seed infection can be reduced by soaking the seeds in ginger, garlic, and neem extracts for 5 min.

11.2.1.4 Rhizoctonia Stem Rot

Economic Importance

A significant barrier to soybean cultivation is seedling disease. One of the world's most significant root and seedling diseases of soybean is Rhizoctonia root rot (Wrather et al. 2001a; Ajayi-Oyetunde and Bradley 2017; Mahmoud and Abo-Elyousr 2014). The disease has been reported to occur in numerous soybean-growing regions across the world, including Brazil, China, India, Illinois, Ohio, Minnesota, and North Dakota in the United States (Wrather and Koenning 2006; Zhao et al. 2005). Seedling diseases were identified as the third-highest cause of yield losses for the soybean crop in the United States and Canada during the 3 years from 2006 to 2009 (Koenning and Wrather 2010). In Iowa, United States, an epidemic of root rot brought on by *R. solani* appeared in 1967. According to reports, the disease prevents the crop's stands from establishing themselves by 50% (Wrather et al. 2001a, b; Xue et al. 2006). Preemergence and postemergence damping-off of seedlings, root and hypocotyl rot, and aerial blight in soybean have all been linked to *R. solani* in Brazil.

Disease Symptoms and Epidemiology

Before the first trifoliate leaf appears, the plant is at its most susceptible. On the hypocotyls of the germinating seed, the infection first appears as a small, sunken, reddish-brown lesion just below the soil surface (Kumar et al. 2019). Such lesions enlarge and girdle the seedling as it germinates, which causes postemergence damping-off of the seedlings. If the seedling is able to survive the infection, the plant's roots grow normally, and it may fully recover from the pathogen infection. The field's afflicted plants appear chlorotic and stunted. The plant may wilt if the soil is not sufficiently moist. Warm temperatures and moist soil favor the development of the disease. Any root damage from stress, trauma, worm activity, insect feeding, or herbicide harm may encourage the growth of root rot (Duke et al. 2007). The pathogen can persist in the soil and plant waste as sclerotia or latent mycelium. It may persist for extended periods of time without a host plant in the soil as a saprophyte. The primary inoculum for the disease's onset during the growing season is provided by these sclerotia and mycelium that are still present in the soil.

Disease Cycle

Disease Management

The environment in the soil is less favorable for the formation of pathogens attributed to proper tillage, which also helps to promote soil drainage. Injuries brought on by nematodes, insects, or herbicides may make the plant more vulnerable to infection. Therefore, effective field management of insect pests and nematodes is crucial, and herbicide damage should be avoided or treated. Soybean cultivars that are resistant should be used for commercial-scale production. To decrease the number of inoculums in the field, crop rotation and crop debris clearance are also crucial (Zhang 2011). Root rot in the field has been observed to be reduced by seed treatment with biocontrol agents like *Trichoderma viride* at 10 g/kg seed (Barpete and Verma 2019). Fungicides from the strobilurin group, such as Azoxystrobin, can be used to treat seeds to help prevent the spread of disease. The management of soybean root rot in fields has also been proven to be successful when seed treatment with carboxin (2 mg/kg seed) and foliar spraying with propiconazole and neem oil (2 mL/L) are used (Kumar et al. 2019).

11.2.1.5 Rust

Economic Importance

One of the most significant diseases that affect soybeans is potentially soybean rust. It has a rapid rate of spread and can defoliate soybean plants and generate leaf spots. When conditions are right, the pathogen can result in yield losses of more than 50% (Miles et al. 2003). In 2004, soybean rust was discovered for the first time in the South of the United States. Every growing season afterward, the disease has moved further north, with modest levels of occurrence in the upper Midwest.

Pathogen

Phakopsora pachyrhizi, also known as the Asian or Australasian soybean rust pathogen, which is now also present in the western hemisphere, and *P. meibomiae*, the so-called New World soybean rust pathogen, which is only present in the western hemisphere, are two closely related fungi that cause rust on soybeans. The two fungi look morphologically identical, with the exception of a few small differences; however, *P. pachyrhizi* is significantly more aggressive on soybeans than *P. meibomiae*. In Central and South America, *P. meibomiae* has not yet been shown to significantly reduce yields (Hernández 2004). The microorganisms that cause soybean rust are obligate parasites, like other rusts, and require a living host to develop and replicate. Under natural circumstances, they can only endure as uredinospores for a brief period of time.

Disease Symptoms and Epidemiology

The signs of soybean rust are most frequently seen on the leaves and begin in the lower canopy. Additionally, petioles, pods, and stems might get lesions (Bromfield 1984). Once the plant begins to blossom, the onset of symptoms happens quickly, and under ideal environmental conditions, defoliation can reach substantial levels. Small, unevenly shaped yellow dots are the initial symptoms of lesions. Lesions are tan to dark reddish brown in hue and grow to be 1/16th to 1/12th of an inch in diameter as the disease progresses. There are one to multiple uredinia in the form of volcanoes within each lesion (spore-producing structures). Plants lose their leaves early and typically develop early as the intensity of the rust increases (APS Press Release 2014). Temperatures between 54 °F and 84 °F (65–80 °F is ideal) and relative humidity above 90% for more than 12 h are favorable for the development of soybean rust. As long as the nighttime temperature falls within the ideal range for disease development, soybean rust can be active in conditions with daytime temperatures as high as 100 °F and probably higher (Hernández 2004). Six hours of nonstop leaf wetness are required for spores to germinate and infect the plant. With longer leaf wetness durations of up to 12 h, infection rates rise. Significant rust formation in South America is associated with wet weather.

Disease Cycle

Epidemics of soybean rust are brought on by airborne inoculum (urediniospores). This disease differs from other rusts in that it has a wide range of potential hosts that could act as sources of inoculum. *P. pachyrhizi* spores are sensitive to ultraviolet light; hence long-distance rust spore migration is likely to take place in storm systems where clouds protect the spores from the sun. Environmental factors affect the development of infection and subsequent epidemics when viable spores have touched down on the leaf surface of a suitable host. Within 7 or 8 days of infection, lesions and pustules containing urediniospores might develop, signaling the start of the subsequent infection cycle. Due to this short life cycle, epidemics of soybean rust can rapidly develop under the correct circumstances, rising from nearly undetectable levels to extremely high ones. Early infections often occur in the lower canopy and are difficult to identify; epidemics may appear to spread much more quickly. In addition to the environment, plant age influences outbreaks of soybean rust. Except in situations where there are substantial inoculum levels early in the season, rust lesions are typically not discovered on soybean until flowering. This might be because plants are more vulnerable to rust as the host reaches the reproductive phases, because spores are better protected from UV light in the lower regions of the canopy, or because the canopy becomes more humid as the canopy closes. Lesions can develop at any stage of growth, although significant increases in disease do not occur until after flowering (Bromfield 1984).

Disease Management

Several cultural activities could help in controlling soybean rust. Changing planting and harvest dates may prevent disease in most regions where rust must be introduced annually for an epidemic to arise. It is possible to prevent rust by planting early and using a cultivar that matures quickly until the crop has either been harvested or has progressed to the point where the disease will have little effect on yield. Additionally, planting dates may be postponed so that the reproductive period is sensitive during dry conditions that do not encourage rust (Dunphy et al. 2006).

Wider row spacing and fewer plants may speed canopy drying in locations where the weather is unfavorable for the formation of rust, so lowering the dew period and preventing or at least slowing disease development. The timing of fungicide application is essential for controlling soybean rust. The initial fungicides (azoxystrobin, difenoconazole, fluquinconazole, epoxiconazole + pyraclostrobin, and tebuconazole) were suggested in emergency cases. The preventive use of azoxystrobin + cyproconazole, tebuconazole, and pyraclostrobin + epoxiconazole had the best results. The most effective treatments, when used therapeutically, were flutriafol and tebuconazole (FAO 1982). The strobilurins prevent the pathogen from respiring through its mitochondria. Strobilurins serve as protective fungicides and must be used before infection begins in order to be successful. Strobilurins can be effective for up to 2 weeks after application, depending on the rate used, but they won't shield freshly emerging leaves (Dorrance et al. 2007a, b).

11.2.1.6 *Phytophthora* Root Rot

Economic Importance

On poorly drained soils, *Phytophthora* root rot is a concern all around the world (Wrather et al. 2001a). While soybean production was beginning to boom, the illness was first discovered in the United States in Ohio in 1951 and Indiana in 1948. However, the causal agent was not identified until 1958 (Dorrance and Gerdemann 1958).

Pathogen

Phytophthora sojae is an oomycete and a close relative of the golden-brown algae (or Straminipila). These species create zoospores, which are swimming spores with tubular flagella. Oomycetes grow in a manner that is akin to fungi since they also produce mycelium, although the mycelium is coenocytic (aseptate, lacking cross walls). Sporangia measure about 40 mm in length and 28 mm in width. Either hyphae are produced directly during germination or a large number of zoospores (more than 20) are produced indirectly during germination. Another sporangium will form in an enlarged and nested pattern from the empty one (Dorrance et al. 2004).

Disease Symptoms and Epidemiology

The oomycete plant pathogen *Phytophthora sojae* is the reason behind *Phytophthora* root and stem rot. When soil conditions are favorable for pathogen development, this water mold can infect seeds, seedlings, and plants at all developmental phases of reproduction. The majority of the time, symptoms appear 1 to 2 weeks after heavy rains and are especially prevalent in poorly drained soils.

In case of seed rot and damping-off: Under moist and warm soil conditions, *Phytophthora sojae* causes seed decay as well as preemergence and postemergence damping-off of soybeans. The ideal temperature for the development of the disease is between 25 °C and 30 °C (77 °F and 86 °F). Replanting is frequently necessary in fields with severe seed rot and preemergence damping-off. As seedlings emerge from the earth, a pale brown soft rot could appear on the roots or the hypocotyl. The seedlings may perish as the roots and/or hypocotyls become colonized.

In case of root and stem rot: The severity of the infection on soybean plants in the vegetative and reproductive stages of growth is closely connected to the amount of resistance in the plant. In particularly susceptible cultivars, *P. sojae* colonization begins in the roots, then spreads several nodes up to the stem. The plant's leaves change yellow, the stem and roots take on a chocolate-brown color, and the entire structure takes on a reddish-orange to orange-brown color. On rare occasions, a lesion will only affect one side of the plant, but it always extends up the plant from the soil line. As the plant withers, the wilted, yellowed leaves adhere to it. During the production season, nearly every plant in the field may die in cultivars with severe susceptibility; cultivars with moderate to low partial resistance may experience stand losses of 20–50%. The stem rot phase does not appear in cultivars with higher partial resistance levels. Light brown roots that are still colonized are present, and in certain circumstances, the plants may be stunted. When cultivars exhibit high partial resistance levels in field circumstances, there are typically no overt signs other than reduced yields (Dorrance et al. 2004).

Disease Cycle

A plant pathogen that lives in soil and plant debris is called *Phytophthora sojae,* and it reproduces as oospores. Oospores can tolerate freezing and extended periods of frigid temperatures, and they can persist for a number of years in a dormant form. *P. sojae* oospores exhibit endogenous dormancy, so even under ideal circumstances, not all of them will germinate simultaneously. Under situations of high soil moisture, oospores grow into mycelia, and this mycelium then forms sporangia and zoospores under ongoing or subsequent saturated soil conditions. The zoospores lose their flagella once they get to a root, encyst, germinate, and then infect the root tissues. In order to produce oospores, *P. sojae* colonizes the root and stem tissues (Fig. 11.3). It is considered that oospores are the direct cause of the root and stem diseases that can affect soybeans all season long. In most areas, the pathogen's inoculum survives as oospores, frequently in large numbers (Dorrance et al. 2004). Due to the nature of oospores, which do not all germinate at once, *P. sojae* infections are thought to be monocyclic, i.e., having only one effective infection cycle every growing season. Low parts of a field or places with poor drainage might experience significant damping-off (Dorrance et al. 2007a, b).

Disease Management

In regions of the field that are poorly drained or inundated, *Phytophthora* root rot is more severe. Any cultural intervention that enhances soil drainage eventually shortens the period during which soils remain saturated, hence reducing the risk of infection. Field drain tile installation and soil tillage, either separately or together, may lessen the prevalence and severity of *Phytophthora* root and stem rot. Rotation and tillage are ineffective management strategies because *P. sojae* oospores can survive in soil for extended periods of time (Dorrance et al. 2007a, b). The only chemical control method utilized to manage *P. sojae* is fungicide seed treatments, which are intended to prevent seed rot and damping-off in the early growing season. Both the extremely similar chemicals mefenoxam and metalaxyl can be employed as seed treatments. Higher rates of seed treatment product are more effective than lower rates for *P. sojae* management. For the management of *Phytophthora* root rot in soybeans, seed treatment with Azoxystrobin 2.5% + Thiophanate Methyl 11.25% + Thiamethoxam 25% FS has been suggested. Trifloxystrobin 13.28% w/w + Penfluren 13.28% w/w FS have also been suggested for soybean seed and seedling rot infections (Ministry of Agriculture and Farmers Welfare 2021).

11.2.1.7 Sclerotinia Stem Rot (White Mold)

Economic Importance

White mold of soybean, also known as Sclerotinia stem rot, is a disease caused by the fungus *Sclerotinia sclerotiorum*. It has become an important soybean disease in parts of the northern region of the United States, Argentina, and Canada. Yield losses usually occur when the incidence of disease is 15% or greater. Yield losses can range from 1.3 to 3.7 bu./A for every 10% increase in disease incidence (Wrather et al. 1997).

Pathogen

The disease is caused by the fungus *Sclerotinia sclerotiorum*, which survives as sclerotia in the soil and can resist adverse conditions.

Symptoms

The symptoms begin during the late reproductive stages, as the pathogen infects the flower after pollination. Initially, wilting and withering of leaves followed by the death of plants is observed. As the symptom advances, white cottony growth of fungus can be seen on the stem, leaves, and pods. On the infected stem, a large lesion starts developing, which gives the stem a bleached appearance (Fig. 11.4). In the later stage, hard black sclerotia of different shapes and sizes are formed outside or inside the stem and pods, which later contaminate the seed during harvesting.

Disease Cycle

As the conditions favor, i.e., close canopy, foggy or cloudy weather, and sufficient moisture and temperature (40–60 °F; 4–16 °C), the sclerotia present in the soil will germinate into cup-shaped fruiting body apothecia, inside which millions of ascospores are formed (Peltier et al. 2012). When these wind-blown ascospores land on flowers under favorable weather conditions, infection is initiated. Infection may also take place through wounds or other injuries. Disease development initially requires continuous surface wetness; once the pathogen is established, a short period of wetness will allow disease development. In the later stage, as the mycelium grows on the plant tissues, sclerotia will form as a resting structure and survive for a long period in the soil. These sclerotia will fall into the soil during harvesting and overwinter there to become inoculum for the next season (Steadman 1983).

Fig. 11.4 : Symptoms of Sclerotinia stem rot. (Image courtesy: Peltier et al. 2012)

Management

The adoption of an integrated management approach is strongly advised since the weather has a significant impact on disease development and it is impossible to predict the weather during flowering at planting time. No cultivars are completely resistant to stem rot, but the selection of tolerant cultivars can be done. Crop rotation with nonhost crops, i.e., small grains and corn, will help to break the pathogen life cycle. Wide row spacing of more than 20 inches helps delay canopy closure. Foliar fungicides Endura (boscalid), Topsin-M (thiophanate-methyl), and Domark (tetraconazole) are available for Sclerotinia stem rot control (Grau et al. 1994). *Coniothyrium minitans*, sold as Contans WG, is a fungal pathogen of *Sclerotinia sclerotiorum*. It parasitizes the sclerotia of the pathogen, thereby helping in reducing the sclerotia load in the soil (Smolińska and Kowalska 2018).

11.2.1.8 Brown Spot or Septoria Leaf Spot

Economic Importance

Brown spot or Septoria leaf spot of soybean is caused by the fungus *Septoria glycines*. The disease was first time reported in the United States in 1923. It may result in an 8–15% yield loss, with around 25–50% of the canopy prematurely defoliating due to this.

Symptoms

The infection starts from lower the leaves and under favorable conditions, it continues to spread to the upper part of the plant. Dark brown spots can be seen on both surfaces of the leaves (Fig. 11.5). In the later stage, the infected leaves appear rusty brown and lead to premature defoliation. During maturity, the brown spots can also be seen on the stem and pods (Sinclair and Hartman 1999; Pataky and Lim 1981).

Fig. 11.5 Symptoms of Septoria leaf spot

Pathogen and Disease Cycle

The disease overwinters on crop debris, and in ideal circumstances, such as high humidity and moderate temperatures (60–85 °F), spore production infects the cotyledons and young leaves. Spore dispersal by wind and rain then causes infection to spread to surrounding soybean plants. On the infected leaves, later fruiting bodies form, serving as an inoculum for later infections of mature leaves, stems, and pods (Cruz et al. 2010).

Disease Management

As most of the soybean cultivars are susceptible to brown spot, some cultivars show higher susceptibility than others. For example, early season cultivars tend to mature when conditions favor disease, so they are highly impacted by brown spot. Ultra-early planting dates (mid- to late-April) can be avoided. Managing weeds and back-to-back consecutive years of planting in the same field will be helpful. In the early phase of disease development, no control is needed or recommended; infected plants will recover with no permanent damage. When the pod filling starts and the environment is conducive to brown spot development, the disease can cause significant yield losses. At that time, fungicides (strobilurin- or triazole-class fungicides) can be applied to prevent premature defoliation and protect crop yield. However, be forewarned that economic returns when using fungicides are highly variable, with profitability only being assured when disease pressure is high (Kentucky 1995; Donald 2012).

11.2.1.9 Downy Mildew on Soybean

Economic Importance

Soybean can be attacked by several diseases, among which downy mildew caused by *Peronospora manshurica* can cause 8% to 14% damage to susceptible U.S. genotypes (Dunleavy 1987).

Pathogen

Downy mildew of soybean is caused by the oomycete fungus *Peronospora manshurica*. The disease is favored by periods of high humidity and moderate temperatures. It can be distinguished from other foliar soybean diseases by the tufts of tan-colored fungal growth on the underside of infected leaves

Symptoms

The initial symptoms of the disease can be seen on the upper surface of the leaf as small, light green spots (not water-soaked), which later enlarge and coalesce into large, brown, irregular spots. During wet and humid conditions, grayish growth can be seen on the lower leaf surfaces. In the later stage, both pods and seeds become infected, with white fungal growth can be seen on the seed surface (Donald 2012).

Disease Cycle and Epidemiology

Downy mildew is caused by the fungus-like organism, *Peronospora manshurica*. The downy mildew fungus overwinters in the field as thick-walled resting spores (oospores) in leaf debris and on seed. The oospores serve as the initial source of inoculum; however, wind dissemination of spores produced on the lower surface of leaf lesions is the most important means of spread within the field and from field to field. Typically, infection requires the presence of extended periods of dew and temperatures between 50 °F and 80 °F (Picinini and Fernandes 2000).

Disease Management

In the majority of cases, control of this disease is not guaranteed and there are no "rescue" treatments available (Donald 2012). The fungicides commonly used on soybean do not control downy mildew. Therefore, for managing the disease, seed quality is of utmost importance, i.e., certified seed and resistant cultivars, where available, can be used. Consecutive planting of soybean in the same field should be avoided.

11.2.1.10 Frogeye Leaf Spot

Economic Importance

The first reports of frogeye leaf spot on soybeans were made in Japan in 1915 (Lehman 1928) and in the United States in 1924 (Melchers 1925). The southern U.S. soybean crop region is where the disease is most prevalent. During Argentina's 2009–2010 soybean season, FLS is thought to have caused losses of roughly $2 billion (Carmona 2011). Due to reduced photosynthetic area, premature defoliation, and decreased seed quality, yield loss from frogeye leaf spot on vulnerable soybean cultivars has been estimated to be as high as 60% in the United States. Due to FLS, yield decreases between 10% and 60% have been documented (Mian et al. 1998). In five southern U.S. states (South Carolina, Georgia, Florida, Mississippi, and Alabama), four pairs of FLS-susceptible versus resistant near-isogenic lines (NILs) were established. Mian et al. (1998) compared the performance of these four pairs. When compared to the corresponding resistant NILs, the yield losses for the four vulnerable soybean cultivars reached 31%. For the cultivars in the study, the average FLS seed yield loss in untreated plots ranged from 30% to 37%. Reduced seed size was the primary cause of FLS's yield reductions.

Pathogen

Cercospora sojina K. Hara is currently recognized as the causal agent of FLS, although earlier literature reported *C. daizu* as the causal agent of this disease (Athow 1987). The primary and secondary inoculum are hyaline conidia of 5 to 7 μm × 39 to 70 μm, which are produced on leaf and stem residues or infested seeds. Conidia are formed on the tips of conidiophores and are pushed aside as the conidiophores continue to grow. The size and shapes of conidia and conidiophores may vary depending on the substrate on which the fungus grows. Conidia can germinate on a leaf surface within an hour of deposition in the presence of water at 25–30 °C, producing one to a few germ tubes from the end cells and sometimes from the intervening cells. In culture media, conidia frequently germinate to form short germ tubes that may produce secondary conidia (Mian et al. 2008). *Cercospora sojina* is a dynamic pathogen with extensive virulence or race diversity (Lin and Kelly 2018).

Disease Symptoms and Epidemiology

Late in the growth season, when prolonged conditions encourage disease development, FLS symptoms can also appear on stems, pods, and seeds. FLS symptoms are most noticeable and typically found on leaves (warm and humid conditions). Even though infections can happen at any stage of soybean development, they tend to happen more frequently throughout the reproductive growth stages (from bloom to maturity) and may even start to manifest earlier in continuous soybean fields or in favorable environmental conditions. Young, developing foliage is easily infected by the disease as leaflets grow. On leaves, lesions start as tiny, gray, wet patches that grow into gray to brown spots with slender, dark reddish-brown borders.

These spots range in size from less than 1 to 5 mm, and their shapes include angular and round.

With time, lesions develop thin, reddish-brown edges and ash gray to light brown central parts. The dots are darker and have light- to dark-gray centers on the underside of the leaves. Conidiophore clusters form in the middle of the lesion during sporulation. Lesions may combine to form larger, irregularly shaped spots in plants that are extremely diseased. Heavy infection rates (>30% severity) can cause leaves to blight and wither quickly before dropping before their time, which can cause severe yield losses (Lin and Kelly 2018).

Spore formation on contaminated leftovers and/or on seedlings resulting from infected seeds is facilitated by warm, humid conditions. Wind and sputtering rain both disperse the spores.

Within 48 h of the onset of apparent symptoms, sporulation can be abundant in plants grown in warm (25–30 °C) and humid environments (>90% relative humidity) (Phillips 1999; Mian et al. 2008).

Disease Cycle

A polycyclic disease, frogeye leaf spot repeats infection, symptom development, and conidia production over the course of the growing season. The primary or first inoculum (the amount of initial inoculum) and the number of disease cycles that take place in the field have an impact on the amount of disease (Lin and Kelly 2018). The fungus survives the winter on infected seeds and leftovers from infected soybeans. No other crop or weed host has been proven to have the illness caused by the virus. Poor germination results from heavily infested discolored seeds, and the percentage of germination may be adversely related to the degree of surface discoloration (Sherwin and Kreitlow 1952; Phillips 1999). Typically, seedlings from infected seeds have lesions on their cotyledons and are weak and stunted. Young leaves may become infected by spores that form on diseased cotyledons. The lesions do not show up on young, growing leaves until almost 2 weeks following the invasion of the host tissue. Seed infection first happens when seeds come into touch with pod lesions, but as seeds mature in the field, infection may spread from seed to seed. During harvest, healthy seeds may pick up spores or mycelium contamination. Air currents and raindrop splashes can both transport conidia over short distances. Under ideal circumstances, secondary infection of leaves, stems, and pods lasts the entire growing season of soybeans (Mian et al. 2008).

Management

The best method for controlling a disease is typically crop resistance. Use high-quality, certified disease-free seeds when planting to prevent the spread of the fungus in the field. Early season infection risk is reduced by seed treatments with fungicides. Crop rotation gives inoculum in the field more time to deteriorate before soybeans are replanted. Soybean residue can contain enough inoculum, even in minute levels, to start an epidemic. A 2-year rotation away from soybeans may be necessary on farms with severe frogeye leaf spot in order to fully break down the residue left over from the previous soybean crop. In fields with high disease pressure, thorough tillage may be required to eliminate the overwintering fungal formations.

If an extended dry spell is predicted in the near future, waiting it out may be more cost-effective because dry weather significantly slows the spread of disease. The best time to apply a fungicide to protect plants from infection and maintain production potential is between the R2 and R5 growth stages. Products with two or more potent active components and a variety of delivery systems are more likely to successfully control the condition. For the treatment of soybean frogeye leaf spot, fluxapyroxad 167 g/L and pyraclostrobin 333 g/L SC have been recommended (Ministry of Agriculture and Farmers Welfare 2021).

11.2.1.11 Damping Off

Economic Importance

Preemergence and postemergence damping-off reduce plant populations throughout all or parts of fields, reducing yields. Stand loss may be so severe that replanting is necessary. Preemergence damping-off, or the death of seedlings before emergence (preemergence damping-off), or postemergence damping-off are two illnesses that are together referred to as "damping-off." These diseases are caused by a range of soil- and seed-borne viruses (postemergence damping-off). The main factor contributing to poor seed quality in the majority of nations that grow soybeans is Phomopsis seed decay (PSD) of the soybean, *Glycine max* (L.) Merrill (Sinclair 1993). PSD has caused substantial financial losses (Baird et al. 2001). In 1994, losses on a global scale were around 0.19 million metric tonnes (MMT) (Kulik and Sinclair 1999). PSD had effects on yields in the United States from 1996 to 2007 that varied between 0.38 and 0.43 MMT (Wrather and Koenning 2009). In 2009, PSD resulted in about 12 million bushels of production losses in 16 states as a result of the predominance of hot and humid weather conditions from pod fill through harvest in the southern United States (Koenning 2010).

Pathogen

Pythium Species

P. ultimum is the species that is both the most abundant and harmful. Young seedlings may die during or soon after emergence from Pythium species, or their seeds may rot. A soft watery rot, wilting, browning, and eventual death will be seen in affected seedlings (Broders et al. 2009).

Phytophthora Species

One of the significant pathogens that cause damping-off is *Phytophthora sojae*. Seedlings can be killed before, during, or after emergence, just like *Pythium*. The difference in moisture requirements between *P. sojae* and *Pythium* most likely accounts for the former disease's less frequent occurrence. In contrast to *Pythium*, which often disappears as seedlings grow, *Phytophthora* can cause plants to wilt and die at any age (Qutob et al. 2000).

Rhizoctonia solani

When the crop is under stress, *R. solani* primarily causes postemergence damping-off. *Phytophthora* and *Pythium* are two fungi that are less dependent on water, yet this one still causes harm. *R. solani* frequently does not harm plants; instead, they outgrow the issue and continue to seem healthy for the rest of the growing season.

Phomopsis

Phomopsis longicolla and *Diaporthe phaseolorum* var. *sojae*, a fungal complex, are responsible for the demise of seeds and seedlings (Xue et al. 2007). Conidiomata with conspicuous necks longer than 200 μm and opening via apical ostioles are black, stomatic, unilocular or multilocular, solitary or aggregated, and pycnidial. They have globular, uniostiolate or multiostiolate locules that can reach a width of 500 μm. Hyaline, simple or typically branching, septate, 3.5–24 × 1–4 μm conidiophores (Sinclair 1993; Hartman et al. 1999).

Miscellaneous Fungi

Many different fungi have the power to damage soybean seeds and seedlings, resulting in preemergence or postemergence damping-off, or both. Fungi, including Fusarium species, *Macrophomina phaseolina, Cercospora sojina* (frogeye leaf spot), and *Cercospora kikuchii* (Cercospora leaf blight).

Disease Symptoms and Epidemiology

Pythium spp.

Pythium oospores can infect seeds within 90 min of sowing because they can germinate in response to exudates from the seed and the root. *Pythium* spp. infection causes damping-off, which is the premature weakening and death of growing seedlings as well as seed rot. Higher soil moisture content and cooler soil temperatures encourage infection (Navi et al. 2019).

Phytophthora

Any stage of the soybean plant's development is susceptible to *Phytophthora* infection. Rapid yellowing and withering, as well as a soft rot and root collapse, are symptoms in immature plants. As the growing season goes on, mature plants typically lose vigor and may even slowly die. On older plants, foliar symptoms include a widespread yellowing of the lower leaves that spreads upward before the plant wilts and dies. The root system is typically badly damaged, resulting in branch and lateral roots that are nearly totally discolored. When split, the interior tissues of tap roots exhibit a tan to brown discoloration in addition to a brown discoloration on the surface (Qutob et al. 2000; Schmitthenner 1988).

Rhizoctonia spp.

By early summer, wilted or dead seedlings are typically seen dispersed throughout the field or in tiny concentrated patches, which are the typical symptoms of this disease. *Rhizoctonia solani* can cause lesions on hypocotyls, root rot, and seed rot (Wrather et al. 2001a, b). When developing plants become infected before emergence, damping-off develops. Lesions that are reddish-brown and sunken develop on immature seedling hypocotyls. The seedling may girdle as a result of the resultant stiff, dry canker, which will similarly collapse symptoms, such as typical sunken, reddish-brown cankers on the lower stem close to the soil surface and uneven or stunted development. These symptoms will be present in older plants or seedlings that survive seedling infection. Older sick plants turn chlorotic and resemble plants with a nitrogen deficit in appearance (Kumar et al. 2019).

Phomopsis spp.

P. longicolla or other *Phomopsis* spp.-infected soybean seeds might range from symptomless to shriveled, elongated, or fractured, and frequently appear chalky-white. Compared to healthy seeds, infected seeds either fail to germinate or germinate more slowly.

Preemergence and postemergence damping-off is brought on by seed infection, and under extreme circumstances, stands can be drastically decreased to the point of decreasing output. After they form, soybean pods are susceptible to infection at any moment. The fungi first colonize the seed coverings before moving on to populate the cotyledons and plumules. Through the funiculus and hilum, mycelia enter the developing seeds and ovules. The fungus eventually invades the radicle and plumule in addition to the cotyledons and seed coat tissues within the seed (Xue et al. 2007; Mengistu et al. 2009).

Disease Management

Use certified seed for a high germination rate and consistent quality. Avoid planting in soils that are chilly or moist, especially when using no-till techniques. If seed quality is subpar or if seeds must be planted in less-than-ideal soil, use a broad-spectrum seed treatment fungicide. Throughout the growth season, keep the soil fertile. Avoid using pesticides, insects, or other physical elements that can harm growing seedlings. Azoxystrobin 2.5% + Thiophanate Methyl 11.25% + Thiamethoxam 25% FS for root rots and damping-off induced by several pathogens in soybean, and Fluxapyroxad 333 g/L FS for Rhizoctonia root rot (Kumar et al. 2019). For collar rot, use carbendazim 25% + mancozeb 50% WS. For soybean seedling infections, use carboxin 37.5% along with thiram 37.5% WS (Ministry of Agriculture and Farmers Welfare 2021).

11.2.2 Category: Viral Diseases

11.2.2.1 Soybean Yellow Mosaic Disease

Economic Importance

SYMD and SMD are the most serious dangers to soybean farming, particularly in the Himalayan foothills, where they show in moderate to severe form in most standard cultivars throughout the growth phase (Singh et al. 2019). Soybean is an economically significant legume crop in which YMD causes losses of 15–75% (Sharma et al. 2014). The disease was first reported in India in 1960 (Nariani 1960), and it is a major constraint to soybean production in Uttarakhand, Punjab, Madhya Pradesh, Delhi, Haryana, Uttar Pradesh, Rajasthan, and Karnataka. It accounts for 30–70% yield loss and increases to 80–100% during its severe incidence (Nene 1972). It is a major constraint to soybean production in Uttarakhand, which can lose up to 100%

of their yield due to SYMD (Naimuddin 2001). The disease causes a 30–70% yield loss, which can rise to 80–100% when the disease is severe (Swathi et al. 2021).

Causal Organism

Mungbean yellow mosaic virus (MYMV), *Mungbean yellow mosaic India virus* (MYMIV), *Dolichos yellow mosaic virus* (DoYMV), and *Horsegram yellow mosaic virus* (HgYMV) are the four species of Begomoviruses that cause yellow mosaic disease in legumes in Southeast Asia (Gupta and Chauhan 2005). *Mungbean yellow mosaic virus* and *Mungbean yellow mosaic India virus*, both belong to the genus *Begomovirus* of the family *Geminiviridae,* cause SYMD in India (Singh and Aravind 2019). Begomoviruses contain circular, single-stranded DNA (ssDNA) genomes that are encapsulated in twinned icosahedral virions that are 18 nm × 30 nm in size (Stanley et al. 2005). The genomes of YMVs that infect soybean have two ssDNA genomic components, DNA A and DNA B, each of which is approximately 2750 nts long. In both virion-sense and complementary-sense strands, the genetic components encode proteins. On the virion sense, DNA A codes for the coat protein (CP) (AV1) and the precoat protein (AV2), whereas on the complementary-sense strand, genes (AC1, AC2, and AC3) code for the replication-associated protein (Rep), transcriptional activator protein (TrAP), and replication enhancer protein (REn), respectively (Table 11.1). AC4 and AC5, two more complementary-sense genes, encode proteins that are symptom determinants and suppressors of RNA silencing, respectively (Li et al. 2015).

Symptom and Epidemiology

The disease is identified by the presence of green and yellow spots on older leaves, as well as full yellowing of new leaves (Singh and Aravind 2019). Variable symptoms such as yellowing, stunting, and puckering with bright yellow mosaic have also been recorded in diverse cultivars, in addition to the normal mosaic symptoms (Ashok 2018). Small yellow color specks on the leaf lamina first appear as YMD symptoms, then transform into mosaic-like irregular green and yellow patches

Table 11.1 Functions of various regions of *Begomovirus* (Rego-Machado et al. 2019)

Gene	Protein	Functions
V1	CP	Whitefly-mediated transmission, virion capsid, entry in nucleus
V2	AV2	Symptom initiation and accumulation of viral DNA
C1	Rep	Replication-associated protein. Viral DNA replication
C2	TrAP	Transcriptional activator for the virus-sense genes; suppresses RNA silencing and other host defense responses. Activate expression of the capsid protein gene
C3	REn	Replication enhancer protein
C4	NSP	Shuttling of viral DNA in nucleus
BV1	Nuclear shuttle protein	Nuclear trafficking (NSP)
BC1	Movement protein (MP)	Cell-to-cell movement and pathogenicity determinant

Fig. 11.6 Symptoms of soybean yellow mosaic mosaic disease

(Fig. 11.6). At maturity, the yellow color regions on the lamina become larger, and the entire leaf turns yellow (Fig. 11.6). Plants that are damaged become unproductive with less flowers and pods (Malathi 2007).

Rugosity is most severe in plants growing at temperatures about 18 °C. At 24 °C to 25 °C, symptoms are less acute, and at 30°C, symptoms are mostly hidden. The amount of time between infection and symptom development is similarly affected by temperature, ranging from 4 days at 29.5 °C to 14 days at 18.5 °C. Soybean crops can be infected by two strains of soybean mosaic virus, SMV mild and SMV severe, which cause moderate and severe symptoms, respectively. It also shows that as the temperature was raised to 30–35 °C, infection dropped progressively (Koranga and Bahukhandi 2021; Domier et al. 2011). Yellow mosaic was first noticed on soybeans in July–August, around 26–54 days after sowing (Gupta and Varma 2015). The weather conditions the previous week had a substantial impact on the peak YMD incidence. Maximum temperature, sunlight hours, and the number of whitefly populations were shown to be significantly connected to YMD, whereas rainfall and evening relative humidity were adversely related. When the following parameters were met, a weather-based rule was developed for predicting the peak of YMD in soybean during the preceding 2 weeks: The number of whiteflies per plant should be greater than 12 with a highest temperature greater than 33 °C and the monsoon in its passive phase for more than a week. When the dry period coincides with a maximum temperature over 33 °C for more than or equal to 5 days, the transmission rate is higher (Srivastava et al. 2021).

Disease Cycle

Paracalyx scubious was an alternative host of the disease, which was spread by whitefly. Other hosts for the pathogen are *Corchorus olitorius*, *Ageratum conyzoides*, *Alternanthera sessilis*, and *Phaseolus trilobus,* which may play a part in pathogen survival, as these weeds were found infected throughout the year (Gupta and Varma 2015).

A single viruliferous whitefly can spread the virus, but ten viruliferous whiteflies per plant are needed for 100% viral transmission. For viral transmission, the minimal acquisition access and inoculation access feeding durations required were 15 min each, whereas 100% virus transmission was reported during acquisition and inoculation periods of 12 h each. With longer acquisition and inoculation durations, the percent transfer rate rose. Pre-acquisition starving was favorably connected with transmission rate, but post-acquisition famine was adversely correlated (Swathi et al. 2021).

Management

The genotypes PK-472, Pb-1, Bragg, NRC-7, VLS-58, JS-72-280, and PK-262 possess some degree of resistance and are used as resistance donors in breeding programs for developing resistant commercial cultivars in soybean (Singh and Aravind 2019). Three genotypes, viz., JS 20–29, JS 20–69, and JS 20–98, were found to be resistant against YMV (Mishra et al. 2020). The antiviral compound Zillon at 6 mL/L applied at 15, 30, and 45 DAS was found to be significantly effective, recording the lowest disease (20.0%) (Amrate et al. 2020).

The intron-spliced hairpin construct expressing the MYMIV coat protein sequence was used to create transgenic soybean plants under the control of the 35S promoter and the ocs terminator and validate the integration of the coat protein gene in separately transformed plants using PCR and Southern hybridization, which were then inoculated with viruliferous whiteflies acquiring the virus. Northern blot analysis found siRNA in transgenic lines infected with viruliferous whiteflies, with a high level of siRNA (21–22 nt), nondetectable viral replicative DNA, and a low quantity of viral transcript for replication, as determined by qRT-PCR. Because it targets the most conserved region controlling whitefly transmission, the transgenic construct can be exploited to generate resistance to begomoviruses in soybean and other crops, and such transgenic approach should be exploited (Kumari et al. 2018).

The silencing of complementary-sense viral genes involved in virus replication is the major focus of efforts to create resistance against the genus Begomovirus. To create resistance to the *Mungbean yellow mosaic India virus,* researchers targeted a complementary-sense gene (AC1) generating Replication initiation Protein (Rep), which is found throughout the Indian subcontinent. Legume host plants treated with infectious constructions of *Mungbean yellow mosaic India virus* isolates as well as an antisense Rep gene construct display viral resistance (Haq et al. 2010). Limiting arthropod vectors hasn't always yielded consistent disease management.

11.2.2.2 Soybean Mosaic Disease

Economic Importance

Soybean mosaic disease, caused by the *Soybean mosaic virus*, has been observed to reduce yields by 8–50% in natural fields and by 100% in severe outbreaks (Lamani et al. 2003). In all soybean-growing locations throughout the world, SMV causes seed quality to deteriorate and severe yield losses of up to 50% (Yang et al. 2014).

Causal Organism

SMD is a member of the Potyvirus genus, which belongs to the Potyviridae family (Singh and Aravind 2019). The *Potyviridae* family is linked to *Soybean mosaic virus*. Using an ELISA technique, several more flexuous, rod-shaped particles are discovered in aggregation with soybean mosaic disease (Benscher et al. 1996). The size range of soybean mosaic virus may be compared to that of members of the *Potyviridae* family, which have a size range of 182–420 nm, typically in the range of 338–371 nm and a breadth of 79–93 nm. The virus is a cytoplasmic form as found in ultrathin leaf sections (Benscher et al. 1993). The thermal inactivation point (TIP) means the virus gets inactivated after being heated at 70 °C for 10 min (Eggenberger et al. 2008) and has a dilution endpoint (DEP) of 1:10,000 with a Longevity In Vitro (LIV) of 7 days at 21 °C, which is room temperature (Domier et al. 2003).

The *Fabaceae, Amaranthaceae, Chenopodiaceae, Passifloraceae, Scrophulariaceae*, and *Solanaceae* plant families are the only ones that SMV may infect. *Glycine soja* (wild soybean) and *Glycine max* are the most usually infected hosts (cultivated soybean). The *Leguminosae* family was systemically infected by the virus, causing mosaic disease of soybean, and the percent transmission of the current virus varies from 70% to 100% depending on the host species (Balgude et al. 2021). Based on their virulence on different genotypes of soybean, a number of SMV isolates with varying pathogenicity and symptomatology have been categorized into seven strain groups from G1 to G7. Mosaic strains outnumber necrotic strains when mixed infection occurs, and both strains outnumber avirulent strains (Chen and choi 2008).

Symptom and Epidemiology

The initial indication of SMD on the mechanically infected soybean plant in the greenhouse is the emergence of yellow vein clearing on fresh trifoliolate leaves. The infected soybean plant also has a downward bending of the leaf edges at the sides and an upward curving of the leaf tip (Brand et al. 1993). Before reaching maturity, the leaves take on a coarse, leathery look and become brittle. When compared to a healthy plant, the development of the infected plant is stunted, and the number of poles is reduced. Mottling happens in both cold and warm temperatures, and the plant gets disguised in hot weather. Localized lesions appear on the leaves as well (Hajimorad et al. 2005). Vein clearing, mosaic mottling, puckering, distortion and bending of leaves, and plant stunting were all seen in the soybean virus isolate. Dolichos bean displayed necrotic local lesions, whereas bean, cowpea, and cluster bean showed mosaic mottling signs (Balgude et al. 2021). Compared to the nodules in healthy plants, root nodules in soybean mosaic virus-infected soybean roots showed a decline in reducing sugars, nonreducing sugars, and starch. Nonreducing sugar ranged from 1.4 to 2.75 mg g^{-1} dry weight, while in healthy and diseased, it ranged from 1.3 to 2.45 mg g^{-1} dry weight. Reducing sugar ranged from 2.0 to 3.39 mg g^{-1} dry weight in healthy, and in diseased, it ranged from 1.86 to 3.00 mg g^{-1} dry weight (Gupta et al. 2010).

Disease Cycle

SMV is a virus that is spread through aphids and seeds. SMV infections are transmitted mostly by seed-borne diseases. SMV is transmitted by contaminated seeds in around 0–43% of cases. The Asian soybean aphid, *Aphis glycines*, may also transmit SMV in a limited way (Halbert et al. 1981). Transmission by insects is a natural and important mode of soybean mosaic virus transmission. The virus is spread by leafhoppers in the case of SMD. In this scenario, the leafhoppers (*Nephotettix nigropictus*) initially feed on the SMV-infected soybean plant before moving on to the healthy soybean plant. The symptoms of the newly infected plant are the same as those of the previously infected plant from which the leafhoppers (*Nephotettix nigropictus*) became virulent (Jones and Vodkin 2013). Between infection and the formation of the septum, there is a 4–14 day incubation period. When an overnight acquisition feed was given to aphids for acquiring *Soybean mosaic virus* (SMV)-infected soybean leaves, the soybean aphid (*Aphis glycines*) was a poor vector. However, when given a 1-min acquisition probe on the same infected leaves used for the feeding treatment, the aphids were more efficient in disease transmission. A set of primers used to amplify a PCR fragment in the coding area of the SMV coat protein was used in an RT-PCR experiment to detect SMV in single soybean aphids. RT-PCR is a technique used to detect nonpersistently transmitted viruses and aid in virus epidemiology (Wang and Ghabrial 2002).

Management

SMV has resulted in a rapid drop in soybean yields, resulting in significant losses for the soybean business, making early diagnosis critical. With a combined convolutional neural network and support vector machine (CNN-SVM) technique for the early identification of SMV, a novel classification method for early SMV was developed, classifying its severity into grades 0, 1, and 2. The feasibility of applying the CNN-SVM model is to identify early SMV infection and present a new avenue for early SMV detection based on hyperspectral images (Gui et al. 2021). This will prevent the primary spread of the disease.

Insecticides used against aphid vectors result in a small reduction in SMV occurrence, similar to other nonpersistently transmitted viruses. Systemic insecticides are used (Pedersen et al. 2007; Johnson et al. 2008). The optimal management options for SMD are the planting of virus-free seeds and deploying resistant soybeans with R gene pyramiding (Shakiba et al. 2012). The five genotypes, viz., PK-472, Pb-1, Bragg, NRC-7, and VLS-58, were resistant to the disease. The use of resistant genotypes is the most desired way to control the disease (Singh and Aravind 2019). Since no SMV-resistant gene has yet been cloned, it is not possible to produce resistant soybean plants by merely inserting the resistance genes. Furthermore, due to the fast development of avirulence/effector genes, R gene-conditioned resistance will be swiftly overcome (Gagarinova et al. 2008; Liu et al. 2011).

11.2.2.3 Bud Blight of Soybean

Economic Importance

The disease is sporadic but causes serious yield losses, especially in southern and south-central Illinois. It can cause a 25–100% reduction in yield. In general, losses are high when infection occurs in young plants or infected seed is used for sowing (Dean 1992).

Symptom and Epidemiology

This disease includes symptoms of bud blight, shepherd's crooking, stunting, wilting, delayed maturation, bud and leaf proliferation, decreased pod quantity, and discolored, deformed, and dead leaves. Plants infected with TRSV may exhibit distorted and discolored petioles, discolored veins in the leaves, and poor root nodulation. Plants with the infection lose their terminal bud and crook (Fig. 11.7). Later, further buds perish and pods fail. The dead areas start to crack. If infection happens in time, the plant doesn't produce many, if any, seeds and stays green until winter. Infected seeds may be the result of later diseases (Sehgal and Thottappilly 1992). Other indications of early infection include reduced leaf size, increased axillary branching, and delayed seed development, all of which contribute to yield loss by lowering the quantity of pods and seeds per plant and resulting in lesser seed weights. Later in the growth season, infection can produce bud blight and necrotic patches on pods, as well as red nodes or necrotic streaks at nodes and delayed maturity. On pods, there are also uneven, black streaks and necrotic patches. Incidence

Fig. 11.7 Symptoms of bud blight

and seed moisture have a substantial positive association, while yield has no significant link. Thrips benefit from hot, dry conditions, which promote the spread of TSV and soybean vein necrosis virus (Rabedeaux et al. 2005).

Disease Cycle

TSV is transmitted through pollen and seeds, and many thrips species can aid in the spread of the virus. Particularly with regard to soybeans, it is unknown how exactly thrips contribute to the propagation of TSV. Reducing thrips populations, eliminating alternative weed hosts from fields and nearby areas, and postponing plantings have all been found to be effective in lowering TSV infection in soybeans in Brazil (Rabedeaux et al. 2005). TSV spreads by seed, thrips, and sap. TSV is carried by *T. tabaci* and *Frankliniella occidentalis*. In the field, naturally infected perennial or annual weeds and seed-borne inoculum propagate TRSV and TSV (Sehgal and Thottappilly 1992). The primary spread, which happened from weeds to soybeans, was distributed randomly. Insects from *Aphanamixis polystachya* moved to soybean plants after colonizing them, multiplied, and spread to nearby plants, giving the disease a clustered distribution pattern.

Management

There is a dearth of knowledge regarding the genetics and origins of the causative virus's resistance. Finding sources of resistance is the first and most important stage in creating resistant strains for effective disease control. Host resistance is a key tool in the management of diseases. KHJB 1, JS 84-1, and JS 71-05 were extremely resistant, whereas JS 84-227, ES 5, JS 2, JS 79-81, and JS 340 were resistant. JS 2 and other JS lines should be used as donors in resistance breeding programs because they have been demonstrated to be resistant to bud blight (Thakur et al. 1998). Controlling vectors is also crucial. The use of a systemic insecticide is advised. In order to manage weeds in soybeans, a combination of cultural, mechanical, and chemical techniques is used. Up until recently, cultural and mechanical weed management were the sole options. The IDM combined method will be very helpful.

11.2.3 Category: Bacterial Diseases

11.2.3.1 Bacterial Pustule of Soybean

Economic Importance

Xanthomonas axonopodis pv. *glycine* is a cause of bacterial pustule, a serious bacterial disease that reduces soybean output by 20–40% globally (Kaewnum et al. 2005). This is one of the most common bacterial diseases of soybeans in Korea also, where it causes significant output losses of up to 19.8% in 2006 and 16.8% in 2007 (Hong et al. 2011).

Causal Organism

One of the most common bacterial diseases of soybean is bacterial pustule produced by *Xanthomonas axonopodis* pv. *glycines* (Hong et al. 2010). Smith (1905) first noticed bacterial pustule in the United States, and Clinton (1915) characterized the causative bacterium. In Japan, *Pseudomonas glycines* (Nakano), Dye et al. (1980) published the first appropriate description of the pathogen, which was later reclassified as *Xanthomonas campestris* pv. *glycines* (Nakano). Hedges (1922) designated it *Bacterium phaseoli* var. *sojense* in the United States, without knowing about the Japanese work. It was renamed *Xanthomonas axonopodis* pv. *glycines* (Xag) after a DNA–DNA hybridization study (Vauterin et al. 1995). The bacteria have one polar flagellum and are Gram-negative rod-shaped. Since, xanthomonadin pigments are present in the bacterial cell walls, colonies on ordinary bacteriological medium appear yellow. The pathogen grows at 35 °C, hydrolyzes starch, generates acid from arabinose, and thrives on the selective medium SX, like other *X. campestris* strains (Schaad et al. 2001). Auxins, extracellular polysaccharides, and biological races are produced by the pathogen (Hartman et al. 1999).

Symptom and Epidemiology

Small, pale green patches with raised pustules, which grow and coalesce into massive necrotic lesions resulting in premature defoliation, are characteristic symptoms of the disease (Narvel et al. 2001). The severity of the bacterial pustule had a significant impact on the weight and yield of 100 seeds, but not on stem length, number of branches per plant, number of pods per plant, or number of seeds per plant (Hong et al. 2011). The first signs are little light green spots with slightly elevated centers that appear on the lowermost surface of the leaf, although some also appear on the top. Cellular hypertrophy and, on rare occasions, hyperplasia cause pustules to develop, and the elevated centers can grow rather big (Hokawat and Rudolph 1993). To avoid misunderstanding, bacterial pustules are surrounded by conspicuous chlorotic yellow halos, whereas rust lesions are tan or brown lesions with a central hole for spore release. Premature defoliation can occur in severe bacterial pustule occurrence. There was a substantial positive and negative relationship between disease spread among plants and maximum temperature and relative humidity in correlation studies. The spread of the disease was adversely correlated with rainfall and nighttime relative humidity (Shankar et al. 2018). The disease manifestation is determined by the susceptible genotype, the pathogen virulence, and favorable climatic circumstances. The transmission of the disease is aided by windy rain and rain splashes. At temperatures of 30–33 °C and extreme humidity, severe infection occurs. The needed temperature for disease development during incubation is critical (Khare et al. 2003). Moderately warm conditions with moderate relative humidity result in severe disease severity, while when the relative humidity is low, the disease severity is decreased (Srivastava and Bais 1986; Khare and Khare 1995).

Disease Cycle

Under natural conditions, *X. campestris* pv. *glycines* has been observed to be seedborne, with a very low proportion of bacterial association with soybean seeds, even when recovered from highly sick plants. The bacterium has been found in all components of seeds, including the seed coat, cotyledons, plumule bud, and radicle (Khare et al. 2003). Bacteria can live in seeds for two and a half years (Graham 1953). The bacteria were found in damaged crop waste as well as collateral hosts such as *Dolichos biflorus*, *Phaseolus vulgaris*, *Phaseolus lunatus*, *Macrotyloma uniflorum*, and *Brunichia cirrhosa* (Elliot 1951). In the field, the disease is spread via rain splashes and wind-blown rain. During threshing, the bacteria infect the seeds as well (Groth and Braun 1986).

Management

The use of resistant cultivars is the most promising one. JS-72-44, JS-75-46, JS-71-05, JS-335, and Shivalic genotypes were free from the disease in Raipur (Chhattisgarh) and should be used in the field (Shankar et al. 2018). It's time to get rid of the alternate host. The seed-borne infection of *X. campestris* pv. *glycines* is eradicated using a hot water seed treatment. By soaking the seed in water at 53 °C for 25 min, it was possible to suppress the seed-borne bacteria (Mishra 1983). The bacterial pustule pathogen was greatly inhibited by tetracycline and streptomycin sulfate. In addition, oxolinic acid application was beneficial in inhibiting pathogens. The pathogen was reported to be inhibited in vitro by streptomycin sulfate + oxytetracycline (18.8 + 1.5%) WP, oxytetracycline (17%) WP, and oxolinic acid (20%) WP. On naturally infected soybeans, foliar treatment with streptomycin sulfate + oxytetracycline WP and oxytetracycline WP had a strong control value (Hong et al. 2010). The best treatment for boosting plant resistance to bacterial pustule disease and soybean crop production was to mix the biological agent biofresh in a solid formulation with soybean litter compost. Plant resistance was induced systemically, as evidenced by increases in salicylic acid of 0.3 mg and peroxidase of 0.07 unit/mL in sample plants (Khaeruni et al. 2018).

11.2.3.2 Bacterial Blight of Soybean

Economic Importance

The yield losses caused by *P. syringae* range from 11% to 20% in soybean crops (Suryadi et al. 2009). Also, the infected plants become dwarfed and die if the disease infects immature plants (Semangun 2004). Severe crop losses have been observed throughout North America, predominantly in the United States, as a result of the infection (Kennedy and Alcorn 1980).

Causal Organism

Bacterial blight of soybean is caused by the bacterium, *P. syringae* pv. *glycinea. Pseudomonas syringae* is a genetic complex of strains that was primarily categorized based on its phenotypic characteristics (Young 2010). Bacterial blight of soybean is one of the most damaging diseases, rendering the crop unproductive and declines the seed quality. Timely detection and diagnosis are critical for predicting disease outbreaks (Surbhi et al. 2023). Following with genetic investigations based on DNA–DNA hybridization and multilocus sequence analysis of housekeeping genes exposed the *P. syringae* complex, including several pathovars (Gardan et al. 1999; Sarkar and Guttman 2004). Parkinson et al. identified seven *P. syringae* phylogroups based on the phylogeny of the rpoD housekeeping gene and previously known strains (Mulet et al. 2010). The bacteria are Gram-negative, aerobic, unicellular rods varying from 0.7 to 1.2 × 1.5 μm with one to many polar flagella. Neither the capsule nor the endospore has been formed. The colonies are round, milky-white, elevated, sparkling white in reflected light, somewhat iridescent in transmitted light, transparent, with a smooth surface and continuous edge. It is a fluorescent pseudomonad that grows in King's B medium and generates a green fluorescent pigment (Gade et al. 2021).

Symptom and Epidemiology

Pseudomonas syringae pv. *glycinea* is a major pathogen that infects soybeans in the field, causing bacterial blight of soybean (Qi et al. 2011). This pathogen has been documented to infect either native soybean or edamame in Jember, Indonesia, causing damage to the stem, leaf and pod (Masnilah et al. 2013). Systemic chlorosis, stunted plants, and elongated leaves were among the symptoms that characterized the "systemic toxaemia" phase of bacterial blight (Gnanamanickam et al. 1982). The young leaves are the most susceptible to infection and manifest itself as lesions, which appear as little yellow to brown spots on the leaf surface (Fig. 11.8). These lesions dry out, become reddish brown to black, and develop a yellowish-green halo around them. Small lesions may develop and coalesce to form larger one, uneven, and necrotic. Older lesions might fall off, leaving diseased leaves with a ragged look. Stem petioles and pods can also be infected by the pathogen. If the pods get infected, it spreads to the seeds (Tereshchenko 1977). *Pseudomonas syringae* pv. *glycinea* PG4180, which generates the phytotoxin coronatine (COR) in a temperature-dependent manner, is best generated at 18 °C, with no detectable synthesis at 28 °C. When the bacteria were cultivated at 18 °C prior to injection, PG4180 induced typical bacterial blight symptoms on soybean plants, but not when obtained from cultures grown at 28 °C. For bacterial invasion via natural plant apertures, preinoculation temperature and phytotoxin production were critical (Budde

Fig. 11.8 Bacterial blight of soybean Symptoms

and Ullrich 2000). Strong wind with high moisture is most favorable with temperatures of 22–26 °C. Cool nights followed by warm days and wind-driven rain are favorable conditions for disease development (Williams and Nyvall 1980).

Disease Cycle

The disease is primarily spread from infested seed and crop debris with cool nights, warm days, and wind-driven rain (Agrios 2005). During rain accompanied by wind, *Pseudomonas glycinea* outbreaks occurred on immature soybean leaves, and lesions were frequently connected with wounded regions. Even when inoculum was available, little disease was developed in the absence of storms or during rain without wind. The importance of wind as a role in disease onset has been demonstrated experimentally, yet only one or two of a plant's youngest open leaves got infected in both real and artificial wind and rain storms (Daft and Leben 1972).

Management

The use of naturally existing antagonists to reduce plant diseases is a promising substitute for traditional protection practices. Detecting the presence of the pathogen is critical for disease management in order to ensure the quality and quantity of soybean output. Several management strategies for bacterial blight disease caused by *Pseudomonas syringae* pv. *glycinea* have been developed to minimize yield loss. One of the most prevalent management techniques for controlling this infection is to employ resistant plants (Selote and Kachroo 2010). Nevertheless, due to the pathogen's tolerance to harsh environments, including host-plant resistance, this approach is less effective. As a result, alternate *P. syringae* management measures, such as the use of bacteriophages through phage treatment, are required to sustain national soybean production (Chan et al. 2013). Four bacteriophages named SK2a, SK2b, SK2c (from Sukorambi), and MGX1 (from Mangli), with double-stranded deoxyribonucleic acid (dsDNA) and different abilities in propagation in the bacterial host, *P. syringae* pv. *glycinea* strain H3, can be exploited with detailed study (Addy and Wahyuni 2016). In the field, the treatment with streptocycline (100 ppm) + copper oxychloride (0.25%) had the lowest mean percent disease incidence with the maximum seed yield (2605 kg/ha) and test weight (14.33 g). Neem plant leaf extract (5% and 10%) had the smallest mean colony diameter and the greatest mean mycelial growth inhibition over the test pathogen, followed by ginger (Jagtap et al. 2012).

The efficacy of a bacterial strain as a biocontrol agent in the field may be related to the ecological similarity between the biocontrol agent and the target pathogen. Therefore, a number of different *Pseudomonas syringae* strains were evaluated for their antagonistic activities in vitro and in greenhouse assays against the target pathogen, *Pseudomonas syringae* pv. *glycinea. Pseudomonas syringae* pv. *syringae* strain 22d/93, has shown considerable promise in suppressing *P. syringae* pv. *glycinea*, causing bacterial blight of soybean. Its efficacy against this pathogen is very repeatable, even in field experiments, and the suppression processes include producing much more siderophores, achromobactin, and pyoverdin than the pathogen (Wensing et al. 2010). Biological control of foliar pathogens through colonization of the host plants with near isogenic or ecologically similar antagonistic strains seems to be a realistic goal (Völksch and May 2001).

References

Addy HS, Wahyuni WS (2016) Nucleic acid and protein profile of bacteriophages that infect *Pseudomonas syringae pv. glycinea*, bacterial blight on soybean. Agric Agric Sci Proc 9:475–481

Agrios GN (2005) Induced biochemical defenses in the hypersensitive response resistance. In: Plant pathology, 5th edn. Academic Press, London, pp 233–243

Ahmad Y, Hameed A, Ghaffar A (2006) Enzymatic activity of fungal pathogens in corn. Pak J Bot 38:1305–1316

Ajayi-Oyetunde OO, Bradley CA (2017) Identification and characterization of *Rhizoctonia* species associated with soybean seedling disease. Plant Dis 101(4):520–533

Amrate PK, Shrivastava MK, Pancheshwar DK (2020) Yield reduction and efficacy of antiviral product against *yellow mosaic virus* disease in soybean. J Crop Sci 8(1):27–34

APS Press Release (2014) Plant pathologists offer soybean rust identification and management tips. https://www.apsnet.org/edcenter/apsnetfeatures/Pages/SoybeanRustIdentification.aspx

Ashok K (2018) Survey on yellow mosaic disease of soybean in Rajasthan. Int J Agric Sci 10(16):6916–6919

Athow KL (1987) Fungal diseases. In: Wilcox JR (ed) Soybeans: improvement, production, and uses, Agronomy Monographs, vol 16, 2nd edn. ASA, CSSA, SSSA, Madison, pp 687–727

Baird RE, Abney TS, Mullinix BG (2001) Fungi associated with pods and seeds during the R6 and R8 stages of four soybean cultivars in southwestern Indiana. Phytoprotection 82:1–11

Balgude YS, Zanjare SR, Suryawanshi AV (2021) Host range studies of the virus causing mosaic disease in soybean. Int J Curr Microbiol App Sci 10(1):1821–1824

Banerjee A, Chandra S, Swer EKP (2014) Aust Plant Dis Notes 9:150

Barpete RD, Verma VK (2019) Management of rhizoctonia root rot disease in soybean in Betul district of Madhya Pradesh. Plant Arch 19(2):2376–2378

Benscher D, Pappu SS, Niblett CL, Rybicki EP, Bird J (1993) Biological and molecular characterization of potyviruses from Passiflora. Phytopathology 83:1422

Benscher D, Pappu SS, Niblett CL, Varón de Agudelo F, Morales F, Hodson E, Lee RF (1996) A strain of soybean mosaic virus infecting *Passiflora* spp. in Colombia. Plant Dis 80(3):258–262

Bhadauria V, Bett KE, Zhou T, Vandenberg A, Wei Y, Banniza S (2013) Identification of lens culinaris defense genes responsive to the anthracnose pathogen *Colletotrichum truncatum*. BMC Genet 14:31

Boufleur TR, Ciampi-Guillardi M, Tikami I, Rogério F, Thon MR, Sukno SA, Júnior NM, Baroncelli R (2021) Soybean anthracnose caused by Colletotrichum species: current status and future prospects. Mol Plant Pathol 22:393–409

Brand RJ, Burger JT, Rybicki EP (1993) Cloning, sequencing, and expression in Escherichia coli of the coat protein gene of a new potyvirus infecting South African passiflora. Arch Virol 128(1):29–41

Broders KD, Wallhead MW, Austin GD, Lipps PE, Paul PA, Mullen RW (2009) Association of soil chemical and physical properties with Pythium species diversity, community composition, and disease incidence. Phytopathology 99:957–967

Bromfield KR (1984) Soybean rust. Monograph 11. American Phytopathological Society, St. Paul

Budde IP, Ullrich MS (2000) Interactions of *Pseudomonas syringae* pv. *glycinea* with host and nonhost plants in relation to temperature and phytotoxin synthesis. Mol Plant 13(9):951–961

Carmona M (2011) Damages caused by frogeye leaf spot and late season disease in soybean in Argentina and control criteria. Trop Plant Pathol 36:1356–1358

Chan BK, Abedon ST, Loc-Carrillo C (2013) Phage cocktails and the future of phage therapy. Future Microbiol 8(6):769–783

Chang KF, Hwang SF, Ahmed HU, Strelkov SE, Harding MW, Conner RL, Turnbull GD (2017) Disease reaction to *Rhizoctonia solani* and yield losses in soybean. Can J Plant Sci 98(1):115–124

Chen PY, Choi CW (2008) Soybean mosaic virus. Characterization, diagnosis and management of plant viruses. Ind Crops 1:389–422

Chen LS, Chu C, Liu CD, Chen RS, Tsay JG (2006) PCR-based detection and differentiation of anthracnose pathogens, *Colletotrichum gloeosporioides* and *C. truncatum*, from vegetable soybean in Taiwan: PCR-based detection of Glomerella/*Colletotrichum* spp. J Phytopathol 154:654–662

Choudhary DK (2011) Plant growth-promotion (PGP) activities and molecular characterization of rhizobacterial strains isolated from soybean (*Glycine max* L. Merril) plants against charcoal rot pathogen, *Macrophomina phaseolina*. Biotechnol Lett 33:2287–2295

Clinton GP (1915) Notes on plant diseases of Connecticut. Conn Agric Exp Stn Annu Rep 1:421–451

Cruz CD, Mills D, Paul PA, Dorrance AE (2010) Impact of brown spot caused by *Septoria glycines* on soybean in Ohio. Plant Dis 94(7):820–826

Daft GC, Leben C (1972) Bacterial blight of soybeans: epidemiology of blight outbreaks. Phytopathology 62(1):57–62

Dean KM (1992) Report on plant disease. University of Illinois. https://ipm.illinois.edu/diseases/rpds/505.pdf

Domier LL, Latorre IJ, Steinlage TA, McCoppin N, Hartman GL (2003) Variability and transmission by Aphis glycines of North American and Asian Soybean mosaic virus isolates. Arch Virol 148(10):1925–1941

Domier LL, Hobbs HA, McCoppin NK, Bowen CR, Steinlage TA, Chang S, Hartman GL (2011) Multiple loci condition seed transmission of Soybean mosaic virus (SMV) and SMV-induced seed coat mottling in soybean. Phytopathology 101(6):750–756

Donald EH (2012) Brown spot of soybean. UK Cooperative Extension Service, PPFS-AG-S-01. https://plantpathology.ca.uky.edu/files/ppfs-ag-s-01.pdf

Dorrance NMJ, Gerdemann JW (1958) Root and stem rot soybean caused by Phytophthora sojae n. sp. Phytopathology 48:201–208

Dorrance AE, Jia H, Abney TS (2004) Evaluation of soybean differentials for their interaction with Phytophthora sojae. Plant Health Prog. https://doi.org/10.1094/PHP-2004-0309-01-RS

Dorrance AE, Mills D, Robertson AE, Draper MA, Giesler L, Tenuta A (2007a) Phytophthora root and stem rot of soybean. Plant Health Instructor. https://doi.org/10.1094/PHI-I-2007-0830-07

Dorrance AE, MA Draper, DE Hershman (2007b) Using foliar fungicides to manage soybean rust. http://www.oardc.ohio-state.edu/SoyRust/index.htm

Duke SO, Wedge DE, Cerdeira AL, Matallo MB (2007) Herbicide effects on plant disease. Pest Manag Sci 18:36–40

Dunleavy JM (1987) Yield reduction in soybeans caused by downy mildew. Plant Dis 71(12):1112–1114

Dunphy J, D Holshouser, D Howle, P Jost, B Kemerait, S Koenning, J Mueller, P Phipps, S Rideout, L Sconyers, E Stromberg, P Wiatrak, A Wood (2006) Managing soybean rust in the mid-Atlantic region. https://www.apsnet.org/edcenter/disandpath/fungalbasidio/pdlessons/Pages/SoybeanRust.aspx

Dye DW, Bradbury J, Goto M, Hayward AC, Lelliott RA, Schroth MN (1980) International standards for naming pathovars of phytopathogenic bacteria and a list of pathovar names and pathotype strains. Rev Plant Pathol 59(4):153–168

Eggenberger AL, Hajimorad MR, Hill JH (2008) Gain of virulence on Rsv1-genotype soybean by an avirulent Soybean mosaic virus requires concurrent mutations in both P3 and HC-Pro. Mol Plant-Microbe Interact 21(7):931–936

Elliot C (1951) Manual of bacterial plant pathogens. Chronica Botanica, Wathan

FAO (1982) Recommended methods for the detection and measurement of resistance of agricultural pests to pesticides. FAO Plant Protect Bull 30:36–71

Gade RM, Koche M, Wasule D, Haque Z (2021) SOYBEAN: seed and seedling blights, rots, bacterial spot and mosaic. In: Khan MR, Haque Z, Ahamad F (eds) Diseases of nationally important field crops. Today & Tomorrow's Printers and Publishers, New Delhi, pp 421–434

Gagarinova AG, Babu M, Poysa V, Hill JH, Wang A (2008) Identification and molecular characterization of two naturally occurring Soybean mosaic virus isolates that are closely related but differ in their ability to overcome Rsv4 resistance. Virus Res 138(1–2):50–56

Gardan L, Shafik H, Belouin S, Broch R, Grimont F, Grimont PAD (1999) DNA relatedness among the pathovars of *Pseudomonas syringae* and description of *Pseudomonas tremae* sp. nov. and *Pseudomonas cannabina* sp. nov. (ex Sutic and Dowson 1959). Int J Syst Evol Microbiol 49(2):469–478

Gnanamanickam SS, Starratt AN, Ward EWB (1982) Coronatine production in vitro and in vivo and its relation to symptom development in bacterial blight of soybean. Can J Bot 60(5):645–650

Graham JH (1953) Overwintering of three bacterial pathogens of soybean. Phytopathology 43(4):189–192

Grau CR, Adee EA, Oplinger ES (1994) An integrated approach to control Sclerotinia stem rot (White Mold) in soybean. In: Proceedings of the integrated crop management conference, Iowa State University, Ames, pp 183–196

Groth DE, Braun EJ (1986) Growth kinetics and histopathology of Xanthomonas campestsris pv. glycines in leaves of resistant and susceptible soybeans. Phytopathology 76:959–966

Gui J, Fei J, Wu Z, Fu X, Diakite A (2021) Grading method of soybean mosaic disease based on hyperspectral imaging technology. Inform Process Agric 8(3):380–385

Gupta GK, Chauhan GS (2005) Symptoms, identification and management of soybean diseases. Technical Bulletin 10. National Research Centre for Soybean, Indian Council of Agricultural Research, Indore

Gupta KN, Varma RK (2015) Ecological traits of yellow mosaic disease in relation to epidemics in soybean. Res Paper Crop 256:125–136

Gupta UP, Manisha S, Uttam G (2010) Influence of Soybean mosaic virus infection on carbohydrate content in nodule of soybean (*Glycine max* L. Merr.). Int J Virol 6(4):240–245

Gupta GK, Sharma SK, Ramteke R (2012) Biology, epidemiology and management of the pathogenic fungus *Macrophomina phaseolina* (Tassi) Goid with special reference to charcoal rot of soybean (Glycine max (L.) Merr.). J Phytopathol 160:167–180

Hajimorad MR, Eggenberger AL, Hill JH (2005) Loss and gain of elicitor function of Soybean mosaic virus G7 provoking Rsv1-mediated lethal systemic hypersensitive response maps to P3. J Virol 79(2):1215–1222

Halbert SE, Irwin ME, Goodman RM (1981) Alate aphid (Homoptera: Aphididae) species and their relative importance as field vectors of soybean mosaic virus. Ann Appl Biol 97(1):1–9

Haq QMI, Ali A, Malathi VG (2010) Engineering resistance against Mungbean yellow mosaic India virus using antisense RNA. Indian J Virol 21(1):82–85

Hartman GL, Sinclair JB, Rupe JC (1999) Anthracnose. In: Compendium of soybean diseases, 4th edn. APS Press, St. Paul, pp 13–14

Hedges F (1922) A bacterial wilt of the bean caused by Bacterium flaccumfaciens nov. sp. Science 55(1425):433–434

Hernández JR (2004) Systematic Botany and Mycology Laboratory, ARS, USDA. Invasive Fungi. Asian Soybean Rust. http://nt.ars-grin.gov/sbmlweb/fungi/index.cfm. Accessed 2 Oct 2007

Hokawat S, Rudolph K (1993) Xanthomonas campestris pv. glycines: cause of bacterial pustule of soybean, Xanthomonas. Chapman & Hall, London

Hong SJ, Kim YK, Jee HJ, Lee BC, Yoon YN, Park ST (2010) Selection of bactericides for controlling soybean bacterial pustule. Res Plant Dis 16(3):266–273

Hong SJ, Kim YK, Jee HJ, Shim CK, Kim MJ, Park JH, Lee BC (2011) Influence of disease severity of bacterial pustule caused by Xanthomonas axonopodis pv. glycines on soybean yield. Res Plant Dis 17(3):317–325

Hymowitz T (1970) On the domestication of the soybean. Econ Bot 24(4):408–421

Islam MS, Haque MS, Islam MM, Emdad EM, Halim A, Hossen QMM, Hossain MZ, Ahmed B, Rahim S, Rahman MS (2012) Tools to kill: genome of one of the most destructive plant pathogenic fungi Macrophomina phaseolina. BMC Genomics 13:1–16

Jagtap GP, Dhopte SB, Dey U (2012) Bio-efficacy of different antibacterial antibiotic, plant extracts and bioagents against bacterial blight of soybean caused by Pseudomonas syringae pv. glycinea. Sci J Microbiol 1(1):1–9

Johnson KD, O'Neal ME, Bradshaw JD (2008) Is preventative, concurrent management of the soybean aphid (Hemiptera: aphididae) and bean leaf beetle (Coleoptera: chrysomelidae) possible? J Econ Entomol 101(3):801–809

Jones SI, Vodkin LO (2013) Using RNA-Seq to profile soybean seed development from fertilization to maturity. PLoS One 8(3):e59270

Joshi D, Singh PK, Adhikari S, Rani S (2018) Screening of soybean germplasm for important disease prevalent in North India. Int J Chem Stud 6(2):2731–2733

Kaewnum S, Prathuangwong S, Burr TJ (2005) Aggressiveness of Xanthomonas axonopodis pv. glycines isolates to soybean and hypersensitivity responses by other plants. Plant Pathol 54(3):409–415

Kaur S (2012) Carbohydrate degrading enzyme production by plant pathogenic mycelia and microscleroia isolates of Macrophomina phaseolina through koji fermantation. Ind Crop Prod 36:140–148

Kennedy BW, Alcorn SM (1980) Estimates of US crop losses to procaryote plant pathogens. Plant Dis 64(7):674–676

Kentucky (1995) Plant disease management guide for soybeans, PPA-10b. https://uknowledge.uky.edu/cgi/viewcontent.cgi?referer=&httpsredir=1&article=1064&context=anr_reports

Khaeruni A, Johan EA, Wijayanto T, Taufik M, Syafar AAR, Sutariati GK (2018) Induction of soybean resistance to bacterial pustule disease (Xanthomonas axonopodis pv. glycines) by rhizobacteria and organic material treatment. IOP Conf Ser 122(1):012052

Khan SN (2007) *Macrophomina phaseolina* as causal agent for charcoal rot of sunflower. Mycopathology 5:11–118

Khare UK, Khare MN (1995) Studies on survival of *Xanthomonas campestris pv. glycines*. Indian Phytopathol 48(2):180–181

Khare UK, Khare MN, Ansari MM (2003) Bacterial pustule disease of soybean-present scenario and future strategies. Soybean Res 1:43–57

Koenning SR (2010) Southern United States soybean disease loss estimate for 2009. In: Proceedings of the Southern soybean disease workers, 37th annual meeting, p 1

Koenning SR, Wrather JA (2010) Suppression of soybean yield potential in the continental United States by plant diseases from 2006 to 2009. Plant Health Prog. https://doi.org/10.1094/PHP2010-1122-01-RS

Koranga N, Bahukhandi KD (2021) Effect of different temperature on host susceptibility to SMV-mild and SMV-severe isolates of soybean (glycine max). In: Advances in environment engineering and management. Springer, Cham, pp 231–235

Kulik MM, Sinclair JB (1999) Phomopsis seed decay. In: Hartman GL, Sinclair JB, Rupe JC (eds) Compendium of soybean diseases. American Phytopathological Society, St. Paul, pp 31–32

Kumar A, Zacharia S, Maurya AK, John V (2019) Effect of fungicides and neem oil on the *Rhizoctonia* root rot of soybean (*Glycine max* L.). Int J Curr Microbiol App Sci 8(1):368–372. https://doi.org/10.20546/ijcmas.2019.801.037

Kumari A, Hada A, Subramanyam K, Theboral J, Misra S, Ganapathi A, Malathi VG (2018) RNAi-mediated resistance to yellow mosaic viruses in soybean targeting coat protein gene. Acta Physiol Plant 40(2):1–12

Lamani MR, Singh KP, Gaur AK, Prasad R (2003) Morphological and biochemical characterization of soybean (Glycine max) cultivars in Central Himalayas. Indian J Agric Sci 73(4):215–218

Lehman SG (1928) Frog-eye leaf spot of soy bean caused by *Cercospora diazu*. J Agric Res 36(9):811–833

Li F, Xu X, Huang C, Gu Z, Cao L, Hu T, Zhou X (2015) The AC 5 protein encoded by Mungbean yellow mosaic India virus is a pathogenicity determinant that suppresses RNA silencing-based antiviral defenses. New Phytol 208(2):555–569

Lin B, Kelly H (2018) Frogeye leaf spot of soybean. Plant Health Instructor. https://doi.org/10.1094/PHI-I-2018-1018-01

Liu JZ, Horstman HD, Braun E, Graham MA, Zhang C, Navarre D, Whitham SA (2011) Soybean homologs of MPK4 negatively regulate defense responses and positively regulate growth and development. Plant Physiol 157(3):1363–1378

Madden LV (1997) Effects of rain on splash dispersal of fungal pathogens. Can J Plant Pathol 19:225–230

Mahmoud AFA, Abo-Elyousr KAM (2014) Genetic diversity and biological control of *Rhizoctonia solani* associated with root rot of soybean in Assiut Governorate, Egypt. J Plant Physiol Pathol 2:4. https://doi.org/10.4172/2329-955X.1000134

Malathi VG (2007) Genetic identity of yellow mosaic viruses infecting legumes and their phylogenetic relationship. Indian Phytopathol 60(2):143

Manandhar JB, Kunwar K, Tribhuwan Singh GL, Hartman GL, Sinclair JB (1985) Penetration and infection of soybean leaf tissues by Colletotrichum truncatum and Glomerella glycine. Phytopathology 75:704–708

Masnilah R, Astono TH, Aini LQ (2013) Karakterisasi bakteri penyebab penyakit hawar daun edamame di Jember. Berkala Ilmiah Pertanian 1(1):10–14

Mathpal M (2016) Epidemiology and management of *Rhizoctonia* aerial blight of soybean (Glycine max L. Merrill). PhD thesis, G.B. Pant University of Agriculture and Technology, Pantnagar, p 221

Melchers LE (1925) Diseases of cereal and forage crops in the United States in 1924. Plant Dis Rep Suppl 40:186

Mengistu A, Castlebury LA, Smith JR, Ray J, Bellaloui N (2009) Seasonal progress of *Phomopsis longicolla* infection on soybean plant parts and its relationship to seed quality. Plant Dis 93:1009–1018

Mian MAR, Boerma HR, Phillips DV, Kenty MM, Shannon G, Shipe ER, Blount AS, Weaver DB (1998) Performance of frogeye leaf spot-resistant and-susceptible near-isolines of soybean. Plant Dis 82(9):1017–1021

Mian MAR, Missaoui AM, Walker DR, Phillips DV, Boerma HR (2008) Frogeye leaf spot of soybean: a review and proposed race designations for isolates of *Cercospora sojina* Hara. Crop Sci 48(1):14–24

Miles MR, Frederick RD, Hartman GL (2003) Soybean rust: is the U.S. Soybean Crop at risk? https://www.apsnet.org/edcenter/apsnetfeatures/Pages/SoybeanRust.aspx

Mishra SP (1983) Studies on Xanthomonas Campestris Pv. Phaseoli-Sojense (Hedges Starr And Burk.) Dye 1978 its detection in soybean seeds and control. Doctoral diss., JNKVV, Jabalpur

Mishra N, Tripathi MK, Tiwari S, Tripathi N, Trivedi HK (2020) Morphological and molecular screening of soybean genotypes against yellow mosaic virus disease. Legume Res 45:1309–1316

Mulet M, Lalucat J, García-Valdés E (2010) DNA sequence-based analysis of the Pseudomonas species. Environ Microbiol 12(6):1513–1530

Naimuddin K (2001) Major viral diseases of pulses and their management. Indian Institute of Pulses Research, Kanpur

Nariani TK (1960) Yellow mosaic of mung (Phaseolus aureus L.). Indian. Phytopathology 13(1):24–29

Narvel JM, Jakkula LR, Phillips DV, Wang T, Lee SH, Boerma HR (2001) Molecular mapping of Rxp conditioning reaction to bacterial pustule in soybean. J Hered 92(3):267–270

Nataraj V, Maranna S, Kumawat G, Gupta S, Rajput LS, Kumar S, Sharma AN, Bhatia VS (2020) Genetic inheritance and identification of germplasm sources for anthracnose resistance in soybean [*Glycine max* (L.) Merr.]. Genet Resour Crop Evol 67:1449–1456

Navi SS, Huynh T, Mayers CG, Yang XB (2019) Diversity of Pythium spp. associated with soybean damping-off, and management implications by using foliar fungicides as seed treatments. Phytopathol Res 1:8. https://doi.org/10.1186/s42483-019-0015-9

Nene YL (1972) A survey of the viral diseases of pulse crops in India, GB Pant Univ Agric Tech, Pantnagar. India J Res Bull 4:191

Pataky JK, Lim SM (1981) Efficacy of benomyl for controlling Septoria brown spot of soybeans. Phytopathology 71(4):438–442

Pedersen P, Grau C, Cullen E (2007) Potential for integrated management of soybean virus disease. Plant Dis 91(10):1255–1259

Pellegrino C, Gilardi G, Gullino ML, Garibaldi A (2010) Detection of phoma valerianellae in lambs lettuce seeds. Phytoparasitica 38:159–165

Peltier AJ, Bradley CA, Chilvers MI, Malvick DK, Mueller DS, Wise KA, Esker PD (2012) Biology, yield loss and control of Sclerotinia stem rot of soybean. J Integr Pest Manage 3(2): B1–B7

Phillips DV (1999) Frogeye leaf spot. In: Hartman GL, Sinclair JB, Rupe JC (eds) Compendium of soybean diseases. American Phytopathological Society, St. Paul, pp 20–21

Picinini EC, Fernandes JMC (2000) Doenças da soja: aspectos epidemiológicos e controle. In: Prasad R (ed) Text book of field crop production. ICAR, New Delhi, p 391

Qi M, Wang D, Bradley CA, Zhao Y (2011) Genome sequence analyses of Pseudomonas savastanoi pv. glycinea and subtractive hybridization-based comparative genomics with nine pseudomonads. PLoS One 6(1):e16451

Qutob D, Hraber PT, Sobral BW, Gijzen M (2000) Comparative analysis of expressed sequences in Phytophthora sojae. Plant Physiol 123:243–254

Rabedeaux PF, Gaska JM, Kurtzweil NC, Grau CR (2005) Seasonal progression and agronomic impact of Tobacco streak virus on soybean in Wisconsin. Plant Dis 89(4):391–396

Rathore V, Sharma HO, Narvariya RK (2021) Growth rate of cost of cultivation of soybean in Maharashtra States of India. Pharma Innov J 10(3):84–89. https://doi.org/10.22271/tpi.2021.v10.i3Sb.5838

Rego-Machado CM, Nakasu EY, Blawid R, Nagata T, Inoue-Nagata AK (2019) Complete genome sequence of a new bipartite begomovirus infecting tomato in Brazil. Arch Virol 164(11):2873–2875

Rubayet MT, Bhuiyan MKA, Jannat R, Masum MMI, Hossain MM (2018) Effect of biofumigation and soil solarization on stemcanker and black scurf diseases of potato (Solanum tuberosum L.) caused by Rhizoctonia solani isolate PR2. Adv Agric Sci 6(3):33–48

Sarkar SF, Guttman DS (2004) Evolution of the core genome of Pseudomonas syringae, a highly clonal, endemic plant pathogen. Appl Environ Microbiol 70:1999–2012

Schaad NW, Jones JB, Chun W (2001) Laboratory guide for the identification of plant pathogenic bacteria, 3rd edn. American Phytopathological Society, St. Paul

Schinke C, Germani JC (2012) Screening Brazilian Macrophomina phaseolina isolates for alkaline lipases and other extracellular hydrolases. Int Microbiol 15:1–7

Schmitthenner AF (1988) Phytophthora rot of soybean. In: Wyllie TD, Scott DH (eds) Soybean diseases of the North Central Region. APS Press, St. Paul, pp 71–80

Sehgal OP, Thottappilly G (1992) Management of virus diseases in soybean. In: Pest management in soybean. Springer, Dordrecht, pp 251–261

Selote D, Kachroo A (2010) RIN4-like proteins mediate resistance protein-derived soybean defense against Pseudomonas syringae. Plant Signal Behav 5(11):1453–1456

Semangun H (2004) Penyakit-Penyakit Tanaman Pangan di Indonesia. Gadjah Mada University Press, Yogyakarta

Shakiba E, Chen P, Gergerich R (2012) Reactions of commercial soybean cultivars from the Mid South to Soybean mosaic virus. Crop Sci 52(5):1990–1997

Shankar G, Koshale KN, Sahu MK, Dantre RK (2018) Screening of soybean genotypes and effect of weather parameter on disease development against bacterial pustule epidemiological study on bacterial pustule of soybean. J Pharmacogn Phytochem 7(1):2826–2829

Sharma ND, Mehta SK (1996) Soybean rust in Madhya Pradesh. Acta Bot Ind 24:115–116

Sharma AN, Gupta GK, Verma RK, Sharma OP, Bhagat S, Amaresan N, Saini MR, Chattopadhayay C, Sushil SN, Asre R (2014) Integrated pest management for soybean. https://niphm.gov.in/IPMPackages/Soyabean.pdf
Sherwin HS, Kreitlow KW (1952) Discoloration of soybean seeds by the frogeye fungus, *Cercospora sojina*. Phytopathology 42(10):568–572
Short GE, Wyllie TD, Bristow PR (1980) Survival of macrophomina phaseolina in soil and residue of soybeans. Phytopathology 70:13–17
Sinclair JB (1989) Compendium of soybean diseases, 3rd edn. The American Phytopathological Society, St. Paul
Sinclair JB (1993) Phomopsis seed decay of soybeans—a prototype for studying seed disease. Plant Dis 77:329–334
Sinclair JB, Hartman GL (1999) Brown spot. In: Hartman GL, Sinclair JB, Rupe JC (eds) Compendium of soy-bean diseases, 4th edn. American Phyto-Pathological Society, St. Paul, p 16
Singh KP, Aravind T (2019) Identification of resistant sources against soybean yellow mosaic and soybean mosaic diseases in field evaluation of soybean (Glycine max (L.) Merr.) genotypes. Indian Phytopathol 72(3):519–521
Singh KP, Aravind T, Bhatt P, Surbhi K (2019) Identification of resistant sources in field evaluation of soybean (Glycine max (L.) Merr.) genotypes against pod blight and Rhizoctonia aerial blight. Int J Curr Microbiol App Sci 8:1823–1827
Smith EF (1905) Bacteria in relation to plant diseases: methods of work and general literature of bacteriology exclusive of plant diseases, vol 1. Carnegie Institution of Washington, Washington, DC
Smolińska U, Kowalska B (2018) Biological control of the soil-borne fungal pathogen *Sclerotinia sclerotiorum*—a review. J Plant Pathol 100(1):1–12
Srivastava SSL, Bais BS (1986) Involvement of pectolytic and cellulolytic enzymes in bacterial leaf pustule of soybean. Farm Sci J 1:82–84
Srivastava AK, Marabi RS, Bal LM (2021) Weather based rules for yellow mosaic disease prediction on soybean in Madhya Pradesh. IJBB 58(5):486–497
Stanley J, Bisaro DM, Briddon RW, Brown JK, Fauquet CM, Harrison BD, Stenger DC (2005) Family geminiviridae. In: Virus taxonomy: eighth report of the International Committee on Taxonomy of Viruses. Academic Press, London, pp 301–326
Steadman JR (1983) White mold—a serious yield-limiting disease of bean. Plant Dis 67(4):346–350
Subedi S, Gharti DB, Neupane S, Ghimire T (2016) Management of anthracnose in soybean using fungicide. J Nepal Agric Res Council 1:29–32
Surbhi K, Singh KP (2020) Influence of weather factors on severity of aerial blight of soybean. Ind Phytopathol. https://doi.org/10.1007/s42360-020-00235-w
Surbhi K, Singh KP, Aravind T, Bhatt P, Jeena H, Rakhonde G (2023) GIS-based survey and molecular detection of bacterial blight of soybean in sub-Himalayan ranges of Uttarakhand, India. Trop Plant Pathol 48:332–346
Suryadi Y, Manzila I, Machmud M (2009) Potensi pemanfaatan perangkat diagnostik ELISA serta variannya untuk deteksi patogen tanaman. Jurnal AgroBiogen 5(1):39–48
Swathi M, Gaur N, Singh K (2021) Virus vector relationship of Yellow mosaic virus and whitefly, Bemisia tabaci (Gennadius) in soybean. Legume Res Int J 1:5
Tereshchenko BA (1977) Soybean cotyledon bacterial disease. Zashch Rast
Thakur MP, Verma KP, Agrawal KC (1998) Characterization and management of bud blight disease of soybean in India. Int J Pest Manage 44(2):87–92
Thind TS (2005) Diseases of field crops and their management. Daya Publication House, New Delhi
Vauterin L, Hoste B, Kersters K, Swings J (1995) Reclassification of xanthomonas. Int J Syst Evol Microbiol 45(3):472–489
Völksch B, May R (2001) Biological control of Pseudomonas syringae pv. glycinea by epiphytic bacteria under field conditions. Microb Ecol 41(2):132–139

Wang RY, Ghabrial SA (2002) Effect of aphid behavior on efficiency of transmission of Soybean mosaic virus by the soybean-colonizing aphid, Aphis glycines. Plant Dis 86(11):1260–1264

Wang J, Zhou P, Shi X, Yang N, Yan L, Zhao Q, Guan Y (2019) Primary metabolite contents are correlated with seed protein and oil traits in near-isogenic lines of soybean. Crop J 7(5):651–659

Wensing A, Braun SD, Büttner P, Expert D, Völksch B, Ullrich MS, Weingart H (2010) Impact of siderophore production by Pseudomonas syringae pv. syringae 22d/93 on epiphytic fitness and biocontrol activity against Pseudomonas syringae pv. glycinea 1a/96. Appl Environ Microbiol 76(9):2704–2711

Williams DJ, Nyvall RF (1980) Leaf infection and yield losses caused by brown spot and bacterial blight diseases of soybean. Phytopathology 70(9):900

Williamson MR, Rothrock CS, Muller JD (2006) First report of *Rhizoctonia* foliar blight of soybean in South Carolina. Plant Health Prog. https://doi.org/10.1094/PHP-2006-1030-01-BR

Wrather JA, Koenning SR (2006) Estimates of disease effects on soybean yields in the United States 2003 to 2005. J Nematol 38:173–180

Wrather JA, Koenning SR (2009) Effects of diseases on soybean yields in the United States 1996 to 2007. Online. Plant Health Prog. https://doi.org/10.1094/PHP-2009-04401-01-RS

Wrather JA, Anderson TR, Arsyad DM, Gai J, Ploper LD, Porta-Puglia A, Yorinori JT (1997) Soybean disease loss estimates for the top 10 soybean producing countries in 1994. Plant Dis 81(1):107–110

Wrather JA, Stienstra WC, Koenning SR (2001a) Soybean disease loss estimates for the United States from 1996 to 1998. Can J Plant Pathol 23:122–131

Wrather JA, Anderson TR, Arsyad DM, Tan Y, Ploper LD, Porta-puglia A, Ram HH, Yorinori JT (2001b) Soybean disease loss estimates for the top ten soybean producing countries in 1998. Can J Plant Pathol 23:115–121

Wrather A, Shannon G, Balardin R, Carregal L, Escobar R, Gupta GK, Tenuta A (2010) Effect of diseases on soybean yield in the top eight producing countries in 2006. Plant Health Prog 10:1094

Xue AG, Cober E, Morrison MJ, Voldeng HD, Ma BL (2006) Effect of seed treatments on emergence, yield, and root rot severity of soybean under *Rhizoctonia solani* inoculated field conditions in Ontario. Can J Plant Sci 87(1):167–174

Xue AG, Morrison MJ, Cober E, Anderson TR, Rioux S, Ablett GR, Rajcan I, Hall R, Zhang JX (2007) Frequency of isolation of species of *Diaporthe* and *Phomopsis* from soybean plants in Ontario and benefits of seed treatments. Can J Plant Pathol 29:354–364

Yadav VK, Khare CP, Tiwari PK, Srivastava JN (2020) Diseases of field crops: diagnosis and management. Apple Academic Press 1:27

Yang HC, Hartman GL (2016) Anthracnose. In: Hartman GL, Rupe J, Sikora EJ (eds) Compendium of soybean diseases and pests. American Phytopathological Society, St. Paul, pp 31–34

Yang XB, Berggren GT, Snow JP (1990) Types of Rhizoctonia foliar blight on soybean in Louisiana. Plant Dis 71:501–504

Yang Y, Lin J, Zheng G, Zhang M, Zhi H (2014) Recombinant soybean mosaic virus is prevalent in Chinese soybean fields. Adv Virol 159:1793–1796

Young JM (2010) Taxonomy of Pseudomonas syringae. J Plant Pathol 92:S1.5–S1.14

Zhang R (2011) Interactions of post emergence herbicides, strobilurin fungicides and *Rhizoctonia* root rot of soybean. MSc thesis, University of Illinois at Urbana-Champaign

Zhao G, Ablett GR, Anderson TR, Rajcan I, Schaafsma AW (2005) Anastomosis groups of *Rhizoctonia solani* associated with soybean root and hypocotyl rot in Ontario and resistance of accession PI 442031 to different anastomosis groups. Can J Plant Pathol 27:108–117

Field Pea Diseases and their Management through an Integrated Approach

12

H. Rajashekara, B. Jeevan, A. R. N. S. Subbanna,
and K. K. Mishra

Abstract

Pulses play an important role in improving the nutritional security of farmers and populations in less developed countries, as well as the sustainability of agriculture in dryland areas worldwide. Of the most important food legumes grown in India, field pea (*Pisum sativum* L.) is grown in the cool season. Biotic stresses are recognized as serious constraints for the production of field pea, which dramatically reduce crop yield and hamper productivity. Among these, fungi and viruses are the major and most important groups affecting all parts of the plant at various growth stages of the crop. According to estimates made in India, diseases alone can cause nearly 10–15% of losses in legume production. This chapter envisages the literature on different fungal diseases of field pea, especially on distribution, the biology of causal agents, symptoms, epidemiology, and their management comprising host plant resistance, cultural, biological, and fungicide use in an integrated way.

Keywords

Field pea · Diseases · Symptoms · Epidemiology · Management

H. Rajashekara (✉) · B. Jeevan · A. R. N. S. Subbanna · K. K. Mishra
Division of Crop Protection, ICAR-Vivekanand Parvatiya Krishi Anusandhan Sansthan, Almora, Uttarakhand, India

V. K. Singh et al. (eds.), *Diseases of Field Crops: Diagnostics and Management*,
https://doi.org/10.1007/978-981-97-6160-9_12

12.1 Introduction

Field pea (*Pisum sativum* L.) or "Dry Pea" is a native of Southwest Asia, popularly known as "Matar" in India. Field pea differs from Garden pea, which is either consumed as fresh or in the form of processed foods. Peas are grown worldwide but are largely confined to temperate regions or cooler seasons in tropical/subtropical regions. The matured seeds are used as whole or split into dal for human consumption. It is also grown as a forage crop for livestock feed since it is highly nutritive and contains approximately 21–25% protein. Pea is also a rich source of amino acids, such as lysine and tryptophan, compared to cereal grains. Besides, it contains 5–20% less trypsin inhibitors than soybeans, making field pea an excellent feed for livestock. It maintains soil fertility through biological nitrogen fixation by *Rhizobium* bacteria prevalent in their root nodules.

India is one of the major producers of field pea in the world and stands at fifth position in the list of major field pea-growing countries, next to France (FAO 2015). Field pea can be grown under different cropping patterns comprising both hills and plains of India. It is capable of utilizing bacterially fixed atmospheric nitrogen, which helps in improving soil fertility by providing nitrogen to the successive crop. It is extensively grown in Uttar Pradesh, Madhya Pradesh, Maharashtra, Uttarakhand, Punjab, Bihar, Gujarat, Haryana, Rajasthan, Karnataka, Andhra Pradesh, Tamil Nadu, West Bengal, Jharkhand, Chhattisgarh, Manipur, and Tripura and contributes around 7% of the world's total produce with a production figure of approximately 7.8 lakh tonnes. Numerous field pea cultivars viz., VL Matar-42, VL Matar-47, Shubhra (IM-9101), Swati (KFPD-24), DDR-27 (Pusa Panna), Alankar, Jayanti (HFP-8712), Aman (IPF5–19), etc., are available in India with an average yield of 600–900 kg/ha. In Uttarakhand, field pea is grown in winter as well as summer seasons. The agro-climatic conditions of Uttarakhand ranging from an altitude of 800–2000 m above mean sea level provide a unique opportunity to grow the pea crop in winter (Tarai area) and as an off-season crop during summer in higher altitudes.

From a global perspective, important diseases include damping off (*Pythium ultimum*), bacterial blight (*Pseudomonas syringae* pv. *pisi*), ascochytosis, powdery mildew, wilt, rust, root rot, downy mildew, pea mosaic, pea streak, and many other minor diseases such as leaf spots, root, collar, and stem rots. In Indian conditions, like other pulse crops, diseases caused by fungi, bacteria, viruses, and nematodes are among the notable risk factors for field pea cultivation. Fungal pathogens are recognized as serious constraints for the production of field pea, which dramatically reduce the crop yield and hamper productivity. They include Fusarium wilt (*F. oxysporum* f. sp. *pisi*), powdery mildew (*Erysiphe pisi*), ascochyta blight, seedling blight, damping off, stem and root rot, septoria leaf spot, rust (*Uromyces viciae fabae*), and downy mildew (*Peronospora viciae*). This chapter highlights the consolidated information on its distribution, yield loss, symptoms, pathogen life cycle, epidemiology, and management of major fungal diseases of field pea (Table 12.1).

Table 12.1 Important fungal diseases of field pea in India by Vishwa and Choudhary (1998)

Diseases	Pathogens
Root rot diseases	
Wilt	*Fusarium oxysporum* f. sp. *pisi*
Root rot	*F. solani* f. sp. *pisi*, *Rhizoctonia solani*, *R. bataticola*
Damping off	*Sclerotium rolfsii*, *Pythium* sp.
Diseases of aerial parts	
Leaf spot	*Alternaria alternate*, *A. brassicae*, *Cercospora pisi*, *Colletotrichum pisi*, *Phyllosticta pisi*, *Helimenthosporium* sp. *Ascochyta pisi*
Anthracnose	*Colletotrichum gloeosporioides*
Powdery mildew	*Erysiphe pisi*, *Oidium erysiphoides*
Rust	*Uromyces viciae fabae*
Downy mildew	*Peronospora viciae*
Blight	*Ascochyta pinodes*
Grey mold	*Botrytis cineria*
Stem rot	*Sclerotinia sclerotiorum*, *Sclerotium rolfsii*
Pod rot	*Phytophthora parasitica*
Pod spot	*Colletotrichum pisi*

12.2 Fusarium Wilt

12.2.1 Distribution

Buxton first reported Fusarium wilt of peas from Europe in the year 1955. There was also some confusion between "wilt" caused by *F. oxysporum* f. sp. *pisi* (*Fop*) and "foot rot" caused by *F. solani* f. sp. *pisi.* and is caused by a combination of both pathogens (Buxton 1955), but *Fop* race 1 causes pea wilt (Jones and Linford 1925). In recent years, its outbreaks have occurred in many pea-growing countries, assuming its epidemic dimensions. In India, wilt is a major problem in Himachal Pradesh (HP), Uttar Pradesh (UP), Haryana, Punjab, Madhya Pradesh (MP), Uttarakhand, Maharashtra, Bihar, West Bengal (WB), and other pea-growing areas.

12.2.2 The Pathogen *Fusarium oxysporum* f. sp. *pisi* (*Fop*)

The disease caused by *Fop* is one of the most destructive, versatile, and widespread pathogens prevalent throughout the world. The pathogen is morphologically characterized by a white to pinkish-white colony with fluffy mycelial growth. The hypha is septate, hyaline, and branched. Conidia are formed on small, compact stroma (sporodochia) called pin notes or directly on mycelium. The pathogen produces two

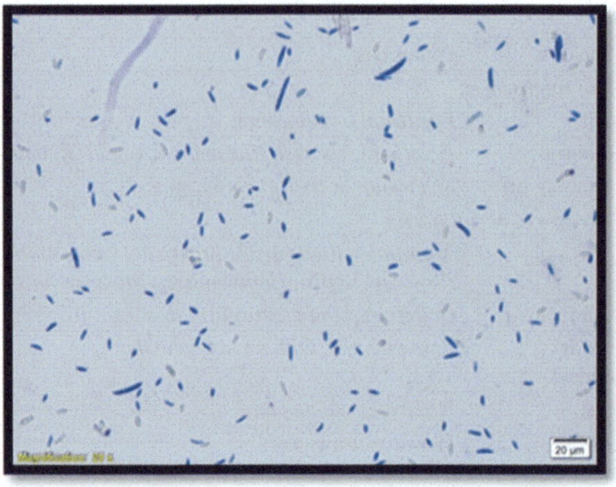

Fig. 12.1 Macro and microconidia of *Fop*

Fig. 12.2 Pea plants infected with *Fop*: (**a**) yellowing of leaves progressing from the base of the plant towards the tip and (**b**) advanced stage of wilting

types of conidia (Fig. 12.1), microconidia (single-celled, oval to comma-shaped) and macroconidia (spindle-shaped hyaline conidia having three to four septations with a distinct pedicellate base). The species is known to produce thick-walled chlamydospores as resting spores that survive in the soil for 10 or more years in the absence of a host. The pathogen is also capable of infecting seeds.

12.2.3 Symptoms

The symptom first appears in the lower leaves, which become pale green, then gradually turn yellow color, and the margins of leaves show downward and inward curling (Fig. 12.2a). A cross-section of the lower stem often reveals brownish-red discoloration of the vascular tissue. Affected plants may be flaccid and turn yellow, starting at the bottom and progressing upward until the entire plant becomes yellow. Finally, the infected plant wilts and dies due to the destruction of feeder roots (Fig. 12.2b).

12.2.4 Disease Cycle

F. oxysporum is known to produce all three kinds of spores: macroconidia, microconidia, and chlamydospores. The pathogen persists in the soil as chlamydospores, especially in the cooler regions. In addition, the pathogen also survives in infected plant detritus as mycelium (Agrios 2005). The pathogen colonizes the host typically by germination of the spore through roots, favored by wounds, and infects the xylem. Further, it moves passively through the plant's vascular system and interferes with water movement, as a result, water flow is retarded in the xylem vessels, which leads to wilting of plants. The secondary spread of the pathogen from infected to healthy plants occurs through irrigation water, infested soil equipment, and infested transplants. Besides there are reports of wind carrying contaminated dust particles that could also be the source of long-distance dispersal of the pathogen under certain environmental conditions (Fig. 12.3) (Rekah et al. 2000).

12.2.5 Epidemiology

Understanding the epidemiology of Fusarium wilt enables us to devise strategies for its management. The presence of susceptible hosts and favorable environmental conditions are necessary for disease development. The disease is favored by warm, dry soil conditions, with a range of temperature from 25 °C to 30 °C, but the pathogen is generally inhibited if the temperature exceeds 33 °C and is not favored below 17 °C.

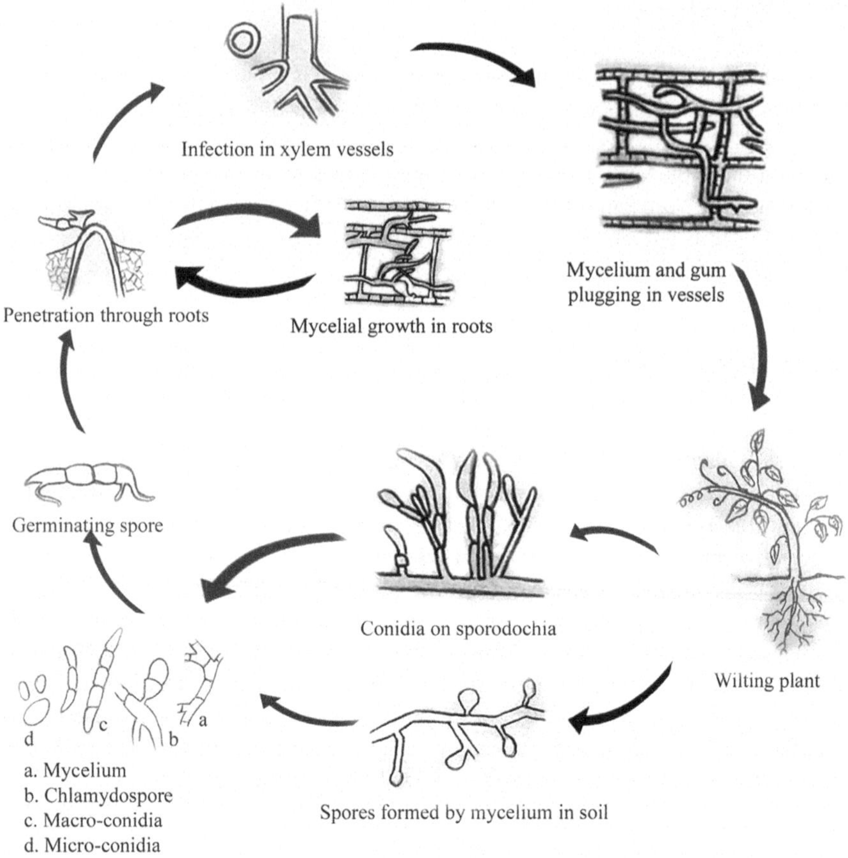

Fig. 12.3 Disease cycle of wilt disease of field pea

12.3 Powdery Mildew (*Erysiphe pisi* (*Erysiphe polygoni*))

Powdery mildew occurs throughout the world and can cause severe yield loss in areas where peas are cultivated (Kraft and Pfleger 2001). It is the most serious and widely distributed disease of field peas in India. Frequent epiphytotics of this disease occur in major pea-growing areas of UP, Bihar, and MP. The disease is more severe in late-season crops. Warm days with cool nights and the formation of dew are favorable conditions for rapid disease development. The disease is more frequently observed under polyhouse conditions in every growing season due to the availability of favorable conditions for disease development. The disease may cause 26–47% losses in pod yield of severely infected crops (Munjal et al. 1963). Often, powdery mildew and rust affect the crop together in UP, causing an average 50% reduction in grain yield.

Fig. 12.4 Symptom of powdery mildew: (**a**) infection on leaves, (**b**) infection on pods, and (**c**) conidial arrangement

12.3.1 Symptoms

The disease appears at all stages of crop growth but more severely after flowering under the prevailing cropping systems of India. It is characterized by the formation of small, diffused, and off-colored spots on the upper surface of the lower part of the leaves. These lesions later appear as a white powdery mass and subsequently cover both surfaces of leaves, petioles, tendrils, stems, and pods. Infected tissues turn brown and necrotic. In severe cases, the entire plant dries prematurely (Fig. 12.4). Reduction in nodulation and nitrogenase activity due to infection is also noticed (Singh and Mishra 1992). The fungus can infect *pisum*, *medicago*, *vicia*, *lupins*, and *lens* spp. Out of three biological forms of the fungus, only the pea form is reported to infect peas (Singh et al. 1994).

12.3.2 Disease Cycle

The fungus overwinters on infected pea crop debris and produces spores that are blown by wind into neighboring crops. Under favorable conditions, the disease may completely colonize the plant in 5–6 days, and once a few plants become infected, it rapidly spreads to adjacent field pea growing areas. The warm temperature range of 15–25 °C with more than 70% relative humidity prevailing for 4–5 days, especially during flowering and pod filling stages, is most favorable for disease development. Rainfall is not favorable for disease development since it washes away the spores from infected plants. Infected seeds are also the source of infection (Fig. 12.5).

Fig. 12.5 Disease cycle of powdery mildew of field pea

12.4 Downy Mildew (*Peronospora pisi*)

Field pea mildew is of wide occurrence globally and is considered a disease of moderate importance. In India, this disease is prevalent in the Indo-Gangetic plains.

12.4.1 Symptoms

It appears on young plants in the form of downy velvety growth on the under surface of the leaf, and the corresponding upper surface will show yellow spots. As the disease progresses, affected leaf tissues turn brownish-grey in color, dry up, and leaves drop prematurely (Fig. 12.6). The disease is incited by *Peronospora viciae* (SYN. *P. pisi*), which is an obligate parasite. The pathogen survives in infected crop debris as oospores and is reported to infect some other hosts such as Lathyrus and *Vicia faba*.

Fig. 12.6 Symptoms of downy mildew: (**a**) yellow spots on leaves, (**b**) drying of leaves, and (**c**) sporangia

12.4.2 Favorable Conditions

Develops quickly in cooler areas with temperatures ranging from 5 to 15 °C, relative humidity of over 90%, and dew conditions lasting for 4–5 days, particularly during the early vegetative stages of the crop growing season.

12.4.3 Disease Cycle

Oospores and Sporangia survive in the soil, and the infected plant debris serves as a primary source of inoculum. The infected seeds are also the other source of primary inoculum. The disease development is favored by high relative humidity (90–100%) and cool weather (5–15 °C) for 4–5 days during the early vegetative stage of the crop. Individual seedlings become infected and act as a secondary source of infection, and rain is the major means of spore dispersal and infection. During the season, heavy dew will enhance sporulation ability. Dry, warm conditions are highly unfavorable for disease development. Systemically infected downy mildew plants will help in disease spread if the conditions are favorable at the stage of crop maturity. If the temperature slightly increases, the plants will escape from the disease incidence without significant yield loss (Fig. 12.7).

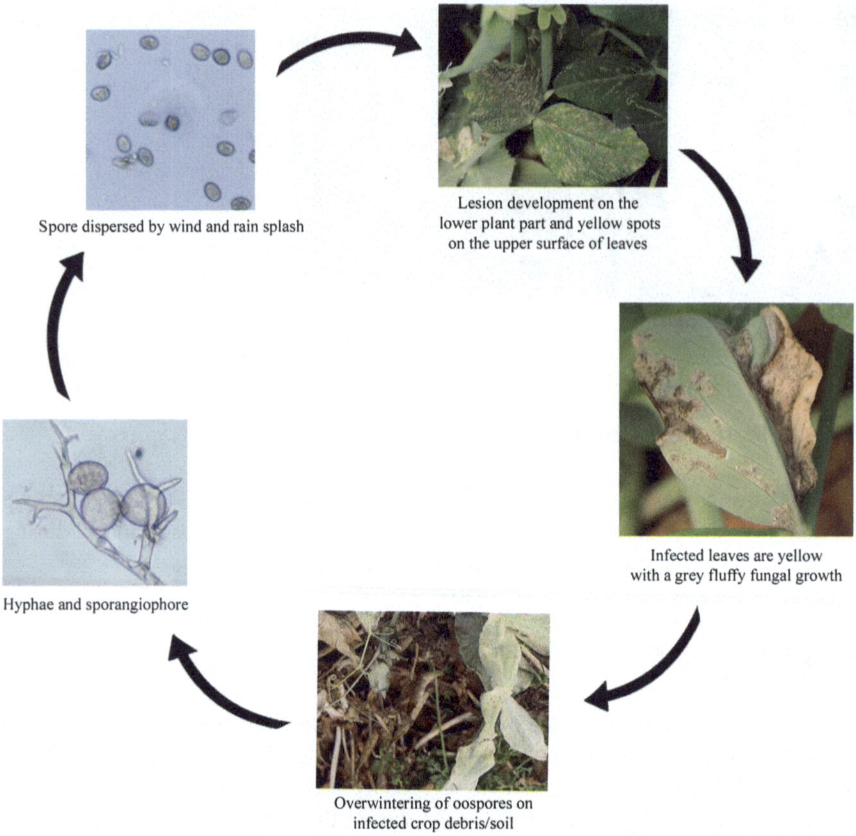

Fig. 12.7 Disease cycle of downy mildew of field pea

12.5 Rust

Field pea rust maintains a moderate status in the world scenario. In India, its occurrence has been erratic, but when it occurs, it devastates the crop. The initial symptoms appear with the development of aecia. The yellow aecia appear first on the undersurface of the leaves, stems, and petioles. The formation of the aecial stage is preceded by a slight yellowing, which gradually turns brown. The uredo pustules are powdery light brown in appearance. All four stages develop on every green part of the host, including the pods. The teleutopustules occur in the same sores as the uredia and develop from the same mycelium.

Rust pathogens are obligate biotrophs. Out of the two species of *Uromyces* infecting peas, only *U. viciae fabae*, which is an autoecious rust, is known to occur in India. The pathogen is known to infect lentil, lathyrus, sweet pea, and faba bean, besides peas. The pathogen survives on crop debris and collateral hosts in the submountains and Indo-Gangetic plains of north India.

12.6 Ascochyta Blight

It is a severe problem in temperate and subtropical zones, including hill areas of India. It is also known as black spot disease. The disease is more destructive when it girdles the stem, and in some locations, pod and seed infection causes downgrading of pea seed quality. Three distinct species, *Ascochyta pinodella*, *A. pisi*, and *A. pinodes*, are known to infect peas. *A. pinodella* causes small, irregular circular, purplish spots on leaves. Lesions sometimes show concentric rings. Almost similar lesions are caused on the stem and pods. It also causes foot rot under humid conditions. Spots of *A. pisi* are tan-colored instead of brown or black. Tiny black fruiting bodies are also visible on the lesions. *A. pinodes* causes blight symptoms on leaves, stems, and pods. *A. pinodella* can be distinguished from *A. pinodes* by larger size conidia, while *A. pisi* produces light buff to flesh-colored spore mass exudates. These pathogens are known to survive through seed and infected crop debris in the soil.

12.7 White Rot

Stem and pod rot caused by *Sclerotinia sclerotiorum* is a disease restricted to cool and humid areas. Under Indian conditions, it is of common occurrence in northern hills and eastern UP and is reported to cause 4.4–30.7% yield loss in eastern UP and 70.3% in the Kullu Valley of Himachal Pradesh. The disease appears at the flowering stage if low temperatures and rain prevail. Typical symptoms are quick rotting of the stem and foliage coupled with white cottony growth of the mycelium. The crop subsequently dries up in patches. Late in the season, black elongated sclerotia of an irregular shape can be seen over and inside the stem. Thick crop canopy and luxurious crop growth favor the disease. The pathogen survives as sclerotia in the infected crop debris.

12.8 Management of Diseases

During recent times, a shift in scientific thinking and practice in the management of field pea diseases has been seen, with greater emphasis on identifying, evaluating, and integrating location-specific components of integrated disease management (IDM). The components of IDM involved in the management of field pea diseases are mentioned below:

1. Host plant resistance (HPR)
2. Disease prediction models for the avoidance of high-risk or disease outbreak
3. Chemical sprays
4. Biological control
5. Cultural practices (sowing dates, crop rotation, etc.)

12.8.1 Cultural Practices

Soil is the main factor harboring the primary source of inoculum for all important diseases either in the saprophytic state or as resting structures in the crop debris. Therefore, sanitation of the field by removing infected plants and crop debris minimizes the inoculum load. Crop rotation with nonhosts for 3–5 years will reduce disease, but careful weed management must be done during this period. Avoid the use of excessive N, as it will encourage disease. The use of calcium nitrate fertilizer instead of ammonium nitrate can reduce disease severity in some soils. In acidic soils, raising the soil pH to 7 can help control the disease. However, these approaches prove fruitful mainly for root rot diseases but fail to give effective control for aerial diseases unless adopted as a large-area approach on a community basis. Similarly, growing early maturing short-duration varieties can also be of practical use in minimizing the losses caused by rust and powdery mildew. Avoiding overcrowding of plants and excessive irrigation before podding helps in reducing the disease spread of downy mildew and rust pathogens. Field pea intercropping with mustard, linseed, wheat, gram, and barley delays the appearance and build-up of powdery mildew. Soil solarization and flooding have been successfully used for the control of Fusarium wilt diseases.

12.8.2 Biological Control

The use of bio-agents can be a suitable approach for the management of wilt disease. Plant growth-promoting rhizobacteria (PGPR), such as Bacillus and Pseudomonas strains, are significant root colonizers (Manikandan et al. 2010), which suppress the soil-borne diseases and also trigger plant defenses. Application of *Trichoderma viride* and *T. harzianum* has significant potential in the control of disease. Besides, the application of neem cake and compost enhances the survivability of bio-agents and also helps in the suppression of soil-borne diseases. Studies on the biological control of field pea diseases have been mainly concentrated on damping off, root rot, and sclerotinia white rot diseases. Biocontrol agents that are found effective against Pythium damping off and Aphanomyces root rot are *Pseudomonas cepacia* and *P. florescens*; seed treatment with these bioagents increases emergence by 61% and 30%, respectively (Parke et al. 1991). The bacteria were highly effective when applied in combination with captan. Control of pea root rot caused by *Rhizoctonia solani* has also been successfully achieved by the use of *Bacillus subtilis* (Hwang and Chakravarty 1992). Singh (1991) reported *Penicillium cyclopium*, *P. sheari*, *Paecilomyces lilacinus*, *Aspergillus niger*, *A. fumigatus*, *Acremonium implicatum*, and *Trichothecium roseum* as antagonists against *Sclerotinia sclerotiorum* from Himachal Pradesh.

12.8.3 Chemical Control

In chemical control, seed treatment (Carbendazim 25% WP + Mancozeb 50% WP) and combinations of soil drenching with azoxystrobin 23 ECWP 0.1% or Carbendazim 25% WP 0.1% can effectively control Fusarium wilt. Wettable sulfur or Karathane was found highly effective in controlling powdery mildew disease. Application of fungicides like propiconazole at 0.1% and hexaconazole at 0.05% were found effective for the management of powdery mildew on field pea (Hiremath and Lal 2018). Metalaxyl and mancozeb were highly effective for downy mildew and rust diseases, respectively.

References

Agrios G (2005) Plant pathology, 5th edn. Elsevier, Amsterdam

Buxton EW (1955) Fusarium diseases of peas. Trans Brit Mycol Soc 38:193

FAO (2015) FAOSTAT. Food and Agriculture Organization of the United Nations, Rome, Italy

Hiremath B, Lal AA (2018) Evaluation of fungicides for management of field pea (*Pisum sativum* L.) powdery mildew caused by *Erysiphe polygoni* DC. Int J Pure App Biosci 6(4):516–520

Hwang SF, Chakravarty P (1992) Potential for the integrated control of Rhizoctonia root rot of *Pisum sativum* using *Bacillus subtilis* and fungicide. Zeitschrift fur planzenkrankheiten und p flanzenschutz 100:308–316

Jones FR, Linford MB (1925) Pea disease survey in Wisconsin. Agricultural Experiment Station of the University of Wisconsin, Madison

Kraft JM, Pfleger FL (2001) Compendium of pea diseases and pests, 2nd edn. APS Press, St. Paul

Manikandan R, Saravanakumar D, Rajendran L et al (2010) Standardization of liquid formulation of *Pseudomonas fluorescens* Pf1 for its efficacy against Fusarium wilt of tomato. Biol Control 54:83–89

Munjal RL, Chenulu VV, Hora TS (1963) Assessment of losses due to powdery mildew (Erysiphe polygoni) on pea. Indian Phytopathol. 19:260–267

Parke JL, Rand RE, Joy AE, King EB (1991) Biological control of Pythium damping-off and Aphanomyces root rot of peas by application of Pseudomonas cepacian and P. fluorescens to seed. Plant Disease. 75:987–992

Rekah Y, Shtienberg D, Katan J (2000) Disease development following infection of tomato and basil foliage by airborne conidia of the soilborne pathogens *Fusarium oxysporum* f. sp. *radicis-lycopersici* and *F. oxysporum* f. sp. *basilici*. Phytopathology 90:1322–1329. https://doi.org/10.1094/PHYTO.2000.90.12.1322

Singh D (1991) Biocontrol of Sclerotinia sclerotiorum by Trichoderma harzianum. Trop Pest Manage 37:374–378

Singh UP, Mishra GD (1992) Effect of powdery mildew (*Erysiphe pisi*) on nodulation and nitrogenase acitivity in pea (*Pisum sativum*). Plant Pathol 41:262–264

Singh RA, Vishwa D, Lal S (1994) The powdery mildew of pea. Indian Institute of Pulse Research, Kanpur

Vishwa D, Choudhary RG (1998) Diseases of pigeonpea and field pea and their management. In: Thind TS (ed) Diseases of field crops and their management. National Agricultural Technology Information Centre, Ludhiana, pp 217–238

13 Major Fungal Diseases of Pulse Crops and their Management to Mitigate its Losses

Raj K. Mishra and Sonika Pandey

Abstract

Pulses play a crucial role in the diets of vegetarians worldwide. India, in particular, stands as the largest producer and consumer of pulses globally. These protein-rich staples contain approximately 25% protein, which is nearly three times the protein content found in cereals. As a result, pulses serve as an essential supplement to cereal-based diets. However, the growth of pulse crops faces numerous challenges, both from biotic and abiotic stresses. Among the biotic stresses, diseases such as wilt, Phytophthora stem blight (PSB), dry root rot, collar rot, Ascochyta blight, powdery mildew, and rust significantly limit yields. These diseases cause substantial economic losses each year and hinder exports due to poor-quality produce. To combat these diseases, chemical pesticides are commonly used worldwide. However, it is widely recognized that the use of agricultural chemicals not only leads to environmental pollution but also poses serious health risks to humans, plants, domestic animals, and wildlife. Consequently, these practices are considered ecologically unacceptable. Therefore, it is imperative to acquire updated knowledge to assess the severity of these problems and develop or refine integrated management strategies. These strategies should incorporate cultural, physical, biological, safer chemical, and molecular approaches to protect crops in a cost-effective and environmentally friendly manner. By adopting these approaches, significant reductions in losses caused by diseases can be achieved, particularly in subsistence farming conditions, while simultaneously enhancing crop yields.

R. K. Mishra (✉) · S. Pandey
Division of Crop Protection, ICAR-Indian Institute of Pulses Research, Kanpur, Uttar Pradesh, India

V. K. Singh et al. (eds.), *Diseases of Field Crops: Diagnostics and Management*,
https://doi.org/10.1007/978-981-97-6160-9_13

Keywords

Pulse crops · Fungal diseases · Losses · Management

13.1 Introduction

Pulse crops are vital for maintaining soil fertility and providing protein to India's large vegetarian population. As the world's largest producer and consumer of pulses, they hold great promise for alleviating widespread protein deficiency and malnutrition amongst the poor. Their cultivation has long been integral to subsistence farming because of their resilience in India's harsh, dry environments. The importance of pulse protein is evident since most Indians avoid meat for religious reasons and rely on pulses as their main protein source. Research and development on managing pulse diseases has focused on developing host resistance and chemical control, often in isolation. Recently, the scientific approach to legume disease management has dramatically shifted to emphasize identifying, evaluating, and integrating region-specific integrated disease management (IDM) components. This chapter attempts to address the major biotic stresses and diseases of key pulses—chickpea, lentil, fieldpea, pigeonpea, mungbean, urdbean—and their integrated management strategies. These globally grown crops suffer substantial losses from numerous biotic and abiotic stresses (Table 13.1).

13.2 Diseases of Pulses and Yield Losses

Pulse crops of both the cool (*rabi*) season—chickpea, lentil, rajmash, and fieldpea—and the warm (*kharif*) season—pigeonpea, mungbean, urdbean, and cowpea—are prone to a large number of fungal, bacterial, phytoplasmal, and viral diseases. This often results in significant economic losses (Nene et al. 1996). Dry root rot, wilt, phytophthora stem blight, powdery mildew, rust, stem rot, and yellow vein mosaic disease are recognized as the most damaging diseases affecting most pulses (Table 13.1). Biotic stresses, mainly insect pests and diseases, can cause over 50% yield loss in pulses. Fusarium wilt in chickpea decreases both seed yield and weight (Haware and Nene 1980), causing 10–15% annual yield loss in India (Singh and Dahiya 1973). However, 70% total crop loss can occur in severe outbreak years (Halila and Strange 1996). Sharma et al. (2016) reported 10% seedling mortality in pigeonpea, chickpea, and lentil due to Fusarium and collar rot. Fusarium wilt in chickpea, pigeonpea, and lentil leads to over 20–25% yield loss across India. Similarly, dry root rot in chickpea causes 10–100% annual yield loss. Stem rot in pulses causes almost 100% yield loss worldwide annually (113). Fusarium wilt in susceptible pigeonpea cultivars can lead to 30–100% yield loss from seedling to maturity (Dhar et al. 2005). Chickpea is most vulnerable to Botrytis gray mold (BGM) infection during flowering and podding, which can lead to complete crop failure (Kannaiyan and Nene 1981). Under endemic conditions, BGM causes up to 50% yield loss. With favorable conditions for infection during vegetative and reproductive growth, losses can reach 100%.

Table 13.1 The major diseases of pulses and their causal agents

S.N.	Crops	Diseases	Causative organisms
1	Pigeonpea	Phytophthora stem blight	*Phytophthora drechsleri* f. sp. *cajani.*
		Wilt	*Fusarium oxysporum* f. sp. *udum* (Butler) Snyd. and Hans.
		Leaf spot	*Cercospora* spp.
		Alternaria blight	*Alternaria alternata*
		Powdery mildew	*Erysiphe polygoni* DC
2	Chickpea	Wilt	*F. oxysporum* f. sp. *ciceri* Matuo and Sato
		Dry root rot	*Rhizoctonia bataticola*
		Stem rot	*Sclerotinia sclerotiorum*
		Rust	*Uromyces ciceris-arietinii (Grogn.)* Jacz. & Boy.
		Wet root rot	*Rhizoctinia solani*
		Botrytis gray mold	*Botrytis cinerea*
3	Lentil	Wilt	*F. oxysporum* f. sp. *lentis*
		Root rot	*Macrophomina phaseolina*
		Rust	*Uromyces fabae*
		Ascochyta blight	*Ascochyta fabae* f. sp. *lentis*
4	Pea	Rust	*U. fabae*
		Powdery mildew	*E. polygoni*
		Root rot	*R. solani*
5	Mungbean and urdbean	Root rot	*M. phaseolina*
		Leaf spot	*Cercospora* spp.

13.3 Major Fungal Diseases

13.3.1 Wilt

Fusarium wilt is a major disease affecting chickpea, pigeonpea, and lentil crops across many countries. This fungal disease, caused by Fusarium species, leads to 20–25% yield losses. Fusarium wilt has been reported in India, Bangladesh, Burma, Ethiopia, Mexico, Pakistan, Syria, Tunisia, Chile, Iran, Nepal, Sudan, the United States, Peru, USSR, Malawi, Spain, Turkey, Italy, and more. The pathogen penetrates the vascular bundles in plant roots, reducing water uptake and causing foliage to wilt. Infected seedlings collapse and lie flat. Older plants show drooping petioles, rachis, and leaflets. Though roots appear healthy externally, internal xylem discolors dark brown when split vertically. In addition to wilting, infected plants may yellow, show xylem necrosis, and have darkened xylem vessels and pith. Pods appear normal but contain smaller, wrinkled, discolored seeds. Seeds from wilted plants may carry the pathogen even if they appear normal. Infested soil and infected seeds spread Fusarium wilt. Recent advances in the symptomatology and management of Fusarium wilt in food legumes have been reviewed, including details on chickpea diseases (Fig. 13.1).

Fig. 13.1 Wilt disease symptoms on pigeonpea, chickpea, and lentil

13.3.2 Phytophthora Stem Blight (PSB)

Phytophthora stem blight (PSB), caused by the pathogen *Phytophthora drechsleri* f. sp. *cajani*, recurs as an economically significant disease of pigeonpea (*Cajanus cajan*). PSB emerges especially when heavy rains fall over short periods and hot, humid weather persists during the crop growing season. PSB is the third most damaging disease of pigeonpea after Fusarium wilt and pigeonpea sterility mosaic disease. Previous reviews have examined the status and research challenges of PSB in pigeonpea (Williams et al. 1975; Kannaiyan et al. 1984). PSB was first reported in India in 1966 (Williams et al. 1968), after which it spread to major pigeonpea growing regions in Asia, Africa, the Americas, Australia, the Dominican Republic, Kenya, Panama, and Puerto Rico (Pal et al. 1970; Williams et al. 1975; Kannaiyan et al. 1984; Wearing and Birch 1988; Nene et al. 1996). Recently, PSB resurgence poses a significant threat to pigeonpea yields in India's Deccan Plateau, irrespective of cropping system, soil type, or cultivar (Pande et al. 2007; Sharma et al. 2006). Although global crop loss data for PSB is unavailable, the disease clearly has the potential for severe damage in susceptible cultivars. PSB deserves attention due to changing rainfall patterns in pigeonpea growing regions. PSB severity largely depends on disease onset in relation to crop growth and disease incidence, both influenced by weather and pathogen levels (Fig. 13.2).

13.3.3 Dry Root Rot

Rhizoctonia bataticola (also known as Macrophomina phaseolina) causes dry root rot (DRR) in chickpea as well as diseases in green gram and black gram via its pycnidial stage. This pathogen is one of the most destructive constraints on chickpea productivity and production in India, causing annual yield losses of 10–20%. R. bataticola is prevalent across the Indian subcontinent, Australia, Ethiopia, Iran, and several other countries (Nene et al. 1991). In India, it is most problematic in the central and southern regions where chickpeas are grown under rainfed conditions.

Fig. 13.2 PSB infected (a) pigeonpea field and (**b**) plants

High daytime temperatures (around 30 °C), coupled with moisture stress during the podding and flowering stages, can rapidly increase disease incidence (Gurha et al. 2003). The disease is more severe in sandy soils and emerges during the podding and flowering stages. Affected plants exhibit drooping petioles and leaflets primarily at the top. The root system loses its lateral roots, and the taproot turns black. When diseased plants are uprooted, the lower taproot often remains in the soil. Dead roots become hard, brittle, and shredded. Dark, minute sclerotial bodies appear on exposed roots or inside the stem base. Vertically splitting discolored stem regions also reveal mycelium or minute sclerotia in the pith. This pathogen is both seed and soilborne with a wide host range, especially among legumes. High daytime temperatures around 30 °C, along with moisture stress during podding and flowering, can rapidly increase disease incidence (Gurha et al. 2003; Singh et al. 2002).

13.3.4 Powdery Mildew

Powdery mildew represents a significant threat to various pulse crops such as fieldpea, rajmash, cowpea, mungbean, and urdbean. Initially, symptoms manifest on the leaves before spreading to other green parts of the plant. The disease is identifiable by the development of white powdery patches on the upper and lower leaf surfaces, as well as on tendrils, pods, and stems. These patches begin as small discolored spots, from which a powdery substance extends in all directions. As the disease progresses, extensive areas of the plant's aerial parts may become covered in these white patches. The powdery mass comprises mycelium and spores of the fungus responsible for the disease. Powdery mildew is a global issue and poses a greater threat compared to other diseases, particularly affecting crops like peas due to its widespread occurrence and impact on a larger host area. The situation exacerbates in dry conditions, with early crop varieties showing relatively less damage. The rapid reduction in yield caused by these diseases is substantial. Severe outbreaks

Fig. 13.3 Powdery mildew infected fieldpea plants parts

typically occur annually towards the end of January and in February, during the pod stage of plant growth. Munjal et al. (1963) estimated that a fully infected crop could experience yield losses ranging from 21% to 31% in pod number and 26–47% in pod weight (Fig. 13.3).

13.3.5 Rust

Rust is caused by *Uromyces* spp., an autoecious fungus that completes its entire life cycle on the same host. The disease typically occurs during the flowering/early podding stage as aecia, which can later develop into secondary aecia or uredia. The resulting aeciospores and uredospores contribute to the further spread of the disease throughout the crop season. Uredia rapidly appear a little later in the crop season, followed by the development of telia (Fig. 13.4). During the summer, the fungus survives as teliospores. Disease development is favored by high humidity, cloudy weather, and temperatures ranging from 20 °C to 22 °C. In the field, affected plants exhibit a visible dark brown or blackish appearance in the form of patches (Mishra et al. 2010). To effectively manage rust, an integrated approach is recommended. This includes controlling volunteer plants during the summer and removing infected plant debris. It is also advisable to use clean seeds without any rust contamination and treat the seeds with a suitable fungicide such as diclobutrazole. Additionally, preventive fungicide sprays of mancozeb (0.25%), Propiconazole (0.1%), and Hexaconazole (0.1%) have been recommended at the early stage of disease

Fig. 13.4 Rust infected (**a**) fieldpea, (**b**) lentil, (**c**) chickpea, and (**d**) rajamash plants

development (Mishra and Pandey 2009). However, the best means of managing rust is through the use of host plant resistance.

13.3.6 Ascochyta Blight

Ascochyta rabiei Labrousse, also known as *Didymella rabiei* (Kovachevski) Arx., is the primary cause of Ascochyta blight, a significant foliar disease in chickpeas. The disease is most prevalent in regions where the chickpea growing season overlaps with cool and moist weather conditions. It is worth noting that the occurrence of this disease has led to a substantial decrease in chickpea cultivation in western Canada, dropping from over 500,000 ha in 2001 to less than 130,000 ha in 2006. Yield losses can reach up to 100% under favorable weather conditions for disease development. Ascochyta blight is seed and stubble-borne, spreading through airborne ascospores and water-splashed conidia during the cropping season. Various regions worldwide, including Australia, Canada, Latin America, southern Europe, the United States, and West Asia, are significantly impacted by this disease. Early symptoms on chickpea plants include epinasty, loss of turgor, and water soaking caused by a toxin secreted by the pathogen. The disease primarily spreads through infected seeds from previous crops. The introduction of Ascochyta blight to countries like Australia, Canada, Iran, and the United States has been attributed to the use of infected seeds. Chickpea seeds that are lightly infected by *A. rabiei* typically show discoloration and have a very low seed weight. However, in cases of severe infection, the pathogen can be observed on the seed coat, cotyledons, and embryo axes (Dey and Singh 1993). The mycelium of *A. rabiei* grows on the seed coat, cotyledons, and emerging seedlings of the infected chickpea seeds (Luthra and Bedi 1932; Lukashevich 1958; Kaiser et al. 1983). Pycnidium serves as a structure for the survival of *A. rabiei* on debris present on the soil surface (Navas-Cortes et al. 1995; Galloway and MacLeod 2003). This structure can also survive as a teleomorph, capable of naturally colonizing infested debris and forming viable pseudopothecia on uninfected parts of the debris. In favorable conditions, pycnidium produces conidia, acting as the primary inoculum for the onset of disease in new crops. The teleomorph, on the other hand, develops on infected chickpea debris that remains in the field and under favorable conditions produces ascospores for the long-distance spread of the disease (Navas-Cortes et al. 1995, 1998).

13.3.7 Botrytis Gray Mold

The *Botrytis cinerea* Pers. ex. Fr. is the causal organism of Botrytis gray mold (BGM), a significant disease affecting chickpea crops. This disease is commonly found in regions characterized by cool, cloudy, and humid weather conditions. During the off-season, the pathogen survives on infected crop debris and seeds. The asexual stage of this necrotrophic fungus is predominant in chickpea crops. When grown on potato dextrose agar (PDA) medium, it initially appears as white and cottony, eventually turning light gray after a few days. The young hyphae of *B. cinerea* are thin, hyaline, and 8–16 μm wide, becoming septate as they age. Conidiophores are light brown, septate, and erect with ramified pseudodichotomous branching. They have slightly enlarged tips with small pointed sterigmata bearing 1–2-celled, hyaline, oval, or globose conidia forming clusters. The size of conidia ranges from 4–25 × 4–18 μm on infected chickpea plants to 4–16 × 4–10 μm on PDA (Nene and Reddy 1987; Pande et al. 2002). According to Joshi and Singh (1969), sporodochia formed by the pathogen on the host surface measure 0.5–5.0 μm in diameter and may develop into hard sclerotial masses. The sporulation of *B. cinerea* is influenced by various factors such as temperature, nutrients, and ecological conditions. Shaw and Ajrekar (1915) were the first to report BGM on chickpea in India, followed by Butler and Bisby (1931). The first epidemic of this disease, resulting in about 95% crop loss, was documented by Carranza (1965) in Argentina. Since then, numerous epidemics caused by BGM have been reported, leading to significant crop losses in various chickpea-growing countries. The disease has raised serious concerns in countries like India, Bangladesh, Nepal, Pakistan, and Australia. BGM has been reported globally on chickpea crops, with several reports from Argentina (Haware and McDonald 1992, 1993; Bakr et al. 1993; Dhar et al. 1993; Karki et al. 1993; Malik et al. 1993; Haware 1998; Pande et al. 2002; Davidson et al. 2004) showing almost complete yield loss (about 100%) under conducive environmental conditions. The disease has been reported in various countries including Canada, Chile, Colombia, Hungary, Mexico, Myanmar, Spain, Turkey, the United States, and Vietnam (Nene et al. 1984; Pande et al. 2002). *B. cinerea* fertilized sclerotia germinate during the sexual stage through the emergence of apothecia that release sexually produced ascospores (Faretra and Grindle 1992). The formation of apothecia requires either two sexually compatible isolates (MAT 1-1 and MAT 1-2) or a pseudo-homothallic isolate (MAT-1/2) (Faretra and Grindle 1992). Although the natural occurrence of sexual structure was not known, it has been reported in India under laboratory conditions (Singh 1997). An antifungal compound produced by Botrytis is known to kill yeast and is commonly used to halt fermentation before sufficient alcohol accumulation in the fermenter. The disease became epidemic in India during 1978–1979, leading to the destruction of approximately 20,000 ha of chickpea (Grewal and Laha 1983; Grewal et al. 1992; Haware 1998). Similarly, the disease occurs almost every year in Nepal with an average yield loss of about 15% (Joshi 1992). In Bangladesh, the disease was first documented in 1981 and later became devastating, resulting in the complete destruction of the crop in 1988 (Bakr and Ahmed 1992). Pod yield is influenced by disease onset, its growth, disease severity, weather conditions, and the level of pathogen inoculums.

13.3.8 Stem Rot/White Rot

Sclerotinia sclerotiorum, also known as Lib. de Bary, is a necrotrophic fungal pathogen that is widely distributed across the globe. It has a diverse range of hosts, including beans, mustard, potatoes, sunflowers, and more. This pathogen infects various broadleaf crops and weeds. Its entry into fields is often facilitated through contaminated seed lots or equipment. The fungus is able to survive in the environment as a mass of fungal hyphae, which is protected by a resilient black rind called sclerotia. Under favorable conditions, such as high humidity levels and the onset of flowering when the crop canopy closes, an apothecium structure forms, resembling an inverted mushroom. This structure produces ascospores that subsequently infect the flowers. Additionally, the hyphae can directly penetrate plant tissues, particularly when there is contact between the plant and soil. The fungus causes tissue death and fills the stem with white hyphae and sclerotia, enabling its long-term survival in the soil for many years (Fig. 13.5).

This particular fungus is a soil-borne fungus that can attack plants at any stage of growth, including young seedlings, mature plants, and even their harvested products. It is a significant disease that affects various leguminous crops such as fieldpea, chickpea, pigeonpea, lentil, soybean, fababean, and alfalfa. The impact of this disease can be devastating, causing almost 100% losses to the crops and resulting in millions of dollars lost annually worldwide. The survival structure of this disease during the off-season is sclerotia, which can remain viable for several years and germinate to produce apothecia. These apothecia release a large number of ascospores that can spread over long distances, serving as the primary source of infection. Managing this disease is challenging due to these characteristics. Many alternatives are being explored for its control, as the use of pesticides can have harmful effects. In India, the disease is regularly reported in certain districts of Himachal Pradesh, leading to significant yield losses. The loss in yield can range from 21% to 47% during the pod-bearing stage of the crop. Limited resistance may be available in some dry pea and lentil cultivars, but no resistance has been observed among chickpea cultivars. Winter-sown lentils are more susceptible to white mold compared to spring lentils due to prolonged wet and cool spring weather conditions.

Fig. 13.5 Stem rot/White rot symptoms on pulses

13.3.9 Collar Rot (*Sclerotium rolfsii*)

Pathogen survives in soil in the form of sclerotia.

13.3.10 Symptoms

Diseases can be observed during the seedling phase, which occurs within 6 weeks after sowing. Impacted seedlings exhibit a yellowing appearance, and young ones may even collapse. When uprooted at the collar region, the seedlings display signs of rotting. The decayed area is covered with whitish mycelial strands. Additionally, a white mycelial coating can be observed on the tap root of completely dried seedlings, and mustard seed-like sclerotia can be seen at the site of infection. Infected seedlings typically appear in small, scattered patches within the field.

13.3.11 Management Strategies

13.3.11.1 Cultural Practices

Healthy seeds, proper farming practices, and a 3-year crop rotation are suggested for disease management. Disease-free seeds can be obtained through surface irrigation in semi-arid areas with high temperatures and low humidity to prevent fungal infection, or through a pedigreed seed program involving isolated seed plots inspected for disease-free seeds. It is important to sow different pulses/legume crops at the optimal time during the rabi, kharif, and spring seasons to achieve maximum yield and minimize disease incidence.

13.3.11.2 Chemical Approaches

Chemical control of wilt disease is not a viable or cost-effective option due to the fact that the pathogen primarily resides in the soil and its chlamydospores can survive for up to 6 years (Haware et al. 1996). Therefore, the most practical and economical method for managing wilt is through the use of host resistance (Govil and Rana 1994). Various systemic and nonsystemic fungicides, such as captan, thiram, and carbendazim, have been utilized by different researchers to reduce seed infections of various diseases (Sindhan and Bose 1981; Trutman et al. 1992). For the management of wilt, foliar sprays of carbendazim, benomyl, zineb, maneb, and mancozeb have been recommended (Chakrabarty and Shyam 1988). Benlate and Bavistin remained in the plant for 15–20 days after seed treatment (Sindhan and Bose 1981). EBI fungicides like triadimefon and triforine were effective against the pathogen under laboratory conditions but were found to be phytotoxic to *P. vulgaris* seedlings when used as seed treatment (Chakrabarty and Shyam 1988). Effective management of this disease was achieved by using carbendazim alone or in combination with thiram as seed treatment, followed by two or three sprays of mancozeb at 45, 60, and 75 days after sowing (Sharma and Sugha 1995). A combination of seed treatment with tebuconazole (0.1%) and foliar sprays of tricyclazole (0.03%) reduced the severity of the disease on both leaves and pods, while the highest seed yield was observed with seed treatment and foliar sprays of carbendazim (Gupta et al. 2000). The utilization of systemic fungicides on a continuous basis has been proven to be linked with the emergence of resistant biotypes (Tu and McNaughton 1980). In a study conducted by Maringoni and Barros (2002), it was observed that all strains of *C. lindemuthianum* collected from *P. vulgaris* fields in Paranapanema Valley, Brazil exhibited insensitivity to benomyl, carbendazim, and thiophanate methyl, indicating the presence of cross-resistance to various benzimidazole fungicides. However, all strains remained sensitive to chlorothalonil. More recently, de Oliveira et al. (2003) reported that strobilurin fungicides such as azoxystrobin and trifloxystrobin effectively controlled this disease, while treatments involving carbendazim and fluquinconazole without rotation were found to be the least efficient. Nevertheless, a combination of systemic fungicides with nonsystemics, along with the rotation of fungicides, can prove to be an effective strategy for managing this disease, rather than relying solely on the continuous use of a single systemic fungicide.

13.3.11.3 Biological Approaches

Various strategies were implemented to combat diseases in pulses by introducing bioagents that thrive in natural substrates within the soil. Soybean bacterial endophytes such as *Bacillus* and *Penibacillus* exhibited their antifungal properties against key soil-borne phytopathogens like *Rhizoctonia bataticola*, *Macrophomina phaseolina*, *Fusarium udum*, and *Sclerotium rolfsii* that commonly infect pulse crops. *Trichoderma* species, among the antagonistic fungi, have been extensively researched as bioagents in the pulse ecosystem. They have proven effective in managing wilts, root rots, collar rot, and stem rot diseases caused by various *Fusarium* spp., *Rhizoctonia solani*, *R. bataticola*, *Sclerotium rolfsii*, *Sclerotinia sclerotiorum*, *Ascochyta*, *Cercospora*, *Alternaria* spp., *Phytophthora* spp., and *Pythium* spp. in pulses and other field crops. Additionally, fungi like *Aspergillus candidus*, *A. niger*, *Paecilomyces lilacinus*, *Glomus mosseae*, *Verticillium chlamydosporum*, *Absidia cylindrospora*, *Coniothyrium minitans*, *Talaromyces flavus*, and *Trichothecium roseum* have been identified to suppress fungal diseases in pulses. Chaudhary et al. (2004) reported that numerous biocontrol agents have effectively managed various phytopathogens in pulses (Table 13.2).

13.4 Conclusion

Previous studies primarily concentrated on resistant sources and chemical management of a limited number of diseases. Presently, the main focus is on recognizing, assessing, and integrating location-specific elements of Integrated Disease Management (IDM). IDM packages for food legumes have been effectively improved and confirmed in collaboration with stakeholders and end users. Despite the creation of various IDM modules to combat economically significant diseases of food legumes, a significant knowledge gap remains between scientists and farmers, especially in developing nations. This is a result of the ever-changing agroecosystems, socio-economic conditions, and the dynamic nature of host-pathogen interactions/co-evolution, which necessitate the enhancement of IDM strategies. Consequently, there is an urgent requirement for continuous improvement, validation, transfer, and adoption of IDM modules. Future endeavors will concentrate on broadening IDM technology by raising farmers' awareness, as well as emphasizing the importance of the quality of crop residues from food legumes used as a crucial component of the existing mixed crop-livestock system. Furthermore, innovative/improved IDM approaches must be established to combat emerging pathogens like Phytophthora blight in pigeonpea and new pathotypes of ascochyta blight in chickpea. Experience from the past few decades has demonstrated that plant viruses cannot be managed solely through preventive measures.

The complex ecology of numerous viruses impacting legume crops, as well as crops in general, necessitates a long-term, sustainable strategy that is economically viable for resource-poor farmers and environmentally friendly. The most effective approach involves utilizing all available measures in an integrated manner. Integrated management has always been and will continue to be the key to controlling

Table 13.2 Biocontrol agents effective against major pathogens of pulse crops

Crop	Pathogens	Effective biocontrol agents
Chickpea	*F. oxysporum* f. sp. *ciceri*	*Trichoderma harzianum, T. viride, T. koningii, T. virens, Gliocladium virens, Pseudomonas* sp, *Bacillus* sp.
	Sclerotium rolfsii	*T. asperellum, T. afroharzianum*
	Rhizoctonia bataticola	
	Rhizoctonia solani	
	Ascochyta rabei	
Pigeonpea	*Fusarium udum*	*T. harzianum, T. viride, G. virens, Bacillus subtilis, T. koningii, T. hamatum, A. niger, Penicillium* sp, *Glomus mosseae, Verticillium chlamydosporum, Glomus mosseae, P, fluorescens*
	Phytophthora drechsleri f. sp. *cajani*	*T. asperellum, T. afroharzianum*
Lentil	*Fusarium oxysporum* f. sp. *Lentis*	*T. viride, T. harzianum, G. virens, Pseudomonas fluorescens*
	Macrophomina phaseolina	*Rhizobium leguminoserum*
	Macrophomina phaseolina	*T. asperellum, T. afroharzianum*
	Ascochyta fabae f. sp. *lentis*	
Fieldpea	*Uromyces fabae*	*T. harzianum, T. viride, T. virens, T. hamatum*
	Erysiphe polygoni DC	*Rhizobium leguminoserum*
	Rhizoctonia solani	
Mungbean/ Urdbean	*Macrophomina phaseolina*	*T. viride, T. harzianum, T. konninghii, T. longibrachiatum, P. fluorescens, Bacillus subtilis, A. niger* and *Pencillium citrinum, T. asperellum, T. afroharzianum*
	Cercospora spp.	

Sources: Chaudhary et al. (2004), Mishra et al. (2018b, 2020, 2021, 2023)

virus-related diseases. Future IDM strategies will take into consideration the impact of climate variability on pathogen dynamics, which contributes to the emergence and resurgence of new strains/pathogens. This underscores the need for refining disease control methods accordingly. High-performance resistance (HPR), fungicides, natural plant products, biofungicides, botanicals, and agronomic practices will remain as potential options for IDM. Furthermore, future strategies will integrate the latest advancements in identifying biocontrol agents and products through biotechnological/breeding interventions to develop disease-resistant cultivars. Simultaneously, there is a growing need to focus on utilizing information technology for disease modeling, creating decision support systems, and leveraging remote sensing to scale up and disseminate IDM technologies.

References

Bakr MA, Ahmed F (1992) Botrytis gray mold of chickpea in Bangladesh. In: Haware MP, Faris DG, Gowda CLL (eds) Botrytis gray mold of chickpea. Summary proceedings of the BARI/ ICRISAT Working Group Meeting. ICRISAT, Patancheru, pp 10–12

Bakr MA, Rahman MM, Ahmed F, Kumar J (1993) Progress in management of Botrytis gray mold of chickpea in Bangladesh. In: Haware MP, Gowda CLL, McDonald D (eds) Recent advances in research in Botrytis gray mold of chickpea. ICRISAT, Patancheru, pp 17–19

Butler EJ, Bisby GR (1931) The Fungi of India. Central Publication Branch, Calcutta

Carranza JM (1965) Wilt of chickpea (*Cicer arietinum* L.) caused by *B. cinerea* (in Spanish). Rev Fac Agron Univ Nacional Plata 41:135–138

Chakrabarty PK, Shyam KR (1988) Evaluation of systemic fungitoxicants against *Colletorichum lindemuthianum*, the incitant of bean anthracnose. Indian J Plant Pathol 6(1):67–70

Chaudhary RG, Sharma N, Prajapati RK (2004) Biological control of soil borne diseases: an updates in pulse crops. In: Ahamad S, Narain U (eds) Ecofriendly management of plant diseases. Daya Publishing House, New Delhi, pp 178–200

Davidson JA, Pande S, Bretag TW, Lindbeck KD, Kishore GK (2004) Biology and management of Botrytis spp. in legume crops. In: Elad Y, Williamson B, Tudzynski P, Delen N (eds) Botrytis: biology, pathology and control. Kluwer Academic Publishers, Amsterdam, pp 295–318

de Oliveira MVM, Vargas FM, Sanchez LMB, Paris W, Frizz A, Haygert IP, Montagner D, Weber A, Cerdotes L (2003) Ruminal degradability and intestinal digestibility of feeds by means of associated technical in situ and mobile nylon bag. Rev. Bras. Zootec. 32(6):2023–2031

Dey SK, Singh G (1993) Resistance to Ascochyta blight in chickpea—genetic basis. Euphytica 68:47–153

Dhar V, Rathi YPS, Tripathi HS, Pal M, Singh G, Haware MP, Upadhyay JP (1993) Multilocational screening for Botrytis gray mold resistance in India. In: Haware MP, Gowda CLL, McDonald D (eds) Recent advances in research on Botrytis gray mold of chickpea. International Crops Research Institute for the Semi-Arid Tropics, Patancheru, pp 14–16

Dhar MK, Kaul S, Sareen S, Koul AK (2005) Plantago ovata: genetic diversity, cultivation, utilization and chemistry. Plant Genetic Resources 3(2005):252–263

Faretra F, Grindle M (1992) Genetic studies of Botryotinia fuckeliana (Botrytis cinerea). In: Verhoeff K, Malathrakis NE, Williamson B (eds) Recent advances in Botrytis research. Proceedings of the 10th international Botrytis symposium. Heraklion, Crete, Greece. Pudoc Scientific Publishers, Wageningen, pp 7–16

Galloway J, MacLeod WJ (2003) *Didymella rabiei*, the teleomorph of *Ascocyta rabiei*, found on chickpea stubble in Western Australia. Australas Plant Pathol 32:127–128

Govil JN, Rana BS (1994) Stability of host plant resistance to wilt (*Fusarium oxysporum* f.sp. *ciceri*) in chickpea. Int J Trop Dis 2(1):55–60

Grewal JS, Laha SK (1983) Chemical control of Botrytis blight of chickpea. Indian J Phytopathol 36:516–520

Grewal D, Baker J, Levy M (1992) "An Experimental Approach to Making Retail Store Environment Decisions," Journal of Retailing, Vol. 68, No. 4, 1992, pp. 445–460.

Gupta SK, Mathew KA, Shyam KR, Gupta DK (2000) Role of fungicides in management of anthracnose of French bean. In: Proceedings of Indian Phytopathological Society-Golden Jubilee International Conference on "Integrated plant disease management for sustainable agriculture", vol II. Indian Phytopathological Society, IARI, New Delhi, pp 705–707

Gurha SN, Singh G, Sharma YR (2003) Diseases of chickpea and their management. In: Kumar S, Singh NB (eds) Masood Ali. Chickpea Research in India. Army Printing Press, Lucknow, India, pp 195–227

Halila HM, Strange R (1996) Identification of the causal agent of wilt of chickpea in Tunisia as Fusarium oxysporum f. sp. ciceris race 0. Phytopathol Mediterr 35:67–74

Haware MP (1998) Diseases of chickpea. In: Allen DJ, Lenne JM (eds) The pathology of food and pasture legumes. ICARDA, CAB International, Wallingford, pp 473–516

Haware, M.P. and McDonald, D. (1992) In: Botrytis. Gray Mold of Chickpea: Summary Proceedings of the. BARIIICRISAT Working Group Meeting. 4-8 March. 1991

Haware MP, McDonald D (1993) Chickpea and pigeonpea diseases: problems and progress. In: Pests and Pest Management in India the Changing Scenario: Lead Papers of the National Seminar on Changing Scenario in Pests and Pest Management in India , 31st Jan. - 1st, Feb. 1992) Hyderabad. India

Haware MP, Nene YL (1980) Influence of wilt at different stages on the yield lossin chickpea. Trop Grain Legume Bull 19:38–40

Haware MP, Nene YL, Natarajan M (1996) The survival of Fusarium oxysporum f. sp. ciceri in the soil in the absence of chickpea. Phytopathologia Mediterranea 35:9–12

Joshi S (1992) Botrytis gray mold of chickpea in Nepal. In: Haware MP, Faris DG, Gowda CLL (eds) Botrytis gray mold of chickpea. ICRISAT, Patancheru, pp 12–13

Joshi MM, Singh RS (1969) A Botrytis gray mould of gram. Indian Phytopathol 22:125–128

Kaiser WJ (1983) et al. Etiology and control of seed decay and preemergence damping-off of chickpea by Pythium ultimum Plant Disease

Kannaiyan J, Nene YL (1981) Influence o f wilt at different growth stages on yield loss in pigeonpea. Trop. Pest Manage. 27:141

Kannaiyan J, Nene YL, Reddy MV, Ryan JG, Raju TN (1984) Prevalence of pigeonpea diseases and associated crop losses in India, Africa, and the Americas. Trop Pest Manage 30:62–71

Karki PB, Joshi S, Chaudhary G, Chaudhary RN (1993) Studies on Botrytis gray mold of chickpea in Nepal. In: Haware MP, Gowda CLL, McDonald D (eds) Recent advances in research on Botrytis gray mold of chickpea. International Crops Research Institute for the Semi-Arid Tropics, Patancheru, pp 11–13

Lukashevich AI (1958) Control measures against ascochytosis of chickpea ... Sci, Rep. Agr. Res. Inst., New Delhi, 1939-40:94–101

Luthra JC, Bedi KS (1932) Some preliminary studies on gram blight with reference to its cause and mode of perennation. Indian Journal of Agricultural Science 2:499–515

Malik BA, Iqbal SM, Bashir M (1993) Present status and future research outlook for Botrytis gray mold—a potential threat to chickpea in Pakistan. In: Haware MP, Gowda CLL, McDonald D (eds) Recent advances in research in Botrytis gray mold of chickpea. International Crops Research Institute for the Semi-Arid Tropics, Patancheru, pp 19–20

Maringoni AC, Barros EM (2002) Occurrence of isolates of Colletotrichum lindemuthianum resistant to benzimidazole fungicides. Summa Phytopathologica 28(2):197:199

Mishra RK, Pandey PK (2009) Efficacy of fungicides against pea rust caused by *Uromyces fabae*. Ann Plant Protect Sci 2:54–55

Mishra RK, Mishra KK, Pandey KK, Singh UP (2010) Biochemical changes in Pea infected by *Uromyces fabae*. Indian Phytopathol 65(2):125–128

Mishra RK, Bohra A, Naimuddin, Kumar K, Sujayanand GK, Saabale PR, Naik SJ, Sarma BK, Kumar D, Mishra M, Srivastava DK, Singh NP (2018b) Utilization of biopesticides as sustainable solutions for management of pests in legume crops: achievements and prospects. Egypt J Biol Pest Control 28:3. https://doi.org/10.1186/s41938-017-0004-1

Mishra RK, Pandey S, Mishra M, Rathore US, Naimuddin, Mohammad A, Krishna K, Singh B (2020) Molecular identification, enzyme profiling and biocontrol potential of *Trichoderma* isolates against wilt disease of major pulse crops. J Food Legumes 33:50–54

Mishra M, Mishra R, Pandey S, Rathore US, Pandey M (2021) Recent Advances in Bio-Intensive Management (BM) of Major Diseases of Pigeon Pea in India. https://doi.org/10.1201/9781003187837-11

Mishra RK, Pandey S, Hazra KK, Mishra M, Naik S, Bohra A, Parihar AK, Rathore US, Naimuddin, Kumar K, Singh B, Singh NP (2023) Biocontrol efficacy and induced defense mechanisms of indigenous Trichoderma strains against Fusarium wilt [*F. udum* (Butler)] in pigeonpea. Physiol Mol Plant Pathol 127:102122

Munjal RL, Chenulu VV, Hora IS (1963) Assessment of losses due to powdery muldew *Erysiphe polygoni* on pea. Indian Phytopath 16:260–267

Navas-Cortes JA, Trapero-Casas A, Jimenez-Diaz RM (1995) Survival of *Didymella rabiei* in chickpea straw debris in Spain. Plant Pathol 44:332–339

Navas-Cortes JA, Trapero-Casas A, Jimenez-Diaz RM (1998) Influence of relative humidity and temperature on development of Didymella rabiei on chickpea debris. Plant Pathol 47:57–66

Nene YL, Sheila VK, Sharma SB (1984) A world list of chickpea (Cicer arietinum) and pigeonpea (*Cajanus cajan* (L.) Millsp.) pathogens. ICRISAT Pulse Pathology Progress Report 32, Patancheru

Nene YL, Reddy MV (1987) Chickpea diseases and their control. In: Saxena MC, Singh KB (eds) The Chickpea. CAB International, Oxon, UK, pp 233–270

Nene YL, Reddy MV, Haware MP, Ghanekar AM, Amin KS (1991) Field diagnosis of chickpea diseases and their control, Information Bulletin no. 28. International Crops Research Institute for the Semi-Arid Tropics, Patancheru

Nene YL, Sheila YK, Sharma SB (1996) A world list of chickpea and pigeonpea pathogens, 5th edn. International Crops Research Institute for the Semi-Arid Tropics, Patancheru

Pal M (1970) et al. Resistance of pigeonpea to Phytophthora blight -1 total phenolic content Indian Phytopathol

Pande S, Singh G, Rao JN, Bakr MA, Chaurasia PCP, Joshi S, Johansen C, Singh SD, Kumar J, Rahman MM, Gowda CLL (2002) Integrated management of botrytis gray mold of chickpea, Information Bulletin No. 61. ICRISAT, Patancheru

Pande S, Sharma M, Rao JN (2007) Etiology, biology and management of diseases of ood legumes. In: Kharkwal MC (ed) Food legumes for nutritional security and sustainable agriculture. Proceedings of the Fourth International Food Legumes Research Conference (IFLRC-IV), 18–22 October 2005, New Delhi, pp 1–15

Sharma PN, Sugha SK (1995) Management of bean anthracnose through chemicals. Indian Phytopathol 48(3):304–307

Sharma M, Pande S, Pathak M, Narayana Rao J, Anil Kumar P, Madhusudan Reddy D, Benagi VI, Mahalinga DM, Zhote KK, Karanjkar PN, Eksinghe BS (2006) Prevalence of phytophthora blight of pigeonpea in the Deccan Plateau of India. Plant Pathol J 22:309–313

Sharma M, Ghosh R, Telangre R, Rathore A, Saifulla M, Mahalinga DM, Saxena DR, Jain YK (2016) Environmental influences on Pigeonpea-Fusarium udum interactions and stability of genotypes to Fusarium wilt. Front. Plant Sci. 7:253

Shaw FJF, Ajrekar SL (1915) The genus Rhizoctonia in India. Mem Dept Agric India Bot Ser 7:117

Sindhan GS, Bose SK (1981) Evaluation of fungicides against anthracnose of French bean caused by *Colletotrichum lindemuthianum*. Indian Phytopath 34(3):325–329

Singh G (1997) Epidemiology of botrytis gray mold of chickpea. In: Haware MP, Lenne JM, Gowda CLL (eds) Recent advances in research on botrytis gray mold of chickpea. ICRISAT, Patancheru, pp 47–50

Singh KB, Dahiya BS (1973) Breeding for wilt resistance in chickpea. In: Symposiumon wilt problem and breeding for wilt resistance in Bengal gram. Indian Research Institute, New Delhi, pp 13–14

Singh BB, Ehlers JD, Sharma B, Freire Filho FR (2002) Recent progress in cowpea breeding. In: Fatokun CA, Tarawali SA, Singh BB, Kormawa PM, Tamo M (eds) Proceedings, the World Cowpea Conference III. Challenges and opportunities for enhancing sustainable cowpea production, International Institute of Tropical Agriculture (IITA), Ibadan, pp 22–40

Trutman O, Paul KB, Cishabayo (1992) Seed treatments increase yield of farmer varietal field bean mixtures in the Central African highlands through multiple disease and bean fly control. Crop Prot 11(5):458–464

Tu JC, McNaughton ME (1980) Isolation and characterization of benomyl-resistant biotypes of the delta race of Colletotrichum lindemuthianum. Can J Plant Sci 60:585–589

Wearing AH, Birch J (1988) Root rot of Pigeonpea. Int. Pigeonpea Newslett. 8:13

Williams FJ, Grewal JS, Amin KS (1968) Serious and new diseases of pulse crops in India in 1966. Plant Dis. Reptr. 52:300e304

Williams FJ, Amin KS, Baldev B (1975) Phytophthora stem blight of pigeonpea. Phytopathology 65:1029–1030

Advances in Nematode Diseases Management in the Field Crops

14

Pankaj, Vishal Singh Somvanshi, and Mansi Machal

Abstract

Nematoda is a phylum that includes extremely small unsegmented roundworms that are plant parasitic. These parasites may be small in size, but they have developed amazing defense mechanisms to take advantage of plants. A small percentage of nematode species have developed specialized adaptations to feed on plant tissues, which can result in a variety of diseases that can have disastrous effects on agricultural and natural ecosystems. Nevertheless, the great majority of nematode species are benign or even helpful to ecosystems. Numerous plant diseases may be caused by nematodes, disrupting natural ecosystems and resulting in substantial losses in the agricultural sector. Nematodes pose a serious threat to plant health and agriculture, necessitating integrated management plans to reduce their effects and preserve ecosystem and crop output.

Keywords

Nematodes · Diseases · Field crops · Management

Pankaj (✉) · V. S. Somvanshi · M. Machal
Division of Nematology, ICAR-Indian Agricultural Research Institute, New Delhi, Delhi, India

V. K. Singh et al. (eds.), *Diseases of Field Crops: Diagnostics and Management*,
https://doi.org/10.1007/978-981-97-6160-9_14

14.1 Introduction

The phylum Nematoda, commonly known as Nemata, falls within the superphylum Ecdysozoa, which comprises moulting animals. It's believed that the earliest nematodes appeared in marine environments around 600 to 550 million years ago during the Cambrian period. Nematodes are microscopic roundworms that inhabit a variety of environments. Nematodes are incredibly adaptable, living in freshwater, marine, and terrestrial environments. Despite their prevalence as the most abundant multicellular organisms on Earth (with an estimated 0.5–10 million species), they are often inconspicuous due to their minuscule size (ranging from 0.25 to 12 mm for plant parasites and free-living nematodes, and 5 cm to 10 m for animal parasites), as well as their typically colorless bodies.

Nematodes are well-known pathogens responsible for a range of diseases affecting humans, animals, and plants. In humans, infections commonly include pinworm (causing itching in the anal area, predominantly affecting children), giant worm (an intestinal parasite leading to anemia and intestinal blockage), hookworms (causing issues like anemia), filarial worm (responsible for conditions like elephantiasis and river blindness), and guinea worm (causing filariasis). Nematodes also impact livestock, affecting animals such as sheep, dogs, and chickens. They can parasitize insects like grasshoppers and mosquitoes, with some being entomopathogenic and beneficial for controlling insect pests in crop fields. In the realm of plant parasitic nematodes (PPNs), economically significant species include root-knot, cyst, seed gall, reniform, citrus, burrowing, stem and bulb, foliar, lesion, and pinewood nematodes. Nematode infestations lead to approximately 14% of global crop losses, amounting to approximately $100 billion annually. It's worth noting that PPNs make up only a small portion (about 15%) of the total described nematode species (around 30,000 species); the vast majority are free-living, nonparasitic organisms that play essential roles in terrestrial and sediment food chains. Plant-parasitic nematodes are of great economic importance. However, because the majority of them are soil-dwelling, they provide one of the most challenging pest issues to identify, demonstrate, and eradicate (Stirling et al. 1998). With the ability to penetrate host plant tissue through the use of a stylet, at least 2500 species of plant-parasitic nematodes have been identified. Free-living nematodes represent approximately 40% of the described species and mainly feed on bacteria, fungi, algae, protozoa, diatoms, and other nematodes. Animal parasites, affecting vertebrates and invertebrates, constitute 44% of the described species, while plant parasites make up the remaining 15%. According to Sasser (1987), plant-parasitic nematodes cause, on average, 12.3% losses annually in 40 major crops at a global level, with losses higher in developing countries (14.6%) than in developed nations (8.8%). According to a comprehensive worldwide assessment, the yearly economic agricultural output losses resulting from plant-parasitic nematodes in principal crops are estimated to be worth USD 173 billion (Elling 2013). According to Kumar et al. (2020), *Heterodera avenae* causes 28.5% yield losses in wheat, *Meloidogyne graminicola* causes 16% yield losses in rice, and *Heterodera zeae* causes 12% yield losses in maize in India.

14.2 Types of Plant Parasitism in Nematode

In general, plant parasitic nematodes may be divided into three groups according to their parasitic behaviors on plants: ectoparasites, semiendoparasites, and endoparasites. The most basic kind of nematodes are ectoparasitic ones, which obtain their food externally by penetrating the cells of root tips and hairs with their stylet, which functions like a spear. A large number of them are specialized sedentary (sessile) ectoparasites (*Hemicycliophora*, *Paratylenchus*, *Xiphinema*, *Longidorus*, etc.) that have evolved to feed sedentarily on deeper cells by penetrating their long stylet for a longer period of time. These grazers and browsers include migratory species like *Tylenchorhynchus*, *Belonolaimus*, *Trichodorus*, *Paratrichodorus*, and many others that move frequently over the root surface. At the feeding sites, they occasionally modify the cells they feed. Endoparasites are nematodes that migrate into the tissues of their hosts. Endoparasites are further divided into two categories: sedentary and migratory. While sedentary endoparasites, such as *Meloidogyne*, *Heterodera*, *Globodera*, etc., are the most advanced parasites among all plant parasitic nematodes that cause economic damage to the crops, migratory endoparasites, such as *Pratylenchus*, *Radopholus*, *Hirschmanniella*, *Bursaphelenchus*, etc., invade roots, stems, leaves, rhizomes, bulbs, etc. and cause extensive damage during their migration through the cortex and other tissues, leading to the formation of cavities and lesions on the necrotic tissues. Sedentary endoparasites invade plant tissues, and their stylet-mediated oesophageal gland cell secretion causes the most complex molecular alterations necessary for them to locate their feeding sites. Semi-endoparasites include several relatively highly adapted plant parasitic nematodes, such as *Tylenchulus* (citrus nematode) and *Rotylenchulus* (commonly known as reniform nematode). They feed by embedding their head and anterior end in the tissue of their hosts, changing their cells there (Table 14.1).

Table 14.1 The diseases caused by nematodes in field crops can be studied under following heads crop-wise

Sr. no.	Crop	Common name	Scientific name
1.	Rice	Rice root nematode	*Hirschmanniella oryzae, H. Mucronata*
		Rice root-knot nematode	*Meloidogyne graminicola*
		White tip nematode	*Aphelenchoides besseyi*
		Rice stem nematode	*Ditylenchus angustus*
		Rice cyst nematode	*Heterodera oryzicola*
2.	Wheat	Seed-gall nematode/ Earcockle disease	*Anguina tritici*
		Tundu disease	*Anguina tritici + Clavibacter tritici*
	Wheat and barley	Cereal cyst nematode/ Molya disease	*Heterodera avenae*
3.	Sorghum, maize and pearl millet	Sorghum cyst nematode	*Heterodera sorghi*
		Maize cyst nematode	*Heterodera zeae*
		Lesion nematode	*Pratylenchus thornei*
		Lance nematode	*Hoplolaimus indicus*
		Stunt nematode	*Tylenchorhynchus vulgaris, T. Brevilineatus, Merlinius indicus*

14.3 Rice

Due to early damage in certain regions and challenges with accurate detection and diagnosis, a few nematode issues that were discovered as significant in rice got less attention. Many nematodes that feed on above-ground plant parts, such as the White tip nematode (*Aphelenchoides besseyi*), Rice stem nematode (*Ditylenchus angustus*), and below-ground feeding nematodes, like Rice root nematode (*Hirschmanniella oryzae*), Rice Root-knot nematode (*Meloidogyne graminicola*), and Rice Cyst nematode (*Heterodera oryzae* and *H. oryzicola*), are known to infect rice. Kumar et al. (2020) estimated that PPNs account for 21.3% of all agricultural losses annually, or Rs. 102,039.79 million (1.58 billion USD). The anticipated losses from 19 different horticultural crops were Rs. 50,224.98 million, of which Rs. 23,272.32 million came from rice alone. In India, it is cultivated on 43.01 million hectares (Mha) and yields around 112.46 million tonnes, according to Pathak et al. (2020). It is well-known that *Heterodera oryzicola* may infect upland rice, resulting in decreased root development and subsequent decrease in production. Upland rice and regions where other cereals are produced in a rice-based cropping system are also affected by *Pratylenchus* spp. and *Hoplolaimus indicus* (Pankaj et al., 2012a). The rice white-tip disease is caused by *Aphelenchoides besseyi.*

14.3.1 Rice-Root Nematode, *Hircshmaniella oryzae*

Currently, there are roughly 35 species in the genus *Hirschmanniella.* Southern USA, China, Japan, and other tropical and subtropical parts of the world are home to almost half of them, which have been reported on rice. Less than ten species of non-rice species may be found in the northern temperate zone and in the same region. Each individual is an obligatory migratory endoparasite of roots, with an affinity for graminaceous roots and adaptation to a wide variety of humid conditions (Loof 1991). Furthermore, weeds and other dicotyledonous plants like cotton serve as hosts. Additionally, aquatic plant roots and leaves, as well as soil near bonsai trees, have been found to have *Hirschmanniella.* It is typical to find many *Hirschmanniella* species on rice.

(a) **Symptoms of rice-root nematode:** The symptoms above ground are not well defined, and they are readily mistaken for a nutritional deficiency. Other symptoms like stunted growth, poor tillering, and reduction in grain weight and ear-head can be observed. As the early root necrosis progresses, the entire root system becomes smaller and browns by the time the crop reaches maturity.
(b) **Life cycle:** Male and female nematodes both penetrate the roots through the epidermis. One nematode's entry creates an opening that other nematodes utilize, resulting in an accumulation of nematodes in the affected region. After entering at the root tip, the nematodes infect the cortical cells. The life cycle takes 30 days to complete under ideal circumstances.

(c) **Spread and survival:** Nematodes may be found in weeds such as *Marsilia, Echinochloa, Cyperus*, and *Monochloria*. Significant nematodes that can infest the next crop can be found in the rice stubbles left behind after harvest.

(d) **Management:**
- Rotation with nonhost crops such as groundnut in kharif or wheat, linseed, potato, cauliflower, mustard, berseem, and chickpea in rabi.
- Extremely arid summer plowings.
- Maintaining soil fertility through fertilizer application.
- Adhering to standard procedures for field sanitation.
- Eliminating weed hosts and wild rice from the field.
- Considering that eggs and young might hibernate in dead roots, this nematode is able to endure the desiccation of rice fields by going into an anhydrobiotic condition, which keeps the nematode populations dormant until the rains arrive. As a result, control strategies have not shown much promise. Also, because rice does not have a high enough economic value in developing nations, some control strategies, including the use of nematicides, are not economically viable.
- In the long run, weed control and rice's natural resistances may prove to be more successful control strategies than varietal resistance; nevertheless, varietal resistance has not been well studied and so remains ineffective.
- Even in severely infected plants, proper cultural techniques can produce an adequate plant stand that enables rice to tolerate some RRN.

14.3.2 Rice Root-Knot Nematode, *Meloidogyne graminicola*

In well-drained soils, *M. graminicola* is a severe pest of upland rice and nurseries worldwide. Patnaik (1969) made the initial observation of this nematode in India. The root-knot nematode, *M. graminicola*, has been identified as a significant pest of rice and is common in the eastern, northeastern, and southern states of India, among other pests and diseases affecting wheat and rice crops. In North-Indian plain zones, such as Jammu, Punjab, HP, Haryana, Delhi, UP, and the Andaman Islands, there have been isolated instances of severe infection of rice nurseries and transplanted crops (Pankaj et al. 2006a, b; Singh and Singh 2009). The life cycle of *M. graminicola* takes 26–51 days to complete, depending on the time of year, and a temperature range of 22–29 °C has been reported to be optimal for the nematode's predominance (Rao and Israel 1973). According to Gaur (1993), several monocot weeds in Delhi, Uttar Pradesh, and Haryana are infected with *M. triticoryzae*, a type of root-knot nematode that affects both rice and wheat.

(a) **Symptoms of rice root-knot nematode:** Rice plants in infected fields have patches of stunted and chlorotic plants, exhibiting reduced vigor, leaf curling, and yellowing. Leaves have less chlorophyll (Swain and Prasad 1988). *Meloidogyne graminicola* induces terminal, spiral, or horseshoe-shaped galls. There are few effective tillers, poor tillering, and a poor crop stand

(Figs. 14.1, 14.2, and 14.3). When grains are insufficiently packed, the size and quantity of spikes decrease, leading to significant yield losses.

(b) **Life cycle:** The life cycle lasts between 26 and 51 days, depending on environmental conditions. The J_2 enters the root in the zone of elongation and migrates

Fig. 14.1 Infested rice plants left in the field after the crop harvest

Fig. 14.2 Patches of stunted and chlorotic rice plants in the field at early stage (left) and late stage (right)

Fig. 14.3 Rice roots showing terminal and horse-shoe shaped galls caused by *Meloidogyne graminicola*

intercellularly. At the feeding location, giant cells are produced, and these cells negatively impact the xylem vessels. When fully grown, the female nematode deposits 200–300 eggs in a gelatinous matrix within the rice roots. In upland conditions, infectious second-stage juveniles of *M. graminicola* infiltrate rice roots just behind the root tips (Rao and Israel 1973). The females lay their eggs mostly in the cortex after developing inside the root system (Roy 1976). Depending on the season of the year, *M. graminicola* takes 26–51 days to complete its life cycle. In India, a temperature range of 22–29 °C is ideal for nematode predominance (Rao and Israel 1973; Singh et al. 2003). However, in Bangladesh, this temperature range is too short for the nematode, with a life cycle of fewer than 19 days (Bridge and Page 1982).

(c) **Management:**
- Soil solarization of the area for nursery beds for 3 weeks in summer.
- Good puddling before transplanting.
- Crop rotation with jute, mustard, and chickpea.
- Using nematode-resistant varieties.
- When used appropriately, biological control utilizing nematode parasites and microorganisms that promote plant development can be a useful alternative to pesticides. When exposed to culture filtrates of *Trichoderma harzianum* (at 100% and 50% concentration), *M. graminicola* juveniles had maximum mortality (>96%). This resulted in more than 90% mortality in *M. graminicola* (Pathak and Kumar 1996). Compared to *T. harzianum*, *Trichoderma virens* was more effective at suppressing the nematode population in host tissue (Pathak and Kumar 2003).

Genetic engineering has proven a promising tool for the development of biotic and abiotic stress tolerance in crop plants. Additionally, genetic engineering leading to transgenic plants harboring nematode resistance genes has demonstrated its significance in the field of plant nematology.

14.3.3 White Tip Nematode, *Aphelenchoides besseyi*

Another plant parasitic nematode that feeds on rice is *Aphelenchoides besseyi*. It is also known as strawberry crimp disease nematode, spring dwarf, rice white tip, and strawberry bud. In many nations that cultivate rice, including Africa, Asia, Eastern Europe, the Pacific Islands, North, Central, and South America, this nematode is found in upland or irrigated rice.

(a) **Symptoms of white-tip nematode:** Infected seeds emerge late in seed beds and produce small seedlings. The upper 2–5 cm leaf tips turn white or pale yellow in the tillering stage, then brown, necrotic, and frayed. Flag leaves are characteristically shortened and twisted at their apical portions (Fig. 14.4). General stunting of the plant accompanies leaf injury. Panicles are smaller, and there are fewer spikelets. Small kernels with distorted shapes are produced by reduced

Fig. 14.4 White tip disease of rice

panicle terminal tissues. If the panicle is sterile, maturation of the panicle is postponed.

(b) **Life cycle:** Beneath the kernel's hull, the nematode is carried in an immature, dormant preadult state. During its quiescent phase, the nematode can live for 2 years. Upon sowing these contaminated seeds, the nematodes emerge, migrate to the leaf and stem development locations, and start ectoparasitic feeding. Once the eggs are deposited in the panicles or leaf axils, several generations take place within a season. Reproduction occurs after the plant grows, and the nematodes rise as the leaf grows. Thin moisture films produced by precipitation, dew, or high humidity are the main source of nematode migration. Both the panicles and the hull contain a significant amount of nematodes. The life cycle lasts 8 days at 23 °C and 10 days at 21 °C.

(c) **Spread and survival:** The main method of dissemination is through infected seeds. Furthermore, the nematode uses irrigation water to spread to nearby crops. Since J_2 cannot survive in the soil, it remains in a dormant form either outside or within the husk. There, it may live for 3 years, but when exposed to grain that has been left on the field, it dies in a couple of months. If rice is grown from shed grain, the nematodes could be able to live from one season to the next.

(d) **Management:**

- Since the nematode survives and spreads in the seed, seed treatment can be used to kill the nematodes. Hot water treatment by presoaking seeds in water (1:2 ratio) overnight, followed by the addition of two volumes of boiling water and continuous stirring for 10 min at 52–55 °C. After treatment, seeds are taken out and dried in shade. These seeds can then either be directly planted or quickly dried for storage.
- Presoaking of infested seeds in plain water or in 0.1% NaCl or KCl solution, followed by sun drying.
- Deep dry summer plowing. Destruction of infected sources of inoculum like seeds, weeds, and debris left after harvest in the field.
- Use certified seeds and burn previous crop stubbles. The most familiar management practice for *A. besseyi* is to use clean seed stocks.

14.3.4 Rice Stem Nematode, *Ditylenchus angustus*

Both the names *Tylenchus angustus* Butler (1913) and *Anguillulina angusta* (Butler) Goodey are synonymous with *Ditylenchus angustus*. The rice stem nematode is the accepted common name for *D. angustus*; however, the organism is also known as the ufra nematode, so named because of the illness it causes in rice (ufra). Because the nematode's field damage mimics that of a lightning strike, Ufra disease is locally known in Bangladesh as "dak pora." In Myanmar, the nematode is known as "akhet-pet," while in Vietnam, it is known as "tiem dot san." In many of the world's irrigated, rain-fed lowland, highland, deepwater, and tidal wetland rice-growing zones, *D. angustus* has adapted to flooded soils and high temperatures (Luc et al. 2005). There have been reports of the nematode in the Philippines, Thailand, Vietnam, Bangladesh, India, Indonesia, Malaysia, and Myanmar. *D. angustus* is primarily a pest of rice, with a restricted geographic range. It may infect many varieties of flooded rice, ranging from lowland irrigated to deepwater. On rice from India, Bangladesh, Thailand, Myanmar, and Vietnam, it has been noted and verified.

(a) **Symptoms of rice stem nematode:** Chlorosis or streaks on the upper leaves of rice plants older than 2 months are the first indications in the field. There are two different types of symptoms: the first is called "swollen ufra," in which the panicles remain enclosed within the leaf sheath, and there is a strong tendency for the stem to branch out in the infected areas; the second is called "ripe ufra," in which the panicle emerges and produces normal grains only near the tip. As the peduncle turns a dark brown color, the lower flowers continue to grow unfertilized. Infected, stunted plants typically exhibit withering leaves. Chlorosis or streaks on the upper leaves of rice plants older than 2 months are the first indications in the field (Fig. 14.5).

(b) **Life cycle:** Butler (1913) and Vuong (1969) studied the biology of *D. angustus*. The infected plants produce masses of preadult fourth-stage larvae (nematode wool) that resemble cotton. In humid weather, they become active, go to the top of the plant, and infect the growing point. The nematode consumes the portion of the stem directly above the upper nodes, the peduncle, and the immature seed

Fig. 14.5 Symptoms of Ufra disease of rice

head epidermal cells in an ectoparasitic way. There are also specks of young leaves rolling. In developing heads, there are many nematodes that range in stage from eggs to adults. It migrates upward as the plant grows to feed on freshly formed tissues, feeding ectoparasitically on the tissues of immature or delicate leaves and leaf sheaths. *D. angustus* is found coiled up in crop residue, dirt, and seeds in a dried form during harvest. Research has shown that *D. angustus* requires high humidity levels and/or protection from plant tissues to withstand desiccation since it lacks the innate ability to regulate the amount of water it loses from its body. When 4–10% of infected seedlings were transplanted in a field, losses ranging from 1.26 to 3.94 t/ha were observed (Mondal and Miah 1989). Within 1 h, nematodes can penetrate a host in water. Older plants and those in areas with low flood water are less easily invaded; therefore, the invasion relies on the age of the plants and the water level.

(c) **Spread and survival:** Plant sections possess cotton-like nematode masses near the end of the growing season. Some nematodes remain on the stubbles even after harvest. As the rainy season approaches, the cottony masses of nematodes in anabiosis spring to life and infect the following crop. In addition, irrigation water helps the nematode to spread. The nematode can withstand almost 15 months of desiccation in cottony masses.

(d) **Management:**
- Plowing of the field during summer.
- Crop rotation with jute and sesame in rice soil. Growing a nonhost crop, such as jute, in rotation with deepwater rice can reduce the incidence of ufra in fields.
- Burning of crop residues to eliminate all infested stem terminals, extending the overwintering period by delayed planting.
- *D. angustus* survives for a limited period; lengthening the overwintering period can reduce primary infection. This can be achieved with deepwater rice by using short-duration cultivars.
- Removal of ratoon rice plants, wild rice, and other host weeds will help prevent the carryover between seasons, and banks between plots may also be beneficial.
- Lowland rice infection might be decreased by immersing young seedlings for 1 week if that is practical. In lowland rice, lowering water levels well below the top of the leaf-sheath collar during the crop reduces the spread of nematodes and the incidence of secondary and tertiary infections.
- Using nematode-resistant varieties. Numerous rice varieties, both lowland and deepwater, have been evaluated against *D. angustus*. In parts of Bangladesh and India, *D. angustus* does not affect the cultivars Padmapani and Digha.

14.3.5 Rice Cyst Nematode, *Heterodera oryzae, H. oryzicola*

A dangerous upland rice root nematode also infects bananas in Kerala. The main field symptoms include decreased yield and tillering, as well as patchy seedling death. Compared to transplanted rice, direct-sown rice sustains greater damage. The infected plants show chlorosis on the leaves and browning on the roots.

(a) **Symptoms of rice cyst nematode:**
The second-stage juvenile (J_2) of this sedentary endoparasite nematode damages rice plants. This deadly rice root nematode also infests bananas in Kerala's uplands. Reduced yield, inadequate tillering, and unequal seedling death are the primary field symptoms. When rice is transplanted, it sustains less harm than when it is seeded directly. The leaves of the affected plants have chlorosis, and the roots turn brown. The nematode causes severe leaf chlorosis, inhibits growth, and decreases the number of tillers on upland rice seedlings 2 weeks after germination. The roots are defined by a darker hue and fewer root hairs. There is an annual fluctuation that is rather noticeable in the field, as well as a tendency for an unequal distribution of disease symptoms.
(b) **Life cycle:** *H. oryzae* has a short life cycle; between 26 and 30 days after infecting the roots, J_2 begin to emerge from the cysts. Around the female body, there is tylose production. One rice growing season may see two or even more generations.
(c) **Management:**
- Avoid movement of seed materials as well as seedlings from infested areas.
- Crop rotation with pulses, vegetables, and oilseeds.

14.3.6 Integrated Management

The following integrated management package against root-knot nematodes in a rice-wheat cropping system.

1. **Nursery-bed treatment**
    (i) Selection of a root-knot nematode-free area for the nursery.
    (ii) Soil solarization of the area for nursery beds, for 3–6 weeks in late April–May.
    (iii) Application of carbofuran or phorate at 0.1 g a.i./ha.
2. **Main field**
    (i) Summer plowing of the main field twice at a 2-week interval
    (ii) Puddling of the main field before transplanting
    (iii) Avoiding direct sowing of rice in the main field
    (iv) Crop rotation with mustard or other nonhost crops
    (v) Late sowing of wheat in late November to mid-December using a suitable variety
    (vi) Green manuring with *Sesbania* spp. or *Crotolaria* spp. if time permits

The other nematode was reported recently by Pankaj et al. (2012b), *Pratylenchus thornei*, from the field of IARI New Delhi.

Lesion nematodes are found all across the world and mostly harm rice that is rainfed and directly sown. In Andhra Pradesh, Assam, Gujarat, Kerala, Orissa, Madhya Pradesh, Rajasthan, Uttar Pradesh, and West Bengal, *P. indicus* and *P. zeae* have been found on rice (Prasad et al. 1987). There have been reports of *P. thornei* causing damage to wheat (Nicol et al. 1999; Nicol and Ortiz-Monasterio 2004), mungbean, and chickpea (Di Vito et al. 1992). Although upland rice has never been affected, the nematode has previously been shown to harm the rice-wheat-legume cropping sequence in India (Dwivedi and Upadhyay 1998). *P. thornei* was initially discovered in rice by Oteifa (1962) in Egypt. Due to *Pratylenchus thornei*'s ease of spread in rice fields with soil and water, there is a significant chance that the country's contaminated area could expand, which would severely reduce rice output losses, especially in situations where management practices are not ideal (Pankaj et al. 2006a, b).

14.4 Wheat

A variety of organisms are capable of infecting wheat, but the most significant ones are bacteria, fungi, and viruses that have an impact on the amount and quality of the crop. The principal diseases that impact wheat crop output include rusts, smuts, and plant parasitic nematodes. Nematodes are estimated to be responsible for $157 billion in agricultural losses globally each year (Abad et al. 2008). The nematode-caused damage has not received sufficient concern. Plant parasitic nematodes that infect wheat crops include the ear cockle nematode, *Anguina tritici*, and the cereal cyst nematode (CCN), *Heterodera avenae*, which is the causal agent of "Molya disease" (Vasudeva 1958). These nematodes are thought to be the main pests and cause an annual loss of Rs. 97.28 million in India (Jain et al. 2007).

Cyst nematodes are highly specialized, sedentary plant endoparasites. Second-stage juveniles (J_2s) induce "syncytia," an extremely complex nematode feeding site, after invading host-plant roots (Jones and Northcote 1972). It serves as a nutritional sink and gives the nematode's constant nutrition supply. The majority of Europe, Australia, Russia, Israel, India, and Pakistan are home to CCN, *H. avenae*. It has been recorded from Rajasthan, Haryana, Delhi, Himachal Pradesh, Madhya Pradesh, Punjab, Uttar Pradesh, and Jammu & Kashmir in India (Kaushal et al. 2007). This nematode can cause a total yield loss of 50–90% in some areas, resulting in an annual loss of Rs. 80 million in Rajasthan alone (Gaur and Pankaj 2009).

14.4.1 Wheat Seed-Gall Nematode, *Anguina tritici*

When farm-saved seed is planted without using conventional seed cleaning techniques, seed gall nematode is commonly seen on small grain cereals. While barley is less susceptible to the nematode in India, wheat, barley, and rye are often infected worldwide (Paruthi and Gupta 1987). In Iraq, barley is also attacked, although the attack is caused by an isolate that does not cause damage to wheat (Al-Tabib et al. 1986).

(a) **Symptoms of seed-gall nematode:** The base of the stems of infected seedlings shows a modest proliferation. The leaves of these seedlings are often folded, crinkled, or twisted, with the tips preserved near the growing point. Severe infection causes stunting and possibly early death of seedlings. Usually exhibiting significant tillering, the infected plants might start producing ears around 30 to 40 days ahead of the healthy plants. The affected ears are usually smaller. The glumes are more widely spaced by the galls, which substitute the kernels. Seed galls are smaller, proportionately shorter, and darker in color than healthy seed.

(b) **Symptoms of tundu disease:** The "tundu" disease is distinguished from ear cockle by the development of a light yellow slime or gum on the surfaces of the young plants' leaves as well as on their abnormal ears. As it dries, this yellow slime hardens, turns brittle, turns brown, and settles on the tissues. Culms infected with the bacteria either perish quickly or grow until they begin to head (Fig. 14.6a–c). In the latter case, the grains are either totally or partially replaced by the bacterial mass, and the growing spike is both smaller and shorter, failing to emerge from the boot-leaf. Usually, the stalk is twisted when the ear exhibits the bacterial symptoms. Ear cockle disease does not cause as much damage as tundu does.

Fig. 14.6 Plants showing symptoms of seed-gall nematode infestation. (**a, c**) Basal swelling in young wheat plants. (**b**) Yellowing and twisting of leaves

(c) **Life cycle:** During harvest, seed galls fall to the ground and break down in the soil, releasing the juvenile in its second stage. Depending on the size of the gall, each seed gall may contain 3000–12,000 J_2. The J_2 penetrate the seedling, proceed to the aerial section, and feed ecto-parasitically on the leaf sheath and growth tip. The plants are invaded by nematodes that transition from an ectoparasitic to an endoparasitic lifestyle during the production of embryonic flowers. In the growing seed, which turns into a gall, the nematode matures into an adult. Thousands of eggs are laid by the females after mating. The eggs hatch after the adult dies. For J_2 in the seed gall, the larva must undergo one moult in the gall. For up to 32 years, the J_2 in seed gall remains alive.

(d) **Host range:** Wheat, rye, *Triticum dicoccum*, and Spelt are susceptible to seed-gall nematode, while oats and barley are immune to the nematode.

14.4.2 Cereal Cyst Nematode, *Heterodera avenae*

Kuhn (1874) initially identified the cereal cyst nematode, *H. avenae*, as a parasite of cereals in Germany, essentially describing it as a biological variation of *H. schachtii*. The physiological and anatomical characteristics that distinguished beetroot eelworm from cereal cyst nematode were eloquently explained by Wollenweber (1924). The most significant nematode of wheat, barley, oats, rye, maize, and other graminaceous plants is the cereal cyst nematode (CCN), *H. avenae*. Assuming the significance of a major limiting factor, *H. avenae* is widely distributed throughout many wheat-growing regions of the world, including Europe, India, Canada, the USA, China, Pakistan, Iran, Saudi Arabia, Israel, Morocco, Tunisia, Pakistan, Libya, Estonia, and Australia (Vasudeva 1958; Krall et al. 1999). In 1958, in the Rajasthani village of Nim ka Thana, in the Sikar District, Vasudeva discovered *H. avenae* on wheat roots for the first time in India. Since then, the agro-climatic zones of northern to central India, which include the states of Punjab, Haryana, Uttar Pradesh, Himachal Pradesh, Jammu and Kashmir, Delhi, and Madhya Pradesh, have come to recognize it as an endemic concern.

Any soil type can be negatively impacted by *Heterodera avenae* if cereal cultivation intensity lies above a particular threshold (Kort 1972). However, light soils sustain more harm. Tropical cereals like sorghum and maize are not hosts for the nematode, while temperate cereals like barley and wheat in India are severely infected by a disease called Molya (Gill and Swarup 1971; Sharma and Swarup 1984). Nematicidal treatments have resulted in yield increases of 4 to 16 times for wheat and barley, respectively, in the Northwestern region of India.

(a) **Symptoms of Molya disease:** Typically, the infected fields seem spotty. With continuing wheat or barley production, such patches rapidly enlarge and may eventually cover the entire field in 3–4 years. The worst periods for damage are the dry seasons. Along with their slower growth and general chlorosis, the infected plants' leaf blades are stiffer, thinner, and narrower than those of healthy plants. Tillering is severely hampered, and culms become thinner and weaker. When the ears do form, they contain a very little number of grains, and the ill plants flower too soon (Figs. 14.7, 14.8, and 14.9). The afflicted roots are short, have several branches, seem bunchy, and commonly include tiny growths that resemble galls. Typically, roots develop bushy and show signs of nematode infection in the form of little swellings. Nematode infestation is confirmed by the appearance of white, glistening females on the roots in January or February.

Depending on the host, different symptoms might arise on roots. Because of their multiplication, the root system becomes shallow, bushy, and bunchy, with modest swelling around the root tip. These plants are effortlessly removed from the soil. On the other hand, the only conclusive proof of nematode infestation is the existence of cysts on the roots (Kort 1972). When *H. avenae* attacks wheat, the plant produces more roots, giving the roots a "bushy-knotted" look and allowing many females to be seen at each knot (Rivoal and Cook 1993). Oat

Fig. 14.7 Above-ground symptoms caused by *H. avenae*

Fig. 14.8 Root system with slight swelling near the root-tips

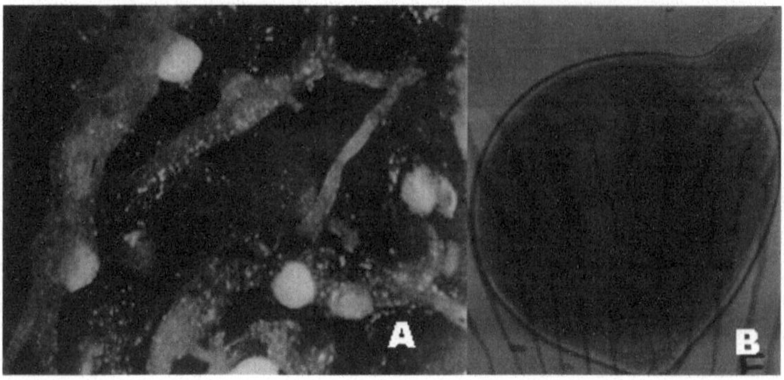

Fig. 14.9 White cysts of *H. avenae* on barley roots (**a**); (**b**) cysts later turn brown at crop maturity

roots are thinner and shorter, whereas barley roots resemble those of afflicted wheat. On the roots of cereals, other *Heterodera* species also seem to cause host-specific symptoms.

(b) **Life cycle:** Dark brown colored lemon-shaped cysts can be recovered from the infested field in large numbers. The cysts can also be recovered from infested field roots and soils. Each cyst may contain 200–300 eggs and larvae. These eggs and larvae remain dormant until the next host crop is available. When the cooler season starts, the infective second-stage larvae start emerging from the cysts. After emergence, the J_2 penetrate the root and remain as sedentary endoparasites. The J_2 develop in 25–30 days and become adults. Lemon-shaped white females can be seen protruding out of the roots about 35 days after penetration. Due to quinone tanning, the white females perish and the body wall hardens, appearing as brown cysts. These cysts may survive harsh environments because they are resilient and leathery. These egg-containing cysts decompose into the soil, where they provide an inoculum supply for the following growing season. After the larvae invade, the nematode takes 9–14 weeks to finish its life cycle.

(c) **Host range:** Wheat, barley, oats, rye, and maize are some of the host plants for the nematode.

(d) **Spread and survival:** The CCN spread through implements, soil adhering to laborers' feet, and also along with irrigation water.

(e) **Management:**

- As the cereal cyst nematode is highly host specific, growing nonhost crops like carrot, fenugreek, onion, mustard, and gram has shown to reduce the nematode population effectively.
- Barley varieties Rajkiran, C-164 and BH 393 and wheat variety Raj Mr-1 are resistant to *Heterodera avenae*.
- The source of earcockle and tundu diseases being the seed material can be removed by using nematode gall-free clean seeds. This can be achieved by adopting dry-cleaning methods like sieving and winnowing or by seed floating in plain water, or preferably, in a 20% brine solution.

- Hot water treatment of seeds can reduce nematode infestation in seeds. The seeds are presoaked for 4–6 h in cold water and then treated for 10 min in hot water at 54 °C.
- Growing nonhost crops for 1 or 2 years is sufficient to eliminate nematodes remaining in the field after harvest.
- Infected plants can be easily detected in early stages and should be removed and destroyed away from the field.

14.5 Sorghum, Maize, and Pearl Millet

(a) **Symptoms of maize cyst nematode:** Results in a decrease in yield, yellowing of the leaves, discoloration, and a bunchy look of the roots.
(b) **Symptoms of sorghum cyst nematode:** Nematodes induce plant yellowing, as well as decreased growth and production.
(c) **Symptoms of lesion nematode:** Plants that have nematodes suffer from yellowing, stunting, and vitality loss. Brown lesions are seen in the root system, which is a common lesion nematode feature. There is significant cortical parenchyma loss, lateral root multiplication, poor grain filling, and a decrease in yield.
(d) **Symptoms of lance nematode:** Nematodes can induce tunnel formation in the cortical area, thickening of cell walls, and lesions in the roots. Additionally, plants may be seen to be stunted and turning yellow.
(e) **Symptoms of stunt nematode:** Nematodes don't produce any particular symptoms. The most common symptoms include plant yellowing and stunting, vitality loss, and decreased production.
(f) **Management:**
   - Summer fallowing and heavy plowing in April/May successfully limit nematode populations.
   - Two years crop rotation with nonhost crops such as pulses, vegetables, and oilseeds.
   - Using nematode-resistant cultivars and avoiding monocropping in nematode-infested fields year after year.
   - Integrated nematode management should be followed in different crops against most of the plant parasitic nematodes (Table 14.2).

The current availability and/or use of resistant cultivars and rootstocks for nematode management reflect the success to date of research efforts in identifying and evaluating resistance sources, incorporating them into commercially acceptable crop selections, and implementing them in management programs. The relatively few crop and nematode combinations where plant resistance or tolerance to nematodes are in use is probably less a reflection of the difficulty of the task and more a reflection of limited research emphasis (Table 14.3). The potential for success in this area seems enormous in light of the essentially untapped sources of resistance to nematodes in a broad range of botanical groups, and the rapid technical advances such as in-embryo rescue, somatic hybridization, and direct gene transfer that

Table 14.2 Integrated nematode management, strategies, and tactics

Strategies	Tactics	Relative efficacy
1. Exclusion	Quarantine	M,H
2. Reduce initial population density	a. Eliminate established foci	L,M
	b. Cultural	
	• Use of clean plant stock	H
	• Crop rotation	L-H
	• Inter- and intra-cropping	L,M
	• Cover/trap crops	L,M
	• Soil amendments	L,M
	• Fallow	L-H
	• Time of planting/harvesting	L,M
	• Farm hygiene and general culture	M,H
	c. Resistance	H
	d. Nematicides	
	• Fumigants	M,H
	• Non-fumigants	L-H
	e. Biological	
	• Natural	L-H
	• Introduced	L-H
	f. Physical	
	• Tillage/root destruction	L-H
	• Heat solarization	L,M
	• Flooding	L,M
3.Suppress nematode reproduction	Resistance horizontal/quantitative	L,M
4. Restrict current crop damage	Tolerant cultivars	L-M
	One or more tactics under 1 and 2	

L low, *H* high, *M* moderate

Table 14.3 Major crops in which resistance to nematodes is available and used for nematode pest management

Crop	Nematode
Beans	*Meloidogyne incognita, M. Javanica, Pratylenchus scribneri*
Carrot	*M. incognita, M. javanica*
Cowpea	*M. incognita*
Cotton	*M. incognita*
Potato	*Globodera rostochiensis, G. pallida*
Soybean	*M. incognita, M. javanica, Heterodera glycines, Rotylenchulus reniformis*
Sweet potato	*M. incognita, M. javanica, M. arenaria*
Small grains (wheat, barley, oats)	*Heterodera avenae, Ditylenchus dipsaci*
Tobacco	*M. incognita, M. arenaria, Globodera* spp.
Tomato	*M. incognita. M. javanica, M. arenaria*
Alfalfa	*M. incognita, D. dipsaci*
Apricot	*Meloidogyne* spp.
Citrus	*Tylenchulus semipenetrans*
Grape	*Meloidogyne* spp., *Xiphinema index*
Walnut	*Meloidogyne* spp.

should promote a more efficient genetic transfer across conventionally difficult biological barriers (e.g., sexual incompatibility, polyploidy, unacceptable gene linkages).

14.6 Recent Advances

Recent advancements in biotechnological approaches have made it possible to incorporate and express indigenous and heterologous proteins from one organism to another. This has brought about a new era of crop improvements after the advent of the so-called "Green Revolution" in the 1960s. Genetic engineering of different crop plants has led to the improvement of both the quality and quantity of the produce, in addition to enhancing tolerance against various stresses. Several disease and pest resistance genes have been introduced into crop plants through genetic engineering.

In the past, several transgenic strategies have been used to enhance nematode resistance in plants. The resistance genes from natural resources have been cloned from different plant species and could be transferred to other plant species. For instance, *Mi* gene from tomato provides resistance against *M. incognita, Gpa*-2 from potato against *Globodera pallida, Hero A* from tomato against *G. rostochiensis*, and $Hs1^{pro\text{-}1}$ from sugar beet (*Beta vulgaris*) against *H. schachtii* (reviewed by Fuller et al. 2008). The overexpression of different protease inhibitors (PIs) such as cowpea trypsin inhibitor (CpTI), PIN2, cystatins, and serine proteases has been used for producing nematode-resistant plants (Lilley et al. 1999). Another main strategy was the targeted suppression of important nematode effectors in plants using the RNA interference (RNAi) approach. Unlike these strategies, some recent researches have suggested that nematode resistance could be enhanced in plants by modifying the expression of particular genes in syncytia (Klink and Matthews 2009; Ali 2012; Ali et al. 2013a, b).

The natural resistance genes could exist in both polygenic as well as single dominant manner. The resistance conferred by host single dominant resistance genes, the *R* genes from plants, interacts specifically with corresponding avirulence (*Avr*) genes in the nematode, resulting in a so-called "gene-for-gene" interaction. This type of interaction initiates a cascade of defense responses in the host plants.

Several natural host resistance genes have been cloned from some plant species and could be transferred to other plant species. For instance, *Mi-1.2* from tomato against *M. incognita* (Milligan et al. 1998; Williamson 1998), $Hs1^{pro\text{-}1}$ from *Beta procumbens* against beet cyst nematode *H. schachtii* (Cai et al. 1997), *Gpa*-2 from potato against potato cyst nematode (PCN, *Globodera pallida*) (Van der Vossen et al. 2000), and *Hero A* from tomato against *G. pallida* and *G. rostochiensis* (Sobczak et al. 2005), and *Cre* loci from *Aegilops* spp. against cereal cyst nematodes in wheat (Williams et al. 1994) are some examples that could be used in the future to develop cyst nematode resistance in crop plants. Transgenic expression of resistance proteins also induces the expression of PR (pathogenesis-related) proteins to establish nematode resistance in plants. Potato roots expressing the *Hero A* gene showed high levels of several salicylic acid (SA)-dependent PR genes in the incompatible interaction with PCN at 3 dpi (Uehara et al. 2010). They confirmed that SA-inducible PR-1 (P4) was a hallmark for the cultivar resistance conferred by *Hero A* against PCN and that nematode parasitism resulted in the inhibition of the SA signaling pathway in the susceptible cultivars. Similar effects were found in the resistant line of hexaploid wheat carrying the *Cre2* gene, which showed upregulation of ascorbate peroxidase coding gene in response to cereal cyst nematode (*H. avenae*) when compared with the expression in the susceptible lines (Simonetti et al. 2010).

RNA interference (RNAi) has emerged as a suitable tool for gene-silencing aimed at functional analysis of different genes by suppressing their expression in a wide variety of organisms, including plant nematodes. In this strategy, the nematodes uptake double-stranded RNA (dsRNA) or short interfering RNAs (siRNAs) from the plants expressing these RNAs, eliciting a systemic RNAi response in nematodes. The transgenic expression of dsRNA targeting a specific nematode effector gene could be handful to suppress the expression of that effector gene, which is crucial for the infection process. There are several review articles recently emphasizing the usefulness and application of RNAi technology to induce nematode resistance in plants by silencing the expression of nematode effectors mainly (Fuller et al. 2008). Most of the aspects of the RNAi application in nematode resistance are reviewed (Lilley et al. 2012). Lilley et al. (2012) have reviewed various features ranging from in vitro assays with *C. elegans* to delivering RNAi *in planta* as an important strategy for crop protection against cyst nematodes.

Research concerning applications of nanotechnology in the field of agriculture has gained significant attention during the recent years (Pankaj et al. 2021). Though a great amount of research data is available in this field, a comprehensive review of this topic covering the application of nanotechnology in different sectors of agriculture is still missing in the literature. Collaborative research among institutes exploring different uses of nanomaterials would be crucial to develop efficient, multifunctional, stable, cost-effective, and target-specific synthetic and natural molecules.

14.7 Conclusion

The policy makers' increasing understanding of plant parasitic nematodes justifies the inclusion of nematodes as pests in the agricultural production scenario. In terms of managing natural resources in relation to agriculture, nematodes also gained prominence. After decades of studying the effects of different nematodes on crops, a comprehensive management schedule package was developed for the benefit of the users. The main challenge to this is the scattered distribution of the nematode population, which makes it difficult to determine the precise extent of the nematode issues and prevents the adoption of the management strategies that the investigator has occasionally recommended. It is crucial to assess the loss accurately. Nematicides are either rarely or infrequently employed in India's nematode management practices. This has also played a role in the nematode population growth over threshold. The breeding program has to place adequate attention on the resistance, which has not yet been utilized in nematode control. The latest biocontrol organisms should be used on farms to create an internal pest management system. India has an abundance of organic materials, which should be encouraged among farmers to practice organic farming.

References

Abad P, Gouzy J, Aury JM, Castagnone-Sereno P, Danchin EG, Deleury E, Wincker P (2008) Genome sequence of the metazoan plant-parasitic nematode meloidogyne incognita. Nat Biotechnol 26(8):909–915

Ali MA (2012) Induction of resistance against cyst nematodes by modifying gene expression in Syncytia. PhD thesis, University of Natural Resources and Life Sciences, Vienna

Ali MA, Abbas A, Kreil DP, Bohlmann H (2013a) Overexpression of the transcription factor RAP2.6 leads to enhanced callose deposition in syncytia and enhanced resistance against the beet cyst nematode *Heterodera schachtii* in Arabidopsis roots. BMC Plant Biol 13:47

Ali MA, Plattner S, Radakovic Z, Wieczorek K, Elashry A, Grundler FM et al (2013b) An Arabidopsis ATPase gene involved in nematode-induced syncytium development and abiotic stress responses. Plant J 74:852–866

Al-Tabib NY, Al-Taae AKM, Nimer SM, Stephan ZA, Al-Bedawi AS (1986) New record of *Anguina tritici* on barley from Iraq. Int Nemat Net Newsl 3:25–27

Bridge J, Page SLJ (1982) The rice root-knot nematode, *Meloidogyne graminicola*, on deep water rice (*Oryza sativa* subsp. indica). Rev Nematol 5:225–232

Butler E (1913) Diseases of Rice, I.—an eelworm disease of rice. Bulletin 34. Agricultural Research Institute, Pusa

Cai DG, Kleine M, Kifle S, Harloff HJ, Sandal NN, Marcker KA et al (1997) Positional cloning of a gene for nematode resistance in sugar beet. Science 275:832–834

Di Vito M, Greco N, Saxena MC (1992) Pathogenicity of Pratylenchus thornei on chickpea in Syria. Nematol Mediterr 20(1):71–73

Dwivedi K, Upadhyay KD (1998) Nematode pests in rice±wheat±legume cropping systems in Central Uttar Pradesh. Nematode pests in Rice±wheat±legume cropping systems. Proceedings of the regional training course. Rice±Wheat consortium paper series, 4, p 60–62

Elling AA (2013) Major emerging problems with minor meloidogyne species. Phytopathology 103(11):1092–1102

Fuller VL, Lilley CJ, Urwin PE (2008) Nematode resistance. New Phytol 180:27–44. https://doi.org/10.1111/j.1469-8137.2008.02508.x

Gaur HS (1993) Population dynamics of root-knot nematode under the wheat—rice cropping system. Ind J Nematol 23:3–4

Gaur HS, Pankaj (2009) Nematodes. In: Handbook of agriculture. ICAR Publications, New Delhi

Gill JS, Swarup G (1971) On the host range of cereal cyst nematode, *Heterodera avenae*, the causal organism of 'molya' disease of wheat and barley in Rajasthan, India. Ind J Nematol 1:63–67

Jain RK, Mathur KN, Singh RV (2007) Estimation of losses due to plant parasitic nematodes on different crops in India. Ind J Nematol 38(2):219–220

Jones MGK, Northcote DH (1972) Nematode induced syncytium: a multinucleate transfer cell. J Cell Sci 10:789–809

Kaushal KK, Srivastava AN, Pankaj, Chawla G, Singh K (2007) Cyst forming nematodes in India—a review. Indian J Nematol 37(1):1–7

Klink VP, Matthews BF (2009) Emerging approaches to broaden resistance of soybean to soybean cyst nematode as supported by gene expression studies. Plant Physiol 151:1017–1022. https://doi.org/10.1104/pp.109.144006

Kort J (1972) Nematode diseases of cereals of temperate climates. In: Webster JM (ed) Economic nematology. Academic, New York, pp 97–126

Krall E, Sturhan D, Muur J (1999) Teraviljadel ja Korsheeintel parasiteerivad tsuste mood ustavad nematoodid Eestis. Teadustoode Kogumik 203:44–48

Kuhn J (1874) Uber das kommen von Rubbennematoden an den wurzehdra halmfruchle. Landav JBR 3:47–50

Kumar V, Khan MR, Walia RK (2020) Crop loss estimations due to plant-parasitic nematodes in major crops in India. Natl Acad Sci Lett 43:409–412

Lilley CJ, Devlin F, Urwin PE, Atkinson HJ (1999) Parasitic nematodes, proteinases and transgenic plants. Parasitol Today 15:414–417

Lilley CJ, Davies LJ, Urwin PE (2012) RNA interference in plant parasitic nematodes: a summary of the current status. Parasitology 139:630–640

Loof PAA (1991) The family Pratylenchidae Thorne, 1949. In: Nickle WR (ed) Manual of agricultural nematology. Marcel Dekker, New York, pp 363–421

Luc M, Sikora RA, Bridge J (eds) (2005) Plant parasitic nematodes in subtropical and tropical agriculture. CAB International, Wallingford

Milligan SB, Bodeau J, Yaghoobi J, Kaloshian I, Zabel P, Williamson VM (1998) The root knot nematode resistance gene *mi* from tomato is a member of the leucine zipper, nucleotide binding, leucine-rich repeat family of plant genes. Plant Cell 10:1307–1319

Mondal AH, Miah SA (1989) Post transplanting nematicidal effect on stubble borne ufra nematode. In: 6th National Botanical Conference, Chittagong (Bangladesh), 18–19 Jan 1989. BBS

Nicol JM, Ortiz-Monasterio I (2004) Effects of the root-lesion nematode, *Pratylenchus thornei*, on wheat yields in Mexico. Nematology 6(4):485–493

Nicol JM, Davies KA, Hancock TW, Fisher JM (1999) Yield loss caused by *Pratylenchus thornei* on wheat in South Australia. J Nematol 31(4):367

Oteifa BA (1962) Species of root-lesion nematodes commonly associated with economic crops in the Delta of the UAR. Plant Dis Rep 46(8):572–575

Pankaj, Sharma HK, Gaur HS, Singh AK (2006a) Effect of zero tillage on the nematode fauna in a rice-wheat cropping system. Nematol Mediterr 34:173–176

Pankaj, Ahlawat JS, Saha M (2006b) Predominant nematode pests of rice nursery in North-Western India. In: 2nd International Rice Congress (9–13 October 2006). ICAR, New Delhi

Pankaj, Ganguly AK, Pandey RN (2012a) Severe damage caused by root-lesion nematode, *Pratylenchus thornei* in aerobic rice in India. Nematologia Meditterrenea 40:67–69

Pankaj, Ganguly AK, Kumar H (2012b) Root-knot nematode, *Meloiodgyne graminicola*: a key nematode pest of rice. Tech Bull. TB–ICN: 90/2012. IARI, New Delhi

Pankaj, Rana VS, Kaushik P, Kumar H, Devaraja KP, Gowda A, Shakil NA (2021) Bioefficacy evaluation of chalcone derivatives against *Meloidogyne graminicola* infecting rice, *Oryza sativa* L. J Environ Sci Health B 56(9):1–8

Paruthi IJ, Gupta DC (1987) Incidence of tundu in barley and kanki in wheat field infested with *Anguina tritici*. HAUJ Res 17:78–79

Pathak KN, Kumar B (1996) Effect of culture filtrates of *Gliocladium virenus* and *Trichoderma harzianum* on the penetration of rice roots by *Meloidogyne graminicola*. Ind J Nematol 25:223–224

Pathak KN, Kumar B (2003) Effect of culture filtrates of *Gliocladium virenus* and *Trichoderma harzianum* on the penetration of rice roots by *Meloidogyne graminicola*. Ind J Nematol 33:149–151

Pathak H, Tripathi R, Jambhulkar NN, Bisen JP, Panda BB (2020) Eco-regional rice farming for enhancing productivity, profitability and sustainability. NRRI research bulletin no. 22. ICAR-National Rice Research Institute, Cuttack

Patnaik NC (1969) All India nematology symposium august, 21-21, 1967, New Delhi. Phytopathology 26:348–349

Prasad JS, Panwar MS, Rao YS (1987) Nematode problems of rice in India. Int J Pest Manage 33(2):127–136

Rao YS, Israel P (1973) Life history and bionomics of *Meloidogyne graminicola*, the rice root-knot nematode. Ind Phytopath 26(2):333–340
Rivoal R, Cook R (1993) Nematode pests of cereals. In: Plant parasitic nematodes in temperate agriculture. CAB International, Wallingford, pp 259–303
Roy AK (1976) Pathological effects of *Meloidogyne graminicola* on rice and histopathological studies on rice and maize. Ind Phytopathol 29:359–362
Sasser JN (1987) A world perspective on nematology: the role of the society. In: Vistas on nematology. Society of Nematologists, Lakeland, pp 7–14
Sharma SB, Swarup G (1984) Cyst forming nematodes of India. New Delhi Ind Cosmo Publ 1:150
Simonetti E, Alba E, Montes MJ, Delibes A, Lopez-Brana I (2010) Analysis of ascorbate peroxidase genes expressed in resistant and susceptible wheat lines infected by the cereal cyst nematode, Heterodera avenae. Plant Cell Rep. 29 1169–1178
Singh SK, Singh KP (2009) A first record of root-knot nematode (*Meloidogyne graminicola)* on wheat in Uttar Pradesh, India. J Mycol Plant Pathol 39(2):340–342
Singh I, Gaur HS, Briar SS, Sharma SK, Sakhuja PK (2003) Role of wheat in sustaining *Meloidogyne graminicola* in rice-wheat cropping system. Int J Nematol 13(1):79–86
Sobczak M, Avrova A, Jupowicz J, Phillips MS, Ernst K, Kumar A (2005) Characterization of susceptibility and resistance responses to potato cyst nematode (*Globodera* spp.) infection of tomato lines in the absence and presence of the broad-spectrum nematode resistance *hero* gene. Mol Plant Microbe Interact 18:158–168
Stirling GR, Nicol JM, Reay F (1998) Advisory services for nematodes pests–operational guide. Rural Industries Research and Development Corporation Publication No. 99/41, p 120
Swain BN, Prasad JS (1988) Chlorophyll content in rice as influenced by the root-knot nematode, *Meloidogyne graminicola* infection. Cur Science 57: 895–896
Uehara T, Sugiyama S, Matsuura H, Arie T, Masuta C (2010) Resistant and susceptible responses in tomato to cyst nematode are differentially regulated by salicylic acid. Plant Cell Physiol 51:1524–1536
Van der Vossen EA, Van Der Voort JN, Kanyuka K, Bendahmane A, Sandbrink H, Baulcombe DC et al (2000) Homologues of a single resistance-gene cluster in potato confer resistance to distinct pathogens: a virus and a nematode. Plant J 23:567–576
Vasudeva RS (1958) Annual report of the Indian Agriculture Research Institute. IARI, New Delhi
Vuong HH (1969) The occurrence in Madagascar of the rice nematodes, *Aphelenchoides besseyi* and *Ditylenchus angustus*. Tech Commun Commonwealth Bureau Helminthol 40:274–288
Williams KJ, Fisher JM, Langridge P (1994) Identification of RFLP markers linked to the cereal cyst nematode resistance gene (Cre) in wheat. Theor Appl Genet 89:927–930
Williamson VM (1998) Root-knot nematode resistance genes in tomato and their potential for future use. Annu Rev Phytopathol 36:277–293
Wollenweber H (1924) The potato *Heterodera* spp. Illustr Landwirtisch Z 12:101

Rhizoctonia: An Emerging Threat to Field Crops

15

Kumari Surbhi, T. Aravind, Krishna Pratap Singh, Pooja Bhatt, and Vaibhav Kumar Singh

Abstract

Rhizoctonia is one of the most important genera of soil-inhabiting fungi responsible for inhibiting plant growth at all stages of development by causing damping off, seedling rot, root rot, collar rot, and wilt. The fungus causes considerable damage to the yield of a variety of crops, including rice, wheat, maize, soybean, and pulses. Due to its distinctive ability to produce durable survival structures, overwinter in soil, and on alternate hosts, the fungus is difficult to manage. The role of seed treatment chemicals in providing protection in the early stages of the crop and foliar sprays to prevent further development and spread of the pathogen is important. An integrated management approach should be followed, which keeps a check on health of the plant at each stage of development.

Keywords

Rhizoctonia · Anastomosis · Diagnosis · Diseases · Field crops

15.1 Introduction

15.1.1 Importance of Soil-Borne Diseases

The yield of agricultural crops is dependent on several abiotic and biotic factors such as environmental variables, soil factors, drought, flood, mineral deficiencies, insect pests, and plant diseases. Among all these factors, plant diseases account for

K. Surbhi · T. Aravind · K. P. Singh (✉) · P. Bhatt
Department of Plant Pathology, College of Agriculture, Govind Ballabh Pant University of Agriculture and Technology, Pantnagar, Uttarakhand, India

V. K. Singh
Wheat Pathology Laboratory, Division of Plant Pathology, ICAR-Indian Agricultural Research Institute, Pusa, New Delhi, Delhi, India

V. K. Singh et al. (eds.), *Diseases of Field Crops: Diagnostics and Management*, https://doi.org/10.1007/978-981-97-6160-9_15

a loss of about 14.1%, which could pose a serious threat to food security in the absence of timely management action (Agrios 2005). There exists a plethora of microorganisms such as fungi, bacteria, viruses, and nematodes which could establish a parasitic relationship with the crop and cause several economically important plant diseases. The soil-inhabiting fungi are of critical importance as they are responsible for inhibiting the plant growth during all crop growth stages by causing diseases such as damping off, seedling rot, root rot, collar rot, wilt, etc. As such, the soil-borne diseases interfere with the expression of the full yield potential of crop cultivars (Panth et al. 2020). Soil-borne diseases caused by the pathogens such as *Pythium, Phytophthora, Verticillium, Fusarium* spp., *Sclerotinia*, and *Rhizoctonia* spp. are responsible for yield loss of about 50–75% in crops including cereals, pulses, vegetables, and horticultural crops (Baysal-Gurel and Kabir 2018; Panth et al. 2020). Among the various soil-inhabiting plant pathogens, the fungus *Rhizoctonia* is of considerable economic importance due to the extensive damage caused in a wide range of crops.

15.1.2 Genus *Rhizoctonia*

Genus *Rhizoctonia* (Division: *Basidiomycota*; Class: *Agaricomycetes*) is a group of filamentous fungi that are not known to produce asexual spores. The genus *Rhizoctonia* was described for the first time by De Candolle in 1815 (Yang and Li 2012). Members of this genus are mostly soil-dwelling saprotrophs and facultative parasites that cause various important diseases in agricultural crops all over the world (García et al. 2006). One of the important and most widely studied species of this genus is *Rhizoctonia solani* Kühn (*Thanatephorus cucumeris* Frank (Donk)), which was reported by Julius Kühn (1888) on diseased potato tubers (Mathpal 2016; Surbhi 2018). The fungus has been known to produce a considerable damage to the yield of a variety of crops worldwide (García et al. 2006). *Rhizoctonia solani* is a ubiquitous soil fungus having a cosmopolitan asexual stage and a rather rare sexual stage.

15.1.3 Morphological Characters of *Rhizoctonia solani*

Genus *Rhizoctonia* was formally recognized by the French mycologist Augustin Pyramus De Candolle in the year 1815. He described the presence of a dolipore septum, absence of clamp connection, septum formation at the point of origin of a branch, constriction at the base of the branch arising at 90°, presence of branching close to the distal septum of cells in young hyphae, absence of conidia, and production of sclerotia with uniform texture (Kumar et al. 2016). The mycelium of *R. solani* gives a silvery appearance when young but turns brown at maturity. The mycelium has clear septation with secondary branches connected at an angle of 90° to the main branch. The diameter of the mycelium varies from 8 to 12 μm, with dark brown to black sclerotia, which are 99.9 to 166.5 μm in diameter. The shape of sclerotia may vary from irregular or flattened to roughly spherical and smooth (Mukhopadhyay and Singh 1984). The mycelium and sclerotia of *R. solani* perpetuate in soil and crop debris and could remain viable for several months to years. *R. solani* is known to produce various different symptoms such as aerial blight (web blight), sheath blight, root and stem rot, and damping off in different crops.

15.1.4 Anastomosis Groups and Host Range of *Rhizoctonia solani*

The species *Rhizoctonia solani* is divided into different groups known as anastomosis groups (AGs) on the basis of anastomosis reaction of its hyphae (Carling et al. 2002). Isolates in which fusion of hyphae leading to an exchange of cytoplasmic and nuclear material is possible are placed under same anastomosis group, while isolates in which hyphal fusion and exchange of cytoplasmic and nuclear material do not occur are placed under different anastomosis groups. The system of anastomosis groups was first proposed by and was developed further by. It is fascinating that members of the different anastomosis groups have been found to have the same teleomorph, i.e., *Thanatephorus cucumeris* (Ajayi-Oyetunde and Bradley 2018). At present, 14 anastomosis groups, i.e., AGs 1-13 and AG-B1, have been identified within the species. Some of these anastomosis groups, viz., AGs 1, 2, 3, 4, 6, 8, and 9, have further been divided into subgroups on the basis of morphological characters, nutritional requirements, sclerotia formation, virulence, host plants infected, biochemical, and molecular characters (Carling et al. 2002; Anees et al. 2010). AG-2, which has been subdivided on the basis of frequency of fusion among the isolates, is an exception to the general rule of classification of other anastomosis groups. Although *R. solani* is a species with a wide host range, the preference for host plants varies for different anastomosis groups of the fungus.

AG-1 has been known to consist of six subgroups, i.e., AG-1-1A, AG-1-1B, AG-1-1C, AG-1-1D, AG-1-1E, and AG-1-1F, which infect maize, soybean, common bean, cabbage, lettuce, and sorghum (Priyatmojo et al. 2001; Kuninaga 2002; Grosch et al. 2004). The second group, AG-2, consists of the subgroups AG-2-1 and AG-2-2, which are known to cause diseases in crucifers, sugar beet, tulip, tobacco, carrot, soybean, maize, canola, wheat, and tomato (Kuninaga et al. 2000; Dorrance et al. 2003; Paulitz et al. 2006; Misawa and Kuninaga 2010; Jaaffar et al. 2016). The host range of AG-3 includes mainly solanaceous crops such as tomato (AG-3TM), tobacco (tobacco), and potato (AG-3PT) (Kuninaga et al. 2000; Woodhall et al. 2007; Misawa and Kuninaga 2010). AG-4 consists of three homogeneous groups (HG), i.e., HG-I, HG-II, and HG-III. The members of AG-4 are known to cause diseases in peanut, potato, tomato, cotton, beans, melon, broccoli, and spinach (Fenille et al. 2002; Kuramae et al. 2003). The host range of AG-5 includes sugar beet, potato, soybean, turf grass and common bean. AG-6, which has been divided into two subgroups, mainly consists of nonpathogenic members usually found in association with orchids and mycorrhiza (Pope and Carter 2001). The host range of AG-7 includes soybean, potato, cotton, and watermelon. The isolates of AG-8 infect barley and wheat and have been divided into five zymogram groups, namely ZG1-1, ZG1-2, ZG1-3, ZG1-4, and

ZG1-5. The members of AG-9 have been found to be pathogenic on potato, lettuce, and carrot. Members of AG-10 are nonpathogenic, while the members of AG-11 could cause diseases on many different crops such as potato, wheat, soybean, cotton, and radish. The isolates of AG-12 are known to cause diseases in radish and cauliflower, while AG-13 has been reported to be nonpathogenic. The fungus is also known to infect and survive on various weed hosts such as *Echinochloa colonum*, *Cyperus rotundus*, *Sesbania exaltata*, *Cynodon dactylon*, etc. (Black et al. 1998).

15.2 *R. solani* as a Pathogen of Major Field Crops

15.2.1 Sheath Blight of Rice

15.2.1.1 Economic Importance

Rice (*Oryza sativa*) is one of the most important cereal crops consumed all over the world. Sheath blight of rice, caused by *Rhizoctonia solani* Kühn anastomosis group AG1-IA, is one of the most destructive diseases of rice and could lead to a loss of about 45–50% in yield of the crop under favorable environmental conditions (Margani and Widadi 2018; Uppala and Zhou 2018). The disease was reported for the first time in 1910 in Japan and thereafter it was known to spread across all the regions with intensive rice cultivation. In India, the disease was first reported in 1963 by Pancer and Chahal (Datta and Vurukonda 2017). According to Rush and Lee (1983), the disease could cause a yield loss of about 20% to 50% in cases where all the leaf sheaths have been infected. Roy (1979) reported that yield loss due to the disease could vary from 10% to 36%, depending on the growth stage of the plant at the time of fungal attack. The disease prevails in areas of high temperature and humidity. Excess application of nitrogenous fertilizer also contributes to the occurrence of the disease. The management of sheath blight is difficult due to the wide host range and high genetic diversity of *R. solani*.

15.2.1.2 Disease Symptoms and Epidemiology

The symptoms of the disease start appearing later in the tillering stage in the form of oval to ellipsoidal green to gray water-soaked spots on the plant near the water line. With the progress of the disease, these spots increase in size and coalesce to form larger grayish spots with a dark brown irregular outline. In case of severe infection, the extensive development of spots has been noticed which spread to the leaves of the plant. The disease softens and weakens the sheath of the plant at the time of tillering, which leads to sagging of tillers. Due to weakening of the stem, water transportation in the plant is hampered, which leads to reduced photosynthetic ability and as such, grain filling is severely affected (Wu et al. 2012; Uppala and Zhou 2018; Singh et al. 2019a, b). The sclerotia from the last crop season could survive in the soil and start fresh infection in the new crop.

A warm temperature varying from 28–32 °C, presence of high humidity of about 95%, and high amount of nitrogen fertilizer are the major contributing factors to the occurrence of disease in the field (Kumar et al. 2009; Singh et al. 2019a, b). In rice infection of *R. solani* is assisted by the production of a toxin known as RS toxin. This toxin is a mixture of four components viz., glucose, mannose, *N*-acetyl glucosamine, and *N*-acetyl galactosamine (Singh et al. 2019a, b). The pathogen is also related with the production of several effector proteins, such as glycosyltransferases, which are important component of virulence of pathogen in rice (Zheng et al. 2013; Singh et al. 2019a, b).

15.2.1.3 Disease Management

The most important component of management practices for sheath blight of rice is the use of proper agronomic practices. Since dense canopy of the crop leads to a warm and humid environment which is favorable for the growth of fungus, dense planting of the crop should be avoided. To avoid excessive vegetative growth of the crop which leads to susceptibility of the crop to *R. solani*, nitrogenous fertilizers should be applied optimally at recommended doses. Many new fungicides such as strobilurins (Azoxystrobin), Hexaconazole, validamycin, and thifluzamide have been reported to be useful in managing the disease (Kumar et al. 2018a). Foliar spray and seed treatment are the most popular methods for application of fungicides for sheath blight (Singh et al. 2019a, b). The application of fungicide should be done after 7 days after the differentiation of panicles until there is 50–75% heading in the crop (Uppala and Zhou 2018). Several biocontrol agents such as *Pseudomonas fluorescens*, *B. megaterium*, and some other Bacillus species have been reported to be effective in reducing the infection of the fungus on rice (He et al. 2002; Kazempour 2004; Radjacommare et al. 2004). Several genetic approaches such as RNA silencing of pathogenesis genes of the fungus and development of transgenic rice cultivars with improved resistance to disease have been found promising for the management of sheath blight (Singh et al. 2019a, b). Crop monitoring systems could play a very important role in early detection and warning for the disease and as such help to reduce the use of chemicals in rice.

15.2.2 Banded Leaf and Sheath Blight of Maize

15.2.2.1 Economic Importance

Banded leaf and sheath blight (BLSB) is one of the most common and destructive diseases infecting maize. It is caused by *Rhizoctonia solani* f. sp. Sasakii (teleomorph: *Corticium sasakii*, syn. *Thanatephorus cucumeris*). It has been known to cause considerable yield losses in maize. In tropical regions where humid conditions prevail, the disease causes rotting of the ears, giving a brownish appearance in which light brown cottony mycelium of the fungus can be seen. Small, rounded, black-colored sclerotia are produced by the fungus on the diseased plant parts. Under favorable environmental conditions, the disease spreads from the sheaths to the developing ears and covers a large area in the form of concentric spots. The disease has been reported all over the world in different countries such as the United States, Venezuela, Germany, England, and China (Hooda et al. 2017). The disease has been reported to cause heavy yield losses to the crop in South and Southeast Asian countries, viz., India, Bangladesh, Pakistan, Nepal, Sri Lanka, Indonesia, Cambodia, Myanmar, Thailand, Laos, Vietnam, Philippines, Taiwan, Malaysia, Korea, and Japan (Singh and Shahi 2012; Hooda et al. 2017). In India, the disease was reported for the first time in 1960 from the Tarai region in Uttar Pradesh. Later, it occurred in epidemic form in Himachal Pradesh (1972) and was also reported from Meghalaya in 1977 (Hooda et al. 2017). The disease has been found to be prevalent in Uttar Pradesh, Rajasthan, Haryana, Himachal Pradesh, Punjab, West Bengal, Orissa, Jharkhand, Madhya Pradesh, Meghalaya, and Assam (Akhtar et al. 2009; Rani et al. 2013; Hooda et al. 2017). Intensive farming practices have led to the spread of the disease to new areas, where it is challenging for the farmers and maize growers to manage the disease. The use of cultural practices and fungicides alone is not adequate for management, and a planned strategy including different management practices is required to avoid losses due to the disease.

High-yielding varieties of maize are under widespread cultivation by farmers. BLSB of maize has the ability to reduce the yield of even the high-yielding varieties considerably. In India, the disease has been reported to cause 24% to 40% losses in grain yield of the crop, which could reach as high as 90% under favorable environmental conditions (Lal et al. 1980). In China, 87.5% and 57.8% yield loss have been reported in the case of hybrid varieties (Sharma 2005). Buddemeyer et al. (2004) reported that the disease resulted in a loss of 37% and 12% in fresh matter of shoot and grain yield, respectively. In the United States, a 42% yield reduction was reported due to the disease by Sumner and Minton (1989) when the soil was infested with a high level of inoculum of the pathogen.

15.2.2.2 Disease Symptoms and Epidemiology

The disease has been reported to appear in 40–50-day-old plants just before the flowering stage (Saxena 2002). The symptoms start as irregular, straw-colored, water-soaked lesions, which rapidly increase in size to form larger discolored areas showing characteristic peculiar alternate dark bands on the base leaf sheaths. Sclerotia of the fungus are also clearly visible on the infected plant parts. The pathogen attacks the leaf sheaths and ears rather quickly, while the tassels of the plant are free from any symptoms (Maiti 1978). Owing to the soil-borne nature of the pathogen, the infection usually starts from the first and second leaf sheath of the plant above the ground and it eventually spreads to the whole plant, reaching the ears, ultimately resulting in ear rot. In ears of the infected plant, light brown colored cottony mycelium of the fungus can be seen along with small black sclerotia, which are round in shape. Severe infection leads to spotting of the stalk rind, weakening and breaking of the plant stalk, and cracking of stylar fibers due to production of horseshoe-shaped lesions around the caryopsis. Due to excessive foliar damage, the photosynthetic ability decreases, which leads to weakening and lodging of the crop (Lu et al. 2012).

The severity of infection of the disease is dependent on the degree of host susceptibility and the prevailing environmental conditions. The disease has been found to be destructive in places where high humidity and high temperature prevail. An average temperature in the range of 25 °C to 30 °C and relative humidity (RH) from 90–100% has been reported to be favorable for the development of the disease (Singh and Shahi 2012). At RH below 70%, the disease spread and development is very slow. Rainfall of 100 mm or more during the first 2 weeks of crop growth could lead to severe infection. The disease severity is also increased in the case of high crop density (Saxena 2002).

Sclerotia of the pathogen could survive in the soil for many years. These overwintering sclerotia in soil and crop debris may come up to the surface of the soil during field preparation and serve as a source of inoculum for primary infection. Active mycelium surviving on alternate hosts, such as weed plants, could also serve for starting new infections of the fungus in the crop (Simon et al. 2014). The fungus may spread in the field by means of irrigation, contaminated soil, and infected crop debris. In the presence of a favorable range of temperature and humidity, the fungus is attracted to chemical stimulants produced by the newly planted crop. Contact with leaf sheaths of infected plants could lead to secondary spread of the disease in healthy plants (Gilligan 2002). The disease covers the whole plant, including the ears, but not the tassels, and the infection of the fungus may continue up to the mid-dough stage of the crop. Occurrence of infection at later stages of the crop may lead to rotting of seeds, seedling blight, and reduced germination percentage of the seeds.

15.2.2.3 Disease Management

The present understanding of inheritance for resistance to banded leaf and sheath blight in maize is ambiguous; thus, there are no stable, widely adapted, and commercial sources of resistance to the disease. As such, for management of the disease, a strategic plan including all possible measures such as cultural, biological, and chemical control needs to be adopted (Singh and Shahi 2012). Cultural practices aimed at managing BLSB focus on reducing humid conditions in the field so that the pathogen is deprived of favorable growth conditions. In order to avoid excess humidity conditions in the field and contact of growing seedlings with excess water, a well-drained field should be selected, and the raised bed method of planting should be followed. High crop density should be avoided as it leads to a humid microclimate in the field. Stripping is another important cultural practice, in which the second and third leaf sheaths are stripped off from 35–40 day-old plants (Mehra et al. 2012). Inter-cropping of maize with leguminous crops such as soybean has also been found to be effective in reducing disease incidence (Hooda et al. 2017).

The biocontrol agents such as *P. fluorescens* and *T. harzianum* not only reduce the incidence of disease in the field but are also known to increase growth of the plant (Sivakumar et al. 2000; Sharma and Saxena 2001; Meena et al. 2003). Seed and foliar treatment with *P. fluorescens*, when used in combination, has been reported to be beneficial in reducing the incidence of sheath blight in maize by 30%. The combined application of *P. fluorescens* and *T. viride* is also effective in decreasing the disease incidence. In the case of chemical control, fungicides that have multiple modes of action against the pathogen, such as inhibition of sclerotial germination, inhibition of mycelial growth, and reduction in the spread of the disease, are helpful for the management of BLSB in maize. Many fungicides, such as benomyl, chloroneb, captafol, edifenphos, iprobenphos, thiophanate methyl, and carboxin, have been found to be effective in decreasing the incidence of disease under field conditions (Singh and Sinha 2004). Some new fungicides, which include the dicarboximide group, strobilurins, and demethylation inhibitors, are also used as sprays for the management of the disease in the field (Krishnam Raju et al. 2008; Amaradasa et al. 2014). An integrated approach using a combination of different components of disease management, including resistant cultivars, cultural practices, organic amendments, fungicides, and bioagents, should be applied for the management of the disease in the field.

15.2.3 Root Rot (Bare Patch) of Wheat and Barley

15.2.3.1 Economic Importance

Root rot, also known as bare patch of cereals, is caused by the fungus *Rhizoctonia solani* (teleomorph: *Thanatephorus cucumeris*) anastomosis group (AG) 8 and is an important disease of cereal grains. The disease appears in the form of patches of stunted wheat in the field. The pathogen is a necrotroph. It invades roots of the seedlings and later leads to the disintegration and collapse of the root cortex structure, which finally gives the characteristic symptom of spear tipping (Schneebeli et al. 2015). Due to root cortex disintegration caused by the action of pathogen, there occurs a reduction in the ability of the plant to access water and nutrients, which ultimately leads to stunting of young seedlings (Paulitz et al. 2002). The disease is prevalent in dryland cropping systems of southern parts of Australia and in the northwest United States. Under dryland cultivation of cereals in Australia, the disease is known to cause an annual loss of about $77 million per year in the yield of wheat and barley crops (Murray and Brennan 2009a; Murray and Brennan 2009b; Barnett et al. 2019). The disease has also been found in wheat fields from Turkey (Unal and Dolar 2012). Bare patch is observed mainly in areas where no-till or minimum tillage practices are used. The activity of pathogen increases in direct-drill method or stubble retention systems of farming, which becomes a major limitation in the adoption of these practices in farming (Paulitz et al. 2002). The patches of stunted plants in wheat or barley fields may be several yards large. The distribution of disease severity is uneven across the field and includes areas of high severity with no plants at all (bare-patches) and areas of reduced plant growth with only 20% crop area (Schillinger and Paulitz 2006; Anees et al. 2010; Handiseni et al. 2013; Barnett et al. 2017).

15.2.3.2 Disease Symptoms and Epidemiology

Formation of circular or oval patches in the field, which consist of stunted or chlorotic plants, is a characteristic symptom of the disease. These stunted plants may produce very less or even no grains at all. The severity of the disease and the expression of symptoms by the infected plants are also dependent on climatic conditions and type of soil as witnessed in countries such as Scotland and the United States. In case of very high disease severity, the stunting of plants is widespread in the field, while low disease pressure may cause minor root rot or stunting of a few individual plants, which are found interspersed within the rows of healthy plants. Apart from this, in case of moderate root damage, the maturation of crop could be delayed by up to 3 weeks.

The symptoms of bare patch become distinctively visible after 3 to 4 weeks of sowing. The patches produced due to the fungus in the field are circular or irregular in shape, ranging in diameter from 0.3 to 200 m. These bare patches could also be found parallel to the direction of rows of other healthy plants and may also change their position and shape in different seasons. Such changes are due to changing environmental conditions, such as soil moisture and temperature, which affect the level and potential of inoculum of the pathogen in the field (Gill et al. 2001).

The fungus survives in plant residues in the upper layers of soil, and after the rainy season, it grows and spreads in the soil, forming a network of hyphae. The growing roots of young seedlings or germinating seeds get infected upon coming in contact with the hyphal network of the fungus. The most favorable temperature for the growth of the fungus lies in the range of 10 °C to 15 °C. *R. solani* is a weak and unspecialized pathogen that can infect young and immature roots of cereal crops, but it can neither attack the mature roots nor the stem tissues. The pathogen starts infection by attacking the primary or seminal roots and prunes away the root tips. Later in the season, the secondary roots and nodal roots are also affected. The rotting of root cortex leads to large patches of dead or stunted plants in the field (Schillinger and Paulitz 2006). The infected roots develop characteristic sharp, reddish-brown points called as the spear tip. The girdling of roots due to the pathogen could make the infected roots look constricted or pinched in appearance.

The pathogen, *Rhizoctonia solani* (AG-8), possesses both saprophytic and parasitic phases, which are not completely independent of each other and may be simultaneously active in the soil. Crop stubble serves as a substrate for the growth of the saprophytic phase of the fungus and also serves for over-summering of the pathogen. The survival of the pathogen is also affected by soil moisture, and higher disease damage has been reported in cases where the infection occurs after dry summers. Rainfall of more than 30 mm in a week during summer could significantly reduce the inoculum of the fungus in the field.

15.2.3.3 Disease Management

To avoid the buildup of inoculums in the field, management practices should be planned early, and the multiplication of the fungus in the field should be prevented in the year before sowing. There are very few cultivars resistant to *R. solani*; as such, major emphasis should be given to cultural control of the disease (Okubara et al. 2009; Mahoney et al. 2016). The seed setting and growth of grassy weeds should be controlled from the first autumn flush by spraying of selective herbicides. Two to three weeks before sowing, the field should be free from weeds. Tillage of the top 5 to 10 cm of soil a fortnight before sowing is effective in reducing the amount of inoculum of the pathogen. In Australia, some of the recently registered fungicides for seed treatment include Rancona® Dimension (Chemtura), EverGol® Prime (Bayer), and Vibrance® (Syngenta), while the fungicide Uniform (Syngenta) has been recommended for in-furrow treatment (Bogacki et al. 2014; Almasudy et al. 2015; Hüberli et al. 2015). However, these fungicides have been found only partially effective in managing the disease.

15.2.4 Aerial Blight of Soybean

15.2.4.1 Economic Importance

Soybean is one of the most important oilseed crops cultivated all over the world. Among the various diseases affecting the productivity and successful cultivation of the crop, aerial blight is of utmost importance. Aerial blight is one of the most destructive diseases of soybean, which leads to considerable losses in yield of the crop (Kumar et al. 2016). The appearance of the disease starts around the last week of August to the first week of September and later in the season may lead to the complete death of the affected plants (Surbhi and Singh 2020). It is found in all tropical and subtropical soybean-growing areas. The disease was reported for the first time from Philippines in 1980. Later, the disease was found to be prevalent in countries all over the world such as India, China, Taiwan, Malaysia, Mexico, North America, South America, Louisiana, Brazil, and Argentina. Aerial blight is an important disease of soybean in India and is widespread in the country. In India, the disease was reported for the first time from Pantnagar in 1967 (Mukhopadhyay and Singh 1984). Later, the disease spread to almost all soybean-growing areas in different states of the country, including Rajasthan, Haryana, Punjab, Uttar Pradesh, Bihar, Sikkim, Madhya Pradesh, and Chhattisgarh (Sharma and Tripathi 2001; Ray et al. 2007).

The pathogen is able to infect the soybean plants at any growth stage and may lead to severe defoliation and huge losses in yield (Crispin and Gallegos 1963). The disease may cause up to 35% reduction in yield of the crop (Crispin and Gallegos 1963). In northern regions of Brazil, damage due to the disease fluctuates owing to the prevailing environmental conditions and crop losses may range from 31% to 60%. In the United States, the disease was reported to cause up to 50% yield loss in the crop. Severe losses of up to 48% and 52% were reported in stand establishment and yield of the crop, respectively (Chang et al. 2017). Infection of *R. solani* in soybean causes a decrease in shoot length and pod formation of the crop. It is also known to reduce the total number of seeds produced and the weight of the seeds. Under humid weather conditions, the disease could cause 35% to 60% losses in crop yield (Patel and Bhargava 1998). Severe occurrences of the disease were recorded during high rainfall periods in Chhattisgarh, Madhya Pradesh, and Uttarakhand (Wrather et al. 2010). The severity of aerial blight is greatly dependent on environmental conditions (Joshi et al. 2018).

15.2.4.2 Disease Symptoms and Epidemiology

Symptoms of aerial blight generally appear on 45–60-day-old soybean plants, and infection starts from the lower or middle parts of the plant and then spread upwards to cover leaves, stems, and pods (Prasad 2005). The symptoms initiate in the form of water-soaked, pale green, small, irregular lesions on the leaflets of the plants. The smaller lesions soon coalesce to form larger irregular spots with reddish-brown margins and tanned grayish centers, which cover the whole leaf (Thind 2005; Williamson et al. 2006). The fungal hyphae could be seen on the infected plant parts in the form of a brownish spider-like web. Small oval to irregular brown to dark-colored sclerotia are present on infected surface of plant. The leaves of severely infected plants wither and die. In the aerial blight phase of the disease, the drooping leaves of the plant stick to the stem portion and pods immediately below them and thus act as a source of infection for initiating infection in pods of the plant. The characteristic brown-colored lesions caused by the pathogen start appearing on petioles and stems at this stage of infection. The infection of pods leads to the production of infected seeds. The development of mycelial growth of the fungus and the production of sclerotia are enhanced in conditions of high humidity (Kumar et al. 2016).

Under field conditions, the disease response of different cultivars varies with respect to severity of symptoms produced on the plants. The susceptible varieties may support the growth of infection cushion of the pathogen by producing chemical stimulants, which are found to be absent in case of resistant and tolerant varieties of soybean. Several cultivars showing a moderately resistant response to the disease are in use, but no cultivar of soybean has been reported to be completely resistant to the disease. Several workers have emphasized the role of weather factors such as temperature, humidity, rainfall, sunshine hours, etc. in the development of the disease over time in the crop season (Mathpal 2016; Joshi et al. 2018; Singh et al. 2019a, b; Surbhi and Singh 2020). An atmospheric temperature in the range of 10–40 °C, with an optimum at 28 °C, is favorable for the development of the disease, while the production of sclerotia of the fungus is inhibited at temperatures below 10 °C. A relative humidity of 85% to 90% supports the growth of *R. solani* in the field (Surbhi and Singh 2020).

15.2.4.3 Disease Management

The most important step while planning the management of aerial blight is the selection of soybean variety. The use of resistant varieties is a cheap, easy and safe method to manage the disease. However, no commercial soybean cultivars with complete resistance or immunity to the infection of the pathogen are still known, but moderately resistant cultivars could be used for cultivation (Palat et al. 2004; Thind 2005). Crop rotation with nonleguminous crops, destruction or burning of crop residues, summer ploughing, and use of recommended doses of fertilizers help in reducing the incidence of the disease (Prasad 2005). Another important management practice is the maintenance of proper spacing between plant rows in the field. A dense canopy could create a humid environment, thus favoring the growth of the pathogen. Also, the chances of spread of infection due to contact of leaves of healthy plant with infected plants or splashing of mycelial fragments due to rain might be more if the canopy is dense.

Under chemical control, fungicides such as Pyraclostrobin (0.1%), Pencycuron (0.1%), and Fluopyram (0.1%) have been reported to be effective in controlling the disease (Mathpal 2016). Seed treatment with Thiram is known to reduce the initial infection by the pathogen (Thind 2005). Two foliar sprays of Tilt (0.1%) at an interval of 15 days have been reported to reduce the severity of the disease in the field. Some other fungicides such as Contaf, Bavistin, and Indofil M-45 have also been reported to give a good control of the disease in the field (Sharma and Tripathi 2001). Biocontrol agents such as *Trichoderma harzianum, Pseudomonas fluorescens*, and *B. subtilis* are also helpful in management of the disease (Mathpal 2016).

15.2.5 Root and Hypocotyl Rot of Soybean

15.2.5.1 Economic Importance

Seedling diseases are a major constraint in the cultivation of soybean. *Rhizoctonia* root rot, caused by *Rhizoctonia solani* Kühn, is one of the most important root and seedling diseases of soybean [*Glycine max* (L.) Merr.] in the world (Wrather et al. 2001; Dorrance et al. 2003; Ajayi-Oyetunde and Bradley 2017). The root and hypocotyl rot of soybean is an important disease affecting the growth of plants in early stages (Mahmoud and Abo-Elyousr 2014). All over the world, various anastomosis groups of *R. solani* including AG-2-2IIIB, AG-4, and AG-5 have been found to cause the seedling diseases of soybean (Fenille et al. 2002; Zhao et al. 2005). AG-3, AG-7, and AG-11 are also known to cause root rots but not as severe as the other groups. The disease has been known to occur in different soybean-growing areas all across the world, including Illinois, Ohio, Minnesota, Kansas, and North Dakota in the United States, Brazil, China, and India (Wrather and Koenning 2006; Zhao et al. 2005). For a period of 3 years from 2006 to 2009, seedling diseases were reported to be the third highest in causing yield losses to soybean crop in the United States and Canada (Koenning and Wrather 2010). During 1967, an epidemic of root rot caused by *R. solani* emerged in Iowa, United States. The disease has been reported to cause a reduction of 50% in stand establishment of the crop (Wrather et al. 2001; Xue et al. 2006). In Brazil, *R. solani* has been reported to cause preemergence and postemergence damping-off of seedlings, root and hypocotyl rot, and aerial blight in soybean.

15.2.5.2 Disease Symptoms and Epidemiology

The most vulnerable stage of the plant is before the development of the first trifoliate leaf. The infection starts as a small, sunken, reddish-brown lesion on the hypocotyls of the germinating seed just near or below the soil line (Kumar et al. 2019).

Such lesions grow in size and cause girdling of the germinating seedling leading to the occurrence of postemergence damping-off of the seedlings (Zhang 2011). If the seedling is able to survive the infection, roots develop normally, and the plant may show complete recovery from the infection of the pathogen. The infected plants in the field appear stunted and chlorotic. In the absence of adequate soil moisture, the plant may wilt. The infected plants may be seen both scattered across the field or in groups. Wet soil and warm conditions are favorable for the development of the disease. Any sort of damage to the roots caused by stress or injury, nematode action, insect feeding, or herbicide damage may foster the infection of root rot (Duke et al. 2007). The pathogen can survive in the form of sclerotia or dormant mycelium in the soil and plant debris. It can live in the soil as a saprophyte and could survive long periods in the absence of any host plant. These sclerotia and mycelium surviving in the soil serve as a primary inoculum for the initiation of the disease in the crop season (Gossen et al. 2016). Temperatures in the range of 26–32 °C, soil moisture of 30–60%, and a soil pH above 6.6 are favorable for the development of the disease. A deficiency of water or nutrients could increase the impact of the disease.

15.2.5.3 Disease Management

Proper tillage helps to improve soil drainage and make the soil environment less favorable for the growth of pathogens. Injuries caused by insect damage, nematodes, or herbicide damage may lead to increased susceptibility of the plant to the pathogen. Thus, proper management of insect pests and nematodes is essential in the field, and herbicide injuries should be avoided or taken care of. For cultivation on a commercial scale, resistant cultivars of soybean should be used. Crop rotation and destruction of crop debris are also important to reduce the amount of inoculum in the field (Zhang 2011). Seed treatment with biocontrol agents such as *Trichoderma viride* @ 10 g/kg seed has also been found effective in reducing the incidence of root rot in the field (Barpete and Verma 2019). Seed treatment with fungicides of the strobilurin group, such as Azoxystrobin, may help reduce the incidence of disease. Seed treatment with Carboxin (2 g/kg seed) and foliar spray of Propiconazole and neem oil (2 mL/L) have also been found effective in managing root rot of soybean in the field (Kumar et al. 2019).

15.2.6 Web Blight of Urdbean

15.2.6.1 Economic Importance

Urdbean [*Vigna mungo* (L.) Hopper] is a valuable pulse crop grown in India. It is also known as black gram. It is cultivated in the *Kharif* season in the country. Several biotic and abiotic factors have a major influence on the productivity of the crop. One of the most important and destructive fungal diseases of urdbean is web blight caused by *Rhizoctonia solani* Kühn (AG 1A). The disease is one of the most important causes of declined productivity of the crop in the country (Dubey and Patel 2001). It has been reported to cause about a 20–30% loss in yield of the crop in

India, depending on the variety used, environmental conditions, stage of the crop, and cultivation practices followed (Dubey and Patel 2001; Shailbala and Tripathi 2007). The disease causes severe damage to the grain quality of the crop. It has been reported from various urdbean cultivating areas of the country, such as Uttarakhand, Uttar Pradesh, Haryana, Punjab, Rajasthan, Bihar, Himachal Pradesh, West Bengal, and Jammu & Kashmir. The disease has also been reported to cause yield losses in urdbean in different countries, including Japan, Sri Lanka, Myanmar, Pakistan, the Philippines, North America, South America, Brazil, Mexico, Argentina, and the West Indies (Kumar et al. 2018b). The disease is difficult to manage due to its wide host range and ability to survive on weed plants (Chahal et al. 2003).

15.2.6.2 Disease Symptoms and Epidemiology

The disease symptoms start in the form of small brown spots on the leaves. These spots are first circular in shape but, after some time, the spots enlarge and their edges become irregular and water-soaked (Singh 2006). The lesions further increase in size and turn from green to reddish-brown or black in color. A fine white-colored fungal mycelium can be seen growing on the lower side of the infected leaves as well as on the young branches and petioles of the infected plant. The mycelium grows on the aerial parts in a spider web-like manner, giving the disease its name. White-colored microsclerotia develop on the infected plant parts. The spread of infection in the collar region of the plant may lead to collar rot. The pathogen is able to infect the pods of the plant at any stage of development, leading to the formation of little irregular specks on the young pods. The spots observed on mature pods are dark brown and circular in shape. Soon, such spots increase in size and cover the whole pods, leading to shriveling of pods and decreased grain quality.

The disease has been known to prevail mainly in tropical and subtropical regions of the world in summer. Warm and wet weather, i.e., a high temperature (26 °C to 32 °C) and high relative humidity (90–100%), is favorable for the growth of the pathogen. Rainfall of about 91–97 mm during the early growth stages of the crop increases the severity of the disease in the field (Sharma and Thripathi 2001). A soil pH in the range of 5 to 8 favors the disease (Kumar et al. 2014b).

R. solani can survive in soil and on plant debris. The pathogen can spread in the field by means of irrigation water, farming equipment, contaminated seeds, and rain splashes. The chemical stimulants released from growing tissues of the young plants attract the fungus. The fungus gains entry into the plant by natural openings or direct penetration of the epidermis. The fungus forms sclerotia in the plant, which survive and cause disease in the next season. The secondary spread of the disease occurs by means of mycelium spread by rain splashes or contact of a healthy plant with an infected plant. Maximum severity of the disease has been reported to occur in 30 to 60-day-old plants (Kumar et al. 2013). Dense crop canopy increases the severity and spread of the disease.

15.2.6.3 Disease Management

Management of web blight in urdbean requires a combination of different methods. Several cultural practices, such as crop rotation, tillage, destruction of crop debris, proper spacing between plant rows, and manipulation of sowing dates, may help to reduce the incidence and severity of the disease. The use of resistant varieties has also been found useful in managing the disease in the field (Kumar et al. 2014a; Neelam et al. 2014). Seed treatment with *Trichoderma viride*, *Rhizobium* sp., and Carboxin or Bavistin (@ 2.5 g/kg seed) has been found to be effective in managing the disease (Dubey and Patel 2001; Shailbala and Tripathi 2010). Several fungicides, such as propiconazole, hexaconazole, tebuconazole, epoxiconazole, and propineb, have also been found effective in managing the disease (Shailbala and Tripathi 2010).

15.2.7 Web Blight of Mungbean

15.2.7.1 Economic Importance

Web blight of mungbean, caused by *R. solani* Kühn, is one of the major diseases responsible for reducing the productivity of the crop. The disease was first reported on mung in 1924 from the Philippines. The disease was later reported from India and Pakistan. The disease is widespread in India and has been reported from all the mungbean growing areas in the country, including Assam, Punjab, Haryana, Madhya Pradesh, Bihar, Rajasthan, Himachal Pradesh, and Jammu and Kashmir. The disease has been reported to cause a reduction of up to 40% in the yield of mungbean (Singh et al. 2013). Web blight is a serious disease of mungbean in Northern India and is known to cause extensive yield damage to the crop.

15.2.7.2 Disease Symptoms and Epidemiology

The pathogen is mainly soil-borne but is also known to spread by air in the form of rain splashes. Although all the above-ground parts (including leaves, petioles, stem, and pods) can be infected by the fungus, the damage done to foliage is most destructive. The fungus infects the plant around second to third week of growth and leads to the death of the young seedlings. The infection of fungus around the collar region of the plant leads to the development of reddish-brown lesions around the hypocotyls, which finally coalesce and girdle the stem, resulting in the death of the seedling (Singh et al. 2013).

On leaves, the symptoms of the disease start appearing in the form of small, water-soaked, greenish spots, which are irregular in shape (Singh et al. 2013). Soon these spots turn dark in color with a tanned center and irregular dark brown margins. As a result of the coalescing of many such spots, the leaf shrivels and dries up. The mycelium of the fungus appears as a spiderweb, and the formation of white microsclerotia takes place in abundant quantities on the infected plant. These sclerotia later turn brown in color. The infection of the fungus also reaches the pods and covers the whole pod under favorable environmental conditions. The pathogen *R. solani* is primarily soil-borne, and it can survive in soil for many years by producing sclerotia. The secondary spread of the disease occurs by the spread of aerial mycelium through rain splashes and the sclerotia formed by the fungus. The prevailing environmental conditions play an important role in the development of the disease. Atmospheric temperature in the range of 26 °C to 32 °C, RH of 90% to 100%, and a soil temperature of 30–33 °C are known to be associated with high severity of the disease. An average rainfall of about 91–97 mm has been found to have a significant effect on the development of the disease in early crop stages (Singh et al. 2013).

15.2.7.3 Disease Management

The use of resistant plant cultivars is a cheap and eco-friendly method of managing the disease in the field. After the selection of a suitable resistant variety, the second most important step in management is following proper cultural practices. The incidence of disease has been found to be higher in early sown crops compared to late sown crops (Singh et al. 2013). Dense planting of the crop should be avoided. A wider row spacing of about 50 cm or more is effective in decreasing the infection of web blight in mung and also improves the yield (Singh et al. 2013). The crop residue should be buried soon after harvesting of the crop, as it may act as a source of inoculum in the field. Crop rotation with a nonhost crop of *R. solani* is effective in managing the disease. Weed plants that may serve as alternate hosts of the pathogen should also be removed and destroyed. The soil should be maintained in a well-drained and well-aerated condition.

Biocontrol agents such as *Trichoderma viride, Trichoderma harzianum*, and *Gliocladium virens* have been found to be effective in the management of web blight (Dubey 2000; Singh et al. 2013; Chauhan 2017; Chauhan et al. 2020). Under field conditions, three foliar sprays of Propiconazole @ 0.1% at an interval of 15 days are effective in managing the disease (Chauhan 2017). Azoxystrobin has been reported to give effective inhibition (91.48%) of the fungal mycelium in in vitro evaluation (Chauhan 2017).

15.2.8 Root Rot of Pulses

15.2.8.1 Economic Importance

India is a major pulse-producing country, contributing about 25% share in the global production of pulses. A wide variety of pulse crops is cultivated in different agro-ecological zones throughout the country. Biotic stress, such as plant diseases, is an important factor responsible for low productivity of pulses in the country. Pulses are heavily infected by different fungal pathogens, which lead to root rot and seedling blight. The major fungal pathogens involved in the root rot complex of pulses include *Fusarium* spp., *Rhizoctonia solani*, and *Pythium* spp. A wide range of pulses, including chickpea, cowpea, mungbean, urdbean, lentil, and French bean, can be infected by *R. solani*. In India, the pathogen has been reported to cause serious yield losses in mung and urdbean (Dubey 2003). The pathogen has been reported to cause around 79% loss in yield was recorded in field pea, 70% in chickpea, 69% in lentil, 88% in bean, and as high as 19% in faba bean (Hwang et al. 2003, 2007; Chang et al. 2004, 2008, 2014).

15.2.8.2 Disease Symptoms and Epidemiology

The most important group of *R. solani*, which has been reported to infect pulses worldwide, is AG-4, but the group AG-2-2 has also been reported to cause diseases in dry beans and faba beans (Conner et al. 2014; Hwang et al. 2003, 2007; Mathew et al. 2012; Sharma-Poudyal et al. 2015; Zhou et al. 2009). Another important group is AG-2-1, which has been reported to cause root rot and seedling death in pea (Sharma-Poudyal et al. 2015). The infection of *R. solani* in the early growth stages of pulse crops leads to rotting of seeds, damping off, and blight of seedlings. In case of fungal infection in older plants, reddish-brown, sunken lesions appear on the epicotyls, hypocotyls, tap root, and base of the stem (Hanson 2005). In case of pea and faba bean, soft rot symptoms have been observed on roots and stem (Hwang et al. 2007). In pea, the pathogen leads to stunting and wilting of the plant and girdles the stem, leading to stem rot (Navarro et al. 2008; Porter et al. 2011). In case of severe infection, the plant can develop adventitious roots adjacent to the surface of soil. The pathogen also leads to a reduction of nodulation in the roots of the infected plant (Chang et al. 2008). The fungus produces small, irregular, brownish-black sclerotia on the infected parts of the plant (Porter et al. 2011). These sclerotia can survive in soil and cause infection in the next season. The fungus is also known to survive in the form of mycelium in infected crop debris (Hanson 2005; Porter et al. 2011).

Atmospheric temperatures in the range of 24 °C to 30 °C favor the hyphal spread of the pathogen and the development of disease symptoms. The severity of disease is greatly affected by the amount of inoculum in the field (Dillard 2001; Hwang et al. 2007). Warm soil temperature is conducive to the development of the disease, and late sowing in such soils has been reported to cause a great reduction in the establishment of seedlings in the field (Chang et al. 2004, 2008; Hwang et al. 2007). Deep sowing of seeds has also been reported to increase the severity of disease in field peas and chickpeas. Low fertility, poor drainage, and high compaction of soil act as predisposing factors for the occurrence of disease in pulse crops (Harveson et al. 2005).

15.2.8.3 Disease Management

Selection of resistant crop varieties for sowing is important for managing the occurrence of disease in the field. Following recommended cultural practices, such as proper depth of sowing, maintenance of proper plant spacing, and proper irrigation and fertilizer management in field, is important for minimizing favorable conditions for the pathogen in the field. The destruction of crop residue, which could serve as a source of inoculum, is important. The rotation of pulse crops with nonhost crops of the fungus to avoid the buildup of pathogen inoculum in the field is also effective in managing the disease. Proper tillage operations should be done to reduce soil compaction and improve drainage of soil. Early sowing of crops could help to escape infection of the pathogen, as infection has been reported to be higher in late-sown crops (Chang et al. 2004, 2008; Hwang et al. 2007). Seed treatment with systemic fungicides and biocontrol agents improves seedling establishment (Abdel-Monaim 2013; Chang et al. 2014). Application of *Rhizobium* has been reported to improve the health of seedlings and plants in chickpeas (Hemissi et al. 2013).

15.2.9 Root Rot of Cotton

15.2.9.1 Economic Importance

Rhizoctonia solani Kühn and *Rhizoctonia bataticola* are serious and destructive pathogens of cotton. The root rot and damping-off caused by *R. solani* is responsible for causing heavy losses during the early growth stages of the crop. The disease has been reported to cause a reduction in plant stand and heavy yield losses in different cotton-producing countries around the world (Yadav 2018). In India, the disease is prevalent in an area of around 1.8 million hectares in the northern cotton-producing states, including Rajasthan, Punjab, and Haryana (Monga and Raj 2016). The disease starts appearing in the month of June and leads to the sudden wilting of plants. The plants infected with the disease become weak and can be easily pulled out from the ground. The severity of the disease increases till July and then slowly declines during August, ultimately ceasing by the end of September (Monga and Raj 2016). During the rainy season, the disease can attain an intensity of up to 90% (Yadav 2018).

15.2.9.2 Disease Symptoms and Epidemiology

With the first appearance of symptoms, the leaves turn slightly yellow leading to bronzing and wilting of leaves. In the field, the disease appears in form of patches of infected plants here and there. These patches get converted into circular diseased patches, and sudden wilting of the infected plants takes place. Sudden wilting of the plants is a characteristic of this disease. The infected plants can be easily pulled out of the ground. The surface of the roots of the infected plant becomes shredded and they turn yellowish in appearance. Favorable soil temperature for the disease varies from 29 °C to 36 °C (Monga and Raj 2016).

15.2.9.3 Disease Management

Seed treatment helps in reducing the incidence of disease in the field. Several fungicides, such as pyraclostrobin, fluxapyroxad, trifloxystrobin, fludioxonil, and carboxin, are effective in managing the disease in the field. Some fungicides can be used singly or in combination as liquid formulations for application in furrows. These include iprodione, azoxystrobin, and a combination of azoxystrobin and mefenoxam (Yadav 2018; Isakeit 2016).

15.2.10 Damping off and Root Rot of Oilseed Rape

15.2.10.1 Economic Importance

Rhizoctonia solani is an important pathogen of rapeseed in Canada, causing severe reductions in germination and establishment of seeds (Hwang et al. 2014). It also leads to symptoms such as damping off, seedling blight, and root rot (wire stem) (Soon et al. 2005). On adult plants of rapeseed, the pathogen causes foot and root rot by girdling the stem base. The root rot caused by *R. solani* has been a major constraint to cultivation of oilseed rape for many years in the Canadian prairies region. The incidence of the disease has been reported to be up to as high as 80–100%. In Finland, Tahvonen et al. (1984) reported a 90% incidence of damping off due to *R. solani* in spring oilseed rape fields. During 2005 and 2006, a 20% reduction in stand establishment of the crop was reported in central Alberta (Hwang et al. 2014). Yield losses due to the disease have been reported to be as high as 20% to 30% in Alberta, Canada (Khangura et al. 1999).

15.2.10.2 Disease Symptoms and Epidemiology

Light brown lesions develop on the taproot and lateral roots, usually at or following the onset of flowering. Lesions become sunken, darken, and may enlarge to girdle the taproot, resulting in premature death of the plants. Girdled plants may have only a short stub of the taproot remaining and are subject to desiccation and lodging. Losses are greatest when high temperatures and dry, windy conditions follow the root damage. This root rot, characterized by a brown girdling of the root and basal stem, causes poor pod formation, pod sterility, loss of seed weight, shriveled seeds, premature senescence of the pods, lodging, and uneven ripening (Hwang et al. 2014).

15.2.10.3 Disease Management

Seed treatment with the fungicides Helix Xtra (thiamethoxam + difenoconazole + metalaxyl + fludioxonil) and Prosper FX (clothianidin + carboxin + trifloxystrobin + metalaxyl) has been reported to increase seedling emergence and yield in canola in field trials in Alberta, Canada (Hwang et al. 2014).

15.3 Conclusion

There exists a vast array of unexplored facts about the taxonomy and population dynamics of the genus *Rhizoctonia*. For a better understanding of the genus and establishment of a more significant taxonomic system, the genetic variability and mechanism of sexual recombination among this group of fungi further need to be evaluated more deeply. Owing to the vast genetic variability and wide host range, the fungus is able to infect many of the economically important field crops, leading to significant reduction in yield and, as such, the income of farmers. Due to its distinctive ability to produce durable survival structures, overwintering in soil and on alternate hosts, the fungus is difficult to manage. Moreover, due to complex mechanisms of inheritance of resistance to *R. solani* in crops, breeding for resistant crop varieties is difficult, and very few commercial cultivars have been known to show absolute resistance to infection by the fungus. As such, to minimize losses caused by the fungus to the crops, the use of suitable cultural practices is of utmost importance. There have been several successful attempts in managing the infection of the fungus through biological control methods. The role of seed treatment chemicals to provide protection in the early stages of the crop and foliar sprays to prevent further development and spread of the pathogen are important. Considering the capability of the fungus to infect crops at any stage of development, an integrated management approach should be followed which keeps a check on the health of the plant at each stage of development.

References

Abdel-Monaim MF (2013) Improvements of biocontrol of damping-off and root rot of faba bean by salicylic acid and hydrogen peroxide. Mycobiology 41:47–55

Agrios GN (2005) Plant pathology, 5th edn. Academic Press, New York

Ajayi-Oyetunde OO, Bradley CA (2017) Identification and characterization of *Rhizoctonia* species associated with soybean seedling disease. Plant Dis 101(4):520–533

Ajayi-Oyetunde OO, Bradley CA (2018) Rhizoctonia solani: taxonomy, population biology and management of rhizoctonia seedling disease of soybean. Plant Pathol 67(1):3–17

Akhtar J, Jha VK, Kumar A et al (2009) Occurrence of banded leaf and sheath blight of maize in Jharkhand with reference to diversity in *Rhizoctonia solani*. Asian J Agric Sci 1:32–35

Almasudy AM, You MP, Barbetti MJ (2015) Influence of fungicidal seed treatments and soil type on severity of root disease caused by *Rhizoctonia solani* AG-8 on wheat. Crop Prot 75:40–45

Amaradasa BS, Lakshman DK, McCall D et al (2014) In vitro fungicide sensitivity of *Rhizoctonia* and Waitea isolates collected from turf grasses. J Environ Hort 32:126–132

Anees M, Edel-Hermann V, Steinberg C (2010) Buildup of patches caused by *Rhizoctonia solani*. Soil Biol Biochem 42(10):1661–1672

Barnett S, Zhao S, Ballard R et al (2017) Selection of microbes for control of Rhizoctonia root rot on wheat using a high throughput pathosystem. Biol Control 113:45–57

Barnett SJ, Ballard RA, Franco CM (2019) Field assessment of microbial inoculants to control *Rhizoctonia* root rot on wheat. Biol Control 132:152–160

Barpete RD, Verma VK (2019) Management of rhizoctonia root rot disease in soybean in Betul district of Madhya Pradesh. Plant Arch 19(2):2376–2378

Baysal-Gurel F, Kabir N (2018) Comparative performance of fungicides and biocontrol products in suppression of *Rhizoctonia* root rot in viburnum. J Plant Pathol Microbiol 9(451):2

Black BD, Griffin JL, Russin JS et al (1998) Weed hosts for *Rhizoctonia* foliar blight in soybeans. Louisiana Agric 41(3):17–18

Bogacki P, Desbiolles J, Correll R et al (2014) Targeted liquid delivery of fungicides: a new tool for Rhizoctonia root rot control. In: Eyre peninsula farming systems summary. Grains Research and Development Corporation, Perth, pp 89–92. https://eparf.com.au/research-project/ep-farming-systems-summary-2014/

Buddemeyer J, Pfahler B, Peterson J et al (2004) Genetic variation in susceptibility of maize to Rhizotonia solani (AG2–2IIIB)-symptoms and damage under field conditions in Germany. J Plant Dis Protect 111:521–533

Carling DE, Kuninaga S, Brainard KA (2002) Hyphal anastomosis reactions, rDNA-internal transcribed spacer sequences, and virulence levels among subsets of *Rhizoctonia solani* anastomosis group-2 (AG-2) and AG-BI. Phytopathology 92(1):43–50

Chahal KS, Sokhi SS, Rattan GS (2003) Investigations on sheath blight of rice in Punjab. Indian Phytopathol 56:22–26

Chang KF, Hwang SF, Gossen BD et al (2004) Effects of soil temperature, seeding depth, and seeding date on Rhizoctonia seedling blight and root rot of chickpea. Can J Plant Sci 84:901–907

Chang KF, Hwang SF, Gossen BD et al (2008) Effects of inoculum density, temperature, seeding depth, seeding date and fungicidal seed treatment on the impact of Rhizoctonia solani on lentil. Can J Plant Sci 88:799–809

Chang KF, Conner RL, Hwang SF (2014) Effects of seed treatment and inoculum density of *Fusarium avenaceum* and *Rhizoctonia solani* on seedling blight and root rot of faba bean. Can J Plant Sci 94:693–700

Chang KF, Hwang SF, Ahmed HU et al (2017) Disease reaction to Rhizoctonia solani and yield losses in soybean. Can J Plant Sci 98(1):115–124

Chauhan S (2017) Etiology and management of web blight disease of mungbean incited by *Rhizoctonia solani* Kuhn. MSc thesis, GB Pant University of Agriculture and Technology, Pantnagar

Chauhan S, Yadav LB, Kushwaha KPS et al (2020) Potential of botanicals and biocontrol agents against *Rhizoctonia solani* Kühn Incitant of web blight disease of Mung bean: an *in vitro* evaluation. Int J Curr Microbiol App Sci 9(6):2627–2636. https://doi.org/10.20546/ijcmas.2020.906.xx

Conner RL, Hou A, Balasubramanian P et al (2014) Reaction of dry bean cultivars grown in western Canada to root rot inoculation. Can J Plant Sci 94:1219–1230

Crispin A, Gallegos CC (1963) Web blight: a severe disease of beans and soybeans in Mexico. Plant Dis Rep 47:1010–1011

Datta A, Vurukonda SSKP (2017) Rice sheath blight: a review of the unsung fatal disease. Trends Biosci 10(45):9216–9219

Dillard HR (2001) *Rhizoctonia* seed and seedling rot. In: Kraft JM, Pfleger FL (eds) Compendium of pea diseases and pests. APS Press, St. Paul, p 9

Dorrance AE, Kleinhenz MD, McClure SA et al (2003) Temperature, moisture and seed treatment effects on *Rhizoctonia solani* root rot of soybean. Plant Dis 87:533–538

Dubey SC (2000) Biological management of web blight of groundnut (*Rhizoctonia solani*). J Mycol Plant Pathol 30(1):89–90

Dubey SC (2003) Integrated management of web blight of urd/mung bean by bio-seed treatment. Indian Phytopathol 56:34–38

Dubey SC, Patel B (2001) Evaluation of antagonists against *Thanatephorus cucumeris* causing web blight of urd and mungbean. Indian Phytopathol 54:206–209

Duke SO, Wedge DE, Cerdeira AL et al (2007) Herbicide effects on plant disease. Pest Manag Sci 18:36–40

Fenille RC, De Souza NL, Kuramae EE (2002) Characterization of *Rhizoctonia solani* associated with soybean in Brazil. Eur J Plant Pathol 108:783–792

Gill JS, Sivasithamparam K, Smettem KRJ (2001) Effect of soil moisture at different temperatures on *Rhizoctonia* root rot of wheat seedlings. Plant Soil 231:91–96

Gilligan CA (2002) An epidemiological framework for disease management. Ad Bot Res 38:1–64

Gossen BD, Conner RL, Chang KF et al (2016) Identifying and managing root rot of pulses on the northern great plains. Plant Dis 100(10):1965–1978

Grosch R, Schneider JHM, Kofoet A (2004) Characterisation of *Rhizoctonia solani* anastomosis groups causing bottom rot in field-grown lettuce in Germany. Eur J Plant Pathol 110:53–62. https://doi.org/10.1023/B:EJPP.0000010137.69498.10

Handiseni M, Brown J, Zemetra R et al (2013) Effect of Brassicaceae seed meals with different glucosinolate profiles on *Rhizoctonia* root rot in wheat. Crop Prot 48:1–5

Hanson LE (2005) *Rhizoctonia* root rot. In: Schwartz HF, Steadman JR, Hall R, Forster RL (eds) Compendium of bean diseases. APS Press, St. Paul, pp 19–20

Harveson RM, Smith JA, Stroup WW (2005) Improving root health and yield of dry beans in the Nebraska panhandle with a new technique for reducing soil compaction. Plant Dis 89:279–284

He QF, Chen WL, Ma ZC (2002) Purification and properties of antagonistic peptide produced by *Bacillus subtilis* A30. Chin J Rice Sci 16(4):361–365

Hemissi I, Mabrouk Y, Mejri S et al (2013) Enhanced defence responses of chickpea plants against *Rhizoctonia solani* by pre-inoculation with rhizobia. J Phytopathol 161:412–418

Hooda KS, Khokhar MK, Parmar H et al (2017) Banded leaf and sheath blight of maize: historical perspectives, current status and future directions. Proc Natl Acad Sci B Biol Sci 87(4):1041–1052

Hüberli D, Conner M, MacLeod B (2015) *Rhizoctonia solani* AG8: new breakthroughs in control and management. Agribusiness Crop Updates. p 24–25. http://researchrepository.murdoch.edu.au/id/eprint/25655/1/Rhizoctonia_solani_AG8.pdf

Hwang SF, Gossen BD, Chang KF (2003) Etiology, impact and control of rhizoctonia seedling blight and root rot of chickpea on the Canadian prairies. Can J Plant Sci 83:959–967

Hwang SF, Gossen BD, Conner RL (2007) Management strategies to reduce losses caused by rhizoctonia seedling blight of field pea. Can J Plant Sci 87:145–155

Hwang SF, Ahmed HU, Turnbull GD (2014) The effect of seed size, seed treatment, seeding date and depth on Rhizoctonia seedling blight of canola. Can J Plant Sci 94(2):311–321

Isakeit T (2016) Managing seedling diseases of cotton. Texas A&M University, College Station. http://cotton.tamu.edu/Nematodes/Management

Jaaffar AKM, Paulitz TC, Schroeder KL (2016) Molecular characterization, morphological characteristics, virulence, and geographic distribution of *Rhizoctonia* spp. in Washington State. Phytopathology 106(5):459–473

Joshi D, Singh PK, Adhikari S et al (2018) Screening of soybean germplasm for important disease prevalent in North India. Int J Chem Stud 6(2):2731–2733

Kazempour MN (2004) Biological control of *Rhizoctonia solani*, the causal agent of rice sheath blight by antagonistics bacteria in greenhouse and field conditions. Plant Pathol J 3:88–96. https://doi.org/10.3923/ppj.2004.88.96

Khangura RK, Barbetti MJ, Sweetingham MW (1999) Characterization and pathogenicity of *Rhizoctonia* species on canola. Plant Dis 83:714–721

Koenning SR, Wrather JA (2010) Suppression of soybean yield potential in the continental United States by plant diseases from 2006 to 2009. Online Publ Plant Health Prog. https://doi.org/10.1094/PHP2010-1122-01-RS

Krishnam Raju S, Kumar VK, Ramabhadra Raju M (2008) Efficacy of tebuconazole against *Rhizoctonia solani*, causal agent of rice sheath blight. Indian J Plant Protect 36:98–101

Kumar KVK, Reddy MS, Kloepper JW et al (2009) Sheath blight disease of rice (*Oryza sativa* L.)—an overview. Biosci Biotechnol Res Asia 6:465–480

Kumar S, Lal M, Garkoti A et al (2013) Standardization of inoculation techniques, plant age and host range of *Rhizoctonia solani*, the incitant of web blight of urdbean. Plant Dis Res 28(1):45–48

Kumar S, Chand G, Tripathi HS et al (2014a) Assessment of yield losses due to web blight and evaluation of urdbeangenotypes for resistance in Tarai region of Uttarakhand. Ann Plant Prot Sci 22:395–397

Kumar S, Kumar A, Chand G et al (2014b) Dynamics of mycelial growth and sclerotia production of *Rhizoctonia solani* Kuhn (AG1-IB) of urdbean. Ecoscan 8:273–277

Kumar T, Dantre RK, Verma KP (2016) Studies on aerial blight of soybean caused by *Rhizoctonia solani* Kuhn. J Plant Dev Sci 8(10):475–482

Kumar P, Ahlawat S, Chauhan R et al (2018a) In vitro and field efficacy of fungicides against sheath blight of rice and post-harvest fungicide residue in soil, husk, and brown rice using gas chromatography-tandem mass spectrometry. Environ Monit Assess 190:503. https://doi.org/10.1007/s10661-018-6897-7

Kumar S, Kumar A, Tripathi HS (2018b) Urdbean web blight and its management strategies-a review. Agric Rev 39(3):210–217

Kumar A, Zacharia S, Maurya AK et al (2019) Effect of fungicides and neem oil on the *Rhizoctonia* root rot of soybean (*Glycine max* L.). Int J Curr Microbiol App Sci 8(1):368–372. https://doi.org/10.20546/ijcmas.2019.801.037

Kuninaga S (2002) Current situation of the taxonomy of the genus Rhizoctonia and the Rhizoctonia species complex. Jap J Phytopathol 68:3–20

Kuninaga S, Carling DE, Takeuchi T et al (2000) Comparison of rDNA-ITS sequences between potato and tobacco strains in *Rhizoctonia solani* AG-3. J Gen Plant Pathol 66:2–11

Kuramae EE, Buzeto AL, Ciampi MB et al (2003) Identification of *Rhizoctonia solani* AG 1-IB in lettuce, AG 4 HG-I in tomato and melon, and AG 4 HG-III in broccoli and spinach, in Brazil. Eur J Plant Pathol 109:391–395

Lal S, Baruah P, Butchaiah K (1980) Assessment of yield losses in maize cultivars due to banded leaf sclerotial disease. Indian Phytopathol 29:129–132

Lu YL, Xu J, Yuan ZM et al (2012) Comparative LD mapping using single SNPs and haplotypes identifies QTL for plant height and biomass as secondary traits of drought tolerance in maize. Mol Breeding 30:407–418

Mahmoud AFA, Abo-Elyousr KAM (2014) Genetic diversity and biological control of *Rhizoctonia solani* associated with root rot of soybean in Assiut Governorate, Egypt. J Plant Physiol Pathol 2:4. https://doi.org/10.4172/2329-955X.1000134

Mahoney AK, Babiker EM, Paulitz TC (2016) Characterizing and mapping resistance in synthetic-derived wheat to *Rhizoctonia* root rot in a green bridge environment. Phytopathology 106:1170–1176. https://doi.org/10.1094/PHYTO-02-16-0055-FI

Margani R, Widadi S (2018) Utilizing bacillus to inhibit the growth and infection by sheath blight pathogen, *Rhizoctonia solani* in rice. E&ES 142(1):012070

Mathew FM, Lamppa RS, Chittem K et al (2012) Characterization and pathogenicity of Rhizoctonia solani isolates affecting Pisum sativum in North Dakota. Plant Dis 96:666–672

Mathpal M (2016) Epidemiology and management of *Rhizoctonia* aerial blight of soybean (Glycine max L. Merrill). PhD thesis, G.B. Pant University of Agriculture and Technology, Pantnagar

Meena R, Rathore L, Mathur RS (2003) Efficacy of biocontrol agents against *Rhizoctonia solani* f. sp. *sasakii* causing banded leaf and sheath blight of maize. J Mycol Plant Pathol 33:310–312

Mehra R, Kamboj MC, Mehla JC et al (2012) Status of maize diseases and their management in Haryana. In: Proceedings of national seminar on sustainable agriculture and food security: challenges in changing climate. CCS HAU, Hisar, p 217

Misawa T, Kuninaga S (2010) The first report of tomato foot rot caused by Rhizoctonia solani AG-3 PT and AG-2-Nt and its host range and molecular characterization. J Gen Plant Pathol 76:310–319

Monga D, Raj S (2016) Root rot disease of cotton and its management. CICR technical bulletin no: 3. Central Institute for Cotton Research, Nagpur

Mukhopadhyay AN, Singh RA (1984) Phaslon Ke Rog. Anuvad and Prakashan Nideshalya, Pantnagar

Murray GM, Brennan JP (2009a) The current and potential costs from diseases of wheat in Australia. Grains Research and Development Corporation Kingston, Australian Capital Territory, Kingston

Murray GM, Brennan JP (2009b) The current and potential costs from diseases of barley in Australia. Grains Research and Development Corporation Kingston, Australian Capital Territory, Kingston

Navarro F, Sass ME, Neinhuis J (2008) Identification and confirmation of quantitative trait loci for root rot resistance in snap bean. Crop Sci 48:962–972

Neelam, Kushwaha KPS, Upadhyay V (2014) Screening of urdbean germplasm for resistance against *Rhizoctonia solani* Kuhn causing web blight disease. Int J Agric Environ Biotechnol 72:293–297

Okubara PA, Steber CM, Demacon VL (2009) Scarlet-Rz1, an EMS-generated hexaploid wheat with tolerance to the soilborne necrotrophic pathogens *Rhizoctonia solani* AG-8 and *R. Oryzae*. Theor Appl Genet 119:293–303

Palat R, Narain U, Singh PN (2004) Screening of soybean varieties against web blight. Ann Plant Prot Sci 2(2):456–457

Panth M, Hassler SC, Baysal-Gurel F (2020) Methods for Management of Soilborne Diseases in crop production. Agriculture 10(1):16

Patel BL, Bhargava PK (1998) Aerial blight of soybean (*Glycine max.*) caused by *Rhizoctonia solani*. Ind J Agric Sci 68(5):277–278

Paulitz TC, Smiley RW, Cook RJ (2002) Insights into the prevalence and management of soilborne cereal pathogens under direct seeding in the Pacific northwest, USA. Can J Plant Pathol 24:416–428

Paulitz TC, Okubara PA, Schillinger WF (2006) First report of dampingoff of canola caused by Rhizoctonia solani AG 2-1 in Washington State. Plant Dis 90(6):829

Pope EJ, Carter DA (2001) Phylogenetic placement and host specificity of mycorrhizal isolates belonging to AG-6 and AG-12 in the *Rhizoctonia solani* species complex. Mycologia 93:712–719

Porter LD, Paulitz T, Chen W (2011) Rhizoctonia seed, seedling, and wet root rot. In: Chen W, Sharma HC, Muehlbauer FJ (eds) Compendium of chickpea and lentil diseases and pests. APS Press, St. Paul, pp 28–29

Prasad R (2005) Textbook of field crop production. ICAR, New Delhi

Priyatmojo A, Escopalao VE, Tangonan NG (2001) Characterization of a new subgroup of *Rhizoctonia solani* anastomosis group 1 (AG-1-ID), causal agent of a necrotic leaf spot on coffee. Phytopathology 91:1054–1061

Radjacommare R, Kandan A, Nandakumar R et al (2004) Association of the hydrolytic enzyme chitinase against *Rhizoctonia solani* in rhizobacteria-treated rice plants. J Phytopathol 152(365):370. https://doi.org/10.1111/j.1439-0434.2004.00857.x

Rani VD, Reddy PN, Devi GU (2013) Banded leaf and sheath blight of maize incited by *Rhizoctonia solani* f.sp *sasakii* and its management—a review. Int J App Biol Pharm Technol 4:52–60

Ray A, Kumar P, Tripathi HS (2007) Evaluation of bio-agents against *Rhizoctonia solani* Kühn the cause of aerial blight of soybean. Indian Phytopathol 60(4):532–534

Roy AK (1979) Susceptibility of rice plant sheath blight at different stages of growth. Kavaka 7:25–26

Rush MC, Lee F (1983) Rice sheath blight: a major rice disease. Plant Dis 67(7):829–832

Schillinger WF, Paulitz TC (2006) Reduction of *Rhizoctonia* bare patch in wheat with barley rotations. Plant Dis 90:302–306

Schneebeli K, Mathesius U, Watt M (2015) *Brachypodium distachyon* is a pathosystem model for the study of the wheat disease *Rhizoctonia* root rot. Plant Pathol 64(1):91–100

Shailbala, Tripathi HS (2007) Current status of research on web blight disease of urdbean: a review. Agric Rev 28:1–9

Shailbala, Tripathi HS (2010) Biological and chemical management of web blight disease of urdbean caused by *Rhizoctonia solani* Kuhn. J Plant Dis Sci 54:121–125

Sharma RC (2005) Banded leaf and sheath blight (Rhizoctonia solani F. Sp sasakii) of maize. In: Zaidi PH, Singh NN (eds) Stresses of maize in tropics. Directorate of Maize Research, New Delhi, pp 159–171

Sharma G, Saxena SC (2001) Evaluation of biocontrol agents against *Rhizoctonia solani* leaf and sheath blight of maize. Ann Plant Prot Soc 9:144–145

Sharma J, Tripathi HS (2001) Biological and chemical control of web blight disease of urdbean. Indian Phytopathol 54(2):267–269

Sharma-Poudyal D, Paulitz TC, Porter LD et al (2015) Characterization and pathogenicity of Rhizoctonia and Rhizoctonia-like spp. from pea crops in the Columbia basin of Oregon and Washington. Plant Dis 99:604–613

Simon TE, Cointe LE, Delarue R et al (2014) Interplay between parasitism and host ontogenic resistance in the epidemiology of the soil-borne plant pathogen *Rhizoctonia solani*. PLoS One 9:e105159

Singh BP (2006) Effect of carbon and nitrogen sources on *Rhizoctonia solani* and its integrated management. MSc thesis, N.D.U.A.T., Kumarganj, Faizabad

Singh A, Shahi JP (2012) Banded leaf and sheath blight: an emerging disease of maize. Maydica 57:215–219

Singh R, Sinha AP (2004) Comparative efficacy of local bioagents, commercial bioformulation and fungicide for the management of sheath blight of rice under glass house conditions. Indian Phytopathol 57:494–496

Singh J, Mishra KK, Singh AK (2013) Current status of web blight of mungbean. Asian J Soil Sci 8(2):495–504

Singh KP, Aravind T, Bhatt P et al (2019a) Identification of resistant sources in field evaluation of soybean (Glycine max (L.) Merr.) genotypes against pod blight and Rhizoctonia aerial blight. Int J Curr Microbiol App Sci 8:1823–1827

Singh P, Mazumdar P, Harikrishna JA et al (2019b) Sheath blight of rice: a review and identification of priorities for future research. Planta 250:1387–1407

Sivakumar G, Sharma RC, Rai SN (2000) Biocontrol of banded leaf and sheath blight of peat based *Pseudomonas fluorescens* formulation. Indian Phytopathol 53:190–192

Soon YK, Klein-Gebbinck HW, Arshad MA (2005) Residue management and crop sequence effects on the yield and brown girdling root rot of canola. Can J Plant Sci 85:67–72

Sumner DR, Minton NA (1989) Crop losses in corn induced by Rhizoctonia solani AG2-2 and nematodes. Phytopathology 9:934–941

Surbhi K (2018) Epidemiology and molecular characterization of soybean genotypes for identification of resistant sources against *Rhizoctonia* aerial blight. MSc thesis, G.B. Pant University of Agriculture and Technology, Pantnagar

Surbhi K, Singh KP (2020) Influence of weather factors on severity of aerial blight of soybean. Ind Phytopathol. https://doi.org/10.1007/s42360-020-00235-w

Tahvonen R, Hollo J, Hannukkala A et al (1984) Rhizoctonia solani damping-offon spring turnip rape and spring rape (Brassiea spp.) in Finland. J Agric Sci Finland 56:143–154

Thind TS (2005) Diseases of field crops and their management. Daya Publication House, New Delhi

Unal F, Dolar FS (2012) First report of *Rhizoctonia solani* AG 8 on wheat in Turkey. J Phytopathol 160:52–54

Uppala S, Zhou X-G (2018) Rice sheath blight. Plant Health Instr. https://doi.org/10.1094/PHI-I-2018-0403-01

Williamson MR, Rothrock CS, Mueller JD (2006) First report of *Rhizoctonia* foliar blight of soybean in South Carolina. Plant Health Prog 7:1–2

Woodhall JW, Lees AK, Edwards SG et al (2007) Characterization of *Rhizoctonia solani* from potato in Great Britain. Plant Pathol 56:286–295

Wrather JA, Koenning SR (2006) Estimates of disease effects on soybean yields in the United States 2003 to 2005. J Nematol 38:173–180

Wrather JA, Stienstra WC, Koenning SR (2001) Soybean disease loss estimates for the United States from 1996 to 1998. Can J Plant Pathol 23:122–131

Wrather A, Shannon G, Balardin R et al (2010) Effect of diseases on soybean yield in the top eight producing countries in 2006. Plant Health Prog 10:1094

Wu W, Huang J, Cui K (2012) Sheath blight reduces stem breaking resistance and increases lodging susceptibility of rice plants. Field Crop Res 128:101–108. https://doi.org/10.1016/j.fcr.2012.01.002

Xue AG, Cober E, Morrison MJ et al (2006) Effect of seed treatments on emergence, yield, and root rot severity of soybean under Rhizoctonia solani inoculated field conditions in Ontario. Can J Plant Sci 87(1):167–174

Yadav R (2018) Integrated Management of Root rot of cotton caused by *Rhizoctonia solani* (Kuhn.). MSc thesis, MPUAT, Udaipur

Yang G, Li C (2012) General description of Rhizoctonia species complex. In: Plant pathology. IntechOpen, London, p 14

Zhang R (2011) Interactions of post emergence herbicides, strobilurin fungicides and *Rhizoctonia* root rot of soybean. MSc thesis, University of Illinois at Urbana-Champaign

Zhao G, Ablett GR, Anderson TR et al (2005) Anastomosis groups of *Rhizoctonia solani* associated with soybean root and hypocotyl rot in Ontario and resistance of accession PI 442031 to different anastomosis groups. Can J Plant Pathol 27:108–117

Zheng A, Lin R, Zhang D et al (2013) The evolution and pathogenic mechanisms of the rice sheath blight pathogen. Nat Commun 4:1424. https://doi.org/10.1038/ncomms2427

Zhou QX, Chang KF, Hwang SF et al (2009) Pathogenicity and genetic diversity of *Rhizoctonia solani* isolates from lupin and other crops in Alberta, Canada. Can J Plant Pathol 31:340–347

Tapping into the Chickpea Root Microbiome for Combating Fusarium Wilt Disease

16

Aakash V. Patel and Mansi Machal

Abstract

Chickpea is a grain legume of global importance. Its production is hindered by both abiotic as well as biotic stresses. Wilt disease caused by *Fusarium oxysporum* f. sp. *ciceris* is one of the main threats to chickpea production, causing tremendous losses worldwide. With advancements in the field of biological control, numerous microorganisms, especially those that inhabit the same habitat as that of the chickpea wilt pathogen, have been evaluated for their antagonistic capabilities. Many bacteria that colonize the chickpea root have been identified for their plant growth-promoting activities too and are known as plant growth-promoting rhizobacteria (PGPR). These microbes have the potential for their inclusion in an integrated wilt management approach in a holistic manner.

Keywords

Chickpea · Microbiome · PGPR · Wilt · Management

16.1 Introduction

Chickpea (*Cicer arietinum* L.) is an important grain legume crop and is known to occupy an important position in different cuisines worldwide. It is an important source of protein and sustains a large portion of the global population by

A. V. Patel (✉) · M. Machal
Wheat Pathology Laboratory, Division of Plant Pathology, ICAR-Indian Agricultural Research Institute, New Delhi, Delhi, India

V. K. Singh et al. (eds.), *Diseases of Field Crops: Diagnostics and Management*,
https://doi.org/10.1007/978-981-97-6160-9_16

complementing the cereal diet. Its consumption is also associated with numerous health benefits (Merga and Haji 2019). Besides this, it is an important rotation crop for biological nitrogen fixation. Despite bearing such importance, the production of chickpea is stagnant due to several abiotic and biotic constraints. Among the biotic factors, the wilt disease caused by *Fusarium oxysporum* f. sp. *ciceris* (FOC) is one of the main culprits for the low productivity of chickpea and is known to cause tremendous yield losses (Singh and Vyas 2021). The most effective control strategy for chickpea wilt is the use of resistant cultivars, but breeding lines for resistance have become difficult due to the existence of several pathogenic physiological races of FOC (Haware and Nene 1982). The sole reliance on chemical fungicides is not beneficial in the long run due to the issues of fungicide resistance in the pathogen. Moreover, the injudicious use of synthetic chemicals can cause environmental degradation and can pose serious health threats. Hence, alternative approaches are needed for developing a holistic wilt disease management program.

The use of biological control agents for the management of plant diseases is not a new concept. Several antagonistic microorganisms, like fungi, bacteria, actinomycetes, etc., are known to control the phytopathogenic *Fusarium* spp. Besides, many antagonistic bacteria that inhabit the rhizosphere of plants are known to possess plant growth-promoting activities and are categorized as plant growth-promoting rhizobacteria (PGPR). These PGPR can act as biostimulants, bioprotectants, and biofertilizers (Van Loon 2007). This chapter focuses on the use of the microbes inhabiting the chickpea rhizosphere, the different biocontrol mechanisms exhibited by them, and how these microbes can be exploited for the effective management of chickpea wilt in a holistic manner.

16.2 *Fusarium* Wilt of Chickpea

Fusarium wilt of chickpea is caused by *Fusarium oxysporum* f. sp. *ciceri* (FOC), one of the few *formae speciales* of monophyletic origin in the *F. oxysporum* complex. It is a soil-borne as well as a seed-transmitted pathogen. Long-term survival in soil and seeds is known to occur through the means of resting spores—chlamydospores. Upon attacking the plants, the mycelia enter the epidermal tissues by invading the roots and gradually extend to the vascular bundles, forming spores—macro and micro conidia—accounting for the asexual reproduction and proliferation in the plant system. The infection in seeds can cause rot and abortion, necrosis, reduction or elimination of germination capacity, as well as plant damage at later stages of plant growth, resulting in the development of the disease as systemic or local infection (Jiménez-Díaz et al. 2015). When attacked, adult plants show typical wilt symptoms, involving drooping of the aerial plant parts. External rotting is not visible in the case of infected roots, but upon vertical splitting, dark brown discoloration of the internal xylem is visible (Nene et al. 1991). Seeds from the infected pods are smaller in size, discolored, and wrinkled. These seeds can be easily detected upon visual examination, but the seeds that look normal may also harbor the wilt pathogen. Plants can be attacked at any growth stage, and the diseased plants appear throughout the field.

The eight races of FOC are identified as 0, 1A, 1B/1C, 2, 3, 4, 5, and 6 by reaction on a set of differential chickpea cultivars (Jiménez-Gasco and Jiménez-Diaz 2003). Races 0 and 1B/1C cause yellowing, whereas races 1A, 2, 3, 4, 5, and 6 induce wilting (Jiménez-Gasco et al. 2001).

16.3 Chickpea Root Microbiome

Plants harbor a wide range of microbes, viz., bacteria, fungi, nematodes, actinomycetes, protozoa, viruses, etc. These microbes can be beneficial or harmful to the host plant. Many microbes are known to exhibit an endophytic lifestyle, and they do not cause any harm to the host plant. The beneficial microbes inhabiting the phyllosphere, rhizosphere, leaf, and root endosphere are known to provide multiple benefits to the host plant (Singh et al. 2023). Many of them are potent antagonists for various plant pathogens. This microbial diversity is governed by various factors, such as the type of host plant, host morphology, weather parameters, employment of crop protection measures like pesticides, topography and soil characters, root exudates, etc.

The main beneficial microbes inhabiting the chickpea root microbiome are bacteria, fungi, and actinomycetes.

The procedure for the characterization of beneficial microbes from the chickpea microbiome, along with their potential biological applications, is depicted in Fig. 16.1.

The details about some of these reported microbes are summarized in Table 16.1.

16.4 Different Mechanisms Employed by Various Rhizospheric Microbes

16.4.1 Nutrient Bioavailability

The symbiotic association of legumes with *Rhizobium* spp. that aids in biological nitrogen fixation (BNF) is a well-known phenomenon (Totey et al. 1997). The chickpea crop also possesses a symbiotic association with *Mesorhizobium* sp. and is known to be an efficient nitrogen-fixing bacterium. Many fungi, such as *Aspergillus* spp., *Penicillium* spp., *Trichoderma* spp., and several bacterial genera, like *Bacillus* spp. and *Pseudomonas* spp., are known to produce organic acids that can solubilize as well as mobilize the nutrients like phosphorus and potassium (Zaidi et al. 2009; Ahmad et al. 2016). Besides, arbuscular mycorrhizae are also getting popular for their plant growth-promoting activities. They are obligate intracellular symbionts and are known to possess vesicles and arbuscles for directing and storing nutrients in the plant roots (Mosse 1973). The best-known example is *Glomus* sp. Many

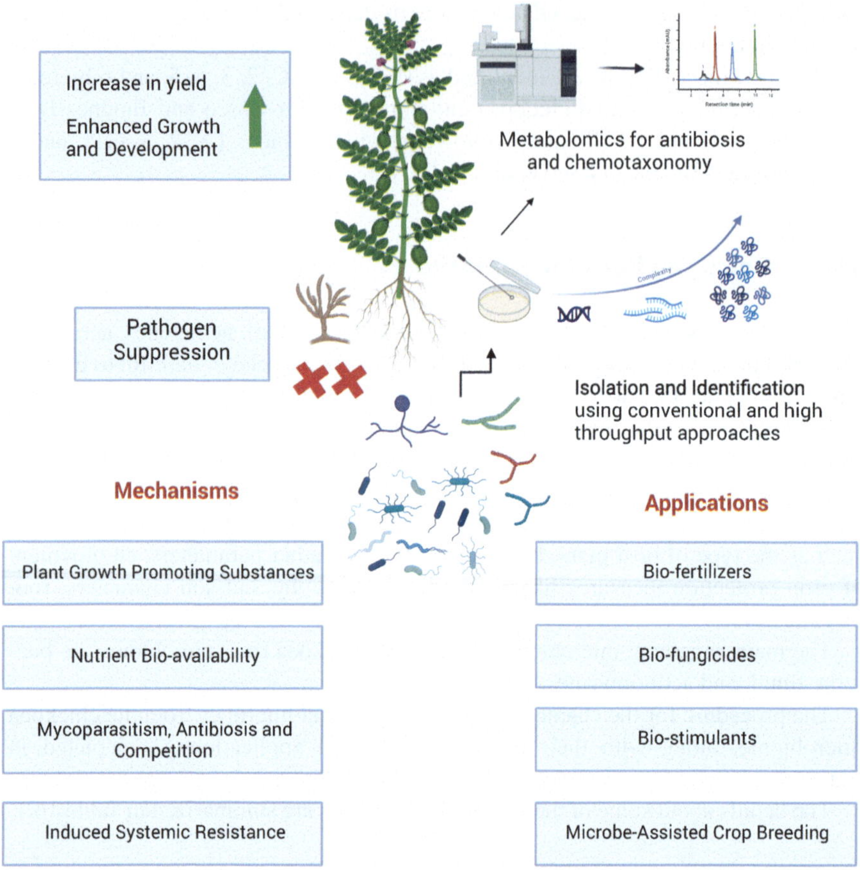

Fig. 16.1 Schematic diagram depicting the process of characterization of chickpea rhizospheric microbes and how their biological attributes can be utilized in sustainable agriculture (Created with BioRender.com)

rhizospheric microbes can also solubilize micronutrients like zinc, silicon, sulfur, etc. (Kumar et al. 2022).

16.4.2 Secretion of Phytohormones

The production of plant growth-promoting substances is indeed a significant ecological service rendered by the rhizospheric microbes. These substances are known to play a vital role in plant growth and development as well as providing resistance upon pathogen attack (Kumawat et al. 2022). Indole Acetic Acid (IAA) is one of the most widely studied plant hormones produced by microbes. It plays a very

Table 16.1 List of some of the beneficial microbes reported from chickpea rhizosphere along with their functions

Microbes	Biological function	References
Bacillus spp., *Paenibacillus* sp., *Pseudomonas* sp.	Antagonism	Elbouazaoui et al. (2022)
Pseudomonas aeruginosa	Antagonism and plant growth promotion	Mozumder et al. (2022)
Priestia megaterium, *Brucella haematophila*, and *Microbacterium paraoxydans*	Antagonism and plant growth promotion	Khanna et al. (2022)
Streptomyces araujoniae	Antagonism and plant growth promotion	Zeyad et al. (2022)
Bacillus siamensis	Antagonism and plant growth promotion	Gorai et al. (2021)
Streptomyces griseus, S. africanus, S. coelicolor	Antagonism and plant growth promotion	Ankati et al. (2021)
Panebacillus polymyxa	Antagonism and plant growth promotion	Gahoi et al. (2021)
Trichoderma asperellum, T. atroviride, T. hamatum, T. harzianum, T. koningii, T. parareesei, T. reesei, T. virens, T. brevicompactum	Antagonism	Poveda (2021)
Pseudomonas fluorescens, Pseudomonas diminuta	Antagonism and plant growth promotion	Nagpal et al. (2021)
Bacillus pumilus, B. subtilis, B. licheniformis, B. safensis, B. cereus	Antagonism and plant growth promotion	Sharma et al. (2018)
T. viride, T. harzianum, T. hamatum, T. viride, T. polysporum, T. koningii	Antagonism	Khan et al. (2014), Patole et al. (2017)
Funneliformis mosseae, Glomus fasciculatum, Rhizophagus irregularis	Antagonism and plant growth promotion	Shukla et al. (2015)
Acaulospora spinose, A. nicolsonii, Gigaspora albida, Glomus ambisporum, G. fasciculatum, G. heterosporum, G. hoi, G. lacteum, G. mosseae, G. clarum	Antagonism and plant growth promotion	Singh et al. (2013)

important role in root development as well as it is also responsible for apical dominance. Besides, ethylene is also an important phytohormone produced by various beneficial microbes that plays a vital role in induced systemic resistance (ISR).

16.4.3 Siderophores

Several bacterial genera, like *Pseudomonas* spp. and *Bacillus* spp., are known to secrete siderophores. These siderophores are low molecular weight compounds, known to chelate iron from oxyhydroxides and can make it available to plants. These siderophores also protect the plants from the toxicity of heavy metals by forming stable complexes with them (Daniel et al. 2022). The fluorescent siderophores produced by *Pseudomonas* spp. are also known to possess antifungal properties. They also play an important role in competition for iron acquisition with plant pathogens.

16.4.4 Cell Wall Degrading Enzymes

16.4.4.1 Chitinases

Chitin is a key component of the fungal cell wall. Depending on the mechanism of chitin degradation, chitinases can be classified as 1,4-β-N-acetylglucosaminidases, endochitinases, and exochitinases (Viterbo et al. 2002).

16.4.4.2 Xylanases

These enzymes degrade β-1,4-xylan into xylose subunits and hydrolyze hemicellulose, which is the major. They help in the degradation of organic matter into nutrients (Beg et al. 2001).

16.4.4.3 Lipases

Lipases are catalysts of esterification and trans-esterification reactions and are known to hydrolyze triacylglycerides into fatty acids, di-acylglycerols, and mono-acylglycerols (Fernandes et al. 2007).

16.4.4.4 Proteases

Proteases are hydrolases that hydrolyze complex proteins into amino acids. They can lyse proteins that do not belong to living cells.

16.4.4.5 β-1,3 Glucanases

These enzymes can hydrolyze the glycosidic bond present in glucan (laminarin), another important constituent of the fungal cell wall.

16.4.5 Induced Systemic Resistance (ISR)

Induced Systemic Resistance is nonspecific in nature and is known to be triggered by various beneficial microbes. It can be considered as a form of plant–microbe interaction that is known to alleviate the plant's ability to defend itself. It involves the ethylene and jasmonic acid pathways. It is responsible for the activation of various defenses in plants, such as cell wall lignification, production of antioxidants and phenolics, upregulation of various defense-related genes, etc. ISR is most widely studied in the case of *Trichoderma* spp., *Bacillus* spp., and *Pseudomonas* spp. Besides, it also gives protection against insects and herbivores. The gene NPR1 (Nonexpresser of Pathogenesis-Related Genes 1) plays a vital role in ISR as it is involved in the crosstalk between Jasmonic Acid and Salicylic Acid Signaling Pathways.

16.4.6 Mycoparasitism

Members of the genus *Trichoderma* are well known for their mycoparasitic activities, which is one of the most widely used mechanisms by them. It mainly involves the host recognition, hyphal attachment, coiling, and drawing of nutrients from the phytopathogenic fungi.

16.4.7 Antibiosis

Rhizospheric microbes are known to produce very secondary metabolites. Many of these metabolic compounds are potent antifungal substances. The most widely known secondary metabolites produced by *Pseudomonas* spp. are 2,4-diacetylphloroglucinol, pyoluteorin, phenazines, pyrrolnitrin, and HCN (Sindhu et al. 2002). Several strains of *Bacillus* spp. can produce antifungal endotoxins (Ahemad and Kilbert, 2014). The well-known antifungal compounds produced by *Trichoderma* spp. are harzianic acid, trichodermin, viridin, gliotoxin, etc.

16.4.8 Ecological Niche Occupation

Successful colonization of the root zone by beneficial microbes restricts the entry of the pathogens as they need to invade the barriers created by the beneficial microbes. Thus, the pathogens have to face competition for space and nutrients. An efficient biocontrol agent should be a rapid colonizer and should adapt well to the root zone habitat.

16.5 Evaluation of Rhizospheric Antagonists

Mainly, in vitro and greenhouse as well as field experiments are employed to evaluate the efficacy of rhizospheric microbes.

16.5.1 In Vitro Experiments

The most common technique for the evaluation of biocontrol agents is the dual culture technique, wherein the pathogen and the bioagent are inoculated on the same petri plate and allowed to grow under ambient conditions. The observed zone of inhibition shows the antagonistic potential of the biocontrol agents. A typical dual culture plate assay of *Fusarium oxysporum* f. sp. *ciceris* with *Pseudomonas* sp. and *Trichoderma* sp. is shown in Fig. 16.2.

These potential biocontrol agents can also be evaluated further in terms of their volatile as well as nonvolatile compounds using various techniques like broth dilution assay, inverted plate assay, disc diffusion assay, spore germination assay using a cavity slide, etc.

The antifungal metabolites of the biocontrol agents can be further characterized using various chromatographic techniques, and the detected compounds can be separated and evaluated using the above-mentioned techniques.

Besides, the microbes can also be evaluated for their plant growth promotion activities, such as the production of phytohormones like IAA, production of siderophores, and solubilization of nutrients like phosphorus, zinc, sulfur, potash, etc. as well as by measuring plant growth attributes.

The plant defense activities can be determined by measuring total phenol content and qualitative enzymatic assays for chitinase, glucanase, peroxidase, etc. Upregulation of several defense-related genes can be quantified using RT-PCR.

Fig. 16.2 Dual culture assay of *Fusarium oxysporum* f. sp. *ciceris* with *Pseudomonas* sp. and *Trichoderma* sp., respectively (Unpublished data)

16.5.2 Greenhouse and Field Evaluation

The in vitro experiments are performed under controlled conditions and might not yield the same kind of results under open field conditions due to various factors that give rise to variable results. Moreover, along with the validation of laboratory results, field trials are also important for the release of these microbes as inoculants for commercial use. The main factor responsible for the efficacy of these microbes is the successful colonization of the plant roots. Mainly, the microbes are applied as seed treatments, seedling dips, or are applied to the soil along with some carrier material or some organic amendments. Various treatments are evaluated using different formulations and combinations, and the most effective ones can be further validated before going for commercialization.

16.6 Applications of Rhizospheric Microbes as Biofertilizers

The beneficial rhizospheric microbes can be applied as a biofertilizer to meet the nutrient requirements of the crops. In recent times, instead of using a single strain, the use of multiple strains as a mixed inoculant biofertilizer is gaining popularity (Aloo et al. 2022). These mixed strains can exhibit different biological mechanisms, thus giving better results than the singly applied microbes. A study has shown the combined effect of a co-inoculant comprising rhizobacteria and Arbuscular Mycorrhizal Fungi (AMF) can have a profound impact on the growth of legumes (Igiehon and Babalola 2017).

With the advancements in culturomics, it has become easy to infer a bacterial species by the development of multiple growth conditions in order to make it suitable for its inclusion in the plant and soil microbiome (Mitter et al. 2021). Bioinformatics has made a significant contribution to the field of plant microbiome. Online databases like KOMODO are a great help in order to obtain the desired microbial inoculum by aiding in the in vitro development of an effective growth media (Oberhardt et al. 2015). Another advancement is the use of bacterial biofilms as a biofertilizer. Biofilms are more resistant to various biotic and abiotic stresses when compared to their planktonic forms (Parween et al. 2017). It has been found that the stress tolerance and nutrient absorption in plants can be significantly enhanced when inoculated with a fungal-bacterial biofilm (Velmourougane et al. 2017).

The development of nanobiofertilizer by combining biofertilizer and nanobiotechnology is indeed a promising approach. The main prospect of using nanofertilizer lies in the fact that microbials and nanomaterials show synergism and can improve the nutrient uptake by plants (Kumari and Singh 2019). They are produced by the reduction of metals/organic compounds by microbial extracts. This process is known as green synthesis. The coating of microbial biofertilizer by the process of nanoencapsulation using organic compounds like chitosan is known to facilitate proper and controlled diffusion along with an enhanced shelf life (Golbashy et al. 2017; Qureshi et al. 2018).

16.7 Microbe-Assisted Crop Breeding

The interactions between beneficial rhizobacteria and plants are important factors governing the agricultural ecosystem. Through the concept of rhizosphere engineering, bioinoculants can be developed by creating a suitable environment for enhanced interactions with the host plant. These inoculants can provide various ecological services like enhanced nutrient uptake, soil bioremediation, biofortification, etc. This approach can give rise to new horizons in the field of agricultural sustainability by mere utilization of the plant–microbe interactions (Chandra et al. 2020).

16.8 Conclusion

Food security and sustainable development are among the biggest challenges that are to be overcome in this twenty-first century. In this regard, numerous approaches are to be explored that not only fulfill the criteria of being environment friendly but also prove to be effective in these times of climate change. One such approach is the use of beneficial microbes associated with plants, which play a significant role in improving crop as well as soil health. Agro-ecosystems possess a great diversity of microbes that may possess beneficial traits. Many of these microbes are still undiscovered or not fully characterized. Hence, efforts are to be made in order to isolate and evaluate beneficial microbes from diverse conditions using a combination of conventional and high-throughput approaches. Emphasis should be made on the detailed study of the interaction of these beneficial microbes with the plants. Also, the biosafety parameters and field evaluations should be the main focus area for the use and commercialization of these microbes or their compounds.

References

Ahmad M, Nadeem SM, Naveed M, Zahir ZA (2016) Potassium-solubilizing bacteria and their application in agriculture. In: Meena VS, Maurya BR, Prakash Verma J, Meena RS (eds) Potassium solubilizing microorganisms for sustainable agriculture. Springer, New Delhi, pp 293–313

Ahemad M, Kibret M (2014) Mechanisms and applications of plant growth promoting rhizobacteria: current perspective. J King Saud Univ Sci 26(1):1–20

Aloo BN, Tripathi V, Makumba BA, Mbega ER (2022) Plant growth-promoting rhizobacterial biofertilizers for crop production: the past, present, and future. Front Plant Sci 13:1002448. https://doi.org/10.3389/fpls.2022.1002448

Ankati S, Srinivas V, Pratyusha S, Gopalakrishnan S (2021) Streptomyces consortia-mediated plant defense against Fusarium wilt and plant growth-promotion in chickpea. Microb Pathog 157:104961. https://doi.org/10.1016/j.micpath.2021.104961

Beg QK, Kapoor M, Mahajan L, Hoondal GS (2001) Microbial xylanases and their industrial applications: a review. Appl Microbiol Biotechnol 56:326–323

Chandra AK, Kumar A, Bharati A, Joshi R, Agrawal A, Kumar S (2020) Microbial-assisted and genomic-assisted breeding: a two way approach for the improvement of nutritional quality traits in agricultural crops. Biotech 10(1):2. https://doi.org/10.1007/s13205-019-1994-z

Daniel AI, Fadaka AO, Gokul A, Bakare OO, Aina O, Fisher S, Burt AF, Mavumengwana V, Keyster M, Klein A (2022) Biofertilizer: the future of food security and food safety. Microorganisms 10:1220. https://doi.org/10.3390/microorganisms10061220

Elbouazaoui A, Sijilmassi B, Maafa I, Allal D, Ahmed S (2022) Biocontrol activity of bacillus, Paenibacillus and pseudomonas against Fusarium wilt of chickpea in Morocco. Acta Agric Scand B Soil Plant Sci 72(1):847–859. https://doi.org/10.1080/09064710.2022.2100819

Fernandes MLM, Saad EB, Meira JA, Ramos LP, Mitchel DA, Kriegeir N (2007) Esterification and trans-esterification reactions catalysed by addition of fermented solids to organic reaction media. J Mol Catal B Enzym 44:8–13

Gahoi P, Omar RA, Verma N, Gupta GS (2021) Rhizobacteria and Acylated Homoserine lactone-based Nanobiofertilizer to improve growth and pathogen defense in *Cicer arietinum* and *Triticum aestivum* plants. ACS Agric Sci Technol 1(3):240–252. https://doi.org/10.1021/acsagscitech.1c00039

Golbashy M, Sabahi H, Allahdadi I, Nazokdast H, Hosseini M (2017) Synthesis of highly intercalated urea-clay nanocomposite via domestic montmorillonite as eco-friendly slow-release fertilizer. Arch Agron Soil Sci 63(1):84–95

Gorai PS, Ghosh R, Mandal S, Ghosh S, Chatterjee S, Gond SK, Mandal NC (2021) *Bacillus siamensis* CNE6- a multifaceted plant growth promoting endophyte of *Cicer arietinum* L. having broad spectrum antifungal activities and host colonizing potential. Microbiol Res 252:126859. https://doi.org/10.1016/j.micres.2021.126859

Haware MP, Nene YL (1982) Races of *Fusarium oxysporum* f.sp. *ciceri*. Plant Dis 66:809–810

Igiehon NO, Babalola OO (2017) Biofertilizers and sustainable agriculture: exploring arbuscular mycorrhizal fungi. Appl Microbiol Biotechnol 101(12):4871–4881. https://doi.org/10.1007/s00253-017-8344-z

Jiménez-Díaz RM, Castillo P, Jiménez-Gasco MM, Landa BB, Navas-Cortés JA (2015) Fusarium wilt of chickpeas: biology, ecology and management. Crop Prot 73:16–27. https://doi.org/10.1016/j.cropro.2015.02.023

Jiménez-Gasco MM, Jiménez-Diaz RM (2003) Development of a specific polymerase chain reaction based assay for the identification of *Fusarium oxysporum* f. sp. *ciceris* and its pathogenic races 0, 1A, 5, and 6. Am Phytopathol Soc 93(2):200–209

Jiménez-Gasco MM, Pérez-Artés E, Jiménez-Díaz RM (2001) Identification of pathogenic races 0, 1B/C, 5, and 6 of *Fusarium oxysporum* f. sp. *ciceris* with random amplified polymorphic DNA (RAPD). Eur J Plant Pathol 107:237–248

Khan M, Ashraf S, Rasool F, Salati K (2014) Field performance of Trichoderma species against wilt disease complex of chickpea caused by *Fusarium oxysporum* f. sp. *ciceri* and *Rhizoctonia solani*. Turkish J Agric For 38(4):447–454

Khanna A, Raj K, Kumar P, Wati L (2022) Antagonistic and growth-promoting potential of multifarious bacterial endophytes against *Fusarium* wilt of chickpea. Egypt J Biol Pest Control 32:17. https://doi.org/10.1186/s41938-022-00516-8

Kumar S, Diksha, Sindhu Satyavir S, Kumar R (2022) Biofertilizers: an ecofriendly technology for nutrient recycling and environmental sustainability. Curr Res Microb Sci 3:100094. https://doi.org/10.1016/j.crmicr.2021.100094

Kumari R, Singh DP (2019) Nano-biofertilizer: an emerging eco-friendly approach for sustainable agriculture. Proc Natl Acad Sci India Sect B Biol Sci:733–741

Kumawat KC, Razdan N, Saharan K (2022) Rhizospheric microbiome: bio-based emerging strategies for sustainable agriculture development and future perspectives. Microbiol Res 254:126901. https://doi.org/10.1016/j.micres.2021.126901

Merga B, Haji J (2019) Economic importance of chickpea: production, value, and world trade. Cogent Food Agric 5(1):1615718. https://doi.org/10.1080/23311932.2019.1615718

Mitter EK, Tosi M, Obregón D, Dunfield KE, Germida JJ (2021) Rethinking crop nutrition in times of modern microbiology: innovative biofertilizer technologies. Front Sustain Food Syst 5:606815. https://doi.org/10.3389/fsufs.2021.606815

Mosse B (1973) Plant growth responses to vesicular-arbuscular mycorrhiza. New Phytol 72(1):127–136

Mozumder AB, Chanda K, Chorei R, Prasad HK (2022) An evaluation of aluminum tolerant *Pseudomonas aeruginosa* A7 for in vivo suppression of Fusarium wilt of chickpea caused by *Fusarium oxysporum* f. sp. *ciceris* and growth promotion of chickpea. Microorganisms 10:568. https://doi.org/10.3390/microorganisms10030568

Nagpal S, Sharma P, Sirari A, Kumawat KC, Wati L, Gupta SC, Mandahal KS (2021) Chickpea (*Cicer arietinum* L.) as model legume for decoding the co-existence of *Pseudomonas fluorescens* and *Mesorhizobium* sp. as bio-fertilizer under diverse agro-climatic zones. Microbiol Res 247:126720. https://doi.org/10.1016/j.micres.2021.126720

Nene YL, Reddy MV, Haware MP, Ghanekar AM, Amin KS (1991) Field diagnosis of chickpea diseases and their control. In: Information Bulletin No. 28. International Crops Research Institute for the Semi Arid Tropics, Patancheru

Oberhardt MA, Zarecki R, Gronow S, Lang E, Klenk HP, Gophna U (2015) Harnessing the landscape of microbial culture media to predict new organism—media pairings. Nat Commun 6:8493. https://doi.org/10.1038/ncomms9493

Parween T, Bhandari P, Siddiqui ZH, Jan S, Fatma T, Patanjali PK (2017) Biofilm: a next-generation biofertilizer. In: Mycoremediation and environmental sustainability. Springer, Cham, pp 39–51

Patole SP, Dhore SB, Pradhan RS, Shankara K (2017) In vitro evaluation of *Trichoderma viride* and *Trichoderma harzianum* against Fusarium wilt of chickpea. Int J Pure App Biosci 5(5):460–464

Poveda J (2021) Biological control of *Fusarium oxysporum* f. sp. *ciceri* and *Ascochyta rabiei* infecting protected geographical indication Fuentesaúco-chickpea by *Trichoderma* species. Eur J Plant Pathol 160:825–840. https://doi.org/10.1007/s10658-021-02286-9

Qureshi A, Singh DK, Dwivedi S (2018) Nano-fertilizers: a novel way for enhancing nutrient use efficiency and crop productivity. Int J Curr Microbiol App Sci 7(2):3325–3335

Sharma A, Kashyap PL, Srivastava AK, Bansal YK, Kaushik R (2018) Isolation and characterization of halotolerant bacilli from chickpea (*Cicer arietinum* L.) rhizosphere for plant growth promotion and biocontrol traits. Eur J Plant Pathol 153:787–800. https://doi.org/10.1007/s10658-018-1592-7

Shukla A, Dehariya K, Vyas D, Jha A (2015) Interactions between arbuscular mycorrhizae and *Fusarium oxysporum* f. sp. *ciceris*: effects on fungal development, seedling growth and wilt disease suppression in *Cicer arietinum* L. Arch Phytopathol Plant Prot 48:240–252. https://doi.org/10.1080/03235408.2014.884831

Sindhu SS, Suneja S, Goel AK, Parmar N, Dadarwal KR (2002) Plant growth promoting effects of *pseudomonas* sp. on coinoculation with *Mesorhizobium* sp. *Cicer* strain under sterile and "wilt sick" soil conditions. Appl Soil Ecol 19:57–64

Singh C, Vyas D (2021) The trends in the evaluation of Fusarium wilt of chickpea. In: Dmitry K (ed) Diagnostics of plant diseases. IntechOpen, London. https://doi.org/10.5772/intechopen.95612

Singh PK, Singh M, Agnihotri VK, Vyas D (2013) Arbuscular Mycorrhizal fungi: biocontrol against Fusarium wilt of chickpea. Int J Sci Res Publ 3:1–5

Singh A, Mazahar S, Chapadgaonkar SS, Giri P, Shourie A (2023) Phyto-microbiome to mitigate abiotic stress in crop plants. Front Microbiol 14:1210890. https://doi.org/10.3389/fmicb.2023.1210890

Totey NG, Khatri PK, Shadangi DK, Bagde M, Pathak HD (1997) Effect of (rhizobium) biofertilizer on the growth of seedlings and germination of seed of D. *Sissoo*. Indian J For 20(1):54–56

Van Loon LC (2007) Plant responses to plan growth-promoting rhizobacteria. In: New perspectives and approaches in plant growth-promoting rhizobacteria research. Springer, Dordrecht, pp 243–254

Velmourougane K, Prasanna R, Saxena AK (2017) Agriculturally important microbial biofilms: present status and future prospects. J Basic Microbiol 57(7):548–573

Viterbo A, Ramot O, Chemin L, Chet I (2002) Significance of lytic enzymes from *Trichoderma* spp. in the biocontrol of fungal plant pathogens. Antonie Van Leeuwenhoek 81:549–556

Zaidi A, Khan M, Ahemad M, Oves M (2009) Plant growth promotion by phosphate solubilizing bacteria. Acta Microbiol Immunol Hung 56(3):263–284

Zeyad MT, Tiwari P, Ansari WA, Kumar SC, Kumar M, Chakdar H, Srivastava AK, Singh UB, Saxena AK (2022) Bio-priming with a consortium of *Streptomyces araujoniae* strains modulates defense response in chickpea against Fusarium wilt. Front Microbiol 13:998546. https://doi.org/10.3389/fmicb.2022.998546

17 Nanoparticle-Based Strategies to Enhance Plant Disease Protection Techniques of Field Crops

Pooja Bhatt, Krishna Pratap Singh, T. Aravind, Kumari Surbhi, and Vaibhav Kumar Singh

Abstract

Plant disease is a deterioration of the natural state of plants that can severely hinder and transform their basic behaviour and function. Plant diseases caused by different pathogens, including bacteria, viruses, and fungi, have been pervasive issues in agriculture for centuries around the world. The present scenario of pest management totally relies on pesticides, such as fungicides, insecticides, and herbicides. Despite the fact that these agrochemicals have many advantages, like quick action, reliability, cost-effectiveness, and high availability, they also show many adverse effects. Development of pest/pathogen resistance, resurgence of the pest population, and harmful effects on non-targeted organisms and the environment are some negative aspects of pesticide application. Nanoscience and technology could be a potential solution for value addition in the current crop production and protection systems. The engineered nanoparticles are designed with desired properties, like pore size, shape, and surface characteristics, so that these can be used to protect the plants from diseases in different ways, like nanoparticles themselves serving as protectants, nanoparticles as a

P. Bhatt
Nuclear Agriculture & Bio-Technology Division, Bhabha Atomic Research Centre, Trombay, Mumbai, Maharashtra, India

K. P. Singh (✉) · K. Surbhi
Department of Plant Pathology, College of Agriculture, Govind Ballabh Pant University of Agriculture and Technology, Pantnagar, Uttarakhand, India

T. Aravind
Department of Plant Pathology, Centurion University of Technology and Management, Bhubaneswar, Odisha, India

V. K. Singh
Wheat Pathology Laboratory, Division of Plant Pathology, ICAR-Indian Agricultural Research Institute, Pusa, New Delhi, Delhi, India

V. K. Singh et al. (eds.), *Diseases of Field Crops: Diagnostics and Management*, https://doi.org/10.1007/978-981-97-6160-9_17

delivery medium either for active compounds such as pesticides or for double-stranded RNA (to provide protection by RNA interference), and nanoparticles as biosensors (nanosensors) for monitoring the plant diseases in large areas through field-sing or remote sensing. As an alternative to current agricultural practices, nanoparticle-based strategies have tremendous potential to improve agricultural practices, but they are also associated with undesirable side effects, such as their distribution in the food chain and environment, risks to human, animal, and plant health, etc. Risk assessment and management are required for understanding and evaluating the behaviour of nanoparticles and their associated risks. To mitigate these risks, microorganism- and plant-derived green nanoparticles have gained more attention in recent years due to their biological or eco-friendly nature and easy production methods. Several researches are being conducted worldwide in order to boost up the significance of nanoparticles and nano-based techniques to make them an essential and fundamental part of smart agricultural systems.

Keywords

Crop protection · Nanoparticles · Plant diseases · Field crops

17.1 Introduction

The world's rapidly increasing population and changes in climate are becoming major challenges to agriculture. By 2050, the population will be estimated at 9.6 billion; to feed this increasing population, about a 70% increase in global food production is required (FAO 2009). While the resources are limited, the production of more food is a big challenge, especially for developing countries where agriculture is the strongest pillar of their economies. Another major issue in the field of agriculture is the indiscreet use of agrochemicals. The consumption of pesticides all over the world is about 2 million tonnes/year (De et al. 2014). This indiscriminate use causes various problems like wastage of pesticides, decrement of soil biodiversity and useful soil microbes, increased chances of pest/pathogen resistance, bioamplification of pesticides, and destruction of the natural habitat of a farmer's friend, earthworms (Tilman et al. 2002). Plant pests and pathogens are also a major obstacle in the way of sustainable food production, with estimated global losses of 20–40% per year (Flood 2010), and the present scenario of pest management totally relies on pesticides, such as fungicides, insecticides, and herbicides. Despite the fact that these agrochemicals have many advantages, like quick action, reliability, cost-effectiveness, and high availability, they also show many adverse effects. Development of pest/pathogen resistance, resurgence of the pest population, and harmful effects on non-targeted organisms and the environment are some negative aspects of pesticide application (Stephenson 2003). Approximately 90% of pesticides applied in different agricultural practices are lost during or post application (Ghormade et al. 2011; Stephenson 2003). To deal with all these challenges, agriculture and its related practices need to become more precise, effective, diversified, and productive. Advancement of technology could be a potential solution for value addition in the current crop production and protection systems (Prasad et al. 2014), which guarantees increased food production, efficient use of limited resources, cost-effectiveness, assurance of the upliftment of farmers' economies and production values, nutrient security, and better ecology (Subramanian and Tarafdar 2011). In much recent literature, nanoscience and technology are reported as promising for sustainable food production (Kagan 2016). 'Nanotechnology' is the science in which material is manipulated or enhanced at the scale of 1–100 nm (known as 'nanoscale') for the fabrication of macroscale products. This technology is also known as 'nanoscale technology', and it utilizes its nanoscale properties to add value in agriculture. This technology is particularly based on minuscule particles known as nanoparticles. Unique properties like catalytic reactivity, broad definite surface area, and quantum confinement (Nel et al. 2006) possessed by these nanoparticles make them interesting even after their extremely small size, challenging to handle and synthesize, and have the potential to open new ways in the field of agriculture. Nano is a Greek word that means 'dwarf', and 1 nm is equal to 10–9 mt. Nanoparticles have dimensions of 1–100 nm (Ball 2002), and they have the ability to fill the gap between atomic/molecular structures and bulk materials. This property makes these particles very significant for scientific studies. Various materials, like metals and metal oxides, lipids, polymers, quantum dots, semiconductors, emulsions, etc., are used to make nanoparticles (Puoci et al. 2008).

The potential applications of nanoparticle-based technology in promoting sustainable agriculture are enormous, including (i) pest management via nanomaterial-based formulations of pesticides; (ii) enhanced delivery systems for controlled and steady release of agrochemicals that allow the gradual uptake of active ingredients and minimize wastage; (iii) nanosensor-enabled remote sensing or field sensing techniques to record pests, diseases, and crop conditions over a large area; and (iv) nanomaterial-based molecular techniques for the development of disease-resistant varieties by improving plant traits (Gogos et al. 2012; Chen and Yada 2011).

To boost crop production, novel, innovative, and efficient technologies are needed that not only increase food production with more nutritive values but also efficiently utilize limited resources. Nanotechnology has made this possible by changing agriculture with nanofertilizers (NFs) and their targeted delivery, by making water management systems more efficient, by making agrochemicals more effective with a lower dose, and by making plants better at using the sun's energy (Ditta et al. 2015). The biofortification of fertilizers with nanoparticles has many advantages, like preventing leaching and volatilization of chemicals, ensuring targeted delivery of active ingredients, reducing fertilizer wastage, etc. Scientists have created such nanoparticles (NPs) that could act as nutrients for plants (nano-based micronutrients) to improve growth, germination rates, yield, and numerous other physiological parameters (Liu and Lal 2015). For example, when soybean seeds were treated with Cu, Co, and Fe nanomaterials, the chlorophyll index increased by 7–15% compared to the control samples, the number of nodules increased by 20–49%, and the yield increased by 16% (Ngo et al. 2014). Researchers achieved another success by enhancing the plant's ability to harvest more sunlight by delivering carbon nanotubes (CNTs) into the chloroplasts; furthermore, these tubes also worked as artificial antennas and captured ultraviolet, green, and near-infrared wavelengths of light (Cossins 2014).

Nanotechnology is emerging as a powerful alternative for dealing with plant diseases. The engineered nanoparticles are designed with desired properties, like pore size, shape, and surface characteristics, so that these can be used to protect plants from diseases in different ways: (a) nanoparticles themselves serve as protectants, (b) nanoparticles act as a delivery medium either for active compounds such as pesticides or double-stranded RNA (to provide protection by RNA interference) via encapsulation, adsorption, or conjugation (Khandelwal et al. 2016), (c) nanoparticles as biosensors, or 'nanosensors', for monitoring plant diseases in large areas through field-sensing or remote sensing, (d) nanomaterial-based gene/DNA transfer techniques are used to develop disease-resistant varieties, and (e) as diagnostic tools. For routine monitoring of plant pathogens and other advantageous microorganisms, tools like quantum dots are commonly used in gene profiling (André Lévesque 2001). The microfabrication technique plays an important role in viral detection because it is cost-effective, easy to operate, and has an improved detection limit (Cheng et al. 2009). This chapter briefly describes how nanoparticle-based strategies make crop protection practices more efficient and advanced for sustainable food production.

17.2 Overview of Diseases of Field Crops

Plant disease is a deterioration of the neutral state of plants that can severely hinder and transform their basic behaviour and function. All plant species, whether wild or cultivated, are prone to disease. Plant disease occurrence can vary seasonally, depending on various factors like climatic conditions, the existence of pathogens, and the variety of the crop cultivated. Plant disease can lead to epidemics and cause annihilation and devastation, which may limit the kinds of crops and types of industry in a specific area. Figure 17.1 depicts the most prevalent diseases in field crops. Field crops constitute major resources of food, fodder, oil, fibre, etc., but plant pests and pathogens cause the loss of these resources. On a global scale, pathogens and pests are causing losses of wheat (10–28%), rice (25–41%), maize (20–41%), potato (8–21%), and soybean (11–32%) (Savary et al. 2019). A few examples of field crops that have historical significance include ergot of rye, wheat, and pearl millet (*Claviceps* spp.), brown spot of rice (*Pyricularia oryzae*), rust of wheat (*Puccinia* spp.), and southern corn leaf blight (*Bipolaris maydis*). These diseases caused epidemics and had an impact on the economies of the affected countries.

Management of plant diseases is the most important part of plant pathology for reducing the economic damage caused by plant diseases. Many methods have been used for decades to do this, such as resistant cultivars, different cultivation processes like crop rotation, intercropping, use of pathogen/disease-free seeds and seedlings, proper cultivation time and plant density, moisture control of fields, and use of chemicals. But now, the latest and most advanced techniques are needed to fight plant diseases. Nanotechnology, one of the most recently developed technologies,

Fig. 17.1 Major diseases of field crops (Source: TNAU Agritech Portal, Crop Protection)

can play a vital role in this area, particularly with the utilization of nanoparticles as an alternative to pesticides (Martinelli et al. 2015).

17.3 Synthesis of Nanoparticles

Different organic or inorganic materials (NMs) are used to synthesize nanoparticles by physical, chemical, and biological methods via two basic approaches: the bottom-up approach and the top-down approach (Murphy 2002). In a top-down approach, the size of the starting material is reduced by using mechanical or chemical means like sonication, high pressure, milling, lithography (Tapaszto et al. 2008), sputtering deposition (Bell et al. 2001), vapour deposition (Kumar and Ando 2010), etc., while in a bottom-up approach, such as sol-gel processing (Yi et al. 1988), laser pyrolysis (Lacour et al. 2007), microemulsion (Darbandi et al. 2005), etc., at the beginning, different constituents of the nanostructure are formed and then amassed to make the final nanoparticles by chemical or biological procedures. Microorganism- and plant-derived green nanoparticles (Table 17.1) have gained more attention in recent years due to their biological or eco-friendly nature and easy production methods (Iravani 2011). The biological synthesis of nanoparticles follows the bottom-up approach, and it is a single-step bio-reduction method with less energy

Table 17.1 Plant- and microorganism-derived nanoparticles

Biological resources	Nanoparticles synthesized	References
Microorganism-based		
Fungi	Ag, TiO_2, CdS	Gaikwad and Ashok (2012), Jha et al. (2009), Prasad and Jha (2010)
Bacteria	Ag, Ag_2O, Au, TiO_2, CdS	Saifuddin et al. (2009), Dhoondia and Chakraborty (2012), Wadhwani et al. (2014), Jha et al. (2009), Prasad and Jha (2010)
Algae	ZnO, Ag, Au, Fe_3O_4	Azizi et al. (2014), El-Rafie et al. (2013), Singaravelu et al. (2007), Mahdavi et al. (2013)
Virus	Au-Co_3O_4	Nam et al. (2006)
Plant-based		
Seed extracts	Ag	Bar et al. (2009b)
Leaf extracts	Ag, Au, Pd	Krishnaraj et al. (2010), Kumar et al. (2011), Yang et al. (2010)
Bulbs	Ag	Rastogi and Arunachalam (2011)
Root extracts	Ag, Au	Velmurugan et al. (2014)
Latex	Ag	Bar et al. (2009a)
Other biological materials-based		
Honey	Pd, C, Pt, Ag, Au	Reddy et al. (2012), Wu et al. (2013), Venu et al. (2011), Haiza et al. (2013), Philip (2009)
Starch	Ag	Vigneshwaran et al. (2006)

consumption. Different biological resources, like bacteria, fungi, algae, actinomycetes, some viruses, plants, etc., have been used for nanoparticle synthesis and are being precisely used in agriculture (Bansal et al. 2014).

17.4 Nanoparticle-Based Strategies to Enhance Plant Disease Protection Techniques

17.4.1 Nanoparticles as Antimicrobial Agents

Nanoparticles are materials that come in a range of nanometers (10 to 100 nm) and can be designed with exclusive properties that differ from their molecular and bulk counterparts (Yang et al. 2008). Nanoparticles such as silver, magnesium oxide, copper, zinc oxide, titanium dioxide, chitosan, carbon-based NPs, etc., have been successfully investigated for their antimicrobial properties (Table 17.2) (Gogos et al. 2012; Kah and Hofmann 2014; Sathiyabama and Parthasarathy 2016; Kim et al. 2018). Thus, NPs can be applied directly to plants as protectants against pests and pathogens in the form of foliage spray, seed treatment, or applied to soil.

17.4.1.1 Silver NPs

Ag NPs are considered the first and most commonly studied nanoparticles with antimicrobial properties (Zinjarde 2012) and are used in plant disease management (Park et al. 2006; Jo et al. 2009; Kim et al. 2012; Mishra et al. 2014). It was observed that the microbe-killing effects of AgNPs were size-dependent (Raza et al. 2017). Broad-range effectiveness, less toxic effects, adaptability to various substrates, high surface-to-volume ratios, unique nanocrystalline structure, and ease of use (Nangmenyi and Economy 2009) make the silver NPs more favourable for use. AgNPs give new hope for controlling plant diseases and have the potential to be utilized as nanopesticides (Alghuthaymi et al. 2015).

First of all, Park et al. (2006) developed a silica-silver composite nanoparticle formulation and validated its antimicrobial activity against various phytopathogens, viz., *Colletotrichum* sp., *Pythium* sp., *Pseudomonas syringae*, *Xanthomonas campestris*, etc. Several studies demonstrated the wider applicability of AgNPs against phytopathogens such as *Bipolaris sorokiniana*, which causes spot blotch disease in wheat (Mishra et al. 2014), *Magnaporthe grisea*, the causal agent of rice blast (Jo et al. 2009), *Fusarium culmorum* (Kasprowicz et al. 2010), *Colletotrichum* sp. (Lamsal et al. 2011), and sclerotium-forming fungi (Min et al. 2009). Aziz et al. (2016) observed the fungicidal effect of biologically synthesized silver nanoparticles on *Candida albicans*, *Fusarium oxysporum*, and *Aspergillus flavus*, which was comparable to the antifungal effect shown by established fungicides (fluconazole, ketoconazole, and amphotericin B).

Several researchers successfully investigated the antibacterial properties of silver NPs against a variety of phytopathogenic bacteria. Velmurugan et al. (2013) demonstrated the antibacterial activity of the PCE (5 mL)-mediated Ag NPs (average size of 5–50 nm) against both gram-negative and gram-positive plant pathogenic

Table 17.2 Antimicrobial activity and mechanism of action of different nanoparticles against various plant pathogens of field crops

NPs	Mechanism of action	Target microorganism	References
Ag NPs	Induction of pits and cavity in the bacterial cell membrane; interact with disulphide or sulfhydryl groups of enzymes and disruption of metabolic processes. Disturbs permeability, respiration, and cell division	*Magnaporthe grisea*	Jo et al. (2009)
		Burkholdaria glumae	Velmurugan et al. (2013)
		Bipolaris sorokiniana	Mishra et al. (2014)
		Yellow mosaic virus of faba bean	Elbeshehy et al. (2015)
ZnO NPs	ROS generation on the surface of the particles; zinc ion release; morphological changes in the presence of ZnO NPs, increase in membrane permeability and NPs accumulation in the cytoplasm	*Cercospora beticola*	Derbalah et al. (2013)
		Fusarium graminearum	Dimkpa et al. (2013)
		TMV in *Nicotiana benthamiana*	Cai et al. (2019)
CuO NPs	Crossing of NPs from the cell membrane and then damaging the vital enzymes	*Curvularia lunata*	Kanhed et al. (2014)
		Rhizopus oryzae	Mageshwari and Sathyamoorthy (2013)
Mg NPS	Damaging the cell membrane and then causing the leakage of intracellular contents and death of the bacterial cells	*Acidovorax oryzae*,	Ogunyemia et al. (2019)
		Ralstonia solanacearum, causing bacterial wilt of tobacco	Cai et al. (2018)
Chitosan NPs	Disruption of cell membrane, inhibition of H+-ATPase activity, inhibition of toxin production, inhibition of the synthesis of messenger RNA and proteins, and blockage of nutrient flow	*Pyricularia grisea*	Sathiyabama and Parthasarathy (2016)
		Fusarium oxysporum f.sp. *ciceri*	Sathiyabama and Parthasarathy (2016)
		Macrophomina phaseolina	Saharan et al. (2013)
SiO_2 NPs	Influencing the cell functions such as cell differentiation, adhesion and spreading	*Magnaporthe grisea*	Liu et al. (2012)
Carbon NPs	Reactive oxygen containing functionalities could inactivate bacteria and react with chitinous compounds and other polysaccharides on the cell walls of fungi, inhibition of water uptake and induction of plasmolysis	*Fusarium graminearum*	Wang et al. (2014a, b)
		Fusarium poae	Wang et al. (2014a, b)
		Xanthomonas oryzae pv. Oryzae	Chen et al. (2013)

bacteria. They have found inhibition zones of around 10 mm for the gram-positive bacterial strain *Bacillus thuringiensis* and 4.2, 8.9, and 8.6 mm for *Xanthomonas oryzae*, *Pseudomonas syringae*, and *Burkholdaria glumae*, respectively (Table 17.2).

The proposed mechanisms of antibacterial action of silver NPs are: destruction of the outer membrane of bacteria (Lok et al. 2007); pits or gaps formation in the membrane, which causes cell fragmentation (Yun et al. 2013; Iavicoli et al. 2013).

The antiviral properties of silver nanoparticles were demonstrated against yellow mosaic virus of faba bean (Elbeshehy et al. 2015), Tomato mosaic virus, and Potato virus Y (El-Dougdoug et al. 2018). The antiviral activity of silver NPs is size-dependent (Lü et al. 2009; Papp et al. 2010). Nematicidal effects are also shown by silver NPs on root-knot nematode (*Meloidogyne* spp.) (Cromwell et al. 2014) and *Caenorhabditis elegans* (Wang et al. 2009).

17.4.1.2 Copper-Based Nanoparticles

The eminent antimicrobial properties of copper and its renowned history of controlling several plant diseases (Elmer and White 2018) make copper nanoparticles a plausible choice for plant disease protection. The first study that shows copper NP can be used as a fungicidal agent in the field was conducted by Giannousi et al. (2013) to suppress *Phytophthora infestans* in tomatoes. The antifungal activity of copper NPs against *Alternaria alternata, Botrytis cinerea, Curvularia lunata, Fusarium oxysporum*, and *Phoma destructiva* is also well documented (Bramhanwade et al. 2016; Kanhed et al. 2014; Ouda 2014; Viet et al. 2016).

CuO nanoparticles show antibacterial activity against plant-pathogenic Xanthomonads (Ocsoy et al. 2013; Strayer-Scherer et al. 2018). Strayer-Scherer et al. (2018) have used engineered Cu NP composites as an alternative to the standard Cu-based bactericidal compounds (as bacteria have already gained resistance to these chemicals) against *Xanthomonas perforans* (which causes bacterial spots). All of the tested nano-copper products significantly reduced disease severity in the greenhouse as well as in the field. The antiviral properties of copper-based NPs are not well documented (Table 17.2), and there is only one study to date that has determined the nematicidal property of nano-copper against *Xiphinema index* on grapes (Darago 2014).

17.4.1.3 Zinc-Based Nanoparticles

The antimicrobial activity of zinc-based nanoparticles against different plant pathogens has been investigated by several workers as an antibacterial agent (Graham et al. 2016; Indhumathy and Mala 2013), as an antifungal agent against a range of fungal pathogens, including *Alternaria alternata, Botrytis cinerea, Fusarium oxysporum, Mucor plumbeus, Penicillium expansum, Pythium* spp., *Rhizoctonia solani, Rhizopus stolonifera*, and *Sclerotinia sclerotiorum* (He et al. 2011; Zabrieski et al. 2015; Sardella et al. 2017; Wani and Shah 2012), and as a nematicide against *M. incognita* (Kaushik and Dutta 2017) and *Caenorhabditis elegans* (Ma et al. 2009). ZnO NPs were shown to be effective in suppressing Xanthomonas leaf spot in rose (Paret et al. 2013), Cercospora leaf blight in sugar beet (Derbalah et al. 2013), Fusarium head blight in sorghum (Dimkpa et al. 2013), etc., which gives hope that ZnO NPs can be used as an alternative to conventional fungicides. In Egypt, Derbalah et al. (2013) conducted a 2-year field study to evaluate and compare the effects of foliar applications of ZnO NPs (at 500 μg/mL), the fungicide tetraconazole, nano-silica, diatomaceous earth, and six bacterial biocontrol agents to suppress Cercospora leaf blight disease of sugar beet. The ZnO NPs were found to be effective, second only to tetraconazole, in increasing root yield and sugar content as well as reducing disease severity. The antiviral activity of ZnO NPs was demonstrated by Cai et al. (2019) against TMV in *Nicotiana benthamiana*, and it also activates plant immunity. Nano-Zn is also used as an antibacterial agent (Graham et al. 2016; Paret et al. 2013). One zinc-based nanoproduct (Zinkicide™) is in the process of registration for use in managing citrus canker disease (Young et al. 2017).

17.4.1.4 Magnesium-Based Nanoparticles

MgO nanoparticles are considered strong antibacterial agents. According to results, MgO NPs damage the cell membrane and cause the leakage of intracellular contents, which in turn leads to bacterial cell death (Jin and He 2011). The antibacterial activity of MgO nanoparticles was evaluated against *Ralstonia solanacearum*, which causes bacterial wilt in tobacco, both in vitro and in vivo. In this study, MgO NPs showed high antibacterial activity against the bacterial pathogen at a very low concentration (250 mg/mL), and this antibacterial property is facilitated by bacterial cell wall and cell membrane destruction, leakage of the intracellular contents, and accumulation of reactive oxygen species (Cai et al. 2018). Magnesium oxide NPs show inhibitory effects on *Acidovorax oryzae* (Ao) strain RS-2, causing Bacterial Brown Stripe (BBS) disease in rice. The MgO NPs reduced the growth of Ao strain RS-2 by 62.9%; in addition, biofilm formation and swimming motility were also significantly reduced (Ogunyemia et al. 2019).

Chen et al. (2020) evaluated the fungicidal activity of MgO nanoparticles against soilborne pathogens of tobacco, *Phytophthora nicotianae* and *Thielaviopsis basicola*, under laboratory and greenhouse conditions. In vitro studies demonstrated that MgO NPs could effectively suppress fungal growth, inhibit spore germination, and retard sporangium development, while a reduction in black shank and black root rot diseases of tobacco by 36.58% and 42.35%, respectively, was recorded in the glasshouse study. Wani and Shah (2012) exhibited that the germination of conidia of *Fusarium oxysporum*, *Rhizopus stolonifera*, *Alternaria alternata*, and *Mucor plumbeus* was inhibited by MgO NPs at very low concentrations (50 μg/mL).

17.4.1.5 Carbon Nanoparticles

Different types of carbon nanomaterials are synthesized, but three categories are well recognized: carbon nanotubes (single-walled or multi-walled), graphene oxides (oxidized and reduced forms), and fullerenes (C60; buckyballs). The use of carbon nanomaterials in other fields like paints, engineering, textiles, medicine, and electronics has been well recognized for decades. In the past few years, the antimicrobial properties of carbon nanomaterials have also been determined against bacteria (Liu et al. 2009; Wang et al. 2013) and fungi (Sarlak et al. 2014; Wang et al. 2014a, b).

Graphene oxide nano-sheets (reduced) were synthesized to check their antifungal activity against three plant pathogenic fungal spp., *Fusarium oxysporum*,

Aspergillus oryzae, and *Aspergillus niger*. All fungal species were inhibited by the tested nanoparticle, and it was suggested that this was due to its sharp edges. The half maximal inhibitory concentration (IC50) values of the tested nanocompound against *A. oryzae*, *A. niger*, and *F. oxysporum* are 100, 100, and 50 μg mL^{-1}, respectively (Sawangphruk et al. 2012). Six carbon-based nanoparticles, viz., graphene oxide, graphene oxide (reduced), activated carbon, single-walled carbon nanotubes, multi-walled carbon nanotubes, and fullerene, were evaluated to find their antifungal activity against two pathogenic *Fusarium* spp., (*Fusarium graminearum* and *Fusarium poae*). Single-walled carbon nanotubes showed the strongest antifungal activity, followed by multi-walled carbon nanotubes, while fullerene and activated carbon were not very effective in inhibiting the growth of fungus (Wang et al. 2014a, b). Graphene oxides also exhibit antibacterial activity against *Xanthomonas oryzae* pv. *oryzae* at a concentration of 250 μg/mL, with a 95% mortality rate, while only a 13.3% mortality rate was recorded when using the bactericide bismerthiazol (Chen et al. 2013).

The mechanisms by which carbon nanomaterials inhibit microbial growth may be the sharp nanoedges of graphene oxides (which allow direct contact with membranes and cell walls of microbes); reactive oxygen-containing functionalities of the graphene oxide could inactivate bacteria and react with chitinous compounds and other polysaccharides on the cell walls of fungi (Berry et al. 2014; Chen et al. 2013; Liu et al. 2009; Sawangphruk et al. 2012); inhibition of water uptake and induction of plasmolysis by the single-walled nanotubes (Wang et al. 2014a, b).

17.4.1.6 Other less Studied Nanoparticles

The nanoparticles of aluminium, cerium, titanium, sulphur, silica, and chitosan are some less studied nanomaterials that can be efficiently utilized in plant disease management (Liu et al. 2012; Filpo et al. 2013; Rao and Paria 2013; Saharan et al. 2013; Shenashen et al. 2016; Adisa et al. 2018). Chitosan nanoparticles have shown antifungal properties against *Pyricularia grisea*, *Alternaria solani*, and *Fusarium oxysporum* at a concentration of 100 μg and the observed inhibition rates were 92%, 87%, and 72%, respectively (Sathiyabama and Parthasarathy 2016). Chitosan nanoparticles also displayed strong antifungal activity against *Macrophomina phaseolina* (Saharan et al. 2013). SiO_2 NPs were found effective in protecting rice varieties against *Magnaporthe grisea* infection (Liu et al. 2012). The nanosized silica-silver nanocomposites showed antifungal as well as antibacterial activity against various tested fungal (*Botrytis cinerea*, *Rhizoctonia solani*, *Magnaporthe grisea*, *Colletotrichum gloeosporioides*, and *Pythium ultimum*) and bacterial pathogens (*Pseudomonas syringae* pv. *syringae* 2440, *Xanthomonas campestris* pv. *vasicatoria*, and *Azotobacter chrococuum*). The antifungal activity exhibited by the nanocomposite at a very low concentration (3 ppm) as compared to the antibacterial activity (at 100 ppm) (Park et al. 2006). Silicon and titanium nanoparticles were successfully investigated to control powdery mildew disease in wheat plants (Farhat et al. 2018). Titanium dioxide (TiO_2) nanoparticles have the ability to destroy bacterial cells in the presence of light, so they can be effectively utilized as antibacterial agents.

17.4.2 Nanobiosensors Enabled Detection and Prediction of Plant Diseases

'Nanobiosensors' or 'nanosensors' are nano-based devices that record biological events. The increasing demand for nanomaterials for sensor development is mainly due to the friendly platform it provides for the association of bio-recognition elements, great electronic conductivity, a large surface area, and plasmonic properties that boost detection power. Quantum dots, nanometals and metal oxides, carbon nanotubes, and graphene are some examples of nanomaterials used for biosensor construction. Several studies demonstrated the use of nanobiosensors in the detection and diagnosis of plant diseases, and the applicability of these devices in the prediction and real-time monitoring of plant diseases was also reported.

17.4.2.1 Plant Disease Detection and Diagnosis through Nanosensors

To facilitate effective disease management practices, timely detection of plant pathogens is very important. Plant disease diagnosis requires techniques that are fast, automated, cheap, portable, precise, reliable, and sensitive. In this regard, NP-based sensors (nanobiosensors), which can diagnose plant pathogens quickly, efficiently, and economically, are in the development phase. Superparamagnetic iron oxide nanoparticles and quantum dots were evaluated to elucidate the relationship between the nanomaterials and *Fusarium* spp., and results indicated that both nanomaterials rapidly interacted with fungal hyphae, labelling the presence of the pathogenic fungus (Rispail et al. 2014). A gold nanoparticle-based biosensor, conjugated with single-stranded oligonucleotides, acted as a probe to identify Ralstonia solanacearum genomic DNA in farm soil (Khaledian et al. 2017). James (2013) developed a polypyrrole (PPy)-based nano-immunosensor conjugated with a pathogen-specific antibody for the detection of cucumber mosaic virus (CMV). The nano-immunosensor's sensitivity was reinforced by decreasing the electrical conductivity and thickness level of the nano-ribbon from 100 to 25 nm. Quantum dot (QD)-based nanosensors were used to detect phytoplasma (Rad et al. 2012) and *Polymyxa betae*, a vector of *beet necrotic yellow vein virus* (BNYVV) (Safarpour et al. 2012). TiO_2 or SnO_2 NPs on screen-printed carbon electrodes (SPCEs) detect p-ethylguaiacol, a fingerprint compound present in plants infected with *Phytophthora cactorum*, and thus can be used to detect the pathogen (Fang et al. 2014). Gold electrodes and Cu-NPs were used to develop a nanosensor to detect the plant pathogenic fungus *Sclerotinia sclerotiorum*. The nanosensor estimates the production rate of salicylic acid, through which the severity and intensity of infection can be predicted (Wang et al. 2010). Singh et al. (2010) used nano-Au-based immune sensors for the detection of karnal bunt disease in wheat through surface plasmon resonance. An efficient and rapid ELISA method is being developed by making use of superparamagnetic NPs that decreased the competition time, coating, and blocking enzymes of the conventional ELISA procedure (Radoi et al. 2008).

17.4.2.2 Nanobiosensors and Plant Disease Prediction

Forecasting plant diseases means predicting the occurrence of the disease in a specified area ahead of time so that suitable protection measures can be undertaken to avoid losses. In the past few years, research has been in progress to develop nanobiosensors for improving phytopathogen detection methods (Rai and Ingle 2012; Esker et al. 2008) and disease prediction or forecasting methods. The nanobiosensors have the potential to sense and quantify minute amounts of biocontaminants such as fungi, viruses, bacteria, toxins, and other biological hazards in agricultural systems. Therefore, these nanobiosensors could be effectively used in smart farming practices. Nanobiosensors can be linked to a GPS (global positioning system) for real-time monitoring of plant diseases and distributed throughout a large farming area for effective monitoring of soil status, weather conditions, and crop health. The combination of biotechnology and nanotechnology in the development of nanosensors will enhance their sensitivity, allowing rapid and early responses to minute ecological changes and plant disease prevalence. Nanosensors will detect plant diseases before visible symptoms appear and thus facilitate their effective control (Rai and Ingle 2012).

17.4.3 Nanoparticles as Delivering Agent

Nanoparticles are now widely used as carriers for different agrochemicals to develop effective and controlled release formulations (CRFs) through absorption, encapsulation, attachment, or entrapment of the active molecules. As carrier nanoparticles, they provide several benefits to the active compounds, like (i) augmented shelf life, (ii) increased solubility of less water-soluble pesticides, (iii) reduction in toxicity levels, (iv) enhanced targeted delivery, and (v) site-specific action on the target pest (Hayles et al. 2017). Furthermore, the nanoparticles as carriers provide other benefits, like increasing the efficiency and stability of pesticides under environmental stress, reducing the number of chemical applications, and reducing toxicity and cost automatically. There are different types of nano-based carriers, such as chitosan NPs, mesoporous silica nanoparticles (MSNs), dendrimers, nanoemulsions, solid lipid nanoparticles (SLNs), nanocrystals, layered double hydroxide (LDH), and hydrogels.

17.4.3.1 Nanoparticles as Carrier for Fungicides and Insecticides

Several studies demonstrated the effective utilization of different nano-based pesticide formulations in agricultural practices. Chitosan-lactide copolymer-NP-loaded Pyraclostrobin was found to be either similar to or less efficient than pyraclostrobin alone at inhibiting the growth of *Colletotrichum gossypii* during the third and fifth days after application. However, increased inhibition was observed 7 days post-treatment compared to the active alone (Xu et al. 2014). The same results were found in another study done by Qian et al. (2011). They investigated nanosized calcium carbonate (nano-CC)-encapsulated validamycin for its controlled release performance and antifungal activity against *Rhizoctonia solani* in an in vitro assay.

Over the period of 1–7 days, the lower efficiency of nano-CC-encapsulated validamycin was observed compared to validamycin alone; however, after 14 days, the nanoparticle formulation showed slightly better results than the active compound alone, demonstrating the efficacy of the nano-based formulation over longer periods of time with slow and controlled release of the active compound. Polymeric NPs loaded with carbendazim were tested successfully with an increased inhibition rate against *Fusarium oxysporum* and *Aspergillus parasiticus* compared to carbendazim alone and found safer for germination and root growth of cucumber, maize, and tomato seeds (Kumar et al. 2017). Wastage of agrochemicals through leaching in soil and water is a major issue; many studies have demonstrated the use of nano based carriers to deal with this problem (Table 17.3). Metalaxyl loaded onto MSNs was found effective to prevent leaching of active metalaxyl in soil; the release rate of free metalaxyl in soil is 76%, while that of encapsulated metalaxyl is 11.5% within a period of 30 days (Wanyika 2013).

Less-water-soluble insecticides have been loaded into chitosan NP-based carriers (Feng and Peng 2012; Zhang et al. 2013; Lu et al. 2013; Campos et al. 2018) and silica NP-based carriers (Wang et al. 2014a, b) successfully to increase their solubility. A specific nanosized particle, OR-GC, was made by annexation of octahydrogenated retinoic acid (OR) and glycol chitosan (GC), and is used as a carrier for the hydrophobic insecticide azadirachtin (AZA). The AZA-OR-GC (25 wt%) conjugate showed controlled release of the integrated AZA insecticide (in phosphate-buffered saline; pH 7.2, 65% of the loaded dose was released in 7 days at 27 °C) with increased inhibition of *Spodoptera litura* ovarian cell lines (Lu et al. 2013). A higher death rate of *Helicoverpa armigera* larvae was recorded when subjected to dendrimer thiamethoxam formulation, while normally *Helicoverpa armigera* is not susceptible to thiamethoxam (Liu et al. 2015). Nguyen et al. (2015) also recorded an increased mortality rate of *S. litura*, when Layered Double Hydroxides (LDH) nanoparticles loaded with anacardic acid (a cashew nutshell liquid extract) were applied either directly to the skin of the insect or on the leaf surface compared to anacardic acid alone (Table 17.3).

Evaporation or volatilization of the active compound, such as essential oils (well known for their insecticidal and fungicidal properties), is also a common problem associated with the loss of pesticide after application. Janatova et al. (2015) successfully encapsulated seven essential oil components (allyl isothiocyanate, carvacrol, cinnamaldehyde, diallyl disulphide, eugenol, thymol, and thymoquinone) into mesoporous silica nanoparticles (MSNs) to investigate their antifungal activity against *Aspergillus niger* infection. When compared to the pure essential oil components, it has been proven that five out of seven nanoencapsulated compounds, show positive effects on antifungal activity with long-term effectiveness and sustained release (Table 17.3). In another study, polyethylene glycol (PEG)-loaded garlic essential oil was sprayed on the harvested rice and then inoculated with red flour beetles (*Tribolium castaneum*). Positive results were recorded after 5 months: an 80% fatality rate of beetles was observed with the nanoparticle essential oil formulation, while only 11% with essential oil alone (Yang et al. 2009).

Table 17.3 Nano-based carrier system for delivery of fungicides, insecticides, genetic material and RNA silencing inducing molecules

Active compound	Nano-based carrier system	Purpose	References
Fungicide			
Tebuconazole	Bacterial ghosts	Antifungal effect against *Erysiphe graminis, Leptosphaeria nodorum, and Pyrenophora teres*	Hatfaludi et al. (2004)
Carbendazim and Tebuconazole	Polymeric and solid lipid nanoparticles (SLNs)	Sustained/controlled release of active compound	Campos et al. (2015)
Carbendazim	Chitosan/pectin	Antifungal activity aginst *Fusarium oxysporum* and *Aspergillus parasiticus*	Kumar et al. (2017)
Metalaxyl	Mesoporous silica nanoparticles (SLNs)	Decrease the leaching of active in soil and water	Wanyika (2013)
Pyraclostrobin	Chitosan–PLA graft copolymer	Inhibitory effect against *Colletotrichum gossypii*	Xu et al. (2014)
Validamycin	Calcium carbonate	Inhibit the growth of *Rhizoctonia solani*	Qian et al. (2011)
Cymbopogon martini essential oil	Chitosan	Antifungal activity against *Fusarium graminearum*	Kalagatur et al. (2018)
Chitosan, saponin, $CuSO_4$	Chitosan/ tripolyphosphate, chitosan-saponin, chitosan-cu	Several phytopathogenic fungi	Saharan et al. (2013)
Insecticide			
Garlic essential oil	Polyethylene glycol (PEG)	Red flour beetle (*Tribolium castaneum*)	Yang et al. (2009)
Chlorfenapyr	Silica	Chinese cotton bollworm (*Helicoverpa armigera*), *Plutella xylostella*	Song et al. (2012)
Azadirachtin	Chitosan	Tobacco cutworm (*Spodoptera litura*) culture ovarian cell lines Sl-1	Lu et al. (2013)
Imidacloprid	Sodium alginate	Leafhopper (Jassids)	Kumar et al. (2014)
Thiamethoxam	Dendrimers	Cotton bollworm cells and larvae (*Helicoverpa armigera*)	Liu et al. (2015)
Azadirachtin	Zinc oxide and chitosan	Groundnut bruchid (*Caryedon serratus O*)	Jenne et al. (2018)
Genetic materials			
DNA	Mesoporous silica nanoparticles (MSN)	Target-specific delivery	Torney et al. (2007)

(continued)

Table 17.3 (continued)

Active compound	Nano-based carrier system	Purpose	References
Plasmid DNA	Carbon nanotubes (CNTs)	Transformation of *Nicotinia tabacum* callus and leaf explants with nptII gene contained in the pGreen 0029	Burlaka et al. (2015)
RNA silencing inducing molecules			
CMV and PMMov derived ds RNA	Layered double hydroxide (LDH) clay nanosheets	Viral protection against cucumber mosaic virus (CMV) and pepper mild mottle virus (PMMoV) using	Mitter et al. (2017)
Aerosolized short interfering RNAs	Perfluorocarbon	Against three aphid species; *Acyrthosiphon pisum, Aphis glycines*, and *Schizaphis graminum*	Thairu et al. (2017)

17.4.3.2 Nanoparticles as Driving Vehicle for Genetic Material and RNA Silencing Inducing Molecules in Plants

Due to the small size of nanoparticles, NP-enabled gene delivery into plant cells is a promising option to develop disease-resistant varieties. The first reported example of nano-based delivery of DNA is the mesoporous silica nanoparticle (MSN), which enabled DNA and chemicals to be transported into isolated plant cells. MSNs were designed with the capping of gold nanoparticles on the pores to avoid leakage of content and release it in a controlled way to the target (Torney et al. 2007). In plants, to transfer DNA, different inorganic nanoparticles (Sokolova and Epple 2008), single-walled carbon nanotubes, organic carbon-based nanomaterials such as fullerenes, single-walled carbon nanohorns, multi-walled carbon nanotubes (Burlaka et al. 2015), and dendrimers were used. Due to their high cellular uptake and strong binding affinity, the surface-coated MSNs were found to be very convenient for the delivery of DNA and SiRNA (Xia et al. 2009). Vijayakumar et al. (2010) have found carbon matrices embedded (as carriers) in gold nanoparticles very useful for the delivery of genetic material. The successful gene transfer by nanoparticle biomolecule conjugate (fluorescent-labelled starch NPs) was reported by Jun et al. (2008). It has been reported that plant cells uptake positively charged NPs more quickly than negatively charged NPs (Cunningham et al. 2018), so chitosan (positively charged) or chitosan-based nanocarriers may pose a promising platform for the delivery of DNA into plant cells.

RNA silencing (RNAi) is a highly conserved mechanism in eukaryotes, associated with growth, development, and host defence against viruses and transposons. This mechanism is also utilized to target different pests of plants, viz., insects, fungi, viruses, and weeds (Baulcombe 2004; Bartel 2004; Burand and Hunter 2013; Maillard et al. 2013; Borges and Martienssen 2015). In plants, RNAi is triggered by dsRNA, which is derived from the target pathogen. Nano-based delivery of dsRNA (target pathogen-derived) is useful to protect plants from several pathogens, especially plant pathogenic viruses. Mitter et al. (2017) have observed successful viral protection against pepper mild mottle virus (PMMoV) and cucumber mosaic virus (CMV) through layered double hydroxide NP-loaded dsRNA called BioClay.

17.5 Nanoparticles Associated Risks and Limitations

As an alternative to pre-existing farming practices, nanoparticle-based strategies have astounding potential to facilitate agricultural practices but have not yet been satisfactorily commercialized in agriculture. Farmers show less interest in nano-based biochemicals due to (a) higher input costs; (b) insufficient economic returns; (c) less interest shown by the government (Mukhopadhyay 2014); (d) a lack of proper knowledge; and (e) disinterest in the public (Mukhopadhyay 2014). It is very essential to make people aware of this science and enhance its acceptability, which could accelerate the discovery of newer applications.

In addition, some side effects are also associated with nanoparticles, such as their distribution in the food chain and environment, associated hazards to human, animal, and plant health, etc. Risk assessment and management are required for understanding and evaluating the behaviour of nanoparticles and their associated risks. Khot et al. (2012) have suggested that non-toxic, stable, biocompatible, and biodegradable nanoparticles with proper wettability and dispersion ability and well-defined toxico-dynamics and toxico-kinetics would be optimal for utilization in agriculture.

17.6 Summary

Plant diseases are one of the primary challenges to global crop production. Conventional plant disease management strategies rely solely on pesticides, which have many disadvantages. To increase crop production, novel, innovative, and effective technologies are required, which not only increase food production with higher nutrient values but also utilize limited resources in an effective manner. Nanotechnology is an emerging field of study with enormous potential for enhancing plant disease management practices. This chapter briefly describes how nanoparticle-based strategies make crop protection practices more efficient and advanced for sustainable food production. Nanoparticles can be used in a variety of ways to protect plants from diseases, such as direct application to the plants as a protectant, as a carrier for different pesticides, as nanobiosensors for the detection and diagnosis of plant diseases, in a wide range of gene profiling applications, for routine monitoring of plant pathogens, etc. In recent years, green nanoparticles have gained popularity because of their eco-friendly nature. Various biological resources, such as bacteria, fungi, algae, actinomycetes, some viruses, plants, etc., have been used for synthesis and are being precisely used in agriculture.

17.7 Conclusion and Future Aspects

Nanotechnology has emerged as one of the most effective and innovative technologies with its capability to inculcate major positive changes in plant disease management techniques through nanopesticides, nanosensors, and nano-based delivery systems. Despite the fact that nanotechnology has diverse advantages in plant protection, it has not yet found a way into farmers' fields. To accelerate the commercialization of nano-based products, high initial investments and large-scale field use are prerequisites. Nanoparticle-based techniques could be further developed for real-time monitoring of plant diseases by using smart nanosensors. In the upcoming times, the plant disease forecasting system will predominantly use nanosensor-linked remote sensing devices and GPS systems to measure field data related to crop health, environmental conditions, and soil conditions in large areas. A network of smart nanosensors will be distributed throughout the large farming area and work like the sense organs of the farming world, thereby providing accurate data that would deliver state-of-the-art intelligence processes and practices. These minute wireless nanosensors have the capability to communicate the information they have sensed, which will be further computed to respond to various parameters such as disease infestation, insect attack, and any abiotic pressure such as nutrient deficiency, drought, etc. (Ingale and Chaudhari 2013). Multiple studies are being conducted on a global scale to increase the significance of nanoparticles and nano-based techniques in order to make them an essential and fundamental component of smart agricultural systems.

References

Adisa IO, Pullagurala VLR, Rawat S, Hernandez-Viezcas JA, Dimkpa CO, Elmer WH, White JC, Peralta-Videa JR, Gardea-Torresdey JL (2018) Role of cerium compounds in *Fusarium* wilt suppression and growth enhancement in tomato (*Solanum lycopersicum*). J Agric Food Chem 66(24):5959–5970

Alghuthaymi MA, Almoammar H, Rai M, Said-Galiev E, Abd-Elsalam KA (2015) Myconanoparticles: synthesis and their role in phytopathogens management. Biotechnol Biotechnol Equip 29(2):221–236

André Lévesque C (2001) Molecular methods for detection of plant pathogens—what is the future. Can J Plant Pathol 23(4):333–336

Aziz N, Pandey R, Barman I, Prasad R (2016) Leveraging the attributes of Mucor hiemalis-derived silver nanoparticles for a synergistic broad-spectrum antimicrobial platform. Front Microbiol 7:1984

Azizi S, Ahmad MB, Namvar F, Mohamad R (2014) Green biosynthesis and characterization of zinc oxide nanoparticles using brown marine macro alga *Sargassum muticum* aqueous extract. Mater Lett 116:275–277

Ball P (2002) Natural strategies for the molecular engineer. Nanotechnology 13(5):R15

Bansal P, Duhan JS, Gahlawat SK (2014) Biogenesis of nanoparticles: a review. Afr J Biotechnol 13(28):2778–2785

Bar H, Bhui DK, Sahoo GP, Sarkar P, De SP, Misra A (2009a) Green synthesis of silver nanoparticles using latex of *Jatropha curcas*. Colloids Surf A Physicochem Eng Asp 339(1–3):134–139

Bar H, Bhui DK, Sahoo GP, Sarkar P, Pyne S, Misra A (2009b) Green synthesis of silver nanoparticles using seed extract of *Jatropha curcas*. Colloids Surf A Physicochem Eng 348(1–3):212–216
Bartel D (2004) MicroRNAs: genomics, biogensis, mechanism and function. Cell 116(2):281–297
Baulcombe D (2004) RNA silencing in plants. Nature 431:356–363
Bell J, Chen Z, Olofinjana A (2001) Synthesis of amorphous carbon nitride using reactive ion beam sputtering deposition with grazing bombardment. Diam Relat Mater 10(12):2184–2189
Berry TD, Filley TR, Blanchette RA (2014) Oxidative enzymatic response of white-rot fungi to single walled carbon nanotubes. Environ Pollut 193C:197–204
Borges F, Martienssen RA (2015) The expanding world of small RNAs in plants. Nat Rev Mol Cell Biol 16:727–741
Bramhanwade K, Shende S, Bonde S, Gade A, Rai M (2016) Fungicidal activity of Cu nanoparticles against *Fusarium* causing crop diseases. Environ Chem Lett 14(2):229–235
Burand JP, Hunter WB (2013) RNAi: future in insect management. J Invertebr Pathol 112(1):S68–S74
Burlaka OM, Pirko YV, Yemets AI, Blume YB (2015) Plant genetic transformation using carbon nanotubes for DNA delivery. Cytol Genet 49(6):349–357
Cai L, Chen J, Liu Z, Wang H, Yang H, Ding W (2018) Magnesium oxide nanoparticles: effective agricultural antibacterial agent against *Ralstonia solanacearum*. Front Microbiol 9:790
Cai L, Liu C, Fan G, Liu C, Sun X (2019) Preventing viral disease by ZnONPs through directly deactivating TMV and activating plant immunity in *Nicotiana benthamiana*. Environ Sci Nano 6:3653–3669
Campos EVR, Oliveira JLD, Da Silva CMG, Pascoli M, Pasquoto T, Lima R, Abhilash P, Fraceto LF (2015) Polymeric and solid lipid nanoparticles for sustained release of carbendazim and tebuconazole in agricultural applications. Sci Rep 5:13809
Campos EV, Proença PL, Oliveira JL, Melville CC, Vechia JF, Andrade DJ, Fraceto LF (2018) Chitosan nanoparticles functionalized with β-cyclodextrin: a promising carrier for botanical pesticides. Sci Rep 8(1):2067
Chen H, Yada R (2011) Nanotechnologies in agriculture: new tools for sustainable development. Trends Food Sci Technol 22:585–594
Chen J, Wang X, Han H (2013) A new function of graphene oxide emerges: inactivating phytopathogenic bacterium, *Xanthomonas oryzae* pv. *Oryzae*. J Nanopart Res 15(5):1658
Chen J, Wu L, Lu M, Lu S, Li Z, Ding W (2020) Comparative study on the fungicidal activity of metallic MgO nanoparticles and macroscale mgo against soilborne fungal phytopathogens. Front Microbiol 11:365
Cheng X, Chen G, Rodrigue WR (2009) Micro- and nanotechnology for viral detection. Anal Bioanal Chem 393:487–501
Cossins D (2014) Next generation, nanoparticles augment plant functions (2014) The incorporation of synthetic nanoparticles into plants can enhance photosynthesis and transform leaves into biochemical sensors. The Scientist, News & Opinion, March 16
Cromwell W, Yang J, Starr J, Jo Y-K (2014) Nematicidal effects of silver nanoparticles on root-knot nematode in Bermuda grass. J Nematol 46(3):261
Cunningham FJ, Goh NS, Demirer GS, Matos JL, Landry MP (2018) Nanoparticle-mediated delivery towards advancing plant genetic engineering. Trends Biotechnol 36(9):882–897
Darago A (2014) The distribution of dagger nematodes species in Hungarian wind regions and newest control options. PhD thesis, University Pannonia Georgikon, Keszthely
Darbandi M, Thomann R, Nann T (2005) Single quantum dots in silica spheres by microemulsion synthesis. Chem Mater 17(23):5720–5725
Dc A, Bose R, Kumar A, Mozumdar S (2014) Targeted delivery of pesticides using biodegradable polymeric nanoparticles. In: De A, Bose R, Kumar A, Mozumdar S (eds) Worldwide pesticide use. Springer, New Delhi, pp 5–6
Derbalah AS, El-Moghazy SM, Godah MI (2013) Alternative control methods of sugar-beet leaf spot disease caused by the fungus *Cercospora beticola* (Sacc). Egypt J Biol Pest Control 23(2):247–254

Dhoondia ZH, Chakraborty H (2012) *Lactobacillus* mediated synthesis of silver oxide nanoparticles. Nanomater Nanotech 2(1):1

Dimkpa CO, McLean JE, Britt DW, Anderson AJ (2013) Antifungal activity of ZnO nanoparticles and their interactive effect with a biocontrol bacterium on growth antagonism of the plant pathogen *Fusarium graminearum*. Biometals 26(6):913–924

Ditta A, Arshad M, Ibrahim M (2015) Nanoparticles in sustainable agricultural crop production: applications and perspectives. Nanotechnology and plant sciences: nanoparticles and their impact on plants:55–75

Elbeshehy EKF, Elazzazy AM, Aggelis G (2015) Silver nanoparticles synthesis mediated by new isolates of *bacillus* spp., nanoparticle characterization and their activity against B*ean yellow mosaic virus* and human pathogens. Front Microbiol 6:453

El-Dougdoug NK, Bondok AM, El-Dougdoug KA (2018) Evaluation of silver nanoparticles as antiviral agent against ToMV and PVY in tomato plants. Middle East J Appl Sci 8(1):100–111

Elmer W, White JC (2018) The future of nanotechnology in plant pathology. Annu Rev Phytopathol 56(1):111–133

El-Rafie HM, El-Rafie MH, Zahran MK (2013) Green synthesis of silver nanoparticles using polysaccharides extracted from marine macro algae. Carbohydr Polym 96(2):403–410

Esker PD, Sparks AH, Campbell L, Guo Z, Rouse M, Silwal SD, Tolos S, Van Allen B, Garrett KA (2008) Ecology and epidemiology in R: disease forecasting and validation. Plant Health Instr. https://doi.org/10.1094/PHI-A-2008-0129-05

Fang Y, Umasankar Y, Ramasamy RP (2014) Electrochemical detection of p-ethylguaiacol, a fungi infected fruit volatile using metal oxide nanoparticles. Analyst 139(15):3804–3810

Farhat MG, Haggag WM, Thabet MS, Mosa AA (2018) Efficacy of silicon and titanium nanoparticles biosynthesis by some antagonistic fungi and bacteria for controlling powdery mildew disease of wheat plants. Int J Agric Technol 14:661–674

Feng B-H, Peng L-F (2012) Synthesis and characterization of carboxymethyl chitosan carrying ricinoleic functions as an emulsifier for azadirachtin. Carbohydr Polym 88(2):576–582

Filpo GD, Palermo AM, Rachiele F, Nicoletta FP (2013) Preventing fungal growth in wood by titanium dioxide nanoparticles. Int Biodeterior Biodegradation 85:217–222

Flood J (2010) The importance of plant health to food security. Food Secur 2:215–231

Food and Agriculture Organization of the United Nations (2009) How to feed the world in 2050. In: Proceedings of the expert meeting on how to feed the world in 2050. FAO, Rome, pp 24–26

Gaikwad S, Ashok B (2012) Green synthesis of silver nanoparticles using *Aspergillus Niger* and its efficacy against human pathogens. Eur J Exp Biol 2:1654–1658

Ghormade V, Deshpande MV, Paknikar KM (2011) Perspectives for nano-biotechnology enabled protection and nutrition of plants. Biotechnol Adv 29:792–803

Giannousi K, Avramidis I, Dendrinou-Samara C (2013) Synthesis, characterization and evaluation of copper based nanoparticles as agrochemicals against *Phytophthora infestans*. RSC Adv 3(44):21743–21752

Gogos A, Knauer K, Bucheli TD (2012) Nanomaterials in plant protection and fertilization: current state, foreseen applications, and research priorities. J Agric Food Chem 60(39):9781–9792

Graham JH, Johnson EG, Myers ME, Young M, Rajasekaran P, Das S, Santra S (2016) Potential of nano-formulated zinc oxide for control of citrus canker on grapefruit trees. Plant Dis 100(12):2442–2447

Haiza H, Azizan A, Mohidin AH, Halin DSC (2013) Green synthesis of silver nanoparticles using local honey. Nano Hybrids 4:87–98

Hatfaludi T, Liska M, Zellinger D, Ousman JP, Szostak M, Jalava K, Lubitz W (2004) Bacterial ghost technology for pesticide delivery. J Agric Food Chem 52(18):5627–5634

Hayles J, Johnson L, Worthley C, Losic D (2017) Nanopesticides: a review of current research and perspectives. In: New pesticides and soil sensors. Academic Press, Cambridge, pp 193–225

He L, Liu Y, Mustapha A, Lin M (2011) Antifungal activity of zinc oxide nanoparticles against *Botrytis cinerea* and *Penicillium expansum*. Microbiol Res 166(3):207–215

Iavicoli I, Fontana L, Leso V, Bergamaschi A (2013) The effects of nanomaterials as endocrine disruptors. Int J Mol Sci 14(8):16732–16801

Indhumathy M, Mala R (2013) Photocatalytic activity of zinc sulphate nano material on phytopathogens. Int J Agric Environ Biotechnol 6(4S):737–743

Ingale AG, Chaudhari A (2013) Biogenic synthesis of nanoparticles and potential application: an ecofriendly approach. J Nanomed Nanotech 4(2):165

Iravani S (2011) Green synthesis of metal nanoparticles using plants. Green Chem 13:2638–2650

James C (2013) Polypyrrole nanoribbon based chemiresistive immunosensors for viral plant pathogen detection. Anal Methods 5(14):3497–3502

Janatova A, Bernardos A, Smid J, Frankova A, Lhotka M, Kourimská L, Pulkrabek J, Kloucek P (2015) Long-term antifungal activity of volatile essential oil components released from mesoporous silica materials. Ind Crop Prod 67:216–220

Jenne M, Kambham M, Tollamadugu NP, Karanam HP, Tirupati MK, Balam RR, Shameer S, Yagireddy M (2018) The use of slow releasing nanoparticle encapsulated Azadirachtin formulations for the management of *Caryedon serratus* O. (groundnut bruchid). IET Nanobiotechnol 12(7):963–967

Jha AK, Prasad K, Kulkarni AR (2009) Synthesis of TiO_2 nanoparticles using microorganisms. Colloids Surf B Biointerfaces 71(2):226–229

Jin T, He Y (2011) Antibacterial activities of magnesium oxide (MgO) nanoparticles against food borne pathogens. J Nanopart Res 13:6877–6885

Jo YK, Kim BH, Jung G (2009) Antifungal activity of silver ions and nanoparticles on phytopathogenic fungi. Plant Dis 93(10):1037–1043

Jun L, Xuan-Ming L, Su-Yao X, Chuen-Yi T, Dong-Ying T, Xuanming L (2008) Preparation of fluorescence starch-nanoparticle and its application as plant transgenic vehicle. J Cent S Univ Technol 15(6):768–773

Kagan CR (2016) At the nexus of food security and safety: opportunities for nanoscience and nanotechnology. ACS Nano 10:2985–2986

Kah M, Hofmann T (2014) Nanopesticide research: current trends and future priorities. Environ Int 63:224–235

Kalagatur NK, Ghosh OSN, Sundararaj N, Mudili V (2018) Antifungal activity of chitosan nanoparticles encapsulated with *Cymbopogon martinii* essential oil on plant pathogenic fungi *Fusarium graminearum*. Front Pharmacol 9:610

Kanhed P, Birla S, Gaikwad S, Gade A, Seabra AB, Rubilar O, Duran N, Rai M (2014) *In vitro* antifungal efficacy of copper nanoparticles against selected crop pathogenic fungi. Mater Lett 115:13–17

Kasprowicz MJ, Kozioł M, Gorczyca A (2010) The effect of silver nanoparticles on phytopathogenic spores of *Fusarium culmorum*. Can J Microbiol 56(3):247–253

Kaushik H, Dutta P (2017) Chemical synthesis of zinc oxide nanoparticle: its application for antimicrobial activity and plant health management. Paper presented at the 109th Annual Meeting of the American Phytopathological Society, San Antonio, TX, August

Khaledian S, Nikkhah M, Shams-bakhsh M, Hoseinzadeh S (2017) A sensitive biosensor based on gold nanoparticles to detect *Ralstonia solanacearum* in soil. J Gen Plant Pathol 83(4):231–239

Khandelwal N, Barbole RS, Banerjee SS, Chate GP, Biradar AV, Khandare JJ, Giri AP (2016) Budding trends in integrated pest management using advanced micro-and nano-materials: challenges and perspectives. J Environ Manag 184:157–169

Khot LR, Shankaran S, Maja JM, Ehsani R, Schuster EW (2012) Applications of nanomaterials in agricultural production and crop protection. Crop Prot 35(C):64–70

Kim SW, Jung JH, Lamsal K, Kim YS, Min JS, Lee YS (2012) Antifungal effects of silver nanoparticles (AgNPs) against various plant pathogenic fungi. Mycobiology 40(1):53–58

Kim DY, Kadam A, Shinde S, Saratale RG, Patra J, Ghodake G (2018) Recent developments in nanotechnology transforming the agricultural sector: a transition replete with opportunities. J Sci Food Agric 98(3):849–864

Krishnaraj C, Jagan EG, Rajasekar S, Selvakumar P, Kalaichelvan PT, Mohan N (2010) Synthesis of silver nanoparticles using *Acalypha indica* leaf extracts and its antibacterial activity against water borne pathogens. Colloids Surf B Biointerfaces 76(1):50–56

Kumar M, Ando Y (2010) Chemical vapor deposition of carbon nanotubes: a review on growth mechanism and mass production. J Nanosci Nanotechnol 10(6):3739–3758

Kumar VG, Gokavarapu SD, Rajeswari A, Dhas TS, Karthick V, Kapadia Z, Shrestha T, Barathy IA, Roy A, Sinha S (2011) Facile green synthesis of gold nanoparticles using leaf extract of antidiabetic potent Cassia auriculata. Colloids Surf B Biointerfaces 87(1):159–163

Kumar S, Bhanjana G, Sharma A, Sidhu M, Dilbaghi N (2014) Synthesis, characterization and on field evaluation of pesticide loaded sodium alginate nanoparticles. Carbohydr Polym 101:1061–1067

Kumar S, Kumar D, Dilbaghi N (2017) Preparation, characterization, and bio-efficacy evaluation of controlled release carbendazim-loaded polymeric nanoparticles. Environ Sci Pollut Res 24(1):926–937

Lacour F, Guillois O, Portier X, Perez H, Herlin N, Reynaud C (2007) Laser pyrolysis synthesis and characterization of luminescent silicon nanocrystals. Physica E Low Dimens Syst Nanostruct 38(1–2):11–15

Lamsal K, Kim SW, Jung JH, Kim YS, Kim KS, Lee YS (2011) Application of silver nanoparticles for the control of *Colletotrichum* species in vitro and pepper anthracnose disease in field. Mycobiology 39(3):194–199

Liu R, Lal R (2015) Potentials of engineered nanoparticles as fertilizers for increasing agronomic productions. Sci Total Environ 514:131–139

Liu S, Wei L, Hao L, Fang N, Chang MW, Xu R, Yang Y, Chen Y (2009) Sharper and faster "nano darts" kill more bacteria: a study of antibacterial activity of individually dispersed pristine single-walled carbon nanotube. ACS Nano 3(12):3891–3902

Liu JB, Gao C, Gao J, Ma JY, Chang HB (2012) Effects of silicon on the ultrastructure of *Oryza sativa* leaves infected by *Magnaporthe grisea*. J South China Agric Univ 33(1):40–43

Liu X, He B, Xu Z, Yin M, Yang W, Zhang H, Cao J, Shen J (2015) A functionalized fluorescent dendrimer as a pesticide nanocarrier: application in pest control. Nanoscale 7(2):445–449

Lok CN, Ho CM, Chen R, He Qy YWY, Sun H, Tam PKH, Chiu JF, Che GM (2007) Silver nanoparticles: partial oxidation and antibacterial activities. J Biol Inorg Chem 12:527–534

Lü JM, Wang X, Marin-Muller C, Wang H, Lin PH, Yao Q, Chen C (2009) Current advances in research and clinical applications of PLGA based nanotechnology. Expert Rev Mol Diagn 9(4):325–341

Lu W, Lu ML, Zhang QP, Tian YQ, Zhang ZX, Xu HH (2013) Octahydrogenated retinoic acid-conjugated glycol chitosan nanoparticles as a novel carrier of azadirachtin: synthesis, characterization, and in vitro evaluation. J Polym Sci A Polym Chem 51(18):3932–3940

Ma H, Bertsch PM, Glenn TC, Kabengi NJ, Williams PL (2009) Toxicity of manufactured zinc oxide nanoparticles in the nematode *Caenorhabditis elegans*. Environ Toxicol Chem 28(6):1324–1330

Mageshwari K, Sathyamoorthy R (2013) Flower-shaped CuO nanostructures: synthesis, characterization and antimicrobial activity. J Mater Sci Technol 29(10):909–914

Mahdavi M, Namvar F, Ahmad MB, Mohamad R (2013) Green biosynthesis and characterization of magnetic iron oxide (Fe_3O_4) nanoparticles using seaweed (*Sargassum muticum*) aqueous extract. Molecules 18(5):5954–5964

Maillard P, Ciaudo C, Marchais A, Li Y, Jay F, Ding S, Voinnet O (2013) Antiviral RNA interference in mammalian cells. Science 342(6155):235–238

Martinelli F, Scalenghe R, Davino S, Panno S, Scuderi G, Ruisi P et al (2015) Advanced methods of plant disease detection. A review Agronomy for sustainable development 35:1–25

Min JS, Kim KS, Kim SW, Jung JH, Lamsal K, Kim SB, Jung M, Lee YS (2009) Effects of colloidal silver nanoparticles on sclerotium forming phytopathogenic fungi. Plant Pathol J 25(4):376–380

Mishra S, Singh BR, Singh A, Keswani C, Naqvi AH, Singh HB (2014) Bio-fabricated silver nanoparticles act as a strong fungicide against *Bipolaris sorokiniana* causing spot blotch disease in wheat. PLoS One 9(5):e97881

Mitter N, Worrall EA, Robinson KE, Li P, Jain RG, Taochy C, Fletcher SJ, Carroll BJ, Lu GQ, Xu ZP (2017) Clay nano sheets for topical delivery of RNAi for sustained protection against plant viruses. Nat Plants 3(2):16207

Mukhopadhyay SS (2014) Nanotechnology in agriculture: prospects and constraints. Nanotechnol Sci Appl 7(2):63–71

Murphy CJ (2002) Materials science: nanocubes and nanoboxes. Science 298(5601):2139–2141

Nam KT, Kim D-W, Yoo PJ, Chiang C-Y, Meethong N, Hammond PT, Chi YM (2006) Virus-enabled synthesis and assembly of nanowires for lithium ion battery electrodes. Science 312(5775):885–888

Nangmenyi G, Economy J (2009) Nonmetallic particles for oligodynamic microbial disinfection. In: Street A, Sustich R, Duncan J, Savage N (eds) Nanotechnol application for clean water. William Andrew, Norwich, pp 3–15

Nel A, Xia T, Mädler L, Li N (2006) Toxic potential of materials at the nanolevel. Science 311(5761):622–627

Ngo QB, Dao TH, Nguyen HC, Tran XT, Van Nguyen T, Khuu TD (2014) Effects of nanocrystalline powders (Fe, Co and Cu) on the germination, growth, crop yield and product quality of soybean (Vietnamese species DT-51). Adv Nat Sci Nanosci Nanotechnol 5(1):015016

Nguyen TNQ, Le VA, Hua QC, Nguyen TT (2015) Enhancing insecticide activity of anacardic acid by intercalating it into MgAl layered double hydroxides nanoparticles. J Viet Environ 6(3):208–211

Ocsoy I, Paret ML, Ocsoy MA, Kunwar S, Chen T, You M, Tan W (2013) Nanotechnology in plant disease management: DNA-directed silver nanoparticles on graphene oxide as an antibacterial against Xanthomonas perforans. ACS Nano 7(10):8972–8980

Ogunyemia SO, Zhanga F, Abdallaha Y, Zhanga M, Wang Y (2019) Biosynthesis and characterization of magnesium oxide and manganese dioxide nanoparticles using *Matricaria chamomilla* L. extract and its inhibitory effect on *Acidovorax oryzae* strain RS-2. Artif Cells Nanomed Biotechnol 47(1):2230–2239

Ouda SM (2014) Antifungal activity of silver and copper nanoparticles on two plant pathogens, *Alternaria alternata* and *Botrytis cinerea*. Res J Microbiol 9(1):34–42

Papp I, Sieben C, Ludwig K, Roskamp M, Böttcher C, Schlecht S, Herrmann A, Haag R (2010) Inhibition of influenza virus infection by multivalent sialic-acid-functionalized gold nanoparticles. Small 6(24):2900–2906

Paret ML, Palmateer AJ, Knox GW (2013) Evaluation of a light-activated nanoparticle formulation of titanium dioxide with zinc for management of bacterial leaf spot on rosa 'Noare'. Hort Sci 48(2):189–192

Park HJ, Kim SH, Kim SJ, Choi SH (2006) A new composition of nanosized silica-silver for control of various plant diseases. Plant Pathol J 22(3):295–302

Philip D (2009) Honey mediated green synthesis of gold nanoparticles. Spectrochim Acta A Mol Biomol Spectrosc 73(4):650–653

Prasad K, Jha AK (2010) Biosynthesis of CdS nanoparticles: an improved green and rapid procedure. J Colloid Interface Sci 342(1):68–72

Prasad R, Kumar V, Prasad KS (2014) Nanotechnology in sustainable agriculture, present concerns and future aspects. Afr J Biotechnol 13(6):705–713

Puoci F, Lemma F, Spizzirri UG, Cirillo G, Curcio M, Picci N (2008) Polymer in agriculture: a review. Am J Agric Biol Sci 3:299–314

Qian K, Shi T, Tang T, Zhang S, Liu X, Cao Y (2011) Preparation and characterization of nano-sized calcium carbonate as controlled release pesticide carrier for validamycin against *Rhizoctonia solani*. Microchim Acta 173(1):51–57

Rad F, Mohsenifar A, Tabatabaei M, Safarnejad MR, Shahryari F, Safarpour H, Foroutan A, Mardi M, Davoudi D, Fotokian M (2012) Detection of *Candidatus Phytoplasma* aurantifolia with a quantum dots FRET-based biosensor. J Plant Pathol 94(3):525–534

Radoi A, Targa M, Prieto-Simon B, Marty J-L (2008) Enzyme-linked immunosorbent assay (ELISA) based on superparamagnetic nanoparticles for aflatoxin M 1 detection. Talanta 77(1):138–143

Rai M, Ingle A (2012) Role of nanotechnology in agriculture with special reference to management of insect pests. Appl Microbiol Biotechnol 94(2):287–293

Rao KJ, Paria S (2013) Use of sulfur nanoparticles as a green pesticide on *Fusarium solani* and *Venturia inaequalis* phytopathogens. RSC Adv 3(26):10471–10478

Rastogi L, Arunachalam J (2011) Sunlight based irradiation strategy for rapid green synthesis of highly stable silver nanoparticles using aqueous garlic (*Allium sativum*) extract and their antibacterial potential. Mater Chem Phys 129(1–2):558–563

Raza MA, Kanwal Z, Rauf A, Sabri AN, Riaz S, Naseem S (2017) Size- and shape dependent antibacterial studies of silver nanoparticles synthesized by wet chemical routes. Nano 6(4):74

Reddy SM, Datta KKR, Sreelakshmi C, Eswaramoorthy M, Reddy BVS (2012) Honey mediated green synthesis of Pd nanoparticles for suzuki coupling and hydrogenation of conjugated olefins. Nanosci Nanotechnol Lett 4(4):420–425

Rispail N, Matteis LD, Santos R, Miguel AS, Custardoy L, Testillano PS, Risueño MC, Pérez-de-Luque A, Maycock C, Fevereiro P, Oliva A, Fernández-Pacheco R, Ibarra MR, de la Fuente JM, MarquinaC RD, Prats E (2014) Quantum dot and super paramagnetic nanoparticle interaction with pathogenic fungi: internalization and toxicity profile. ACS Appl Mater Interfaces 6(12):9100–9110

Safarpour H, Safarnejad R, Tabatabaei M, Afshin M, Fatemeh R, Marzieh B, Fatemeh S, Fatemeh H (2012) Development of a quantum dots FRET-based biosensor for efficient detection of *Polymyxa betae*. Can J Plant Pathol 34(4):507–515

Saharan V, Mehrotra A, Khatik R, Rawal P, Sharma SS, Pal A (2013) Synthesis of chitosan based nanoparticles and their in vitro evaluation against phyto-pathogenic fungi. Int J Biol Macromol 62:677–683

Saifuddin N, Wong CW, Yasumira AAN (2009) Rapid biosynthesis of silver nanoparticles using culture supernatant of bacteria with microwave irradiation. Eur J Chem 6(1):61–70

Sardella D, Gatt R, Valdramidis V (2017) Physiological effects and mode of action of ZnO nanoparticles against postharvest fungal contaminants. Paper presented at the 109th Annual Meeting of the American Phytopathological Society, San Antonio, TX, August

Sarlak N, Taherifar A, Salehi F (2014) Synthesis of nanopesticides by encapsulating pesticide nanoparticles using functionalized carbon nanotubes and application of new nanocomposite for plant disease treatment. J Agric Food Chem 62(21):4833–4838

Sathiyabama M, Parthasarathy R (2016) Biological preparation of chitosan nanoparticles and its in vitro antifungal efficacy against some phytopathogenic fungi. Carbohydr Polym 151:321–325

Savary S, Willocquet L, Pethybridge SJ, Esker P, McRoberts N, Nelson A (2019) The global burden of pathogens and pests on major food crops. Nat Ecol Evol 3(3):430–439

Sawangphruk M, Srimuk P, Chiochan P, Sangsri T, Siwayaprahm P (2012) Synthesis and antifungal activity of reduced graphene oxide nanosheets. Carbon 50(14):5156–5161

Shenashen M, Derbalah A, Hamza A, Mohamed A, Safty SE (2016) Antifungal activity of fabricated mesoporous alumina nanoparticles against root rot disease of tomato caused by *Fusarium oxysporium*. Pest Manag Sci 73(6):1121–1126

Singaravelu G, Arockiamary JS, Kumar VG, Govindaraju V (2007) A novel extracellular synthesis of monodisperse gold nanoparticles using marine alga, Sargassum wightii Greville. Colloids Surf B Biointerfaces 57(1):97–101

Singh S, Singh M, Agarwal VV, Kumar A (2010) An attempt to develop surface plasmon resonance based immune-sensor for Karnal bunt (*Tilletia indica*) diagnosis based on the experience of nano-gold based lateral flow immune-dipstic test. Thin Solid Films 519(3):1156–1159

Sokolova V, Epple M (2008) Inorganic nanoparticles as carriers of nucleic acids into cells. Angew Chem Int Ed Engl 47(8):1382–1395

Song M-R, Cui S-M, Gao F, Liu Y-R, Fan C-L, Lei T-Q, Liu D-C (2012) Dispersible silica nanoparticles as carrier for enhanced bioactivity of chlorfenapyr. J Pestic Sci 37(3):258–260

Stephenson GR (2003) Pesticide use and world food production: risks and benefits. ACS Publications, Washington, DC, pp 261–270

Strayer-Scherer AL, Liao YY, YoungM RL, Vallad GE, Santra S, Freeman JH, Clark D, Jonees JB, Paret ML (2018) Advanced copper composites against copper-tolerant *Xanthomonas perforans* and tomato bacterial spot. Phytopathology 108(2):196–205

Subramanian KS, Tarafdar JC (2011) Prospects of nanotechnology in Indian farming. Indian J Agric Sci 81:887–893

Tapaszto L, Dobrik G, Lambin P, Biro LP (2008) Tailoring the atomic structure of graphene nanoribbons by scanning tunnelling microscope lithography. Nat Nanotechnol 3(7):397–401

Thairu M, Skidmore I, Bansal R, Nováková E, Hansen T, Li-Byarlay H, Wickline S, Hansen A (2017) Efficacy of RNA interference knockdown using aerosolized short interfering RNAs bound to nanoparticles in three diverse aphid species. Insect Mol Biol 26(3):356–368

Tilman D, Cassman KG, Matson PA, Naylor R, Polasky S (2002) Agricultural sustainability and intensive production practices. Nature 418:671–677

Torney F, Trewyn BG, Lin VSY, Wang K (2007) Mesoporous silica nanoparticles deliver DNA and chemicals into plants. Nat Nanotechnol 2(5):295–300

Velmurugan P, Lee SM, Iydroose M, Lee KJ, Oh BT (2013) Pine conemediated green synthesis of silver nanoparticles and their antibacterial activity against agricultural pathogens. Appl Microbiol Biotechnol 97(1):361–368

Velmurugan P, Anbalagan K, Manosathyadevan M, Lee K-J, Cho M, Lee S-M, Park J-H, Oh S-G, Bang K-S, Oh B-T (2014) Green synthesis of silver and gold nanoparticles using *Zingiber officinale* root extract and antibacterial activity of silver nanoparticles against food pathogens. Bioprocess Biosyst Eng 37(10):1935–1943

Venu R, Ramulu TS, Anandakumar S, Rani VS, Kim CG (2011) Bio-directed synthesis of platinum nanoparticles using aqueous honey solutions and their catalytic applications. Colloids Surf A Physicochem Eng Asp 384(1–3):733–738

Viet PV, Nguyen HT, Cao TM, Hieu LV (2016) Fusarium antifungal activities of copper nanoparticles synthesized by a chemical reduction method. J Nanomater 2016(1):1957612

Vigneshwaran N, Nachane RP, Balasubramanya RH, Varadarajan PV (2006) A novel one-pot 'green' synthesis of stable silver nanoparticles using soluble starch. Carbohydr Res 341(12):2012–2018

Vijayakumar PS, Abhilash OU, Khan BM, Prasad BLV (2010) Nanogold-loaded sharp-edged carbon bullets as plant-gene carriers. Adv Funct Mater 20(15):2416–2423

Wadhwani SA, Shedbalkar UU, Singh R, Karve MS, Chopade BA (2014) Novel polyhedral gold nanoparticles: green synthesis, optimization and characterization by environmental isolate of *Acinetobacter* sp. SW30. World J Microbiol Biotechnol 30(10):2723–2731

Wang H, Wick RL, Xing B (2009) Toxicity of nanoparticulate and bulk ZnO, Al_2O_3 and TiO_2 to the nematode *Caenorhabditis elegans*. Environ Pollut 157(4):1171–1177

Wang Z, Wei F, Liu S-Y, Xu Q, Huang J-Y, Dong X-Y, Yu J-H, Yang Q, Zhao Y-D, Chen H (2010) Electrocatalytic oxidation of phytohormone salicylic acid at copper nanoparticles-modified gold electrode and its detection in oilseed rape infected with fungal pathogen *Sclerotinia sclerotiorum*. Talanta 80(3):1277–1281

Wang X, Liu X, Han HY (2013) Evaluation of antibacterial effects of carbon nanomaterials against copper-resistant *Ralstonia solanacearum*. Colloids Surf B 103(1):136–142

Wang X, Liu X, Chen J, Han H, Yuan Z (2014a) Evaluation and mechanism of antifungal effects of carbon nanomaterials in controlling plant fungal pathogen. Carbon 68:798–806

Wang Y, Cui H, Sun C, Zhao X, Cui B (2014b) Construction and evaluation of controlled-release delivery system of Abamectin using porous silica nanoparticles as carriers. Nanoscale Res Lett 9(1):2490

Wani AH, Shah MA (2012) A unique and profound effect of MgO and ZnO nanoparticles on some plant pathogenic fungi. J Appl Pharm Sci 2(3):40–44

Wanyika H (2013) Sustained release of fungicide metalaxyl by mesoporous silica nanospheres. J Nanopart Res 15:1831

Wu L, Cai X, Nelson K, Xing W, Xia J, Zhang R, Stacy AJ, Luderer M, Lanza GM, Wang LV, Shen B, Pan D (2013) A green synthesis of carbon nanoparticles from honey and their use in real-time photoacoustic imaging. Nano Res 6(5):312–325

Xia T, Kovochich M, Liong M, Meng H, Kabehie S, George S, Zink JI, Nel AE (2009) Polyethyleneimine coating enhances the cellular uptake of mesoporous silica nanoparticles and allows safe delivery of siRNA and DNA constructs. ACS Nano 3(10):3273–3286

Xu L, Cao L-D, Li F-M, Wang X-J, Huang Q-L (2014) Utilization of chitosan-lactide copolymer nanoparticles as controlled release pesticide carrier for pyraclostrobin against *Colletotrichum gossypii*. Southw J Dispers Sci Technol 35(4):544–550

Yang W, Peters JI, Williams RO (2008) Inhaled nanoparticles—a current review. Int J Pharm 356(1–2):239–247

Yang F-L, Li X-G, Zhu F, Lei C-L (2009) Structural characterization of nanoparticles loaded with garlic essential oil and their insecticidal activity against *Tribolium castaneum* (Herbst) (*Coleoptera*: *Tenebrionidae*). J Agric Food Chem 57(21):10156–10162

Yang X, Li Q, Wang H, Huang J, Lin L, Wang W, Sun D, Su Y, Opiyo JB, Hong L, Wang Y, He N, Jia L (2010) Green synthesis of palladium nanoparticles using broth of *Cinnamomum camphora* leaf. J Nanopart Res 12(5):1589–1598

Yi G, Wu Z, Sayer M (1988) Preparation of Pb (Zr,Ti)O_3 thin films by sol gel processing: electrical, optical, and electro-optic properties. J Appl Phys 64(5):2717–2724

Young M, Ozcan A, Myers ME, Johnson EG, Graham JH, Santra S (2017) Multimodal generally recognized as safe ZnO/nanocopper composite: a novel antimicrobial material for the management of citrus phytopathogens. J Agric Food Chem 66(26):6604–6608

Yun H, Kim JD, Choi HC, Lee CW (2013) Antibacterial activity of CNT-Ag and GO-Ag nanocomposites against gram-negative and gram-positive bacteria. Bull Korean Chem Soc 34(11):3261–3264

Zabrieski Z, Morrell E, Hortin J, Dimkpa C, McLean J, Britt D, Anderson A (2015) Pesticidal activity of metal oxide nanoparticles on plant pathogenic isolates of *Pythium*. Ecotel 24(6):1305–1314

Zhang J, Li M, Fan T, Xu Q, Wu Y, Chen C, Huang Q (2013) Construction of novel amphiphilic chitosan copolymer nanoparticles for chlorpyrifos delivery. J Polym Res 20(3):107

Zinjarde S (2012) Bio-inspired nanomaterials and their applications as antimicrobial agents. Chronicles Young Sci 3(1):74

Molecular Marker-Based Approaches for Crop Improvement

18

Pooja Garg, Anamika Kashyap, Kunal Tanwar, Sujata Kumari, Shikha Tripathi, Jyoti Sharma, Pham Thi Thu Ha, Ashish Kumar Gupta, Mahesh Rao, and Sanjay Singh

Abstract

Molecular markers have become one of the most important genomic tools and are being used extensively for crop breeding and improvement programs. Genomics research is giving new insights into statistical approaches that can be used in breeding programs, increasing efficiency and precision. Techniques like Random Amplified Polymorphic DNA (RAPD), Amplified Fragment Length Polymorphism (AFLP), Restricted Fragment Length Polymorphism (RFLP), Simple-Sequence Repeats (SSRs), Sequence-Tagged Sites (STSs) account for the reliability of marker-assisted selection (MAS) of agronomically important traits. Genetic variations can be harnessed from cultivars and wild species using recent genetic approaches like Advanced-Backcross QTL (AB-QTL) analysis, Introgression Libraries (ILs), and Multi-Parent Advanced Generation Intercross (MAGIC) populations. These variations or genes of interest can be introduced through breeding approaches such as Marker Assisted Recurrent Selection (MARS) or Marker Assisted Back Crossing (MABC). This chapter provides an introduction to some recent advancements in biotechnology and the use of molecular marker-based approaches for crop selection and improvement.

Keywords

Molecular markers · Crop improvement · MAS · Genomic tools

P. Garg · A. Kashyap · K. Tanwar · S. Kumari · S. Tripathi · J. Sharma · A. K. Gupta · M. Rao (✉)
ICAR-National Institute for Plant Biotechnology, Pusa Campus, New Delhi, India

P. T. T. Ha
Genomic Research Institute and Seed, Ton Duc Thang University, Ho Chi Minh City, Vietnam

S. Singh
Uttar Pradesh Council of Agricultural Research (UPCAR), Lucknow, Uttar Pradesh, India

V. K. Singh et al. (eds.), *Diseases of Field Crops: Diagnostics and Management*,
https://doi.org/10.1007/978-981-97-6160-9_18

18.1 Introduction

Genetic markers are genetic differences between individual organisms or species, differing in DNA base sequence at any locus or any site. A genetic marker exhibits a polymorphic trait, easily and reliably identified and readily followed in segregating generations. They can be represented as signs or flags, but they cannot be represented as target genes themselves. An ideal genetic marker should be multiallelic and polymorphic based on which individuals could be classified into more than two groups. It should be codominant for discriminating between heterozygotes and homozygotes. Also, they should not be epistatic and pleiotropic. Genetic markers should be abundantly distributed in the entire genome evenly, and they should not show variation with respect to the environment (Singh and Singh 2015). Over time, the discovery of molecular markers is described in Fig. 18.1.

Molecular markers are a type of genetic marker revealing variation at any site in DNA. Molecular markers remain in close proximity to genes on the same chromosomes; they stay linked as further generations of plants are produced. These are identifiable, and specific DNA sequences are transmitted through the standard law of inheritance. Usually, molecular markers seem to have no biological effect and are considered landmarks that are constant in the genome (Mandal 2018). DNA markers can be visualized by gel electrophoresis or detected using radioactive probes, revealing genetic differences among individuals of the same or different species. These markers, which differentiate among two or more species/genotypes, are called *polymorphic markers*. Polymorphic markers can be divided into codominant and dominant depending on their discriminating properties between heterozygotes and homozygotes. Dominant markers can be identified by their presence or absence in the genome, whereas codominant markers can be determined based on the presence of different amplicon sizes in the genomes being analyzed. Codominant markers are multiallelic and dominant markers that identify the single dominant alleles (Singh and Singh 2015).

This book chapter will discuss molecular markers and their potential role in crop improvement as marker-assisted breeding.

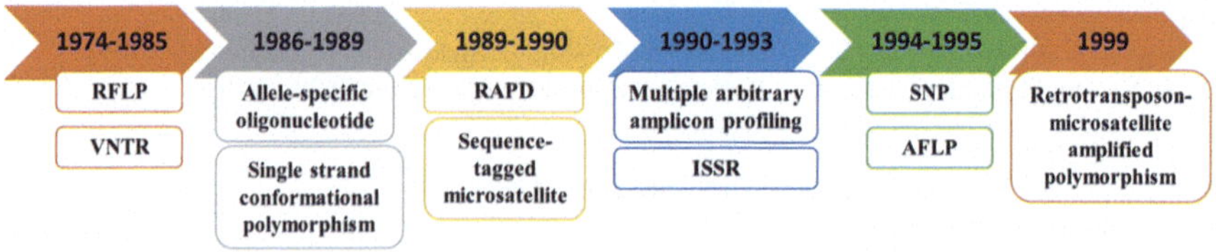

Fig. 18.1 Timeline of significant discovery of different molecular markers

18.2 Conventional Breeding and Molecular Breeding

In 1980–1990, varieties of molecular markers were developed for crop plants (Xu 2010). Because of the expeditious development of DNA markers followed by the improvement of molecular assays, marker-assisted breeding (MAB) is now a regular practice. Molecular markers are widely used for germplasm evaluation, gene discovery, genetic mapping, and improvement of crops, etc., proving that molecular-based technology is a useful and efficient way to induce genetic changes in important agronomic traits of crop plants (Xu 2010; Jiang 2013). Molecular breeding is highly advantageous as compared to conventional breeding procedures:

1. MAB may select the seedling stage for all traits and reduce the time required.
2. It is possible to perform selection under any environmental conditions. This is required to improve specific traits, e.g., stress tolerance and disease resistance expressed under favorable environments.
3. Pleiotropic and polygenic traits can be recognized and selected simultaneously in MAB; therefore, it is useful for gene pyramiding. But in the conventional phenotypic selection, it is impossible to discern these individual loci or genes because of the epistatic effect of genes on one another.
4. Genotypic studies done by molecular markers may be faster and more precise than traditional phenotypic studies. Thus, MAB may be more productive and useful in terms of resources, time, and efforts.

In public and private sectors, research and applications of MAB in plants have been increased, particularly since the 2000s (Xu 2010; Jiang 2013; Babu et al. 2004). MAB has a vital contribution to the future development of plant breeding, but it is not perfect and has some shortcomings. It is a great challenge for MAB to improve economically relevant agronomic traits like quality and yield, which have complex inheritance patterns, even with the recently established Genome-Wide Selection (GWS) (Jiang 2013; Nakaya and Isobe 2012). MAB is not generally or inevitably helpful from the standpoint of a plant breeder (Jiang 2013). The following limitations and/or obstacles remain in the application of molecular techniques for plant breeding:

1. The precision of the QTL mapping study is critical for the effectiveness of MAS, especially for more complicated traits like yield. The efficiency of QTL detection is low with a population number less than 200 during the simulation and experimental studies (Kearsey 1998).
2. In some cases, weak linkage results in recombination between the marker and gene. This may occur even if there is a strong linkage because of the inaccuracy of data obtained from a one-time QTL mapping experiment (Sharp et al. 2001). Thus, to evaluate the accuracy of a marker in predicting phenotype, it must be validated.
3. Markers are expected to behave as diagnostic tools for traits in a broad range of breeding sources, but in application, this may not always be the case. For exam-

ple, *the Sr2* gene, a wheat SSR marker responsible for resistance against stem rust, was found to be diagnostic for all cultivars but not for Australian cultivars. So it cannot be used for the introgression of the highly vulnerable cultivars. Although some crops have a great number of usable markers, specific chromosomal sites harboring an important QTL or gene can be challenging to find polymorphic markers for these sites (Brumlop and Finckh 2011).

4. It has been discovered that QTLs from one mapping population may not be useful in another background. It could be the result of a minor influence of an allele introduced into elite cultivars. In some situations, the use of a QTL is useless for introgression into superior varieties. In some circumstances, due to epistasis, the effect of a QTL can differ in different genetic backgrounds (Li et al. 2003a).
5. The cost of adopting MAS is too high as compared to traditional phenotypic selection due to the benefit ratio of MAS, the methods used for phenotypic evaluation, field cost, glasshouse facility, and human resources cost. It is necessary to make substantial initial investments for equipments, including recurring expenses. Licensing cost is also included in MAS (Dreher et al. 2003).
6. A large percentage of reported markers are not used in practical applications. Therefore, to utilize them in breeding programs on a large scale, it is mandatory to resolve different logistical, economical, practical, and genetic obstacles (Xu et al. 2004).
7. Interactions between genotype and environment also affect genetic studies and hinder predictions. Marker-assisted backcross concept was used by Ribaut and Ragot (2007) in maize to enhance drought tolerance and found that the successful marker selection for quantitative traits depends on many other factors like epistasis, G × E interactions, and the authenticity of plant phenotyping.

Therefore, molecular breeding is not a substitute for traditional breeding but rather a supplement to it. MAB's high price and technical requirements will always be key hindrances to widespread adoption in upcoming times, particularly for prospering nations (Ribaut et al. 2010; Collard and Mackill 2008). Therefore, combining MAB and traditional breeding programs will be a good technique for future crop science advancements.

18.3 Molecular Markers in Crop Improvement

The types of genetic markers have already been described in Fig. 18.2. Many of these used in crop improvement are discussed here in detail. Table 18.1 can be referred to for examples of such markers.

18.3.1 Restriction Fragment Length Polymorphism (RFLP)

RFLP is a hybridization-based marker and was first used by Botstein et al. (1980) for human genome mapping. RFLPs were first used to identify DNA

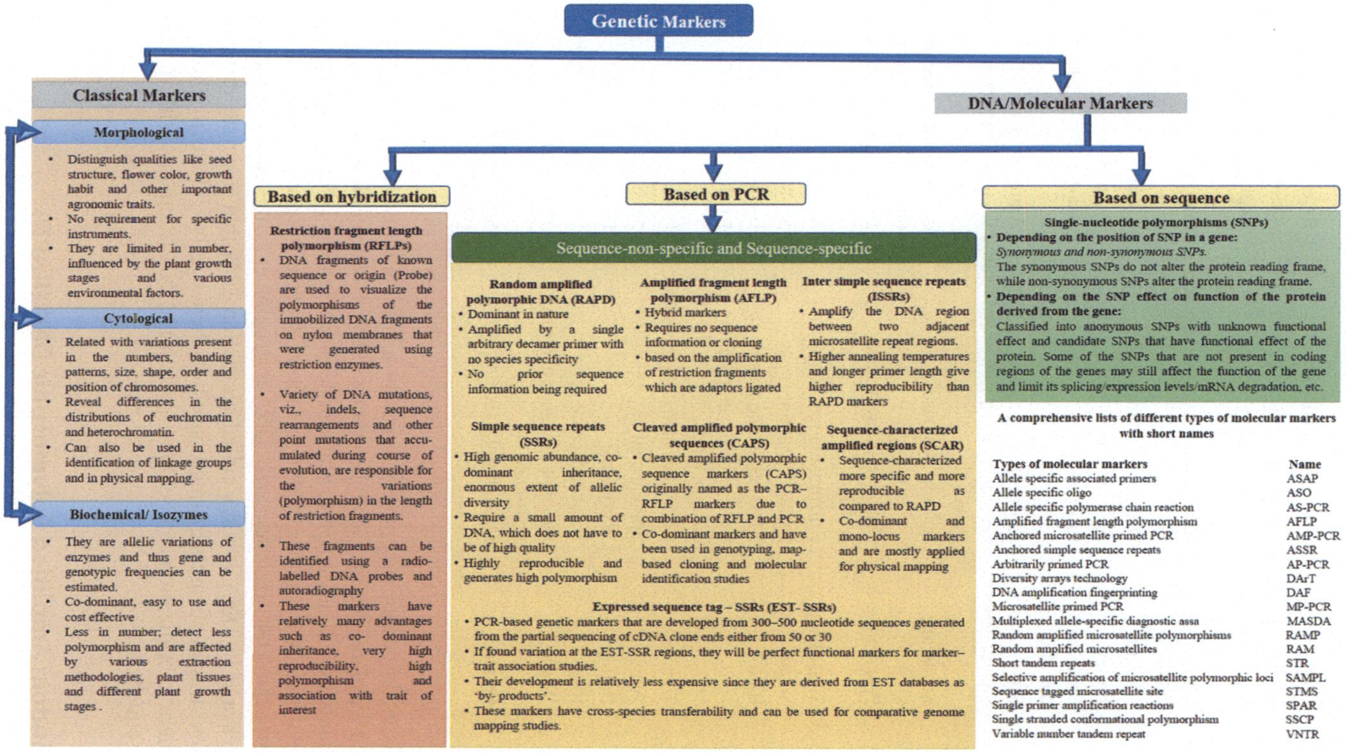

Fig. 18.2 Types of genetic markers: classical markers are divided into three types: morphological, cytological, and biochemical markers; DNA/molecular markers are classified based on the methods of detection into three categories: based on hybridization, based on PCR, and based on sequence

Table 18.1 Summary of major molecular markers used for various associated traits in major crops

Molecular marker	Crop	Trait associated with	References
AFLP	Barley	Water stress tolerance	Altinkut et al. (2003)
AFLP to SCAR	*B. juncea*	Seed coat color	Negi et al. (2000)
CAPS, SCAR	Chinese cabbage	Resistance to clubroot disease	Piao et al. (2004)
	Tomato	Disease resistance	Arens et al. (2010)
RAPD	Barley	Spot blotch	Tyagi et al. (2007)
	Rapeseed mustard	Erucic acid	Jourdren et al. (1996)
	Rice	Blast resistance gene	Araújo et al. (2002)
RFLP	Sour cherry	QTL analysis of flower and fruit traits	Wang et al. (2000)
EST-derived RFLP	Sugarcane	Sucrose synthase	da Silva and Bressiani (2005)
SCAR	Pearl millet	Downy mildew disease resistance	Jogaiah et al. (2014)
	Sugar beet	Rhizomania resistance	Norouzi et al. (2017)
	Sugarcane	Drought tolerant	Srivastava et al. (2012)
SNP	Rice	Blast resistance gene	Hayashi et al. (2004)
		Root traits development in drought stress	Krishna et al. (2018)
	Tomato	Root-knot nematode resistance	Devran et al. (2016)
	Wheat	Spike density	Ma et al. (2018)
	Barley	Waxy gene	Domon et al. (2004)
SSR	Rice	Blast resistance gene	Fjellstrom et al. (2006)
	Wheat	Drought tolerance and rust resistance	Mavi et al. (2018)
	Potato	Drought tolerance	Schumacher et al. (2021)

polymorphisms for genetic mapping of a mutation in adenovirus serotypes and were later adopted for plant genomes (Singh and Das 2009). However close in phylogeny, any two organisms will always have some differences in their DNA sequences due to some mutations. With the use of this technique, these differences can be visualized with respect to DNA fragment size. DNA restriction enzymes catalyze endonucleolytic cleavage in DNA at sites called restriction sites that generate different lengths. These fragments are separated based on their size by gel electrophoresis. Later, Southern hybridization can be done to identify DNA fragments with specific nucleotide sequences.

18.3.2 Random Amplified Polymorphic DNA (RAPD)

Polymorphic markers can be rapidly generated using random primers in a PCR, a method described by Williams et al. (1990). Here, prior knowledge of specific nucleotide sequences is not required, as single primers of random sequences are used to amplify genomic DNA. Polymorphism can be detected by visualizing the bands on the agarose gel. The polymorphic markers can then be used to develop genetic linkage maps. This saves the time and cost of obtaining sequence information (Williams et al. 1990).

18.3.3 Amplified Restriction Fragment Length Polymorphism or Selective Restriction Fragment Amplification (AFLP)

Vos et al. (1995) developed a DNA fingerprinting technique in which selective restriction fragments are amplified from a total digest of genomic DNA. The method for AFLP can be described in three steps. First, the genomic DNA is digested at restriction sites using a pair of restriction enzymes, a rare cutter and a frequent cutter, to limit the number of fragments and ease selective amplification. Then, template DNA is generated by ligating adapters to the ends of the obtained fragments. Second, selective amplification of restricted fragments is done by attaching selective nucleotides to the 3′ ends of PCR primers. The third and last step is to analyze amplified fragments by gel electrophoresis and visualization of bands (Vos et al. 1995).

18.3.4 Sequence Characterized Amplified Regions (SCAR)

Michelmore et al. (1991) described a method in 1991 where RFLP probes or RAPD primers are used to screen the differences in a segregating population and detect markers in a genomic region rapidly. Martin also described a similar method in which isogenic lines rapidly identify sequences associated with the gene of interest (Martin et al. 1991). SCAR is more specific and reproducible than RAPD. In this technique, 24 long nucleotide primers are synthesized for a specific locus linked with the RAPD marker. Further, amplification and detection are done via PCR and gel electrophoresis (Paran and Michelmore 1993).

18.3.5 Single Nucleotide Polymorphism (SNP)

SNPs are more abundant in the genome than SSRs and are inherited stably (Landegren et al. 1998). They are widely used in many genetic applications such as Genotype-by-sequencing (GBS), phylogeny analysis, population structure, genetic purity, genetic diversity, construction of linkage maps, QTL mapping, genomic selection (GS), MABC, and MARS. SNPs can be classified as synonymous and non-synonymous SNPs. Some examples of crop improvement using SNP markers have been given in Table 18.1.

18.3.6 Simple Sequence Repeats (SSR) or Microsatellites

Microsatellites have been developed as molecular markers after the work of many scientists who have established the role of these regions in hypervariability and polymorphism (Litt and Luty 1989; Tautz 1989; Weber and May 1989). SSRs rich in AT repeats are highly polymorphic and suitable for evaluating allelic variations within closely related cultivars of rice (Akagi et al. 1997). SSR markers are PCR-based and are widely used as they are codominant, efficient, readily available, cost-effective, and have high diversity in crop plants.

18.4 Potential Applications in Plant Sciences

18.4.1 Phylogeny and Evolution

Molecular markers play a significant role in retrieving information about phylogeny and evolution (Wei et al. 2010). Phylogenetic studies are generally conducted using chloroplast genome sequences, which are more stable (Dong et al. 2012). Microarray-based expression analysis and cluster analysis were conducted to study maize phylogeny for the transcription factor WRKY (Wei et al. 2012). Phylogenetic relationships in *Brassica* and *Raphanus sativus* were conducted using nuclear and chloroplast DNA (Qi and Zhang 2012).

18.4.2 Heterosis Investigation

The supremacy of the F_1 hybrid over the two crossed parents in yield, performance, etc., is termed heterosis. Positive heterosis is defined when the F_1 hybrids are greater than their parents, and if F_1 hybrids are less than their parents, then it is said to be negative heterosis (Comings and MacMurray 2000). Various molecular studies related to genetic diversity and heterosis were conducted in different crops like wheat (Martin et al. 1995), maize (Betran et al. 2003), and rapeseed (Yu et al. 2005). Recently, SSR markers were used to detect the heterotic pattern in rice (Wu et al. 2013).

18.4.3 Cultivar Identity/Hybrid Testing

Markers can confirm the true identity of individual plants. In cereal hybrid production, a high level of genetic purity is essential for exploiting heterosis. The quality "grow-out tests" are complex and time-consuming, so SSR and STS markers have been used in hybrid rice for purity confirmation (Yashitola et al. 2002).

Table 18.2 Molecular markers used for diversity analysis and linkage studies in different crops

Molecular marker	Crop	Work done	References
RAPD, RFLP	Peach	Genetic linkage map	Rajapakse et al. (1995)
RFLP	Soybean	Mapping for fatty acid content	Diers and Shoemaker (1992)
SSR	Sorghum	Diversity analysis	Anas and Yoshida (2004)
	Melon	Linkage studies	Chiba et al. (2003)
	Pearl millet	Genetic diversity	Budak et al. (2003)
	Tomato	Genetic diversity	Dhaliwal et al. (2013)
EST-SSR	Sugarcane	Transferability, genetic diversity	Ul Haq et al. (2016)

18.4.4 Genetic Diversity Analysis/Population Genetics

Advancement in molecular markers and genome genetic diversity extensively deals with germplasm. Assessment of genetic diversity is very useful in the relative study of plant evolution and their provisional genomics and in understanding the population structure (Nadeem et al. 2018). Mostly RAPD, AFLP, and SSR markers are used in population genetics studies (Brown and Epifanio 2003). In the characteristic analysis of modern rice varieties, an SSR marker based on genetic diversity is used (Shakil et al. 2015). For the study of genetic diversity among the population of Geum reptans with a crossbreed clonal plant, RAPD-PCR is used (Pluess and Stöcklin 2004). More examples of genetic diversity and map construction are given in Table 18.2.

18.4.5 Genetic Linkage Map Construction

A genetic linkage map is considered a road map for explaining the distance and location between markers. Genetic linkage maps act as an important tool in genetic and genomic research. DNA markers can be used for the construction of genetic linkage maps (Khanal et al. 2015). A genetic map of 2275 markers covering 1521.6 cM has been constructed for rice (*Oryza sativa* L.) (Harushima et al. 1998).

18.4.6 Mapping of Quantitative Trait Loci (QTLs)

Most of the agronomically important traits are quantitative and polygenic in nature. The region of the chromosome that contains the gene for these quantitative traits is termed as Quantitative trait locus (QTL). In QTL mapping, molecular markers are used to identify the genes that can harm the trait of interest. These traits are categorized as qualitative and quantitative. Qualitative traits describe discontinuous variations, while quantitative traits explain continuous variations. QTL mapping is an effort to describe the genetic foundation of variation in complex traits (Kearsey 1998). In *P. haitanensis*, 15 QTLs associated with six economic traits were identified by constructing a high-density genetic linkage (Xu et al. 2015). Similarly, a genetic linkage map was constructed for Sponge Gourd, and QTLs for two interspecific reproductive isolation traits were analyzed (Wu et al. 2016).

18.4.7 Mapping of Mutations

A genetic map is more informative if more mutations are mapped; if more genes that are localized along a chromosome, the map will be more detailed in such case. Genetic mapping of the mutated gene is done by isolating and cloning the gene with respect to the normal gene and recognizing its programmed protein.

18.4.8 Map-Based Cloning of Genes

Map-based cloning, or positional cloning, is a method to recognize the genetic basis of a mutant phenotype by studying linkage to markers of known location. Cloning of genes based on the map associated with the soybean maturity locus E3 (Watanabe et al. 2009), isolation of the *Pto* gene in tomato (Martin et al. 1993), and identification of the *a-1* gene in rice (Yoshimura et al. 1996) are some examples of positional cloning. *MOC-1*, a gene responsible for rice tillering, was isolated using this approach (Li et al. 2003b).

18.4.9 Marker-Assisted Selection (MAS)

MAS makes it easy to select recessive alleles in their heterozygous state with the help of codominant markers such as SNP and SSR. Selfing or test crossing is not required here and thus saves time and accelerates the breeding process. MAS is increasing in importance in upbringing programs for simply inherited traits, allowing to pace up the breeding processes (Francia et al. 2005).

18.4.10 Marker-Assisted Pyramiding

Gene pyramiding is commonly used in plant breeding nowadays. It can be used in self-pollinated crops to sum up several desirable genes into one genotype of interest. Nowadays, marker-based gene pyramiding can be used as a method to develop inbred lines with improved traits. The most common and wide use of gene pyramiding is to develop stable resistant plants against diseases by integrating multiple disease resistance genes into that plant. Gene pyramiding mainly consists of three main basic steps: selecting parents with the desired genes, successive backcrossing with one of the parents, and the accumulation of desired genes into a single genotype.

Table 18.3 Different techniques used in crop improvement

Technique used	Crop	Trait	References
MABC	Bread wheat	Leaf rust resistance	Yadawad et al. (2017)
	Wheat	Rust resistance	Singh et al. (2018)
MARBB	Rice	Bacterial blight and blast disease	Sagar et al. (2018)
MAS	Soybean	Kunitz trypsin inhibitor	Kumar et al. (2011)
	Rice	Bacterial leaf blight	Perumalsamy et al. (2010)
MAS/RAPD	*B. napus*	Male fertility restorer gene	Hansen et al. (1997)
MAS/STMS	Rice	Bacterial blight resistance	Basavaraj et al. (2010)

MAS marker assisted selection, *MABC* marker assisted back cross, *MARBB* marker-assisted restricted backcross breeding

Hittalmani et al. (2000) has developed the rice blast resistance gene, which originated from three parents with the help of RFLP and STS markers. Castro et al. (2003) suggested the combination of a single stripe rust gene and two QTLs in quantitative resistance.

18.4.11 Marker-Assisted Backcross Breeding (MABC)

In plant breeding techniques, backcrossing is important as it is used to incorporate several genes into an elite or adapted variety. Like MAS, DNA markers used in MAB can help in the selection of desirable crop traits efficiently (Table 18.3). MAB has been used in bacterial blight resistance in rice with STS markers (Singh et al. 2001) and the Sub1 QTL selection technique by phenotyping traits and SSR (Mackill et al. 2007).

18.4.12 Molecular Markers in Nano Era

Nanoparticles are serving as nano-carriers in the agricultural industry for transferring of genes or herbicides and chemicals targeting a particular part of the plant (Zakaria Mohammed et al. 2015). Increasing applications of nanoparticles have been found to induce genotoxicity in plants. In the direct mechanism of inducing genotoxicity, nanoparticles pass through the cell and nucleus membranes and interact with DNA either by mechanical or chemical binding. This DNA damage and modification can be detected using molecular markers. Liu et al. (2012) used RAPD for identification of DNA damage and mutations in Arabidopsis cd-treated shoots.

18.5 Conclusion

With the advancement of technology and the development of NGS, various molecular markers like RAPD, AFLP, RFLP, SSRs, STSs, SCAR, CAPS, etc. can be employed as efficient tools for the improvement of traits such as disease resistance, stress tolerance, nutrient uptake efficiency, bio-fortification, and other traits in crops. These markers account for the reliability of MAS for agronomically important traits. Genetic variations can be harnessed from cultivars and wild species using recent genetic approaches. These variations or genes of interest can be introduced through breeding approaches such as MABC, MARBB, or MARS (Marker-Assisted Recurrent Selection).

References

Akagi H, Yokozeki Y, Inagaki A et al (1997) Highly polymorphic microsatellites of rice consist of AT repeats and classification of closely related cultivars with these microsatellite loci. Theor Appl Genet 94(1):61–67. https://doi.org/10.1007/s001220050382

Altinkut A, Kazan K, Gozukirmizi N (2003) AFLP marker linked to water-stress-tolerant bulks in barley (*Hordeum vulgare* L.). Genet Mol Biol 26(1):77–82. https://doi.org/10.1590/S1415-47572003000100013

Anas, Yoshida T (2004) Sorghum diversity evaluated by simple sequence repeat (SSR) markers and phenotypic performance. Plant Prod Sci 7(3):301–308. https://doi.org/10.1626/pps.7.301

Araújo LG, Prabhu AS, Filippl MC (2002) Identification of RAPD marker linked to blast resistance gene in a somaclone of rice cultivar Araguaia. Fitopatol Bras 27(2):181–185. https://doi.org/10.1590/S0100-41582002000200010

Arens P, Mansilla C, Deinum D et al (2010) Development and evaluation of robust molecular markers linked to disease resistance in tomato for distinctness, uniformity, and stability testing. Theor Appl Genet 120(3):655–664. https://doi.org/10.1007/s00122-009-1183-2

Babu R, Nair SK, Prasanna BM et al (2004) Integrating marker-assisted selection in crop breeding–prospects and challenges. Curr Sci 87:607–619

Basavaraj SH, Singh VK, Singh A et al (2010) Marker-assisted improvement of bacterial blight resistance in parental lines of Pusa RH10, a superfine grain aromatic rice hybrid. Mol Plant Breed 26(2):293–305. https://doi.org/10.1007/s11032-010-9407-3

Betran FJ, Ribaut JM, Beck D et al (2003) Genetic diversity, specific combining ability, and heterosis in tropical maize under stress and non-stress environments. Crop Sci 43(3):797–806. https://doi.org/10.2135/cropsci2003.7970

Botstein D, White RL, Skolnick M et al (1980) Construction of a genetic linkage map in man using restriction fragment length polymorphisms. Am J Hum Genet 32(3):314–331

Brown B, Epifanio J (2003) Nuclear DNA. In: Hallermann EM (ed) Population genetics: principles and applications for fisheries scientists. American Fisheries Society, Bethesda, pp 458–472

Brumlop S, Finckh MR (2011) Applications and potentials of marker-assisted selection (MAS) in plant breeding: final report of the F+ E project "applications and potentials of smart breeding" (FKZ 350 889 0020)-on behalf of the Federal Agency for Nature Conservation. Deutschland

Budak H, Pedraza F, Cregan PB et al (2003) Development and utilization of SSRS to estimate the degree of genetic relationships in a collection of pearl millet germplasm. Crop Sci 43(6):2284–2290. https://doi.org/10.2135/cropsci2003.2284

Castro AJ, Capettini F, Corey AE et al (2003) Mapping, and pyramiding qualitative and quantitative resistance to stripe rust in barley. Theor Appl Genet 107(5):922–930. https://doi.org/10.1007/s00122-003-1329-6

Chiba N, Suwabe K, Nunome T et al (2003) Development of microsatellite markers in melon (*Cucumis melo* L.) and their application to major cucurbit crops. Breed Sci 53(1):21–27. https://doi.org/10.1270/jsbbs.53.21

Collard BC, Mackill DJ (2008) Marker-assisted selection: an approach for precision plant breeding in the twenty-first century. Philos Trans R Soc Lond Ser B Biol Sci 363(1491):557–572. https://doi.org/10.1098/rstb.2007.2170

Comings DE, MacMurray JP (2000) Molecular heterosis: a review. Mol Genet Metab 71(1):19–31. https://doi.org/10.1006/mgme.2000.3015
da Silva JA, Bressiani JA (2005) Sucrose synthase molecular marker associated with sugar content in elite sugarcane progeny. Genet Mol Biol 28(2):294–298. https://doi.org/10.1590/S1415-47572005000200020
Devran Z, Göknur A, Mesci L (2016) Development of molecular markers for the *Mi-1* gene in tomato using the KASP genotyping assay. Hortic Environ Biotechnol 57(2):156–160. https://doi.org/10.1007/s13580-016-0028-6
Dhaliwal MS, Jindal SK, Gaikwad AK et al (2013) Genetic diversity analysis and DNA fingerprinting of elite chilli pepper lines using SSR markers. Int J Veg Sci 19(3):207–216. https://doi.org/10.1080/19315260.2012.707762
Diers BW, Shoemaker RC (1992) Restriction fragment length polymorphism analysis of soybean fatty acid content. J Am Oil Chem Soc 69(12):1242–1244. https://doi.org/10.1007/BF02637690
Domon E, Yanagisawa T, Saito A et al (2004) Single nucleotide polymorphism genotyping of the barley waxy gene by polymerase chain reaction with confronting two-pair primers. Plant Breed 123(3):225–228. https://doi.org/10.1111/j.1439-0523.2004.00970.x
Dong W, Liu J, Yu J et al (2012) Highly variable chloroplast markers for evaluating plant phylogeny at low taxonomic levels and for DNA barcoding. PLoS One 7(4):e35071. https://doi.org/10.1371/journal.pone.0035071
Dreher K, Khairallah M, Ribaut JM et al (2003) Money matters (I): costs of field and laboratory procedures associated with conventional and marker-assisted maize breeding at CIMMYT. Mol Plant Breed 11(3):221–234. https://doi.org/10.1023/A:1022820520673
Fjellstrom R, McClung AM, Shank AR (2006) SSR markers closely linked to the *pi-z* locus are useful for selection of blast resistance in a broad array of rice germplasm. Mol Plant Breed 17(2):149–157. https://doi.org/10.1007/s11032-005-4735-4
Francia E, Tacconi G, Crosatti C et al (2005) Marker assisted selection in crop plants. Plant Cell Tissue Organ Cult 82:317–342. https://doi.org/10.1007/s11240-005-2387-z
Hansen M, Halldén C, Nilsson NO et al (1997) Marker-assisted selection of restored male-fertile Brassica napus plants using a set of dominant RAPD markers. Mol Plant Breed 3(6):449–456. https://doi.org/10.1023/A:1009647417936
Harushima Y, Yano M, Shomura A et al (1998) A high-density rice genetic linkage map with 2275 markers using a single F2 population. Genetics 148(1):479–494. https://doi.org/10.1093/genetics/148.1.479
Hayashi K, Hashimoto N, Daigen M et al (2004) Development of PCR-based SNP markers for rice blast resistance genes at the *Piz* locus. Theor Appl Genet 108(7):1212–1220. https://doi.org/10.1007/s00122-003-1553-0
Hittalmani S, Parco A, Mew TV et al (2000) Fine mapping and DNA marker-assisted pyramiding of the three major genes for blast resistance in rice. Theor Appl Genet 100:1121–1128. https://doi.org/10.1007/s001220051395
Jiang GL (2013) Molecular markers and marker-assisted breeding in plants. Plant Breed Lab Fields 3:45–83. https://doi.org/10.5772/52583
Jogaiah S, Sharathchandra RG, Raj N et al (2014) Development of SCAR marker associated with downy mildew disease resistance in pearl millet (*Pennisetum glaucum* L.). Mol Biol Rep 41(12):7815–7824. https://doi.org/10.1007/s11033-014-3675-7
Jourdren C, Barret P, Horvais R et al (1996) Identification of RAPD markers linked to the loci controlling erucic acid level in rapeseed. Mol Plant Breed 2(1):61–71. https://doi.org/10.1007/BF00171352
Kearsey MJ (1998) The principles of QTL analysis (a minimal mathematics approach). J Exp Bot 49(327):1619–1623. https://doi.org/10.1093/jxb/49.327.1619
Khanal R, Navabi A, Lukens L (2015) Linkage map construction and quantitative trait loci (QTL) mapping using intermated vs. selfed recombinant inbred maize line (*Zea mays* L.). Can J Plant Sci 95:1133–1144. https://doi.org/10.4141/cjps-2015-091
Krishna GK, Vishwakarma C, Thomas P et al (2018) Association between ABA- and drought-mediated regulation of root traits and identification of potential SNPs in genes

for root development in rice. Indian J Genet Plant Breed 78(1):48–58. https://doi.org/10.5958/0975-6906.2018.00006.8

Kumar V, Rani A, Mourya V et al (2011) Marker-assisted selection for development of kunitz trypsin inhibitorfree soybean varieties: I. Parental polymorphism survey using SSR markers. Indian J Genet Plant Breed 71(4):372–376

Landegren U, Nilsson M, Kwok PY (1998) Reading bits of genetic information: methods for single-nucleotide polymorphism analysis. Genome Res 8(8):769–776. https://doi.org/10.1101/gr.8.8.769

Li ZK, Yu SB, Lafitte HR et al (2003a) QTL× environment interactions in rice. I Heading date and plant height. Theor Appl Genet 108(1):141–153. https://doi.org/10.1007/s00122-003-1401-2

Li X, Qian Q, Fu Z et al (2003b) Control of tillering in rice. Nature 422:618–621. https://doi.org/10.1038/nature01518

Litt M, Luty JA (1989) A hypervariable microsatellite revealed by in vitro amplification of a dinucleotide repeat within the cardiac muscle actin gene. Am J Hum Genet 44(3):397–401

Liu W, Sun L, Zhong M et al (2012) Cadmium-induced DNA damage and mutations in Arabidopsis plantlet shoots identified by DNA fingerprinting. Chemosphere 89:1048–1055. https://doi.org/10.1016/j.chemosphere.2012.05.068

Ma J, Sun M, Yang C et al (2018) Development and validation of markers for spike density QTL, qsd.Sau-7A from tibetan semi-wild wheat (*Triticum aestivum* ssp. tibetanum). Indian J Genet Plant Breed 78(1):11–18. https://doi.org/10.5958/0975-6906.2018.00002.0

Mackill DJ, Collard BCY, Neeraja CN et al (2007) QTLs in rice breeding: examples for abiotic stresses. In: Rice genetics. World Scientific Publishing, Singapore, pp 155–167. https://doi.org/10.1142/9789812708816_0011

Mandal L (2018) Potential applications of molecular markers in plant. Curr Trends Biomed Eng Biosci 12(4):85. https://doi.org/10.19080/CTBEB.2018.12.555844

Martin GB, Williams JGK, Tanksley SD (1991) Rapid identification of markers linked to a pseudomonas resistance gene in tomato by using random primers and near-isogenic lines. Proc Natl Acad Sci 88(6):2336–2340. https://doi.org/10.1073/pnas.88.6.2336

Martin GB, Brommonschenkel SH, Chunwongse J et al (1993) Map based cloning of a protein kinase gene conferring disease resistance in tomato. Science 262:1432–1436. https://doi.org/10.1126/science.7902614

Martin JM, Talbert LE, Lanning SP et al (1995) Hybrid performance in wheat as related to parental diversity. Crop Sci 35(1):104–108. https://doi.org/10.2135/cropsci1995.0011183X003500010019x

Mavi GS, Bazzer SK, Kaur S et al (2018) Marker assisted stacking of drought tolerance QTL and rust resistance genes in an elite Indian wheat genotype. Indian J Genet Plant Breed 78(1):1–10. https://doi.org/10.5958/0975-6906.2018.00001.9

Michelmore RW, Paran I, Kesseli RV (1991) Identification of markers linked to disease-resistance genes by bulked segregant analysis: a rapid method to detect markers in specific genomic regions by using segregating populations. Proc Natl Acad Sci 88:9828–9832. https://doi.org/10.1073/pnas.88.21.9828

Nadeem AM, Nawaz AM, Shahid MQ et al (2018) DNA molecular markers in plant breeding: current status and recent advancements in genomic selection and genome editing. Biotechnol Biotechnol Equip 32(2):261–285. https://doi.org/10.1080/13102818.2017.1400401

Nakaya A, Isobe SN (2012) Will genomic selection be a practical method for plantbreeding? Ann Bot 110(6):1303–1316. https://doi.org/10.1093/aob/mcs109

Negi MS, Devic M, Delseny M (2000) Identification of AFLP fragments linked to seed coat colour in *Brassica juncea* and conversion to a SCAR marker for rapid selection. Theor Appl Genet 101(1–2):146–152. https://doi.org/10.5958/0975-6906.2017.00042.6

Norouzi P, Mahmoudi SB, Darabi S et al (2017) Confirmation of some SCAR molecular markers linked to rhizomania resistance gene (*Rz1*) in sugar beet. Indian J genet plant Breed 77(2):312–315. https://doi.org/10.5958/0975-6906.2017.00042.6

Paran I, Michelmore RW (1993) Development of reliable PCR-based markers linked to downy mildew resistance genes in lettuce. Theor Appl Genet 85(8):985–993. https://doi.org/10.1007/BF00215038

Perumalsamy S, Bharani M, Sudha M et al (2010) Functional marker-assisted selection for bacterial leaf blight resistance genes in rice (*Oryza sativa* L.). Plant Breed 129(4):400–406. https://doi.org/10.1111/j.1439-0523.2009.01705.x

Piao ZY, Deng YQ, Choi SR et al (2004) SCAR and CAPS mapping of CRb, a gene conferring resistance to *Plasmodiophora brassicae* in Chinese cabbage (*Brassica rapa* ssp. pekinensis). Theor Appl Genet 108(8):1458–1465. https://doi.org/10.1007/s00122-003-1577-5

Pluess AR, Stöcklin J (2004) Population genetic diversity of the clonal plant Geumreptans (Rosaceae) in the Swiss Alps. Am J Bot 91(12):2013–2021. https://doi.org/10.3732/ajb.91.12.2013

Qi XH, Zhang MF (2012) Molecular phylogenetic studies on members of the brassica and Raphanus genera based on the nuclear ribosomal internal transcribed spacer, the chloroplast *trn*L intron, *trn*LF, and cpSSRs. J Hortic Sci Biotechnol 87(2):149–156. https://doi.org/10.1080/14620316.2012.11512846

Rajapakse S, Belthoff LE, He G et al (1995) Genetic linkage mapping in peach using morphological, RFLP and RAPD markers. Theor Appl Genet 90(3–4):503–510. https://doi.org/10.1007/BF00221996

Ribaut JM, Ragot M (2007) Marker-assisted selection to improve drought adaptation in maize: the backcross approach, perspectives, limitations, and alternatives. J Exp Bot 58(2):351–360. https://doi.org/10.1093/jxb/erl214

Ribaut JM, De Vicente MC, Delannay X (2010) Molecular breeding in developing countries: challenges and perspectives. Curr Opin Plant Biol 13(2):213–218. https://doi.org/10.1016/j.pbi.2009.12.011

Sagar V, Krishnan SG, Mondal KK et al (2018) Development of basmati rice genotypes with resistance to both bacterial blight and blast diseases using marker assisted restricted backcross breeding. Indian J Genet Plant Breed 78(1):36–47. https://doi.org/10.5958/0975-6906.2018.00005.6

Schumacher C, Krannich CT, Maletzki L et al (2021) Unravelling differences in candidate genes for drought tolerance in potato (*Solanum tuberosum* L.) by use of new functional microsatellite markers. Genes 12(4):494. https://doi.org/10.3390/genes12040494

Shakil SK, Sultana S, Hasan MM et al (2015) SSR marker based genetic diversity analysis of modern rice varieties and coastal landraces in Bangladesh. Indian J Biotechnol 14:33–41

Sharp PJ, Johnston S, Brown G et al (2001) Validation of molecular markers for wheat breeding. Aust J Agric Res 52(12):1357–1366. https://doi.org/10.1071/AR01052

Singh KM, Das BK (2009) Molecular markers: It's application in crop improvement types of molecular markers. J Crop Sci Biotechnol 12(4):169–181. https://doi.org/10.1007/s12892-009-0124-6

Singh BD, Singh AK (2015) Marker-assisted plant breeding: principles and practices. Springer, New Delhi. https://doi.org/10.1007/978-81-322-2316-0

Singh S, Sidhu JS, Huang N et al (2001) Pyramiding three bacterial blight resistance genes (*xa5, xa13 and Xa21*) using marker-assisted selection into indica rice cultivar PR106. Theor Appl Genet 102:1011–1015. https://doi.org/10.1007/s001220000495

Singh A, Jaiswal JP, Badoni S (2018) Enhancing rust resistance in wheat through marker assisted backcross breeding. Indian J Genet Plant Breed 78(1):19–25. https://doi.org/10.5958/0975-6906.2018.00003.2

Srivastava MK, Li CN, Li YR (2012) Development of sequence characterized amplified region (SCAR) marker for identifying drought tolerant sugarcane genotypes. Aust J Crop Sci 6(4):763–767. https://doi.org/10.3316/informit.363108953113553

Tautz D (1989) Hypervariability of simple sequences as a general source for polymorphic DNA markers. Nucleic Acids Res 17(16):6463–6471. https://doi.org/10.1093/nar/17.16.6463

Tyagi K, Kumar U, Prasad LC et al (2007) Inheritance and identification of RAPD markers for resistance to spot blotch of barley caused by *Cochliobolus sativus* L. Indian J Genet Plant Breed 67(4):309–314. https://doi.org/10.1590/S1415-47572008000400021

Ul Haq S, Kumar P, Singh RK et al (2016) Assessment of functional EST-SSR markers (sugarcane) in cross-species transferability, genetic diversity among Poaceae plants, and bulk segregation analysis. Genet Res Int 2016:7052323. https://doi.org/10.1155/2016/7052323

Vos P, Hogers R, Bleeker M et al (1995) AFLP: a new technique for DNA fingerprinting. Nucleic Acids Res 23(21):4407–4414. https://doi.org/10.1093/nar/23.21.4407

Wang D, Karle R, Iezzoni AF (2000) QTL analysis of flower and fruit traits in sour cherry. Theor Appl Genet 100(3–4):535–544. https://doi.org/10.1007/s001220050070

Watanabe S, Hideshima R, Xia Z et al (2009) Map based cloning of the gene associated with the soybean maturity locus E3. Genetics 182(4):1251–1262. https://doi.org/10.1534/genetics.108.098772

Weber JL, May PE (1989) Abundant class of human DNA polymorphisms which can be typed using the polymerase chain reaction. Am J Hum Genet 44(3):388–396

Wei C, Wang L, Yang Y et al (2010) Identification of an S n allele in *Oryza rufipogon Griff.* and its effect on embryo sac fertility. Chin Sci Bull 55(13):1255–1262. https://doi.org/10.1007/s11434-010-0154-y

Wei KF, Chen J, Chen YF et al (2012) Molecular phylogenetic and expression analysis of the complete WRKY transcription factor family in maize. DNA Res 19(2):153–164. https://doi.org/10.1093/dnares/dsr048

Williams JGK, Kubelik AR, Livak KJ et al (1990) DNA polymorphisms amplified by arbitrary primers are useful as genetic markers. Nucleic Acids Res 18(22):6531–6535. https://doi.org/10.1093/dnares/dsr048

Wu JW, Hu CY, Shahid MQ et al (2013) Analysis on genetic diversification and heterosis in autotetraploid rice. Springer Plus 2(1):439. https://doi.org/10.1186/2193-1801-2-439

Wu H, He X, Gong H et al (2016) Genetic linkage map construction and QTL analysis of two interspecific reproductive isolation traits in sponge gourd. Front Plant Sci 7:980. https://doi.org/10.3389/fpls.2016.00980

Xu Y (2010) Molecular breeding tools: markers and maps. Mol Plant Breed:21–58. https://doi.org/10.1079/9781845933920.0000

Xu Y, Beachell H, McCouch SR (2004) A marker-based approach to broadening the genetic base of rice in the USA. Crop Sci 44(6):1947–1959. https://doi.org/10.2135/cropsci2004.1947

Xu Y, Huang L, Ji D et al (2015) Construction of a dense genetic linkage map and mapping quantitative trait loci for economic traits of a doubled haploid population of *Pyropiahaitanensi*s (Bangiales, Rhodophyta). BMC Plant Biol 15(1):228. https://doi.org/10.1186/s12870-015-0604-4

Yadawad A, Gadpale A, Hanchinal RR et al (2017) Pyramiding of leaf rust resistance genes in bread wheat variety DWR 162 through marker assisted backcrossing. Indian J Genet Plant Breed 77(2):251–257. https://doi.org/10.5958/0975-6906.2017.00033.5

Yashitola J, Thirumurgan T, Sundaram RM et al (2002) Assessment of purity of rice hybrid using microsatellite and STS markers. Crop Sci 42:1369–1373. https://doi.org/10.2135/cropsci2002.1369

Yoshimura S, Umehara Y, Kurata N et al (1996) Identification of a YAC clone carrying the Xa-1 allele, a bacterial blight resistance gene in rice. Theor Appl Genet 93:117–122. https://doi.org/10.1007/BF00225736

Yu CY, Hu SW, Zhao HX et al (2005) Genetic distances revealed by morphological characters, isozymes, proteins and RAPD markers and their relationships with hybrid performance in oilseed rape (*Brassica napus* L.). Theor Appl Genet 110(3):511–518. https://doi.org/10.1007/s00122-004-1858-7

Zakaria Mohammed B, Belik Alexei A, Liu C-H et al (2015) Prussian blue derived nanoporous iron oxides as anticancer drug carriers for magnetic-guided chemotherapy. Chem Asian J 10:1457–1462

19 Role of ICT and Artificial Intelligence in Disease Diagnosis, Forecast, and Management

S. N. Islam, Sudeep Marwaha, Chandan Kumar Deb, and Md Ashraful Haque

Abstract

Information Communication Technology (ICT) and Artificial Intelligence (AI) play a pivotal role in disease diagnosis, forecast, and management in agriculture. Referred to as Agriculture Information Technology (AIT), this integration of IT and agriculture has revolutionized traditional farming, enhancing productivity and sustainability. The article explores the development of national databases for crop-specific diseases and insects, expert systems like "EXOWHEM," "EXPSS," "AgriDaksh," "PulsExpert," "AMRAPALIKA," "Rice Expert System," "Tomato Expert System," and "Rapeseed-Mustard Expert System" for integrated pest management, and various AI-based tools for diagnosing and managing agricultural diseases. Additionally, it discusses the importance of mobile technology in providing farmers with real-time information. Overall, this article highlights the potential of ICT and AI in reshaping agriculture, offering solutions to disease and insect management, and empowering farmers with knowledge for improved crop yields.

Keywords

Expert system · Artificial intelligence · National database · Exowhem · Wheat expert system

S. N. Islam (✉) · S. Marwaha · C. K. Deb · M. A. Haque
ICAR-Indian Agriculture Statistical Research Institute, New Delhi, Delhi, India
e-mail: sn.islam@icar.gov.in

V. K. Singh et al. (eds.), *Diseases of Field Crops: Diagnostics and Management*,
https://doi.org/10.1007/978-981-97-6160-9_19

19.1 Introduction

The emergence of the computer is one of the most significant intellectual developments of the twentieth century. It is broadly defined as the theory and practice of representation, data processing, information sharing and processing, communication, expert and information systems, artificial intelligence, etc. (Collins et al. 2021; Sarker 2021). The recent explosive growth in computing power and connectivity is reshaping the relationship between people and organizations. The information and learning revolution in information technology has transformed the global communication system. It has revolutionized almost every aspect of life, right from engineering to agriculture. However, its contribution to agricultural is still very low and needs attention (Islam et al. 2019).

Incidence of disease/pest is a common phenomenon in any agricultural production system. Disease-causing pathogens could attack each and every crop field if they get favorable growing conditions (Marwaha et al. 2012). Disease attack is one of the important factors that affects the yield of agricultural crops. Diseases are the result of the combined action of some distinct factors that include pathogen attacks on a susceptible plant under suitable environmental conditions that favor the infection process (Strange 2003). Each disease needs different environmental conditions; a few require a water film on leaf surfaces, while others are inhibited by the presence of moisture or water. In general, diseases need specific environmental conditions before they become a threat and make it an economic problem. Factors influencing disease development and its intensity in plants include hybrid/variety genetics, plant growth stages, weather conditions (e.g., temperature, rain, wind, hail, etc.), single versus mixed infections, and genetics of the pathogen populations (Chakraborty and Newton 2011). Variation inherent in these factors makes disease diagnosis and forecasting very difficult at the early stages of incidence on individual plants as well as at the early stages of an epidemic (Jeger et al. 2021). However, at some stage of incidence, many disease symptoms do become diagnostic with a reasonable confidence level. They can be considered as diagnosis rules based on these symptoms. Mechanisms for identifying the disease affecting the crop have been developed through standard disease management approaches (Fang and Ramasamy 2015; Mahlein 2016). Extensive research has been done to identify the disease, the pathogen that causes the disease, and insect pests that attack agricultural crops. The management process of many of them has been standardized and may be stored in databases and rules.

ICT encompasses a wide range of technologies, including mobile devices, sensors, remote monitoring, and data analytics. When integrated with AI, which encompasses machine learning and deep learning techniques, this combination offers the agricultural sector the potential to revolutionize the way plant diseases are diagnosed, forecasted, and managed (Li 2021; Eli-Chukwu 2019). By harnessing the capabilities of AI, such as pattern recognition, data analysis, and predictive modeling, ICT can empower farmers and researchers to make strategic decisions in real-time, ultimately leading to more efficient and sustainable agriculture practices (Kujawa and Niedbała 2021). In this article, we will explore the multifaceted role of

ICT and AI in the realm of plant disease management. We will delve into the various applications, ranging from automated disease detection and early warning systems to the precise management of resources, which can significantly enhance agricultural outcomes. Furthermore, we will discuss the potential benefits and challenges associated with these technologies, as well as the ethical considerations surrounding their implementation.

19.2 IT Applications in Crop Protection

Information Communication Technology, which is known as the constituent of computers, information processing, computer networks, and information transmission on computer networks, a new terminology can be conceived with a name Agriculture Information Technology (AIT). This new technology integrates Information Technology with agriculture (Poulton et al. 2010). Let us explore the opportunity attached with this technology, its application, and its future that gives a new direction to our agriculture and encourages entrepreneurs to invest substantially in this area.

Information communication technology has transformed the world scenario significantly and has been more noticeable now. This technology is a step forward in transforming traditional agriculture to a modern smart agriculture, eventually contributing to the significant improvement in agricultural production, management, and extension Islam et al. 2019. This way it helps farmers in acquiring the right information at the right time and place, which is essential for improving the efficiency and viability of small and marginal holdings (Dwivedi et al. 2021).

In this new paradigm for increasingly uncertain and complex situation of agricultural disease management, dynamically improving performance of related outcomes are the key points of how intelligent minds apply intelligent technologies to leverage strategic challenges transforming them to opportunities by effective disease and insect management. There has been significant use of mobile application as an ICT tool that has changed the communication process in agricultural. Its introduction in agricultural brings weather and market information in planning operations, monitoring plant health and other relative aspects. The introduction of mobile phone-based applications has provided new services and applications in the agriculture sector. Mobile-based applications bring dynamic information related to weather, market, demand, and scope. Studies reveal that mobile phones have a positive impact in bringing information relevant to issues covering agricultural production and in harnessing its full potential (Bhavnani et al. 2008).

19.2.1 Development of National Database on Disease and Insects

Incidence of disease/pest is a common phenomenon, and its management is a real challenge of any agricultural production system. There are various diseases and insects that attack various crops when they get favorable environments. Crop-specific research institutions have identified such diseases and insects that attack the crops have been well documented. Mechanisms for identifying the disease affecting the crop have been developed through standard disease management approaches. A national database of all these insects/diseases can be developed and made accessible to stakeholders and may be considered as an information bank. The database should carry the complete details of the insect pest with its full taxonomy using some relational database management system like Oracle, SQL Server, or MySQL, etc. The database should be dynamic in nature so that they can be easily updated, eventually new insect/pest or a pathogen should easily find its place in the system. A user interface should be provided so that authorized users can have access to the system. There should be an expert interface through which information of new disease, insect, pathogen, and other information can be added to the system. We need plant disease models for developing such systems. A plant disease diagnostic model may be considered as a mathematical description of the interaction between pathogens, pant host, and environmental condition that become the reason for diseases. The outcome of such models give a numerical index of risk associated with the diseases and its severity and predicting inoculum development. A model can be represented in the form of tables, rules, graphs, or a simple equation.

Many information systems have been developed that keep databases at their backend. A crop-pest weather database for rice and cotton has been developed by ICAR-CRIDA. It documents weekly pest records (34,472) for 11 insect pests and diseases in rice and 13 insect pests and diseases in cotton, along with corresponding weather across 12 important locations across India. However, we need such systems for all crops and diseases associated with these crops. One integrated system may be developed that covers crop-wise disease information, its environmental situation, pathogens, and management. The developed system should find a place at ICAR data center available at ICAR-IASRI.

19.2.2 Development of Expert Systems and Decision Support Systems

Expert systems are computer programs that acquire knowledge from one or more domain experts and store it in their knowledge base so that it may be used in the future. The knowledge is acquired from the domain expert through a series of conversations between the knowledge engineer and the domain expert. The system developed applies the acquired knowledge to solve the problem submitted by the user. A farmer/user presents a specific problem to the system that uses the stored knowledge acquired from the domain expert to solve it. Expert systems have the following components:

- Knowledge base
- Knowledge base editor
- User's interface
- Expert interface

- Working memory
- Inference engine

19.2.2.1 Knowledge Base

The knowledge base is one of the important components of an expert system used to store the acquired information and data from experts about the domain. It is represented in the form of production rules and relational tables. There are two types of knowledge applied while developing such systems. One is the facts about the domain that have been established over the years and available in the public domain are kept in databases. The other is the heuristics or rules of thumbs that experts apply while taking decisions. Such thumb rules acquired from experts are stored in the form of rules. The accuracy of problem-solving is based on the quality of information acquired from the experts based on heuristics and facts.

19.2.2.2 Knowledge Base Editor

Knowledge base editors are the user-friendly interfaces that give power to a specialist for adding new information and updating the existing one. In an ideal system, all the information may be updated based on the expert's judgment. It is a facility and interface developed for the knowledge engineer and experts to update the information in the knowledge base. Knowledge engineers are the players that collect information from the domain experts, transform them in the form of rules, and enter them into the knowledge bases.

19.2.2.3 The User and Expert Interface

The interface for the user is designed and developed to interact with the developed system to deliver a solution to the problem that is submitted by this interface. Expert Interface helps in entering new knowledge into the system. Knowledge engineers update the information through this interface. Provisions have been made for the experts as well to add new information and validate the information added by the knowledge engineers. Both interfaces are developed on the same basis, which is to make the expert system user-friendly as much as possible. These interfaces are meant to acquire and deliver knowledge regarding various aspects of the domain knowledge.

19.2.2.4 Working Memory

Working memory is one of the prime components of expert systems meant to keep the information temporarily for some specific tasks and functions. It keeps the information, which has been derived from the problem at hand. During processing, some data in terms of variables is transferred into the working memory. We also store some processed data in the working memory as and when a query-based process is initiated. This information is retained during the whole process. A distinct working memory variable is created for each query raised by a user.

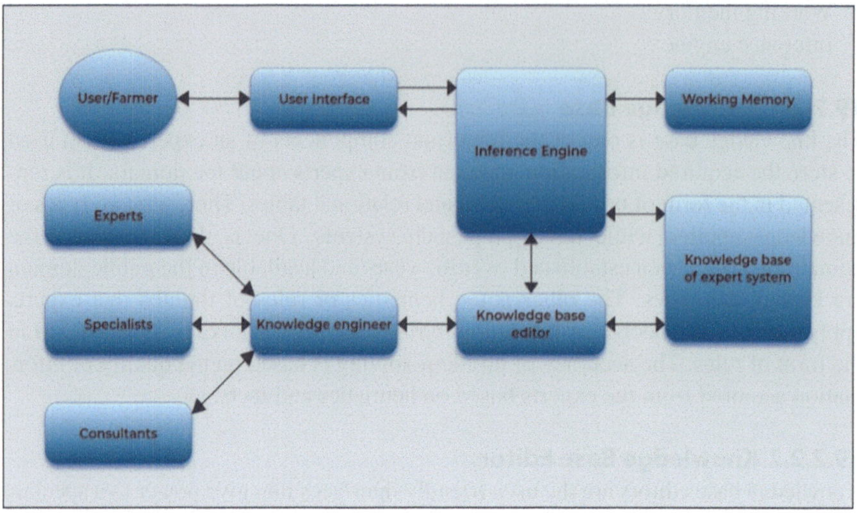

Fig. 19.1 Expert system architecture

19.2.2.5 Inference Engine

Expert systems get their strength and potential from the inference engine through its capacity to draw conclusions and make inferences from the premises obtained from a user. The inference engine makes an expert system intelligent that gives appropriate solutions to the problem submitted. As and when the knowledge is represented as rules, the domain expert should know the way of applying these rules in the proper order. The inference engine interrelates the facts and rules in the knowledge base. On the basis of the problem submitted by the user, the inference engine interacts with the knowledge base, applies algorithms, and then searches for the most suitable rules to be applied. The architecture of the system is depicted in Fig. 19.1.

19.2.2.6 Knowledge Engineering and Inference Mechanism

The knowledge engineering allows the interpretation of the processed information, which is kept in the databases for diagnosing, designing, and making the right analysis of a problem. This way, the system gets rules and factual data from the domain experts through conversation. This information is finally organized in a suitable format or structure designed specifically for storing them as rules and tables in the knowledge base. The knowledge and information stored in the knowledge base are extracted by the inference engine to find a solution. The representation in the knowledge base is done by tools developed for storage. The inference engine is the intellect of the system that takes up the problem, analyzes it, applies algorithms, does comparison and matching, and delivers the solution to the user.

19.2.3 Expert Systems for Integrated Pest Management

IPM is a scientific process that involves common sense and a natural way of treating and controlling pests. It is an integration of multiple methods that comprise evaluation, decision, and control for pest management. Farmers aware of the intensity adopt a four-tiered approach. They have the three basic approaches mentioned below.

Step 1: First of all, a threshold is set, which is the point that indicates about the point of pest population or environmental conditions indicating the need of control and management of pests. It is the point when the pest becomes an economic threat and the situation is critical.

Step 2: Many pest species are beneficial and many of them are innocuous. IPM initiatives are taken up to monitor and identify pests so that the right management practice may be applied in conjunction with action thresholds. This minimizes and prevents the excessive use of pesticides.

Step 3: IPM initiatives work to manage the crop, farm field, lawn, and indoor spaces to prevent pest attacks before they become a threat. IPM uses control methods like crop rotation, pest-resistant varieties, and planting pest-free root stock. These methods are cost-efficient, effective, and have no health hazards.

19.2.3.1 DIARES-IPM

DIARES IPM is an expert system developed for integrated pest management based on diagnostic advisory rules. This system has been developed for Solanaceous crop, which is an automated identification system that helps nonexperts to identify diseases and insects, including beneficial insects. It also identifies nutritional deficiencies and suggest the appropriate management. The purpose of this system is as an education and extension tool to help with the diagnosis and management of pests associated with vegetable crops. It covers economically important diseases, insects (noxious and beneficial insect), and nutritional deficiencies affecting the crop. This knowledge has been kept in the knowledge base of the system (Mahaman et al. 2003).

19.2.3.2 Soypest

Soypest (Soybean Expert System) is a fuzzy web-based system developed for integrated pest management. The system has been developed in JAVA language for providing online decision support to Soybean growers. The system works in four phases: symptom-based identification, morphology-based identification, pest image-based identification, and IPM-based management (Saini et al. 2002).

19.2.4 Expert Systems for Disease Identification and Management

There are many Expert and Information Systems developed for crop management, disease and insect identification and other agriculture related aspects. A few of them has been discussed below.

19.2.4.1 Expert System on Wheat Crop Management (EXOWHEM)

An expert system on wheat crop management has been developed for the wheat-growing farmers of India by ICAR-IASRI, New Delhi Islam et al. (2012). The system guides farmers as an expert, recommends a variety, diagnoses a disease, identifies insects, and provides appropriate solution to them. The knowledge is drawn through the information given by the user applying artificial intelligence principles and the rules supplied by the domain experts. The system has been developed using web-based technology, is platform-independent, and may be accessed by any web browser. The inference engine, which is the process and decision maker of the expert system, has been developed in JavaScript and VBScript using Microsoft visual studio which is an integrated development environment. The inference procedure has been taken up in three phases. In the first stage, it selects the stage of the crop, affected plant part, and symptom. In the next stage, it identifies color and matches the disease through its image and executes the query to finally identify the disease. There is an intermediate memory known as working memory where the outcome is often kept for comparing with the rules and facts stored in the knowledge base. The system looks for a consistent match based on major and minor symptoms of the diseased plant part. It finally compares with the image of the probable disease affecting the crop. The result is given with a certainty factor.

This system can be taken up as a model and have to be replicated for other cereal crops. The diagnostic approach for a disease has been given in Fig. 19.2. A user diagnoses a disease by selecting the stage of the plant, the affected part, and major symptoms appear on a plant. This module helps the farmers in identifying the

Fig. 19.2 Diagnosing wheat disease (Source: From the Site of Wheat Expert System)

disease with which the crop is infected and suggests treatment and control measures to minimize the yield loss.

19.2.4.2 Expert System on Seed Spices (EXPSS)

EXPSS is an expert system developed for seed spices by ICAR-IASRI, New Delhi, in collaboration with ICAR-NRCSS, Ajmer Islam et al. (2011). It has organized a large amount of knowledge on seed spices cultivation in its knowledge base. The information stored is related to varieties, diseases, insects, and weeds. This system provides information on four major and six minor seed spices. A farmer/user may have information on variety selection, disease diagnosis, insect and weed identification, and cultural practices.

The system has an exclusive plant protection module. Through queries, the system narrows the choice and shows images of probable diseases. The Plant Protection module of the system is meant for diagnosing diseases and identifying insects and suggests control measures for Cumin, Coriander, Ajwain, Fennel, Fenugreek, Nigella, Anise, and Celery. Seed spices crops are affected by many diseases mainly caused by fungal pathogens. Wilt, Root Rot, Blight, and Powdery Mildew are the important diseases which cause considerable yield losses in seed spices. Management strategies for the diseases have been developed and are being practiced for commercial cultivation to minimize yield losses. An integrated disease management strategy should be employed for the effective management of seed spices diseases. The system takes a diagnostic approach for identifying a disease by asking questions, showing images. The queries are based on affected plant part, stage at which symptoms appear, minor symptoms associated with major symptom, and color of the affected part. Based on the input provided by the user, shortlisted and the most probable disease with which the crop can be shown, and finally the disease can be identified by the user. Expert Systems and Decision Support Systems can be developed that can help in diagnosing a disease, identifying an insect, or a weed in a user-friendly manner. Figure 19.3 shows diagnosed Cumin Blight Disease.

19.2.4.3 AgriDaksh

AgriDaksh is a holistic system that helps the agricultural domain expert to create expert systems in their respective domain Marwaha (2017). AgriDaksh provides user-friendly interfaces to domain experts with minimal knowledge in developing expert systems (Yadav et al. 2012). It is one of the pioneer systems in agriculture that provides the facility to create expert systems. AgriDaksh provides modules like "Knowledge Model Creation," "Problem Identification," "Knowledge Retrieval," and "Ask Questions to experts and Administration." A couple of expert systems have already used AgriDaksh, i.e., Maize AgriDaksh, Mushroom Agridaksh, Tobacco Agridaksh (Rao et al. 2016), etc. AgriDaksh provides ontology-based disease diagnosis, variety selection, and insect identification. Integrated pest management is also a part of the AgriDaksh (Arora et al. 2015). Apart from the domain expert, AgriDaksh provides facilities for the farmers to send queries to experts of different domains. AgriDaksh collected and resolved thousands of questions globally. The Home Page of the System is shown in Fig. 19.4.

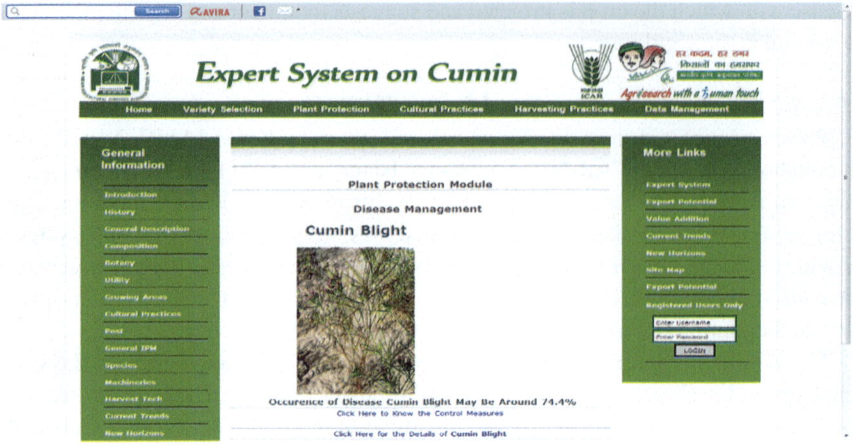

Fig. 19.3 Diagnosing cumin blight (Source: Expert System on Seed Spices)

Fig. 19.4 Home page maize AgriDaksh (Source: Home Page of Agridaksh)

19.2.4.4 Pulse Expert System

Pulse Exp. System has been developed by Devraj and Jain (2011) for the diagnosis and management of pulse diseases. Pulse Exp. is a diagnostic system developed for farmers and extension workers, which helps them in identifying diseases of important pulse crops in India, viz., Pigeon pea, Chickpea, Mung bean, and Urdbean, and suggests the appropriate control measures. The knowledge acquisition system of PulsExpert provides interfaces to the knowledge engineers and experts of the domain for entering and structuring the knowledge. The knowledge base has been developed and designed after examining the structural type of the knowledge from multiple sources like literature, books, databases, farmers, extension workers, etc.

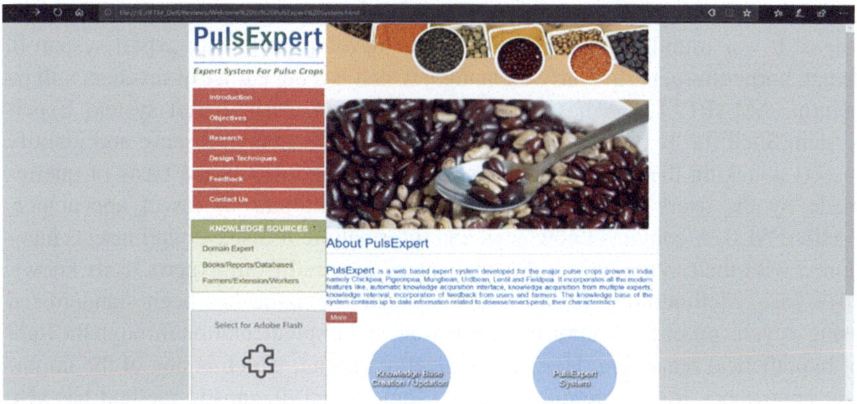

Fig. 19.5 Home page pulse expert system

For a specific crop, knowledge may be entered by one or more experts through the knowledge acquisition system that integrates the knowledge to build up a consistent knowledge base. Figure 19.5 is the Home Page of Pulse Expert System.

PulsExpert contains current knowledge about 19 important diseases affecting pulses right from the seedling stage till they attain maturity. The system provides a user-friendly interface to farmers/users. The question may be asked in textual as well as pictorial forms. The order of questions is decided dynamically depending upon the answers of previous questions given by the farmer. Based on user's response, PulsExpert diagnoses the crop diseases with a confidence factor. It also suggests an appropriate remedy comprising adoption of right cultural practices, chemical control, selecting the right variety, and application of nutrients, etc. The system has been designed and developed for disease identification and management of pulse crops. It has implemented an automated knowledge-acquiring system for getting information from experts. It allows multiple experts to enter knowledge, which integrates dynamically to enrich the system's knowledge base. It has a user-friendly interface for farmers/extension workers. A certainty factor based on symptoms is also given with the diagnosed disease.

19.2.4.5 Mango Expert System

Prasad et al. (2006) developed an expert system known as AMRAPALIKA on Mango Crop for identifying 14 important pests of the crop that includes eight diseases and six insects. Amarpalika has been developed for diseases like Anthracnose, Bacterial Spot, Die Back, Powdery Mildew, Black Spot, Sooty Mould, Red Rust, and Malformation. It helps in identification of insects like Mites, Mealy Bug, Red Ants, Shoot Borer, White Ants, and Fruit Fly. Mango covers 2.3 million ha (70% of fruit-growing area) of growing with a total production of around 11 million tonnes amounting to 57% of total world production (Negi 2000). India is one of the significant contributors to the total production of mango; however, per capita production is low as compared to other mango-growing countries. We lose 20–22% of the total production of this fruit during harvesting and by wastage at various postharvest stages. It was found to be important that the development of an expert system in Indian horticulture, particularly on mango, which is a big earner of revenue, will be helpful. AMARPALIKA has been developed using a rule-based system Expert System Shell for Text Animation (ESTA) to diagnose mango disease and identify insects attacking the crop. The system does the diagnosis on the basis of queries made by the users in the context of disease symptoms or insect appearance. AMRAPALIKA contains knowledge and information of diseases and insects mentioned with their symptoms and management acquired from experts in its knowledge base. Methodology for solving the diagnostic issue has been standardized using a cyclic process of capturing, deduction, and implementation through the light of hypothetical reasoning. However, logical reasoning, which is one of the important approaches for problem-solving and logic-based diagnostic method has also been found to be useful. This expert system applies models of pictorial symptoms for the classification of rules. It has been expressed in ESTA as its syntax rules. The rule-based approach has been applied in several other diagnostic systems successfully. The rule-based technique has also been applied through backward chaining for knowledge representation. The backward chaining method is applied due to the fact of this being goal-oriented that is similar to the reasoning process of the diagnosis. A similar approach has been made to integrate nutrient deficiency in the model with the knowledge base of AMARPALIKA.

19.2.4.6 Rice Expert System

An expert system on rice crop for diagnosing and management of rice disease has been developed by Sharma et al. (2012). This rule-based system has been developed using the well-known shell ESTA. The system has been designed and developed for the diagnosis of a few important diseases that infect the rice crop. The rice expert system comprises of a user interface, knowledge base, and inference engine. This system provides various modules that include decision support for identification of diseases and interfaces for knowledge enhancement and management of the identified insects and diseases. The response is given by the system based on the queries related to disease symptoms. Sarma et al. has applied ESTA for the development of the expert system which is a language for logical programming and integrating the structured knowledge bases. ESTA has tightly binded the knowledge base that is having symptoms of the crop and its facts with the inference engine. A database of images, containing details including symptoms, has also been developed for the system to make inferences and extend decision support to rice-growing farmers. It responds to the user through the interfaces based on inference mechanism transformed into algorithms. This system has been validated through multilevel testing with domain experts and users with their own datasets. The potential of any expert system is entirely dependent on the wealth of knowledge available in its knowledge base that has been obtained from the domain expert. The process taken up in developing the knowledge base is a multistep task that aims in developing a mechanism for identification of the disease. Diagnostic approach of common diseases of rice has been acquired from the domain expert of rice diseases in the development process.

19.2.4.7 Tomato Expert System

Babu et al. (2010) developed a web-based Tomato Expert System for the tomato-growing farmers in India. It is an advisory system that is aimed to help tomato-growing farmers with tomato cultivation. It has the knowledge of its various species, diseases, and many other details on tomato cultivation. The system is a collaborative venture developed with the help of eminent agriculture experts in the area of tomato crops. It comprised a team of designers and domain experts, computer engineers and programmers. This expert system has two main components: one is the tomato information system and the other is an expert system which is advisory in nature. In the advisory system, a user has the opportunity to interact with the expert system. The user has to answer the questions raised through the interfaces of the web-based expert system. Based on the user's query, the expert system gets the specific disease or pest and suggests its control measure. The information system part provides the user with all the static information about species, diseases including viruses, and pests of tomato fruit and the plant along with their disease symptoms. This system also provides the management and control practices for the diagnosed disease. This system has been developed using MySQL at the backend and JavaServer Pages (JSP) at the frontend. The symptoms have been validated using ID3 techniques. This web application has the feature that provides the latest information on tomato crops to the users through the user's interface regarding diseases, viruses, and their control measures. The expert system part contains important modules, which are Information Systems on tomato crop and Tomato Advisory System, and other services related to web applications. It has an additional feature that links to other agricultural information and expert systems. The system has web directory services, articles, and a discussion forum that have been provided on the site to help tomato-growing farmers. It helps the experts with updated findings in the area of tomato crop production.

19.2.4.8 Rapeseed-Mustard Expert System

A web-based expert system for disease identification in rapeseed mustard in India has been developed by Kumar et al. (2008). The system helps in diagnosing the diseases of rapeseed mustard, like White Rust, White Rot, Downy Mildew, Powdery Mildew, and Alternaria Blight, which may be effectively managed through this system. It has been estimated that yield losses due to Alternaria Blight and White Rust cause a significant yield loss in mustard. It has been observed that rot of mustard has also become damaging in the last few years. The disease causes damage up to 40%, resulting in yield losses where the incidence is high that leads to discourage the farmers of tomato crop (Chakraborty and Newton 2011). Panels of experts on rapeseed mustard have been consulted many times. While developing, published information was reviewed to collect the knowledge about the diseases affecting the crop. The appearance of the disease has been divided into six significant parts like root, leaves, flower, pods, or the plant as a whole. The developers have collected

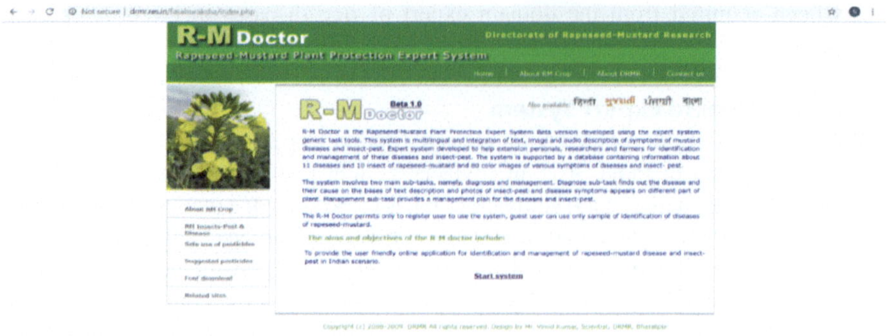

Fig. 19.6 Home page of rapeseed expert system (Source: Home Page of Rapeseed Mustard Expert System)

high-quality images and video clips of diseases with their symptoms. The system has been designed with basic components like user interface, knowledge base, and inference engine. System development takes place in multiple phases. Standard design methodology and implementation of the knowledge base dealt with information like data on characteristics of diseases with photographs, and symptoms of various diseases affecting various parts of the plant. Information on the management of these diseases has been collected from the pathologists and published literature. The coding has been done with the object-oriented programming language. The front-end language used for the development is Visual Basic 6.0, with MS Access-2000 as the backend for developing the software. Figure 19.6 is the image of the Home Page for the Rapeseed Expert System.

19.2.5 Development of Mobile Applications

Mobile communication has transformed the communication system in the past few years. The evolution of tablets and smartphones has extended the use of mobile communication, giving a platform for many applications that run on these devices. These applications are called as "apps" may easily be installed on smartphones and tablets. It administers an economical way to access the Internet and retrieve information related to many aspects of life, including agriculture. Mobile apps have become popular in the agriculture sector, opening new avenues for farmers who have access to smartphones. It has potential for further advancement in acquiring farm inputs, farm management, market information, and many other aspects (Chaurasia et al. 2020).

A large number of mobile apps have been developed for Indian farmers by various organizations doing research in agriculture that includes ICAR institutes and State Agricultural Universities. Applications with unique features like weather alerts and price trends of various agricultural commodities are available. There are many apps available to help farmers in identifying insects and diagnosing plant diseases.

There are 300 agricultural-related mobile apps developed by ICAR institutes that have been compiled together and brought to one platform by ICAR-IASRI. It has been named KISAAN 2.0 (Krishi Integrated Solution for Agri Apps Navigation) and works as an aggregator Android mobile app. It is envisaged to help e-agriculture and to drive smartphone-based agriculture in India. It has been developed with the aim of making farming convenient for Indian farmers. KISAAN 2.0 app facilitates farmers to access agriculture-based knowledge about crops, horticulture, livestock, fisheries, natural resource management, agricultural engineering, agricultural education, and agricultural extension through a single interface in multiple languages. This app will change the approach of Indian farmers in availing information on advanced agricultural operations, varieties, seeds, and disease and pest management. It also covers livestock as per their requirements.

This app contains a few disease-related apps also. ICAR-IASRI has developed two AI-based exclusive apps for diagnosing plant and animal diseases, compatible with all Android devices (required Android 5.0+) and may be installed on PC and Mac. National Image base for Livestock Disease (NIBLD) and National Image Base for Plant Protection (NIBPP) are two comprehensive mobile apps for collecting, validating, annotating, securely storing, and identifying plant and animal disease and pest images. These apps provide different levels and types of users for administrators, domain experts, and data science professionals. The presence of different levels and types of users makes the app modular, making a massive number of functionalities manageable and avoiding inter-account conflicts.

19.2.5.1 Modules in NIBPP and NIBLD App

(a) **Disease and Pest identification module (DPM):** Disease and Pest identification module (DPM) is the prime important module of NIBPP/NIBLD. The module has two different manifestations, one for the tester account and another for the farmers. The tester account uses this module to test the model developed. The farmer's account uses this module in the field condition to identify crop-wise diseases along with the management practices.

(b) **User management Module (UMM):** This module is created for managing the users of different types and levels. There are five types of users that may be created through image uploads with metadata, i.e., basic information and location in the form of latitude and longitude.

(c) **Expert Forum for Farmers (FEF):** This module has been developed for nearly real-time communications between farmers and the domain experts. This module provides a discussion forum for farmers to write crop-related problems with an image upload facility.

(d) **Dashboard:** The Dashboard module has been integrated with many of the account types. In different dashboards, a holistic view of the activities and data status can be seen. The dashboard has been created on a requirement basis.

19.2.5.2 Architecture of the Mobile Apps

The general architecture of the mobile app has been depicted in Fig. 19.7. The mobile application consists of three layers: Data Access layer, Presentation layer, and Business layer (Haque et al. 2018). These layers are discussed below:

Business Layer: This layer consists of three important components, which are business logic, validations, and calculations related to the acquired data. A website may communicate and interact with data through data access layer directly, but it normally goes through the Business Layer. The Business Layer is most vital as it validates the input conditions before calling a method from the data layer. This ensures the data input is correct before processing and ensures the correctness of the output as well.

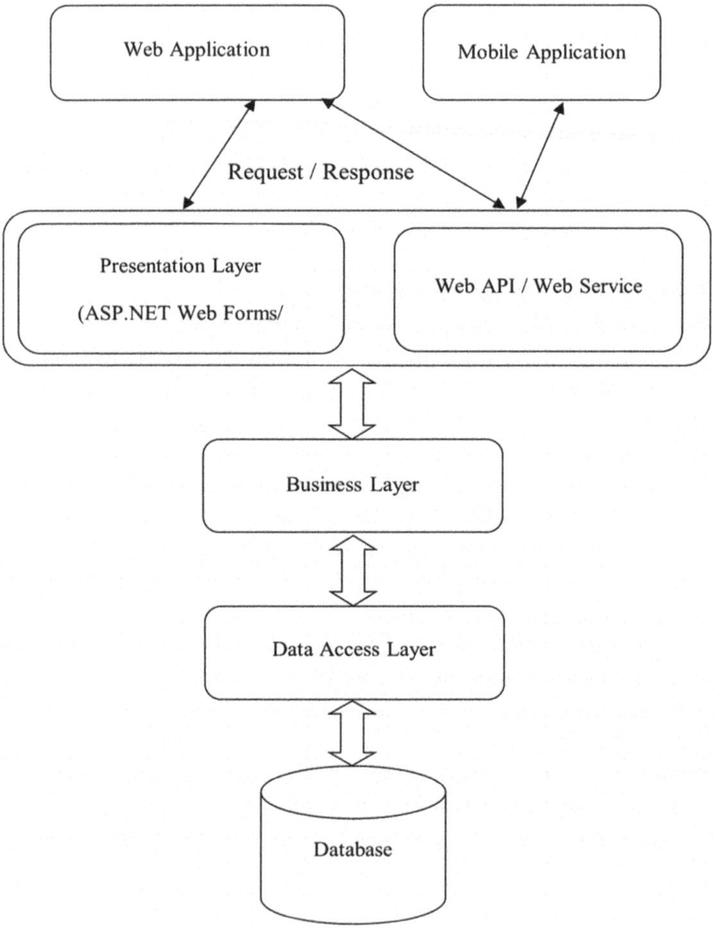

Fig. 19.7 Mobile app architecture

Data Access Layer: This layer comprises the methods that help the Business Layer to connect the data and perform desired actions, whether to return data or to manipulate it through actions like insert, update, delete, etc.

Presentation Layer: This layer provides interface to the user and contains pages with extensions like .aspx or Windows Forms. Through this layer, data is presented to the user or input is taken from the user. The ASP.NET web portal/website or Windows Forms application (the UI for the project) is called the Presentation Layer.

APIs: Web API of ASP.NET is a powerful framework for building HTTP-based services which may be accessed by different applications on different platforms such as windows, web, mobile, etc. It has the same functionality as ASP.NET MVC web application, except that it sends data as a response instead of html view. It is RESTful HTTP-based services. Web API is built on top of the ASP.NET application that supports ASP.NET request/response pipeline. During the process, Web API maps HTTP verbs to method names like GET, PUT, POST, and DELETE. It supports multiple formats of response data that has Built-in support for BSON, JSON, and XML formats (Fig. 19.8).

19.2.6 Application of Artificial Intelligence

Artificial intelligence (AI) is a branch of computer science that studies the intelligence exhibited by machines in a similar manner as an intelligent human can think. As indicated by John McCarthy, father of AI, "It is the science and engineering of making intelligent machines, especially intelligent computer programs" (McCarthy 1987). The primary goal of AI is to program the computer for specific characteristics such as learning, planning, knowledge, reasoning, problem-solving, and so forth. Machine learning is a subclass of AI that uses statistical techniques to enable

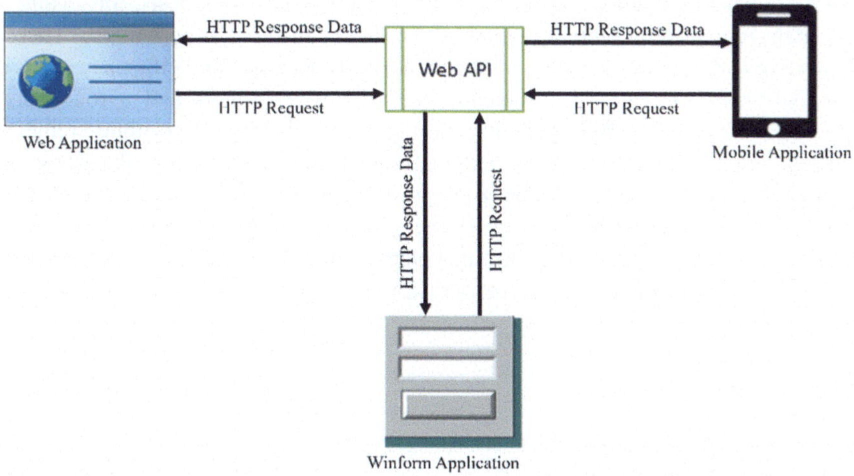

Fig. 19.8 Data exchange between web portal and presentation layer using web API

the machine to perform some particular tasks. It follows the "learning by example" standard. The learning methods may be categorized into three types: (a) supervised learning algorithms supplied with marked data and the desired output, whereas (b) unsupervised learning algorithm works on unlabeled data and recognizes the patterns from the feedback and information, and (c) reinforcement learning algorithm allows machines and programming specialists to conclude the ideal behavior automatically inside a particular context to amplify its performance. AI is a very extensive research field, where machines show cognitive potential and capabilities such as learning behaviors, proactive interaction with the environment, deduction, and inference. It works on computer vision, knowledge representation, speech recognition, problem-solving, perception, and many other areas. More colloquially, AI represents any activity where machines mimic intelligent behaviors usually exhibited by humans. Artificial intelligence takes vision from many areas of mathematical sciences that include computer science, mathematics, and statistics. Artificial Intelligence techniques are being applied in the identification of insects, disease, and weeds (Haque et al. 2021; Marwaha et al. 2023; Haque et al. 2023). Artificial Intelligence technology, along with image processing, object detection, machine learning algorithms, and computer vision, is widely used and analyzed. It has been proven to be effective in different agricultural activities and exclusively in the identification of pests/diseases and their management (Haque et al. 2022a, b; Gole et al. 2023).

19.3 Conclusion

Information Communication Technology is the technology of the future; if we could integrate it with agriculture, it will prove worthy for the farmers and the future of agriculture. The integration of Information and Communication Technology (ICT) and Artificial Intelligence (AI) in the field of plant disease diagnosis, forecasting, and management represents a pivotal advancement in modern agriculture. These innovative technologies have not only enhanced the accuracy and efficiency of disease detection but have also revolutionized the way farmers and agronomists approach plant health. By leveraging the power of AI-driven algorithms, remote sensing, and data analytics, we are better equipped to anticipate outbreaks, mitigate crop losses, and ensure sustainable food production. As we look to the future, the continued collaboration between researchers, farmers, and technology developers will be essential in harnessing the full potential of ICT and AI to safeguard our global food supply and promote a more resilient and productive agricultural sector.

References

Arora A, Saha LK, Marwaha S, Jain R, Jha AK (2015) Online system for integrated Pest management on tomato in AGRIdaksh. In: 2015 2nd international conference on computing for sustainable global development (INDIACom). IEEE, p 1125–1129

Babu MP, Murty NR, Narayana SVNL (2010) A web based tomato crop expert information system based on artificial intelligence and machine learning algorithms. Int J Comput Sci Inform Technol 1(1):6–15

Bhavnani A, Chiu RWW, Janakiram S, Silarszky P, Bhatia D (2008) The role of mobile phones in sustainable rural poverty reduction. World Bank Group, Washington, DC

Chakraborty S, Newton AC (2011) Climate change, plant diseases and food security: an overview. Plant Pathol 60(1):2–14

Chaurasia H, Islam SN, Priyadarshini S, Kumar R, Khan MA (2020) Development of android based crop advisory application for seed spices. Int J Curr Microbiol AppSci 9(3):904–913

Collins C, Dennehy D, Conboy K, Mikalef P (2021) Artificial intelligence in information systems research: a systematic literature review and research agenda. Int J Inf Manag 60:102383

Devraj, Jain R (2011) PulsExpert: an expert system for the diagnosis and control of diseases in pulse crops. Expert Syst Appl Int J 38(9):11463–11471

Dwivedi YK, Hughes L, Ismagilova E, Aarts G, Coombs C, Crick T et al (2021) Artificial intelligence (AI): multidisciplinary perspectives on emerging challenges, opportunities, and agenda for research, practice and policy. Int J Inf Manag 57:101994

Eli-Chukwu NC (2019) Applications of artificial intelligence in agriculture: a review. Eng Technol Appl Sci Res 9(4):4377–4383

Fang Y, Ramasamy RP (2015) Current and prospective methods for plant disease detection. Biosensors 5(3):537–561

Gole P, Bedi P, Marwaha S, Haque MA, Deb CK (2023) TrIncNet: a lightweight vision transformer network for identification of plant diseases. Front Plant Sci 14:1221557

Haque MA, Marwaha S, Arora A, Choubey AK, Paul RK (2018) Mobile application for information retrieval on Pest and disease in crops. In: Proceedings of research frontiers in precision agriculture. AFITA, Mumbai, pp 437–440

Haque MA, Marwaha S, Arora A, Paul RK, Hooda KS, Sharma A, Grover M (2021) Image-based identification of maydis leaf blight disease of maize (Zea mays) using deep learning. Indian J Agric Sci 91(9):1362–1367

Haque MA, Marwaha S, Deb CK, Nigam S, Arora A, Hooda KS et al (2022a) Deep learning-based approach for identification of diseases of maize crop. Sci Rep 12(1):6334

Haque MA, Marwaha S, Arora A, Deb CK, Misra T, Nigam S, Hooda KS (2022b) A lightweight convolutional neural network for recognition of severity stages of maydis leaf blight disease of maize. Front Plant Sci 13:1077568

Haque MA, Marwaha S, Deb CK, Nigam S, Arora A (2023) Recognition of diseases of maize crop using deep learning models. Neural Comput Appl 35(10):7407–7421

Islam SN, Khan MA, Kakani RK, Krishna Kant AO, Anwer MM (2011) Expert system for effective selection of seed spices varieties. Int J Seed Spices 1:88–94

Islam S, Kundu S, Shoran J, Sabir N, Sharma K, Farooqi S et al (2012) Selection of wheat (Triticum aestivum) variety through expert system. Indian J Agric Sci 82(1):39

Islam SN, Agarwal H, Sharma JP, Farooqi MS (2019) Role of IT in entrepreneurship development in agriculture. In: Entrepreneurship in livestock and agriculture. CBS Publishers and Distributors, New Delhi, pp 454–459

Jeger M, Beresford R, Bock C, Brown N, Fox A, Newton A et al (2021) Global challenges facing plant pathology: multidisciplinary approaches to meet the food security and environmental challenges in the mid-twenty-first century. CABI Agric Biosci 2(1):1–18

Kujawa S, Niedbała G (2021) Artificial neural networks in agriculture. Agriculture 11(6):497

Kumar V, Lehri S, Sharma AK, Meena PD, Kumar A (2008) Image based rapeseed-mustard disease expert system: an effective extension tool. Indian Res J Ext Educ 8(3):10–13

Li D (2021) Application of artificial intelligence and machine learning based on big data analysis in sustainable agriculture. Acta Agric Scand B Soil Plant Sci 71(9):956–969

Mahaman BD, Passam HC, Sideridis AB, Yialouris CP (2003) DIARES-IPM: a diagnostic advisory rule-based expert system for integrated pest management in Solanaceous crop systems. Agric Syst 76(3):1119–1135

Mahlein AK (2016) Present and future trends in plant disease detection. Plant Dis 100(2):1–11

Marwaha S (2017) AGRIDAKSH-A tool for developing online expert system. Proceedings of AIPA 2012, New Delhi

Marwaha S, Chand S, Saha A (2012) Disease diagnosis in crops using content-based image retrieval. In: 2012 12th international conference on intelligent systems design and applications (ISDA). IEEE, p 729–733

Marwaha S, Deb CK, Haque MA, Naha S, Maji AK (2023) Application of artificial intelligence and machine learning in agriculture. In: Translating physiological tools to augment crop breeding. Springer Nature Singapore, Singapore, pp 441–457

McCarthy J (1987) Generality in artificial intelligence. Commun ACM 30(12):1030–1035

Negi SS (2000) Himalayan forests and forestry. Indus Publishing

Poulton C, Dorward A, Kydd J (2010) The future of small farms: new directions for services, institutions, and intermediation. World Dev 38(10):1413–1428

Prasad R, Ranjan KR, Sinha AK (2006) AMRAPALIKA: an expert system for the diagnosis of pests, diseases, and disorders in Indian mango. Knowl-Based Syst 19(1):9–21

Rao NS, Marwaha S, Ravisankar H, Sivaraju K, Saha A, Deb CK (2016) Expert system for the identification of diseases in tobacco (Nicotiana tabacum). J Basic Appl Eng Res 3(6):561–563

Saini HS, Kamal R, Sharma AN (2002) Web based fuzzy expert system for integrated pest management in soybean. Int J Inf Technol 8(1):55–74

Sarker IH (2021) Machine learning: algorithms, real-world applications and research directions. SN Comput Sci 2(3):160

Sharma R, Singh H, Gill MS (2012) Development of an image based expert system for identification of rice diseases and their management. Plant Dis Res 27(2):158–161Strange RN (2003) Introduction to plant pathology. John Wiley & Sons, Chichester

Yadav VK, Marwaha S, Kumar S, Kumar P, Kaul J, Parihar CM, Supriya P (2012) Maize AGRIdaksh: a farmer friendly device. Indian Res J Ext Educ 12(3):13–17

Role of Plant Quarantine Under the WTO-SPS Agreement

20

Kavita Gupta, Sunil Chandra Dubey, and Jameel Akhtar

Abstract

India, with its agrarian economy at the forefront of national economic progress, faces mounting challenges in sustaining its agricultural ecosystems amidst escalating production demands and limited agricultural resources. This delicate balance is imperiled by a confluence of factors: pests, diseases, and harmful pesticides. Their collective impact not only diminishes crop yields but also poses significant threats to life and the environment. Natural calamities, such as invasions by migratory pests like locusts and armyworms, inflict extensive damage on diverse food crops, resulting in complete production losses across Africa and Asia. Additionally, invasive alien pests and endemic counterparts perpetuate chronic losses on an annual or seasonal basis, exacerbating the strain on agricultural systems.

Keywords

Biosecurity · Phytosanitary measures · Transboundary movement · Plant genetic resources

20.1 Introduction

20.1.1 Plant Quarantine Scenario in India

Plant quarantine is a government endeavor enforced through legislative measures to regulate the introduction of planting materials, including seeds, plant products, soil, living organisms, etc., in order to prevent the inadvertent introduction of pests,

K. Gupta (✉) · S. C. Dubey · J. Akhtar
ICAR-National Bureau of Plant Genetic Resources, Pusa Campus, New Delhi, Delhi, India
e-mail: kavita.gupta@icar.gov.in

V. K. Singh et al. (eds.), *Diseases of Field Crops: Diagnostics and Management*,
https://doi.org/10.1007/978-981-97-6160-9_20

pathogens, and weeds harmful to the agriculture of a country/state/region and, if introduced, prevent their establishment and further spread (Kahn 1989).

Plant quarantine is a way the government makes rules to control what gets brought into a place like a country, state, or region. This includes things like seeds, plant products, soil, and living things. The goal is to stop harmful pests, diseases, and weeds from getting in and spreading in agriculture. This is super important for countries like India, where a lot of the economy relies on farming. In India, people started paying attention to plant quarantine back in the early 1900s. The government made it compulsory to clean imported cotton to stop a bug called the Mexican cotton boll weevil from coming in. The idea behind these rules is that it's better to stop pests from getting in at the start, even if it's a bit inconvenient, rather than trying to deal with them later, which can cost a lot of money and effort. In 1914, the Indian government made a law called the DIP Act for plant quarantine. Since then, they've made different rules to control what plants and plant stuff can come in from other countries or even from one state to another within India. The most recent rule, called the Plant Quarantine (Regulation of Import into India) Order (2003), replaced an older one from 1989 called the Plants, Fruits, and Seeds (Regulation of Import into India) Order.

Quarantine measures are very important for countries like India because they depend a lot on agriculture. India started taking quarantine seriously back in the early 1900s when the government made it compulsory to fumigate cotton imports to stop bugs from other places, like the Mexican cotton boll weevil, from causing trouble. The idea behind this is that it's better to deal with some inconvenience upfront, like fumigation, rather than dealing with the big costs of pest control later on. Seeing how crucial it is, the Indian government passed the DIP Act in 1914, which has been updated over time. This law restricts bringing plants and plant stuff from other countries or even from one state to another within India (this is called domestic quarantine). The latest rule, the Plant Quarantine (Regulation of Import into India) Order (2003), replaced an older one from 1989 called the Plants, Fruits, and Seeds (Regulation of Import into India) Order. Three categories of materials are being imported under the PQ Order, 2003:

- Bulk consignments for consumption
- Bulk consignments of seeds/planting materials for sowing/planting
- Samples of germplasm in small quantities for research purposes

There are seven Regional Plant Quarantine Stations in places like Amritsar, Bengaluru, Chennai, Kandla, Kolkata, Mumbai, and New Delhi. In addition, there are 71 more Plant Quarantine Stations managed by the Directorate of Plant Protection, Quarantine, and Storage, Faridabad. These stations check and give clearance to consignments coming into the country in the first two categories. The Plant Quarantine Order lists the plants and planting materials that are:

(i) Prohibited for import into India (Schedule IV)
(ii) Permitted for import with additional declarations only through authorized institutions (Schedule V)

(iii) Permitted for import with additional declarations and special conditions (Schedule VI) under restricted conditions

To bring planting materials into the country, we need two important papers: an import permit and a phytosanitary certificate. The ICAR-NBPGR, New Delhi, being the nodal organization, has been authorized under the PFS Order to issue import permits and process germplasm material, including genetically modified planting materials imported by both the public and private sectors for research purposes and for quarantine clearance. Of all the three categories of planting materials being brought into the country, the riskiest is bulk imports for sowing/planting. It's hard to thoroughly check and treat them because there's so much, and the area needed for planting is huge.

Quarantine processing is often restricted to smaller samples derived from them, and based on the results of these samples, the whole consignment is rejected/detained or released after giving fumigation treatment or fungicidal/insecticidal treatments. Bulk consignments meant for consumption pose lesser hazards. However, small samples meant for research purposes are of immense quarantine importance. Since these samples usually comprise of germplasm material, wild relatives, or landraces of a crop, they are more likely to carry diverse biotypes/races/strains of the pest/pathogen. In the case of true seeds, risks are generally higher with deep-seated infections than with surface-borne contamination. However, the import of vegetative propagules presents a much higher order of risk than true seeds.

Usually, only small samples are checked, and based on those results, the whole batch might be rejected, held, or released after being treated with chemicals. Bulk consignments meant for consumption are less risky, but small samples for research are still of immense quarantine importance. They often carry different types of pests or diseases because they're made up of genetic material or wild versions of plants. We need to keep our agriculture safe from these pests without stopping people or stuff from coming in from other countries. One big job of plant protection groups is to find out about pests from other places that could really mess up farming here.

The potential impacts caused by diseases and pests introduced through the international movement of planting materials and agricultural products are well documented (Nath et al. 1988). The Irish famine of 1845, which led to mass migration from Europe, was due to the near-total failure of the potato crop caused by the late blight pathogen (*Phytophthora infestans*) brought from Central America. Powdery mildew (*Uncinula necator*), downy mildew (*Plasmopara viticola*), and the root-eating aphid (*Viteus vitifoliae*) of grapevines were introduced successively into France from North America in the mid-nineteenth century. These pests caused severe damage to grapevines and nearly destroyed the country's wine industry. Coffee rust (*Hemileia vastatrix*) appeared in Sri Lanka in 1875, causing production to drop from 400 million pounds in 1870 to just 5 million in 1889. The disease reached India in 1876 from Sri Lanka, and within 10 years, severely impacted the coffee industry in South India. In 1875, Germany banned potato imports from the United States to prevent the spread of the Colorado potato beetle (*Leptinotarsa decemlineata*). In 1898, most European countries banned American imports to keep out the San Jose scale (*Quadraspidiotus perniciosus*). Among the first countries to implement comprehensive plant quarantine regulations were France, Germany, the United States, the United Kingdom, and Australia.

In India, severe yield losses of the *Cicer arietinum* crop in Haryana, Madhya Pradesh, Punjab, and nearby areas occurred during 1981–82 due to the introduction of a virulent strain of *Ascochyta rabiei* from the Middle East. Similarly, Bunchy top of banana, introduced into India from Sri Lanka, caused annual losses of over Rs 4 crores to banana crops (Khetarpal and Nath 1998). The feared Golden nematode (*Heterodera rostochiensis*) was introduced from the United Kingdom with exotic seed material in the 1960s and has heavily infested potato tubers in the Nilgiris. A harmful weed, *Lantana camara*, was introduced in 1809 from Central America and has spread to wastelands and pastures, displacing native species. These examples clearly show that imported seeds or planting materials, especially in bulk, without proper quarantine measures, may introduce and establish quarantine pests in new areas, severely damaging crop production and the economy of a nation.

In India, plant quarantine operations are carried out by the Directorate of Plant Protection, Quarantine, and Storage, which works under the Ministry of Agriculture and Farmers Welfare. The administrative structure of plant quarantine is shown in the organizational chart (Fig. 20.1).

The creation of the Plant Quarantine (Regulation of Import into India) Order (2003) (referred to as "the PQ Order") highlights the main plant quarantine issues for the Government of India. These are to:

1. Prevent the entry and spread of harmful exotic pests by regulating the import of plants and plant products through appropriate policies and legal measures

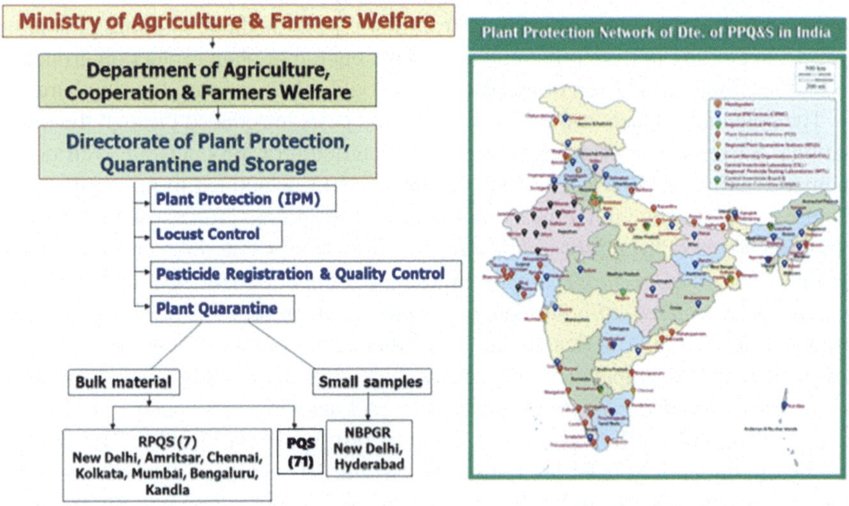

Fig. 20.1 Organizational setup of the Directorate of Plant Protection, Quarantine, and Storage in India

2. Support India's agricultural exports with reliable export certification
3. Enable safe international trade in agriculture by helping producers, exporters, and importers with comprehensive and trustworthy phytosanitary certification

Before the PQ Order was notified in November 2003, the existing plant quarantine laws and regulations dated back to 1914. Until recently, regulations in India were based on the Destructive Insects and Pests Act, 1914, which was created to prevent the entry and spread of harmful pests affecting crops, and the Plant, Fruits, and Seeds (Regulation of Import into India) Order, 1989. Other rules were established for regulating the import of live insects (1941), fungi (1943), and cotton (1972).

The New Seed Policy, 1988, aimed to give Indian farmers access to the best domestic and imported seeds and planting materials. The Plants, Fruits, and Seeds (Regulation of Import into India) Order, 1989, which controlled and regulated the import of plants, plant materials, and similar items into India, was based on postentry quarantine (PEQ) checks. It has now been replaced by the PQ Order of 2003.

20.1.2 Role of ICAR-NBPGR in Plant Quarantine

ICAR-NBPGR is the primary institution responsible for the domestic and international exchange of plant genetic resources (PGR) in India. Under the Plant Quarantine (Regulation of Import into India) Order (2003), it has been authorized to manage the quarantine processing of germplasm and transgenic planting materials imported for research by both the public and private sectors. The ICAR-NBPGR Regional Station at Hyderabad oversees the quarantine inspection and clearance of germplasm for the International Crops Research Institute for the Semi-Arid Tropics in Patancheru, Telangana, and other institutes in South India. ICAR-NBPGR has established advanced laboratories and a greenhouse complex for postentry quarantine growing. Additionally, a CL-4 level containment facility has been set up for processing transgenics.

At ICAR-NBPGR, an effective strategy has been employed to intercept numerous pests of significant economic and quarantine importance found on exotic material, many of which are not yet reported in India. These include insects such as *Acanthoscelides obtectus* in *Cajanus cajan*, *Anthonomus grandis* in *Gossypium* spp., and *Ephestia elutella* in *Macadamia* and *Vigna* spp.; fungi like *Peronospora manshurica* on soybean, *Fusarium nivale* on wheat, barley, and Aegilops, and *Uromyces betae* on sugar beet (Singh et al. 2018); bacteria such as *Xanthomonas campestris* pv. *vesicatoria* on tomato (Dev et al. 2012); and viruses including barley stripe mosaic virus in barley, cherry leaf roll virus on French bean, cowpea mottle virus in cowpea, pea seed-borne mosaic virus in broad bean, and tomato black ring virus on cowpea and French bean (Khetarpal et al. 2006). Additionally, nematodes such as *Rotylenchus minutus* in *Hypoxis hemerocallidea* (Khan et al. 2012), *Heterodera schachtii*, *Ditylenchus dipsaci*, *D. destructor*, and *Rhadinaphelenchus cocophilus* have been intercepted in soil clods and plant debris accompanying plant materials.

20.1.3 Procedure for Import of PGR

Research institutions in the public and private sectors wishing to import plants or planting materials must request an Import Permit (IP) from ICAR-NBPGR, which is non-transferable. A Phytosanitary Certificate is also legally required and is issued by the exporting country. It certifies that the consignment has been inspected according to the importing country's standards and is free from quarantine pests. Upon arrival, the plant material is carefully processed. If pests are found, all efforts are made to salvage the material. If the material cannot be salvaged, it is incinerated. If post-entry quarantine (PEQ) examination of the imported material is needed, it is conducted at PEQ greenhouse facilities at ICAR-NBPGR in New Delhi, its Regional Station in Hyderabad, ICRISAT, or at the indenter's PEQ growing facility if mentioned in the IP application form.

20.1.4 Methods for Detection and Interception of Pests in Quarantine

ICAR-NBPGR imports around 100,000 samples of germplasm and international trial material each year for research conducted by both the public and private sectors. The Division of Plant Quarantine at ICAR-NBPGR has established systematic procedures for pest interception (Fig. 20.1): visual and stereo-binocular examination to detect smut and bunt balls, seed galls, ergot sclerotia, and rust pustules/spores on seeds; washing tests for rusts and downy mildews; blotter methods for fungi and bacteria; and electron microscopy (EM), serological (ELISA), and molecular (PCR-based) tests to identify seed-borne pathogens, including viruses. Infected materials are either incinerated or salvaged using physical, chemical, and mechanical techniques, depending on the type of pest detected before release. Quarantine processing has shown that seed-borne pests can affect seed quality and germination, spread new pest strains or races to new areas, and lead to the development of epidemics. Therefore, detailed quarantine examinations with specialized tests are performed to ensure the detection and identification of fungal, bacterial, viral pathogens, insects, mites, nematodes, and weeds of quarantine importance in seeds and other planting materials.

During quarantine processing, seeds and plant materials are inspected for unwanted weed seeds, plant debris, soil clods, insect and mite pests, plant-parasitic nematodes, and pathogens such as fungi, bacteria, viruses, and phytoplasma (Bhalla et al. 2018). The examination procedures for different sample sizes and numbers of seeds per sample vary between bulk and small samples, but the techniques for pest detection remain consistent.

External feeders and incidental insect pests infesting seeds are easily detected visually, either with the naked eye or using a magnifying glass or stereoscopic binocular microscope. Fungal infections are indicated by discolored and malformed seeds and fungal structures on or within the seed surface. Yellow discoloration around the hilum suggests a bacterial infection.

Specialized/advanced tests for detecting different groups of pests are as follows:

20.1.4.1 Detection of Fungal and Bacterial Pathogens

- Detecting surface-borne pathogens involves examining the suspension after seed washing. Seeds are shaken in water, and the resulting suspension is analyzed under a compound microscope for spores of downy mildew, powdery mildew, and other fungi.
- The blotter method is used to detect many fungal pathogens that produce mycelial growth and fruiting structures during incubation, as well as bacterial pathogens. Seeds are placed on moist filter paper in plastic Petri dishes (110 mm) and incubated at 22 ± 1 °C with 12-h alternating light and dark cycles for 7 days to observe the growth of associated pathogen(s).

20.1.4.2 Detection of Seed-Transmitted Viruses

- Seeds known or suspected to carry seed-transmitted viruses are sown in insect-proof postentry quarantine glasshouses or environment-controlled screen houses. Seedlings displaying viral symptoms are uprooted and burned. Only produce from healthy plants is released to requestors.
- An infectivity test is conducted to detect the presence of viruses by inoculating leaf extracts from symptomatic seedlings onto indicator hosts. This method can reveal symptomless or latent infections in grow-out tests.
- The enzyme-linked immunosorbent assay (ELISA), a relatively simple, rapid, and sensitive technique, is used to simultaneously test a large number of samples.
- Leaf extracts from infected seedlings are examined under a transmission electron microscope to reveal the shape and size of virus particles.

20.1.4.3 Detection of Insects and Mites

- X-ray radiography detects seed infestations by phytophagous chalcidoids, bruchids, and other insects that show no external symptoms. Crops such as pulses, oilseeds, tree species, and some grasses must undergo X-ray radiography for germplasm import (Bhalla et al. 2002). Seeds with insects, identified on the X-ray screen using a real-time system, are hand-picked, and healthy seeds are released to the requester.
- The transparency method detects infestations in small seeds like sesame. Seeds are boiled in a lactophenol solution (a mixture of lactophenol, lactic acid, distilled water, and glycerin in a 2:2:2:1 ratio) for 1–2 h, depending on seed hardness. This process renders the seeds transparent, revealing any insect infestations.

20.1.4.4 Detection of Plant-Parasitic Nematodes

- Soaking seeds suspected of carrying seed-borne nematodes in water overnight softens the seed coat, allowing nematodes to be teased or crushed out into the water. Examining the accompanying soil can reveal viable nematodes, especially ectoparasites and cysts of certain species.

20.1.5 Methodology of Salvaging Infested/Infected/Contaminated Material

Various methods are employed for salvaging infected seeds and other planting materials, as outlined below:

Mechanical Cleaning: Soil clods, plant debris, weeds, discolored, deformed, and shriveled seeds are manually cleaned by handpicking.

Hot Water Treatment: This method involves using different temperature and time combinations to eliminate pathogens like fungi, bacteria, and nematodes. Seeds are treated in a hot water tank fitted with heaters, a stirrer, a thermostat, and a contact thermometer to control the water temperature. Brassica spp. seeds receive mandatory hot water treatment to eradicate seed-borne pathogens.

X-ray Radiography: Used to separate insect-infested seeds without external symptoms from healthy ones. Seeds are laid in a single layer and exposed to X-rays. Infested seeds are easily identified on the developed X-ray film and are hand-picked. Although impractical for bulk material, this method, combined with fumigation, is effective for salvaging infested seeds.

Fumigation: One of the most effective methods for eliminating insects, mites, and nematodes in quarantine. It can be done at atmospheric pressure or under vacuum conditions. Atmospheric fumigation uses aluminum phosphide at 2 g/m^3 or an ethylene dichloride-carbon tetrachloride mixture (3:1) at 320 mg/L at 30 °C for 48 h. Vacuum fumigation is performed in specially designed chambers, enhancing fumigant penetration through tightly packed or internally infested material. Common fumigants include ethylene oxide-carbon dioxide mixture, hydrogen cyanide gas, and methyl bromide.

Pesticidal Treatment: The most practical method in quarantine for controlling surface-feeding insects and nematodes. Seed dressing with a combination of systemic and nonsystemic fungicides or antibiotics effectively eliminates seed-borne fungi and bacteria.

Alcohol Wash: To eliminate externally seed-borne spores of safflower rust (*Puccinia carthamii*) and sunflower rust (*P. helianthi*), contaminated seeds are taken in a test tube with ethyl alcohol and a pinch of sand. The mixture is stirred with a mechanical stirrer for 30 s, separating spores from the seed surface into the alcohol.

Postentry Quarantine Growing: Chemically treated seed material is grown in isolation for one crop season to detect seed-borne pathogens. Healthy seeds from uninfected plants are then released to the indentors.

Tissue Culture Technique: With around 1300 seed-transmitted pests affecting various plants, physical, chemical, and mechanical methods are supplemented by tissue culture, the safest technique for eliminating pests from germplasm. Shoot tip culture raises pathogen-free plants from infected germplasm. This technique relies on culturing in vitro abscised meristems to obtain pathogen-free progeny from infected plants. It is particularly useful for eliminating viruses from valuable coconut material.

These methods, ranging from mechanical and chemical treatments to advanced tissue culture techniques, ensure the effective management of seed-borne pathogens and pests, safeguarding the quality and health of imported germplasm (Fig. 20.2).

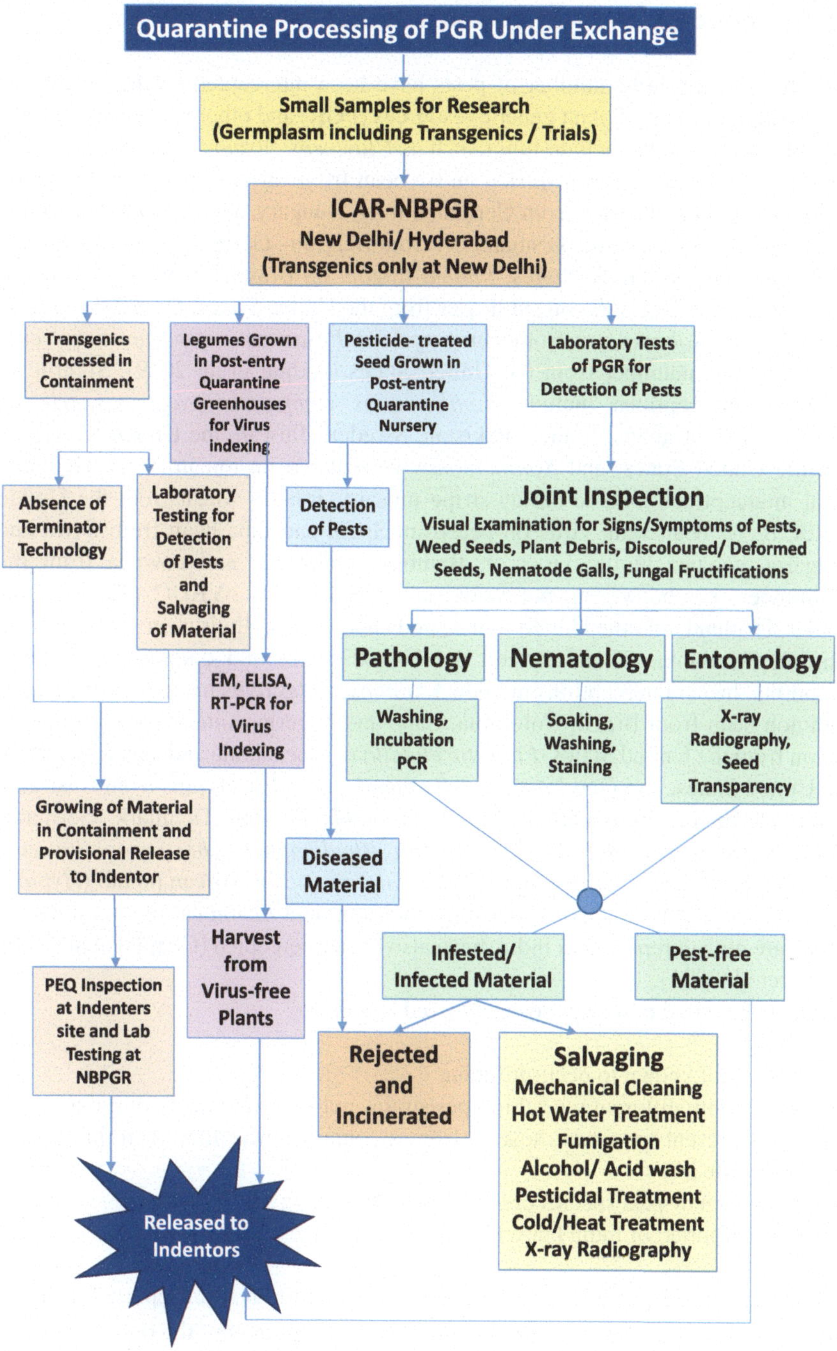

Fig. 20.2 Flowchart of quarantine processing of planting material

20.1.6 Interceptions in Imported Material

Over the years, a large number of pests have been intercepted during quarantine processing of imported plant genetic resources (PGR) and other research materials. Notable interceptions include fungi such as *Claviceps purpurea* in wheat and barley seeds, *Peronospora manshurica* on soybean from various countries, *Fusarium nivale* on wheat and barley from Germany, Italy, Hungary, Sweden, and the United Kingdom, *Uromyces betae* on sugar beet from Belgium, Germany, Italy, the United Kingdom, and the United States, *Phoma lingam* on Brassica from several countries, *Diaporthe longicolla* on sunflower from the United States, *Diaporthe eres* on black locust (*Robinia* sp.) from Hungary, and *Fusarium oxysporum* f. sp. *cucumerinum* on cucumber from the United States (Akhtar et al. 2019). Significant bacterial interceptions include *Xanthomonas campestris* pv. *campestris* on Brassicas from Canada, France, Pakistan, Sweden, Taiwan, the United Kingdom, and the United States, and *Xanthomonas vesicatoria* on tomato from Thailand. Viral interceptions include barley stripe mosaic virus on wheat from the United States, broad bean stain virus on pea from Spain and faba bean from Syria and Bulgaria, cowpea mottle virus on Bambara groundnut and cowpea from the Philippines, raspberry ringspot virus on soybean from AVRDC (Taiwan), Sri Lanka, Thailand, and the United States, and cherry leaf roll virus on soybean from Taiwan, Sri Lanka, Thailand, and the United States and common bean from Colombia. Insect interceptions include *Acanthoscelides obtectus* in pigeon pea and common bean from Brazil, Colombia, Italy, and Nigeria, *Anthonomus grandis* in cotton from the United States, *Ephestia elutella* in Macadamia and Vigna spp. from the United States, and *Quadrastichodella eucalyptii* in Eucalyptus from Australia. Nematode interceptions include *Heterodera schachtii* from Denmark, Germany, and Italy, *Ditylenchus dipsaci*, *D. destructor*, *Rhadinaphelenchus cocophilus*, etc., in soil clods and plant debris, and *Rotylenchus minutus* in African potato (*Hypoxis hemerocallidea*). Weeds like *Cichorium spinosum* and *Echinochloa crus-pavonis*, which are not yet reported in India, have also been intercepted (Chalam et al. 2023; Dubey et al. 2021).

The intercepted pests can be categorized as follows:

(i) Those not known to occur in India.
(ii) Those with different races/biotypes/strains not known to occur in India.
(iii) Pests present on a new host or from a country where they were previously unreported.
(iv) Entirely new pest species previously unreported in science.
(v) Pests reported in India but with a wide host range.

Interceptions, especially of pests and their variability not yet reported in India, underscore the importance of quarantine in preventing the introduction of destructive exotic pests. The third and fourth categories of pests are unexpected in samples based on risk analysis, which is literature-based. Since no records are available on these pests or their hosts, their presence is unexpected and significant from a quarantine perspective. The last category, pests with a wide host range, is critical and could become invasive if they find suitable biotic and abiotic environments (Khetarpal and Gupta 2008). Such interceptions highlight the success of quarantine measures; otherwise, these pests could have entered the country and severely impacted plant biodiversity and agriculture.

In the current international context, quarantine specialists play a crucial role not only in promoting and facilitating the export and import of agricultural goods for their respective nations but also in protecting the environment from invasive alien species. The importance of quarantine has increased significantly under the World Trade Organization (WTO) regime. Adopting appropriate techniques and strategies for pest detection and diagnosis is vital for ensuring pest-free germplasm exchange and transparency in international trade. This approach is considered the best strategy for managing the transboundary movement of pests.

20.1.7 International Developments

The World Trade Organization (WTO), established on January 1, 1995, serves as the legal and institutional backbone of the multilateral trading system. It sets the primary contractual obligations that guide countries in forming and enforcing domestic trade laws and regulations. This organization acts as a platform for the evolution of trade regulations among countries through collective discussion and negotiation. The WTO's scope encompasses a wide range of commercial activities and trade policies. The WTO Agreement comprises around 29 distinct legal texts, covering diverse areas such as services, government procurement, rules of origin, and intellectual property. The WTO Agreement on Agriculture aims to ensure greater equity in agricultural trade, while the Agreement on Trade-Related Aspects of Intellectual Property Rights (TRIPS) seeks to enhance competitive conditions involving ideas and inventions. The Agreement on Technical Barriers to Trade ensures that technical regulations and certification procedures do not become trade obstacles (http://www.wto.org).

The Agreement on the Application of Sanitary and Phytosanitary (SPS) Measures is directly linked to plant protection science and significantly impacts phytosanitation and trade. The SPS Agreement seeks to mitigate health-related barriers to market access by promoting the "establishment, recognition, and application of common sanitary and phytosanitary measures by different Members." The main advantage of utilizing common international standards is that they offer necessary health protection based on scientific evidence while simultaneously enhancing trade flows. Implementing SPS measures at both national and international levels demands extensive scientific research and data collection to benefit stakeholders and promote safe trade (Khetarpal and Gupta 2002).

20.1.8 Sanitary and Phytosanitary Measures

Sanitary and Phytosanitary (SPS) measures are defined as any action undertaken to:

(a) Protect the health and life of animals or plants within a Member State's territory from the risks posed by the introduction, establishment, or spread of pests, diseases, disease-carrying organisms, or disease-causing organisms. This includes regulations and controls that prevent the infiltration of harmful biological agents that can compromise the well-being of local flora and fauna.
(b) Safeguard human or animal health within the Member State's territory from dangers associated with food, beverages, or foodstuffs contaminated by additives, toxins, or disease-causing organisms. This involves monitoring and regulating the presence of harmful substances in consumables to ensure they are safe for consumption and do not pose health risks.
(c) Protect human health within the Member State from diseases transmitted by animals, plants, or their products, or from the entry, establishment, or spread of pests. This includes measures to control zoonotic diseases and ensure that the trade and movement of biological materials do not compromise human health.
(d) Prevent or mitigate other forms of damage within the Member State's territory from the entry, establishment, or spread of pests. This broad category includes actions to protect ecosystems, agriculture, and natural resources from the adverse impacts of invasive species and other harmful organisms.

These measures are essential for maintaining health standards and preventing biological threats from undermining public health, agriculture, and ecosystems.

20.1.9 Existing International Organizations for Developing Standards

20.1.9.1 Food Safety Standards

To achieve international standards for food safety, the WTO has designated the Codex Alimentarius Commission (CAC) as the responsible authority. The CAC, based in Rome, is a joint subsidiary of the Food and Agriculture Organization (FAO) of the United Nations and the World Health Organization (WHO). Under the SPS Agreement, Codex is tasked with evaluating and harmonizing international food safety standards. The commission develops methodologies for food additives, veterinary drugs, pesticide residues, and contaminants. Additionally, it establishes methods of analysis and sampling and creates codes and guidelines for hygienic practices. This comprehensive framework ensures that food safety measures are scientifically grounded and globally consistent, thus facilitating safer trade and consumer protection.

20.1.9.2 Animal Health and Zoonosis Standards

Zoonoses, diseases transmissible from animals to humans, are a significant concern in international trade. The World Organization for Animal Health (OIE), headquartered in Paris, develops the standards, guidelines, and recommendations for managing these diseases. The OIE produces manuals on various animal diseases, diagnostic procedures, vaccination protocols, disease control measures, disinfection techniques, and certification processes. By setting these standards, the OIE helps to prevent the spread of zoonotic diseases and ensures that animal health measures do not become unnecessary trade barriers.

20.1.9.3 Plant Health Standards

The International Plant Protection Convention (IPPC), based in Rome, is crucial for trade-related plant health issues. Recognized by the WTO under the SPS Agreement, the IPPC is the authoritative body for International Standards for Phytosanitary Measures (ISPMs). The IPPC Secretariat develops guidelines and recommendations to protect plants from pests and diseases that could impede trade. These measures ensure that plant health standards are based on scientific evidence and promote safe, efficient trade by minimizing the risk of pest introduction and spread.

20.1.9.4 The "Three Sisters"

The CAC, OIE, and IPPC, often referred to as the "Three Sisters," are integral to the SPS framework. They regularly observe and contribute to SPS Committee meetings and provide expert advice to WTO dispute settlement panels. For areas not covered by these organizations, other relevant international bodies open to WTO members may develop appropriate standards, guidelines, and recommendations as identified by the SPS Committee.

20.1.9.5 Pest Risk Analysis (PRA)

The WTO agreements emphasize the importance of updating quarantine and plant health services to enhance the health status of exports. Countries are encouraged to incorporate Pest Risk Analysis (PRA) into their regulatory decision-making processes for imports and exports. The FAO, in collaboration with various regional plant protection organizations, developed ISPM Part 1, which provides guidelines for pest risk analysis. This was published in February 1996 under the SPS Measures provisions. These standards, guidelines, and recommendations are essential for ensuring that regulatory actions related to international trade are based on scientific evidence and risk assessment.

PRA is a critical tool for determining the entry status of imported agricultural commodities. It helps quarantine services evaluate the risk of pest entry and establish appropriate safeguards. The entry status, which combines a country's laws, regulations, guidelines, policies, and inspector decisions at ports of entry, must be scientifically justified. If the perceived risk is high, the entry status should be conservative, while a low perceived risk should result in a more liberal entry status. Effective safeguards can influence this status by mitigating the risk of pest introduction.

20.1.9.6 Implementing PRA

A PRA conducted according to international standards allows national planners to allocate resources effectively to areas facing the highest risk, thus protecting agriculture and the environment from exotic pests. PRA provides a structured approach to assessing and managing risk by weighing various factors and adopting biologically sound strategies for pest risk mitigation. It consists of both risk assessment and risk management, enabling quarantine services to justify new regulations, phytosanitary guidelines, operational procedures, the entry status of imported articles, and resource allocations. This approach ensures that regulatory actions are transparent and have the least negative impact on international trade while maintaining the necessary level of protection.

20.1.9.7 The IPPC's PRA Framework

The IPPC's PRA framework involves three stages: Initiation, Risk Assessment, and Risk Management. Each stage requires thorough documentation to support review or dispute resolution. The PRA must clearly document the sources of information and the rationale behind management decisions regarding mitigation measures. This transparency ensures that all actions taken are scientifically justified and aligned with international standards.

International organizations play a pivotal role in developing standards for food safety, animal health, and plant health, which are essential for safe and efficient international trade. The Codex Alimentarius Commission, the World Organization for Animal Health, and the International Plant Protection Convention are central to this effort, providing the necessary frameworks and guidelines to harmonize and implement SPS measures globally. Pest Risk Analysis, as promoted by the FAO and IPPC, ensures that regulatory decisions are based on scientific evidence and risk assessment, fostering safer trade practices. These collaborative efforts by international organizations ensure that SPS measures protect human, animal, and plant health while facilitating international trade.

20.1.10 Compliance with International Developments

In an effort to modernize, upgrade, and enhance its outdated plant quarantine system, the Government of India has recently approved a new Plant Quarantine Order. This initiative aims to align India's regulatory framework with the International Plant Protection Convention (IPPC) and the tenets of the SPS Agreement of the WTO. The new Order represents a significant step toward harmonizing India's plant quarantine regulations with internationally accepted standards. This harmonization process includes updating the legal and administrative frameworks governing quarantine activities, import and export inspections, on-field pest surveillance, treatment standards, and certification methodologies.

To safeguard its agriculture from the threat of alien pests, India now requires pest risk analysis for all plant and plant material imports. This measure ensures that the risks associated with the introduction of non-native pests are carefully assessed and mitigated. Furthermore, efforts are underway to enhance the export certification process, ensuring that Indian phytosanitary certificates are robust and credible, providing assurance to importing countries that their plant imports are free from quarantine and regulated pests, including alien species.

20.1.11 Development and Enforcement of National Quarantine Legislation

Several factors necessitated the development of new legislative provisions for plant quarantine in India:

1. **Liberalized Trade in Agriculture**: The WTO Agreements opened up economic opportunities but also introduced new pest risks due to increased international trade in agricultural commodities.
2. **Diverse Agroclimatic Zones**: India's varied agricultural produce and surplus food production positioned it to expand its agricultural trade significantly. However, this expansion required the establishment of credible national standards for phytosanitary activities to comply with stringent international regulations.
3. **Gaps in Existing Legislation**: The existing Plants, Fruits, and Seeds Order of 1989 needed updating to regulate the import of germplasm, genetically modified organisms, transgenic plant material, and live insects and fungi (including biocontrol agents).
4. **Facilitating Safe Global Trade**: There was a need to fulfill India's legal obligations under international agreements and facilitate safe global agricultural trade.
5. **Protecting Farmers and National Interests**: Protecting the interests of Indian farmers by preventing the entry, establishment, and spread of destructive pests, vectors, and alien species was essential.
6. **Protecting National Plant Life and Environment**: Safeguarding the national plant life and environment from the threats posed by invasive species was crucial.
7. **Preserving Biodiversity**: It was necessary to protect national biodiversity from invasive alien species.

The need was felt for the incorporation in regulation of additional special declarations for the freedom of import commodities from quarantine and alien pests. These declarations would be based on standardized pest risk analysis, particularly for seeds and planting materials.

20.1.12 Salient Features of the PQ Order

The Plant Quarantine (PQ) Order (2003), introduced several significant changes and enhancements to India's plant quarantine activities:

1. **Expanded Scope:** The PQ Order has expanded the scope of plant quarantine activities by incorporating additional definitions and making pest risk analysis a precondition for imports.
2. **Prohibition and Restrictions:** It places a prohibition on the import of commodities contaminated with weeds and/or alien species. The import of packaging materials of plant origin is restricted unless treated.
3. **Regulation of Specific Imports:** The new Order includes provisions for regulating the import of:
   - Soil, peat, sphagnum moss
   - Germplasm, genetically modified organisms, and transgenic material for research
   - Live insects, microbial cultures, and biocontrol agents
   - Timber and wooden logs
4. **Classification of Agricultural Imports:** Agricultural imports are now classified into four categories:
   (a) Prohibited plant species
   (b) Restricted species, importable only by authorized institutions
   (c) Restricted species requiring additional declarations of freedom from quarantine pests and specific treatment certifications
   (d) Plant material imported for consumption or industrial processing, permitted with normal phytosanitary certification
5. **Permit Requirements**: Importing seeds, including flower seeds, propagating material, and mushroom spawn cultures, now requires permits. The new Order specifies additional declarations for the import of 698 agricultural commodities, listing over 700 quarantine pests and 57 weed species.
6. **Increased Entry Points**: The number of notified points of entry has increased from 59 to 130, facilitating better regulation and inspection of imported agricultural commodities.
7. **Certification Fees and Inspection Charges**: The Order rationalizes the structure of certification fees and inspection charges, ensuring a more streamlined process.

20.1.13 Structure of the PQ Order

The PQ Order of 2003 is comprehensive, including 15 clauses, 16 forms for various plant quarantine regulatory functions, and 12 schedules that describe various aspects and conditions for importing agricultural articles (plants and plant products) into India.

20.1.14 Key Clauses and Schedules

- **Clauses**: The 15 clauses cover all aspects of plant quarantine regulation, from definitions to operational procedures.

- **Forms**: The 16 forms facilitate various regulatory functions, including applications for import permits and phytosanitary certificates.
- **Schedules**: The 12 schedules detail conditions and requirements for importing different agricultural commodities, ensuring compliance with quarantine regulations.

20.1.15 Implementation and Impact

The Plant Quarantine Order marks a pivotal shift in India's approach to managing plant health and biosecurity. By aligning its regulations with international standards, India aims to protect its agriculture and environment from the threats posed by invasive pests and diseases. The requirement for pest risk analysis ensures that all imports are thoroughly assessed, minimizing the risk of introducing harmful organisms. This proactive approach not only safeguards domestic agriculture but also enhances India's credibility as a reliable trading partner capable of meeting stringent international phytosanitary standards.

India's Plant Quarantine Order of 2003 is a comprehensive framework designed to modernize the country's plant quarantine system, ensuring compliance with international standards and enhancing the safety and efficiency of agricultural trade. By implementing rigorous pest risk analysis and improving phytosanitary certification processes, India aims to protect its agricultural sector and biodiversity from invasive pests and diseases. This modernization effort demonstrates India's commitment to fulfilling its international obligations and promoting safe global trade, ultimately benefiting its farmers, environment, and economy.

Schedule	Description
Schedule I	Points of entry for imports of plants/plant materials and other articles
Schedule II	List of inland container depots and container freight stations for import of plants and plant products
Schedule III	List of foreign post offices for the import of plants and plant products
Schedule IV	List of plants/planting materials and countries from which imports are prohibited, along with justification
Schedule V	List of plants and plant materials for which imports are restricted and permissible only by authorized institutions, with additional declarations and subject to special conditions
Schedule VI	List of plants/plant materials permitted for import with additional declarations and special conditions
Schedule VII	List of plants/planting materials for which imports are permissible on the basis of phytosanitary certificate issued by the exporting country, inspection conducted by inspection authority and fumigation, if required, including all other general conditions
Schedule VIII	List of quarantine weed species
Schedule IX A	Inspection fees
Schedule IX B	Fumigation/disinfection/disinfestation/supervision charges
Schedule X	List of permit-issuing authorities for import of seeds, plants, plant products, and other articles

Schedule	Description
Schedule XI	List of inspection authorities for certification of postentry quarantine facilities and inspection of growing plants
Schedule XII	Quantities of seeds permitted for trial purposes/accession to the gene bank of the National Bureau of plant genetic resources

As of March 28, 2024, the Plant Quarantine (PQ) Order has undergone 123 amendments aimed at revising definitions and clarifying specific queries raised by quarantine authorities worldwide. These amendments include updated lists of crops under Schedules VI and VII, and quarantine weed species under Schedule VIII. The revised Schedule VI now lists 703 crops/commodities, Schedule VII includes 298 crops/commodities, and Schedule VIII features 57 quarantine weed species. The PQ Order incorporates "Additional/Special Declarations" for import commodities to be free from quarantine pests, based on pest risk analysis (PRA) following international norms, particularly for seeds and planting materials (https://pqms.cgg.gov.in/10a4244b-74f1-47fb-85ea-b7508787199d).

1. **Phytosanitary Measures**: Designed to prevent the global spread of harmful pests, these measures are based on justified scientific principles, with PRA as the cornerstone.
2. **Applicability**: The provisions extend to packaging and transportation methods to ensure comprehensive phytosanitary control.
3. **Inspection and Certification**: These provisions align with Article IV of the IPPC, ensuring rigorous inspection and certification processes.
4. **Phytosanitary Certificates**: Issued in accordance with Article V of the IPPC, these certificates meet the plant quarantine requirements of importing countries and are issued after careful inspection and necessary treatments.
5. **Capacity Enhancement**: Emphasis is placed on developing and training staff to enhance capacity.
6. **Transparency**: The new Order is publicly accessible on the Ministry of Agriculture and Farmers Welfare's website, ensuring transparency and uniform application to all exporting countries.

20.1.16 PQ Order 2003 in Harmony with the IPPC

India's PQ Order aligns with the International Plant Protection Convention (IPPC) framework in several critical ways:

1. The phytosanitary measures are designed to prevent the global spread of noxious pests and are based on justified scientific principles, with pest risk analysis as the cornerstone.
2. Provisions have been made applicable to packages and transportation.
3. Inspection and certification provisions accord with Article IV of the IPPC.

4. Phytosanitary certificates are in the format of Article V of the IPPC and according to the plant quarantine requirements of the importing country. They are to be issued after careful inspection and the required treatment.
5. There is emphasis on capacity enhancement, development, and training of staff.
6. The new Order has been hosted on the website of the Ministry of Agriculture and Farmers Welfare (and is available at www.agricoop.nic.in) and made accessible to everyone. It is transparent and applies uniformly to all exporting countries or parties.

20.1.17 Need for Modernizing National Plant Quarantine System

The establishment of the WTO and the government's liberalization policies have opened up significant opportunities for material flow. However, the quarantine setup needs upgrading to prevent the globalization of pests. Increased imports of transgenic planting materials and the stipulations of the WTO's SPS Agreement pose additional challenges for quarantine personnel.

To bring the National Plant Protection Organization up to international standards, there is a need to upgrade manpower, infrastructure, and capabilities. Establishing the National Plant Quarantine Information System is crucial for developing a national database on PRA, Pest-Free Areas (PFAs), quarantine methodologies, policies, and related issues. Research projects must be reprioritized based on critical gaps identified to meet export and import requirements. Immediate areas of focus include:

- **Survey and Surveillance:** Conduct surveys and surveillance of diseases/pests of national and international importance to gather endemic pest data.
- **Epidemiological Data:** Generate comprehensive epidemiological data on significant pests to establish tolerance limits, develop PRA, and identify PFAs as per WTO norms.
- **Diagnostic Protocols:** Develop diagnostic protocols using molecular techniques to detect exotic pests and their variants, which are not easily detectable but pose significant economic concerns.
- **Identification Tools:** Create digitized keys for identifying exotic pests.
- **Biosafety Risks:** Evaluate biosafety risks associated with importing transgenics and beneficial biocontrol agents.
- **GM Contamination Detection:** Develop methods to detect the presence of transgenic/GM contamination in unknown samples during quarantine processing.
- **Development of an Accessible Information Platform**: There is a pressing need to create an accessible platform for information on biosecurity for policymakers, administrators, and industry groups. The current website (http://plantquarantineindia.nic.in/PQISPub/html) provides a national database on legislation, quarantine procedures, and methodologies. However, a more comprehensive Internet-based portal, similar to the International Portal for Food Safety and

Animal and Plant Health, is required. This portal would facilitate communication among countries in these sectors, ensuring easy accessibility to information.

- **Establishment of a National Institute for Emerging and Exotic Plant Pests**: To address emerging pest issues effectively, the government has proposed establishing a National Institute for Emerging and Exotic Plant Pests. This institute would focus on developing an early warning system for pest outbreaks, models for pest/disease forecasting and risk analysis, novel diagnostic approaches for detecting pests, and strategies for enhancing resistance durability. The institute could work in a network mode, facilitating research by less-equipped laboratories and boosting research on many regional and chronic plant diseases.
- **Enhancing Awareness and Domestic Quarantine Regulations**: Raising awareness about biosecurity among the scientific community and the general public is essential for generating respect for regulations in the interest of national security. The domestic quarantine regulations, which cover eight destructive pests (fluted scale, San José scale, coffee berry borer, codling moth, banana bunchy top and mosaic viruses, potato cyst nematode, potato wart, and apple scab), have largely failed, as these pests have spread to neighboring states and different parts of the country. Extensive eradication measures and surveys are needed to identify PFAs, a task required to fulfill international commitments under the SPS Agreement.
- **Need for Organized Plant Quarantine Services at the State Level**: Unlike the United States and Australia, India lacks organized plant quarantine services at the state level. It is essential to establish plant quarantine check-posts at interstate borders, inland container depots, and railroads, manned by trained personnel, to control the movement of planting materials, transport carriers, and containers. This would ensure effective pest control and protection of agricultural production.

Recent introductions of pests, such as the fall armyworm (*Spodoptera frugiperda*) in maize, rugose spiralling whitefly (*Aleurodicus rugioperculatus*) in 2018, root-knot nematode of guava (*Meloidogyne enterolobii*) in 2017, white rust (*Puccinia horiana*) in chrysanthemum in 2016, tomato pinworm (Tuta absoluta) in 2014, Jackbeardsley mealybug (*Pseudococcus jackbeardsleyi*) in 2012, and papaya mealybug (*Paracoccus marginatus*) in 2007, highlight the dangers of increased international travel and trade, exposing the country to exotic pests harmful to agriculture.

Case Study—Wheat Import: The complexities of plant protection and risk analysis during trade are well illustrated by a case study on wheat import. When India considered importing wheat, a thorough pest risk analysis was conducted to assess potential threats from exotic pests. This analysis involved evaluating the presence of pests in the exporting country, assessing the risk of pest introduction through wheat imports, and determining the efficacy of mitigation measures. By adhering to international standards and conducting detailed risk assessments, India ensured that the wheat imports did not pose a significant threat to domestic agriculture.

The Government of India's efforts to modernize and harmonize its plant quarantine system with international standards reflect its commitment to safeguarding agriculture and biodiversity from the threat of invasive pests. The amendments to the PQ Order, alignment with IPPC standards, and emphasis on pest risk analysis underscore the proactive approach taken to enhance biosecurity. Establishing a comprehensive information platform, developing a National Institute for Emerging and Exotic Plant Pests, and raising awareness about biosecurity are critical steps toward achieving this goal.

Moreover, addressing the spread of pests through effective domestic quarantine regulations and establishing organized plant quarantine services at the state level are essential for protecting the country's agriculture. By adopting these measures, India aims to enhance its agricultural trade, fulfill international obligations, and ensure the safety and productivity of its agricultural sector.

20.1.18 Conclusion

India's national plant quarantine system has significantly evolved since the enactment of the Destructive Insects and Pests Act in 1914. Nevertheless, there is much to learn from the more advanced quarantine frameworks of developed nations such as Australia, Europe, the United States, and New Zealand. In today's globalized world, modernizing this mechanism is crucial not only to shield crops from invasive alien pests but also to protect the economy. Enhanced quarantine measures can ensure greater agricultural productivity and higher farm incomes, thus providing a robust defense against potential biosecurity threats and fostering economic stability and growth.

References

Akhtar J, Chalam VC, Kumar P, Kiran R, Dubey SC (2019) Plant quarantine—a phytosanitary requirement for disease free import of plant genetic resources in India. In: Mishra KK, Subbanna ARNS, Rajashekara H, Pattnayak A (eds) Integrated pest management in major crops. J. Staley. ICAR-Vivekananda Parvatiya Krishi Anusandhan Sansthan, Almora, pp 153–163

Bhalla S, Chalam VC, Singh B, Gupta K, Dubey SC (2018) Biosecuring plant genetic resources in India: role of plant quarantine. ICAR-National Bureau of Plant Genetic Resources, New Delhi

Bhalla S, Kapur ML, Lal B, Verma BR, Singh C (2002) Quarantine risk associated with exchange of plant genera carrying hidden infestation. Indian Journal of Plant Genetic Resources 15:160–163

Chalam VC, Gupta K, Singh MC, Khan Z, Akhtar J, Gawade BH, Kumar P, Kumari P, Meena BR, Maurya AK, Meena DS, Sadhana (2023) 75 Quarantine pests intercepted ensuring crop biosecurity. ICAR-National Bureau of Plant Genetic Resources, New Delhi

Dubey SC, Gupta K, Akhtar J, Chalam VC, Singh MC, Khan Z, Singh SP, Kumar P, Gawade BH, Kiran R, Boopathi T, Kumari P (2021) Plant quarantine for biosecurity during transboundary movement of plant genetic resources. Indian Phytopath 74:495–508

Kahn RP (1989) Plant protection and quarantine: selected pest and pathogens of quarantine significance, vol 2. CRC Press, Boca Raton

Khan Z, Mahamood M, Lal A, Akhtar J, Parakh DB (2012) Interception of an exotic plant parasitic nematode (*Rotylenchus minutus*) on African potato (*Hypoxis hemerocallidea*) from Swaziland. Indian J Agric Sci 82(12):1090–1092

Khetarpal RK, Nath R (1998) Evolving national plant quarantine system: harmonization of current quarantine regulations with sanitary and phytosanitary (SPS) agreement. In: National Dialogue (ISPGR Dialogue 1998): issues on plant genetic resources, December 1–2, 1998. NBPGR, New Delhi

Khetarpal RK, Gupta K (2002) Implications of sanitary and phytosanitary agreement of WTO on plant protection in India. Ann Rev Plant Pathol 1:1–26

Khetarpal RK, Gupta K (2008) Plant quarantine in India in the wake of international agreements: a review. Scientific publishers (India), Jodhpur. Rev Plant Pathol 4:367–391

Khetarpal RK, Lal A, Varaprasad KS, Agarwal PC, Bhalla S, Chalam VC, Gupta K (2006) Quarantine for safe exchange of plant genetic resources. In: Singh AK, Saxena S, Srinivasan K, Dhillon BS (eds) 100 years of PGR management in India. National Bureau of Plant Genetic Resources, New Delhi, pp 83–108

Nath R, Mathur VK, Verma BR (1988) Plant quarantine activities in the Indian context. In: Paroda RS, Arora RK, Chandel KPS (eds) Plant genetic resources—Indian perspective. NBPGR, New Delhi, pp 470–480

Plant Quarantine (Regulation of Import into India) Order (2003) The Gazette of India Extraordinary, Part II Section 3 Sub-section (ii) No. 1037 Published by Authority

Singh B, Akhtar J, Kandan A, Kumar P, Chand D, Maurya AK, Agarwal PC, Dubey SC (2018) Risk of pathogens associated with plant germplasm imported into India from various countries. Indian Phytopath 72(1):91–102

The manufacturer's authorised representative in the EU is Springer Nature Customer Service Centre GmbH, Europaplatz 3, 69115 Heidelberg, Germany. If you have any concerns regarding our products, please contact ProductSafety@springernature.com

Printed and bound by CPI Group (UK) Ltd, Croydon, CR0 4YY
01/10/2026
02225941-0013